E DU

D1238207

Particle Astrophysics

Other books by the same author:

The Weak Interaction in Nuclear, Particle and Astrophysics
K Grotz and H V Klapdor
ISBN: 0 85274 312 2 (hardback), 0 85274 313 0 (paperback)

Non-accelerator Particle Physics
H V Klapdor-Kleingrothaus and A Staudt
ISBN: 0 7503 0305 0

Particle Astrophysics

H V Klapdor-Kleingrothaus

Max-Planck-Institut für Kernphysik, Heidelberg

K Zuber

University of Dortmund

Translated by S M Foster and B Foster

Institute of Physics Publishing
Bristol and Philadelphia

British Library Cataloguing-in-Publication Data

A catalogue record for this book is available from the British Library.

ISBN 0 7503 0403 0

Library of Congress Cataloging-in-Publication Data

Klapdor-Kleingrothaus, H. V. (Hans Volker), 1942–
[Teilchenphysik ohne Beschleuniger. English]
Particle astrophysics / H.V. Klapdor-Kleingrothaus, K. Zuber ;
translated by S.M. Foster and B. Foster.
p.   cm.
Includes bibliographical references and index.
ISBN 0-7503-0403-0 (akl. paper)
1. Nuclear astrophysics. 2. Particles (Nuclear physics)
I. Zuber K. II. Title.
QB464.K53 1997
523.01'972--dc21   97-22009
                                                                                        CIP

Simultaneous German edition, *Teilchenphysik ohne Beschleuniger* published by B G Teubner GmbH, Stuttgart

Published by Institute of Physics Publishing, wholly owned by The Institute of Physics, London

Institute of Physics Publishing, Dirac House, Temple Back, Bristol BS1 6BE, UK

US Editorial Office: Institute of Physics Publishing, The Public Ledger Building, Suite 1035, 150 South Independence Mall West, Philadelphia, PA 19106, USA

Typeset in TEX using the IOP Bookmaker Macros
Printed in the UK by J W Arrowsmith Ltd, Bristol

Contents

Preface

The last two decades have seen an explosion of development in the fields of particle physics, astrophysics and cosmology. Particle physics has become a crucial tool in the quest for a deeper understanding of the universe. The arrival of the grand unified theory (GUT) of particle physics has allowed us to trace the development of the universe back to the earliest points in time. Despite the fact that astrophysical and cosmological processes require energies beyond the reach of accelerators, at least in the foreseeable future, such energies could allow the realization of part of the wealth of exotic particle-theoretical predictions: baryogenesis, inflation, production of exotic particles such as monopoles, axions, cosmic strings and many others. Supersymmetric particles (neutralinos), which have just become accessible to observation, are candidates for cold dark matter and could help to explain the origins and large scale structure of the universe. Neutrinos are candidates for hot dark matter and properties of neutrinos influence the explosion mechanisms of supernovae. Astrophysical neutrino sources help in the investigation of the properties of neutrinos, which have a key function for the structure of theories of elementary particles. Astrophysically produced axions could probe the strong CP problem of quantum chromodynamics (QCD).

Thus the new field of *particle astrophysics* was born. In this area attempts are made to understand some of the fundamental problems of modern physics from two opposing approaches. The enormous growth in this field makes it increasingly difficult not only to trace the development of the subject in the literature but also to enter into this area of research, at least to newcomers.

The aim of this book is to give an outline of the essential ideas and basic lines of development. These include the close connections between topics of the micro- and macro-cosmos. We attempt to give an insight into the variety of unanswered or still developing theoretical and experimental problems. We try to show that in particle astrophysics new and unusual observations can be expected almost every day.

In contrast to some of the existing, excellent monographs, for example *The Early Universe*, texts by G Boerner and by E W Kolb and M S Turner, we have tried to address a larger circle of readers in this book. However, we do not claim to provide a comprehensive and self-contained treatment of all the topics discussed. As far as the references are concerned, we have not attempted to provide a comprehensive list, but rather have tried to include overview articles in addition to the fundamental papers. In order to be as up-to-date as possible we have also included many preprints, which can be easily accessed electronically via preprint servers on the World Wide Web.

We thank Professors Ch Wetterich (Institute for Theoretical Physics, University of Heidelberg) and I Appenzeller (Astronomical Observatory and University of Heidelberg) for valuable suggestions and discussions. We are indebted to Dr I Krivosheina (Radiophysical Institute, Nizhnij Novgorod, Russia) for helpful discussions and assistance with illustrations. We thank Mrs C Klehr and Mrs V Traeumer for their untiring technical assistance in the generation of the figures. KZ thanks S Helbich for her patience and support.

We are indebted to Dr Brian Foster (University of Bristol) and also to Mrs S M Foster for the translation from the German original, published by B G Teubner GmbH, Stuttgart, and Drs Jim Revill, Peter Binfield and Ms Lucy Williams of Institute of Physics Publishing for their faithful and efficient collaboration in the publication of this English edition.

H V Klapdor-Kleingrothaus
K Zuber
July 1997

Acknowledgments

The authors and publisher gratefully acknowledge permission to reproduce previously published material, including many figures and tables taken from extensive journal literature, as granted by authors and publishers, and as indicated by corresponding citations in captions. They have attempted to trace the copyright holders of all material reproduced from all sources and apologize to any copyright holders whose prior permission might not have been obtained.

The figures and tables are reproduced from the cited sources, as referenced in the bibliography, by permission of the copyright holders acknowledged below.

Scientific American (©1992): figure 3.5(*a*) from [Cha92].

Science (©1988, American Association for the Advancement of Science): figure 13.17 from [Woo88]; (©1995): figure 7.10 from [Sco95]; (©1996): figure 1.12(*b*) from [Tau96].

Addison-Wesley Publishing Company: figures 3.7, 3.9, 3.11, 4.3, 4.5, 6.12, 6.15, 7.1, 7.11, 9.7 from [Kol90]; figures 1.13(*a*), 1.13(*b*) from [Qui83].

Deutscher Taschenbuch Verlag (dtv) GmbH & Co. KG: figure 13.1 from [Her80].

The University of Chicago Press: figures 6.2, 6.3, 8.2, 12.1, 12.2, 14.3 from [Rol88].

Spektrum Akademik Verlag: figures 6.1(*a*), 6.1(*b*), 6.1(*c*) from [Sex87].

ESO: figures 13.14(*a*), 13.14(*b*) from [Buh87].

New Scientist, IPC Magazines Ltd: figure 7.11 from [Hen91].

CERN Information Services: figure 1.10 from [Per87].

Nature (©1991 Macmillan Magazines Ltd): figures 6.6, 6.7(*a*), 6.7(*b*) from [Dre91b]; (©1993): figure 9.4 from [Alc93].

Sky & Telescope Magazine (©1991): figure 6.1(*d*) from [Arp91]; (©1993): figure 13.4 from [Fil93]; figure 6.8(*b*) from [Sky93]; (©1995): figures 13.11(*a*), 13.11(*b*), 13.23 from [Hay95]; figure 8.20 from [Sky95].

Physics Today (©1987): figure 6.10 from [Sch87b].

Carnegie Institute Publication of Washington: figures 6.1(*a*), 6.1(*b*), 6.1(*c*) from [San94].

Chapter 1

The standard model of particle physics

1.1 The building blocks of matter

The discovery of the electron at the end of the last century spelled the end of the long-standing theory that the atom was the smallest element of matter. Nils Bohr developed his atomic model on the basis of scattering experiments, like those done with electrons by Lenard or with α-particles by Geiger, Marsden and Rutherford. The atom consists of an atomic nucleus that is only one ten-thousandth the size of the atomic radius and, as became apparent later, consists of neutrons and protons. Around it the electrons form an atomic shell and provide electric neutrality. No difference has been found between neutrons and protons with regard to the nuclear force, which is also called the strong interaction, and therefore they are called *nucleons*. Research using cosmic rays and experiments with accelerators resulted in the discovery of vast quantities of new and seemingly elementary particles, which led to the conclusion in the 1950s that perhaps protons and neutrons were also constructed of even smaller particles (see figure 1.1). These are known as *quarks* and today we are aware of six different types (flavours): namely up (u), down (d), strange (s), charm (c), bottom (b) and top (t) quarks. Indeed all particles which experience the strong force, the *hadrons*, can be constructed either out of three quarks (*baryons*) or out of a quark–anti-quark pair (*mesons*). So, for example, the proton is made of a combination of uud-quarks and the neutron of udd-quarks. There are six known particles which do not experience the strong interaction called *leptons*. Apart from the electron these are the muon and the tau and their associated electrically neutral, massless electron-, muon- and tau-neutrinos. These particles are grouped into *families* or *generations* according to their increasing mass. The corresponding members of different families (see figure 1.2) are distinguished only by their gravitational interaction due to their different mass; with respect to other interactions they behave identically. Table 1.1 shows the elementary particles with their quantum numbers. Only the first family is needed to construct normal matter. *All particles used to create matter are fermions*, which means they have spin 1/2 and are subject to the Pauli exclusion principle and therefore

1

must not have identical quantum numbers. Already at this point some questions present themselves: are the above particles really the most elementary building blocks, or do perhaps substructures (i.e. preons) exist? Are there more than three generations? Are neutrinos really massless, as has been suspected until now? We will deal with these points in more detail in chapter 2.

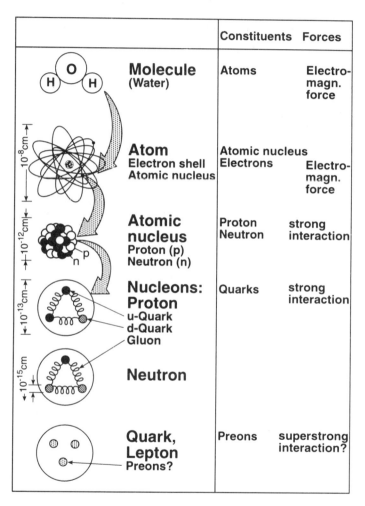

	Constituents	Forces
Molecule (Water)	Atoms	Electro-magn. force
Atom Electron shell Atomic nucleus	Atomic nucleus Electrons	Electro-magn. force
Atomic nucleus Proton (p) Neutron (n)	Proton Neutron	strong interaction
Nucleons: Proton u-Quark d-Quark Gluon	Quarks	strong interaction
Neutron		
Quark, Lepton Preons?	Preons	superstrong interaction?

Figure 1.1. The building blocks of matter. During the last few decades atoms and nuclei have been resolved into ever smaller units. Typical orders of magnitude and the dominant forces are included in the figure (from [Loh81]).

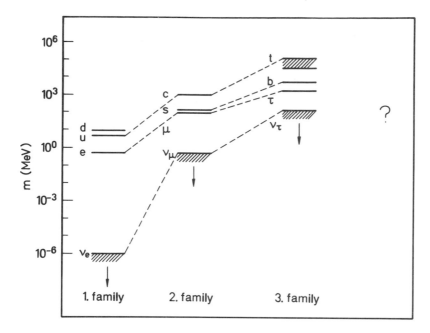

Figure 1.2. The mass spectrum of the known elementary fermions. The dashed lines connect the corresponding particles of the different families. Thus far there is no theoretical explanation for either the absolute values of the mass or of the mass splittings (from [Gro89, Gro90]).

1.2 The fundamental interactions

In modern physics we are aware of four fundamental forces, namely the strong interaction (colour force), electromagnetism, the weak interaction and gravitation. The strongest force, the colour force, acts between the quarks; its long-distance component gives rise to the well known nuclear force. The latter has to be seen as a kind of residual interaction, in analogy to the Van der Waals force between molecules. The coupling strength has an order of magnitude of one (see chapter 2). Next follows electromagnetism, whose force can be expressed in terms of Sommerfeld's fine structure constant $\alpha = \frac{e^2}{4\pi} \approx \frac{1}{137}$. The weak interaction is characterized at low energies by the Fermi constant G_F, which is specified in units of proton mass ($G_F \sim m_P^{-2}$). Among other things it is responsible for β-decay. The weakest force by far is gravitation, characterized by Newton's gravitational constant G. A comparison of individual interactions is shown in table 1.2. The fact that these coupling constants are energy dependent will present the possibility of the unification of the forces later on (see figure 1.3).

The range of the various interactions is as variable as their strength. While

Table 1.1. (*a*) Properties of the quarks. *I* isospin, *S* strangeness, *C* charm, *Q* charge, *B* baryon number, *B** bottom, *T* top. (*b*) Properties of leptons. L_i flavour-related lepton number, $L = \sum_{i=e,\mu,\tau} L_i$.

(*a*) Flavour	Spin	*B*	*I*	I_3	*S*	*C*	*B**	*T*	*Q[e]*
u	1/2	1/3	1/2	1/2	0	0	0	0	2/3
d	1/2	1/3	1/2	−1/2	0	0	0	0	−1/3
c	1/2	1/3	0	0	0	1	0	0	2/3
s	1/2	1/3	0	0	−1	0	0	0	−1/3
b	1/2	1/3	0	0	0	0	−1	0	−1/3
t	1/2	1/3	0	0	0	0	0	1	2/3

(*b*) Lepton	*Q[e]*	L_e	L_μ	L_τ	*L*
e^-	−1	1	0	0	1
ν_e	0	1	0	0	1
μ^-	−1	0	1	0	1
ν_μ	0	0	1	0	1
τ^-	−1	0	0	1	1
ν_τ	0	0	0	1	1

Table 1.2. Phenomenology of the four fundamental forces and the hypothetical GUT interaction.

Interaction	Strength	Range *R*	Exchange particle	Example
Gravitation	$G_N \simeq 5.9 \times 10^{-39}$	∞	Graviton?	Mass attraction
Weak	$G_F \simeq 1.02 \times 10^{-5} m_p^{-2}$	$\approx m_W^{-1}$ $\simeq 10^{-3}$ fm	$W^\pm$, Z^0	β-decay
Electro-magnetic	$\alpha \simeq 1/137$	∞	γ	Force between electric charges
Strong (nuclear)	$g_\pi^2/4\pi \approx 14$	$\approx m_\pi^{-1}$ ≈ 1.5 fm	Gluons	Nuclear forces
Strong (colour)	$\alpha_s \simeq 1$	confinement	Gluons	Forces between the quarks
GUT	$M_X^{-2} \approx 10^{-30} m_p^{-2}$ $M_X \approx 10^{15}$ GeV	$\approx M_X^{-1}$ $\approx 10^{-16}$ fm	*X*, *Y*	*p*-decay

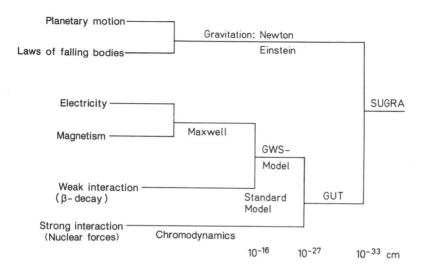

Figure 1.3. The phenomenological fundamental forces and the attempts to unify them. The Glashow–Weinberg–Salam (GWS) model which unifies the electromagnetic and weak interactions, together with quantum chromodynamics (QCD), is known as the standard model of particle physics. Grand unified theories (GUTs) which also include gravitation, are known as supergravity theories (SUGRA) (following [Wes87, Gro89, Gro90]).

gravitation and electromagnetism possess a $1/r$-potential and therefore have an infinite range, the effect of the strong interaction is restricted to nuclear dimensions. The range of the weak interaction is even smaller and variations from a point-like interaction are only visible at high energies. Finally, the GUT force, the original force of grand unified theories, has a range which is lower by several orders of magnitude. The differing ranges of the forces can be understood within the framework of a field theoretical representation as a reflection of the differing masses of the exchanged particles (table 1.3). In the quantum field theoretical representation each of the interactions is carried by these exchange particles (see figure 1.4). In accordance with Heisenberg's uncertainty principle, more massive particles can be produced only for short periods, and can therefore travel only a short distance. Since electromagnetism and gravitation have an infinite range, the photon and the graviton, the as yet hypothetical exchange particle carrying the gravitational interaction, are massless. The massless gluons are carriers of the strong interaction and the massive W and Z bosons are the exchange particles corresponding to the weak interaction. The fact that the strong interaction, despite having massless gluons, has only a limited range is due to the gluons themselves carrying colour charge (see section 1.5.1).

A critical length scale is reached at about 10^{-33} cm, the so-called Planck length, as here a quantum theoretical description of gravity is required. This becomes necessary when two characteristic length scales of a particle, the

Compton wavelength and the Schwarzschild radius (see chapter 2), become of
the same order of magnitude.

Table 1.3. Properties of the exchange bosons.

Boson	Interaction	Spin	Mass (GeV/c^2)	Colour charge	Electric charge	Weak charge
Gluons	Strong	1	0	yes	0	no
γ	Electromagn.	1	0	no	0	no
$W^\pm$; Z^0	Weak	1	80.4; 91.2	no	± 1; 0	yes
Graviton	Gravitation	2	0	no	0	no
X; Y	GUT	1	$\sim 10^{15}$	yes	$\pm 4/3$; $\pm 1/3$	yes

All force carrying particles are bosons, and possess spin 1, with the
exception of the graviton which has spin 2. As bosons therefore these particles
are not subject to the Pauli exclusion principle. An interaction can be described
by vertices in a space-time diagram, which is called a Feynman diagram. In
each case two vertices are needed to describe an interaction (figure 1.5). The
exchange bosons do not show up directly, they are called virtual particles.

1.3   Quantum numbers and symmetries

In quantum mechanics conserved quantities correspond to operators O, which
commute with the Hamiltonian operator H, i.e. the commutator is zero

$$[H, O] = HO - OH = 0. \qquad (1.1)$$

From this relation it follows that eigenstates ψ of H exist which are
simultaneously eigenstates of O.

$$O|\psi\rangle = q|\psi\rangle \qquad (1.2)$$

where q is an eigenvalue of the eigenstate ψ of both H and O. Conserved
quantities imply the invariance of the equations of motion under particular
symmetry transformations (see section 1.4).

There are two different kinds of symmetry. Firstly there are those connected
with space-time symmetries, such as translation and rotational invariance. Such
symmetries are called *'external'* symmetries.   As an example, translation
invariance implies momentum conservation.   At the same time there are
symmetries concerning internal degrees of freedom of the wavefunction, such
as phase transformations (i.e. multiplication of ψ by $e^{i\alpha}$). Such symmetries
are called *'internal'* symmetries; we will return to them later in more detail
(see section 1.4).   The symmetries are also characterized as continuous (e.g.

translation) and discrete transformations (e.g. spatial reflection through the origin). *Continuous* symmetries can be described by real numbers and lead to *additive* quantum numbers, while *discrete* symmetries are described by integers and lead to *multiplicative* quantum numbers. We will now consider some conserved quantities in more detail.

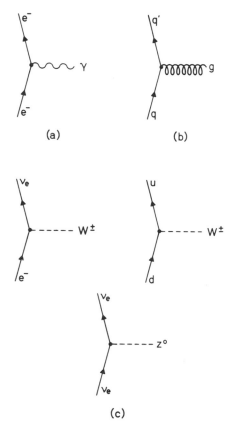

Figure 1.4. Elementary fermion-field quantum vertices for (*a*) electromagnetic, (*b*) strong (colour) and (*c*) weak interactions.

1.3.1 The electric charge Q

The conservation of electric charge $q = e$ is a consequence of quantum electrodynamics. Its conservation results in the stability of the electron, which could otherwise decay by processes such as

$$e^- \rightarrow v_e + \gamma \tag{1.3}$$

$$e^- \rightarrow v_e + v_e + \bar{v}_e. \tag{1.4}$$

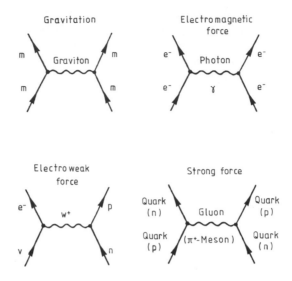

Figure 1.5. Feynman diagrams for the four fundamental interactions. The interaction proceeds via the exchange of the field quanta characteristic of the particular interaction.

Experimental checks for these decay modes come for example from double beta-decay experiments with Ge detectors (see section 2.4.2). The first decay mode would be characterized by the existence of a mono-energetic 255 keV photon, while the second decay channel in the case of germanium produces a signal at 11 keV. This results from the decay of a K-shell electron and the signal given by X-ray quanta emitted during the refilling of the hole thereby created. The present experimental value for the lifetime of the electron from the process (1.3) is [Bal93, Aha95]

$$\tau_e > 3.7 \times 10^{25} \text{ years} \quad (68\% \quad \text{confidence level}) \tag{1.5}$$

and for the pure neutrino decay (1.4) is [Reu91, Aha95]

$$\tau_e > 4.3 \times 10^{23} \text{ years} \quad (68\% \quad \text{confidence level}). \tag{1.6}$$

A different situation with respect to charge non-conservation can occur in nuclei. Normal β-decay is forbidden for two nuclei whose mass difference is less than the electron mass, but charge non-conserving transitions could take place [Kuz66]. The ^{71}Ga–^{71}Ge system is an example of this. Decays of this type would be

$$^{71}\text{Ga} \rightarrow {}^{71}\text{Ge} + X \quad \text{with} \quad X = \gamma, \nu\bar{\nu} + \text{exotic decays.} \tag{1.7}$$

If one interprets the results of the gallium solar neutrino detectors (see chapter 12) in this context as charge-violating, a half-life for this pair of isotopes of

$$T_{1/2}(^{71}\text{Ga} \rightarrow {}^{71}\text{Ge}) > 2.4 \times 10^{26} \text{ years} \tag{1.8}$$

can be obtained [Bar80, Bal93]. Should the electric charge really not be conserved, it should be possible in principle to measure it by the electrostatic charging of macroscopic objects. For example, to avoid the Earth becoming positively charged by protons after electron decay, a limit on the electron lifetime of

$$\tau_e > 3 \times 10^{21} \text{ years} \tag{1.9}$$

independent of the decay channel is required [Dol81]. Astrophysical considerations also give constraints on the lifetime of the electron which are partly many orders of magnitude higher (lifetimes greater than 10^{35} years) than those which can be obtained from laboratory data. However, they are subject to several uncertainties [Ori85]. Of course strong theoretical arguments exist against the non-conservation of electric charge [Oku78]. Radiative decay would be connected to catastrophic bremsstrahlung in the form of the emission of 10^{14}–10^{21} photons.

1.3.2 Parity P and charge conjugation C

Parity, P, is an example of an inner, discrete, symmetry transformation. The parity transformation P corresponds to spatial reflection of the coordinates of a physical state at the origin. For a scalar wavefunction (for example a solution of the Schrödinger equation) then

$$P\psi(\boldsymbol{x}, t) = \psi(-\boldsymbol{x}, t). \tag{1.10}$$

Since $P^2\psi = \psi$, it follows that the eigenvalues are either $\pi = +1$ (even parity) or $\pi = -1$ (odd parity). Since parity commutes with angular momentum $\pi = (-1)^l$, where l are eigenvalues of the angular momentum operator. Experimentally, parity is seen to be conserved in the strong and electromagnetic interactions provided that particles are also allocated an intrinsic parity (see e.g. [Qui83, Per87]).

The weak interaction is the only interaction which does not conserve parity, as was discovered in 1956 in experiments on the β-decay of cobalt. A sample of ^{60}Co at a temperature of about 0.01 K was placed in a magnetic field to align the nuclear spins. The angular distribution of the electrons emitted in the decay was examined [Wu57]. The observed intensity had an angular distribution of the form

$$I(\theta) = 1 + \delta\left(\frac{\boldsymbol{\sigma} \cdot \boldsymbol{p}}{E}\right) = 1 + \delta\frac{v}{c}\cos\theta \tag{1.11}$$

where θ is the angle between the nuclear spin vector and the direction of emission of the electron. By examining the distribution for both possible directions of the ^{60}Co relative to the magnetic field it could be seen that the electrons were emitted preferentially in the direction opposite to the nuclear spin, which implies that $\delta = -1$ for electrons. This result is a clear indication that parity is not conserved, since it implies that the expectation value of a pseudoscalar quantity is not equal to zero. In this case the quantity in question is

$$\Delta(\theta) = \lambda(\theta) - \lambda(180° - \theta) \tag{1.12}$$

where $\lambda(\theta)$ is the probability that the momentum of the emitted electron lies at an angle θ with respect to the spin of the parent nucleus. The application of the parity operator reverses the direction of the momentum while leaving the nuclear spin unchanged. The angle θ becomes

$$\theta \rightarrow 180° - \theta \tag{1.13}$$

so that

$$\Delta(\theta) \rightarrow \lambda(180° - \theta) - \lambda(180° - (180° - \theta)) = -\Delta(\theta). \tag{1.14}$$

Since the electrons are emitted preferentially opposite to the spin direction it can be seen that the expectation value of the pseudoscalar $\Delta(\theta)$ is non-zero.

As momentum changes under the parity operation but spin does not, left-handed particles are changed to right-handed ones and vice-versa under parity transformations,

$$P|e_L\rangle = |e_R\rangle \tag{1.15}$$

$$P|e_R\rangle = |e_L\rangle. \tag{1.16}$$

Here left- and right-handed are defined via the spin direction relative to the momentum. The expectation value of the spin in the direction of the momentum is defined as the helicity h with the helicity operator

$$h = \frac{\boldsymbol{\sigma} \cdot \boldsymbol{p}}{|\boldsymbol{p}|}. \tag{1.17}$$

For electrons the helicity is identical to the longitudinal polarization, i.e. $h = -v/c$. The helicity operator h is not relativistically invariant for massive particles. For massless neutrinos and anti-neutrinos however helicity is a conserved quantity. The helicity operator h has then eigenvalues $h = -1$ and $h = +1$.

There therefore exists a fundamental asymmetry between right and left in nature because of the weak interaction. Parity is in fact maximally violated, since only left-handed neutrinos exist and there are no right-handed neutrinos.

The operation of charge conjugation C applied to a wavefunction ψ changes all its associated charges, i.e. all additive quantum numbers, but leaves quantities such as momentum and spin unchanged. Therefore charge conjugation implies a conversion of a particle into the corresponding *anti-particle* and vice-versa:

$$C|e_L^-\rangle = |e_L^-\rangle^C = |e_L^+\rangle. \tag{1.18}$$

Charge conjugation is also not conserved by the weak interaction. Therefore β-decay results in a preference for left-handed electrons and right-handed positrons. Neutrinos play a special role here. Only left-handed neutrinos have been experimentally observed, i.e. neutrinos with spin aligned opposite to the direction of flight. These are labelled as ν_L. Right-handed neutrinos (spin along the direction of flight) have so far not been observed. In the case of anti-neutrinos exactly the opposite is found. Here only the right-handed anti-neutrino, $\bar{\nu}_R$, is found. Strictly speaking however, the right-handed anti-neutrino $\bar{\nu}_R$ is not the charge-conjugate particle to the left-handed neutrino. Since spin and momentum do not change under charge conjugation,

$$(\nu_L)^C \neq \bar{\nu}_R \tag{1.19}$$

because the charge-conjugate particle should also be left-handed. Instead particle and anti-particle are connected by the *CP* operation:

$$(\nu_L)^{CP} = \bar{\nu}_R \tag{1.20}$$

This can in principle be interpreted in two different ways:

(i) The neutrino is its own charge-conjugate,

$$(\nu_L)^C = \nu_L \tag{1.21}$$

and also

$$(\bar{\nu}_R)^C = \bar{\nu}_R \tag{1.22}$$

The two states ν_L and $\bar{\nu}_R$ then form a *Majorana neutrino*. Other particles identical to their charge-conjugate state are, for example, the photon and the π^0.

(ii) All four states are independent of each other, and $(\nu_L)^C$ and $(\bar{\nu}_R)^C$ are new, as yet unobserved, particles, such that

$$(\nu_L)^C \neq \nu_L \tag{1.23}$$

and

$$(\bar{\nu}_R)^C \neq \bar{\nu}_R \tag{1.24}$$

In this case the neutrino would be a *Dirac neutrino*.

The Majorana description is only possible for neutrinos, since all other fundamental fermions can be clearly distinguished as particles or anti-particles through their electric charge (see figure 1.6) (see e.g. [Gro89, Gro90, Kay89, Boe92]).

The question as to which of the two possibilities actually applies to the neutrino could be solved from experimental data on neutrinoless double beta-decay, since this process is only possible for Majorana neutrinos (see for example [Gro89, Gro90, Kay89, Kla95]).

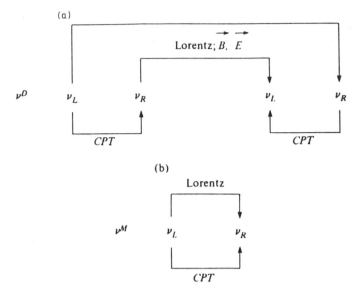

Figure 1.6. The difference between Dirac and Majorana neutrinos. Whereas for left-handed neutrinos, precession in E and B fields in connection with the CPT theorem can lead to four distinct states in the case of Dirac neutrinos, in the case of Majorana neutrinos there are only two. In the limit of massless neutrinos and without right-handed weak currents, the distinction becomes meaningless (from [Boe92]).

1.3.3 CP conjugation

1.3.3.1 CP invariance

Whereas the operations P and C are not necessarily conserved separately, their combination seemed to be well conserved. Consider the reaction

$$\pi^+ \rightarrow e^+ \nu_{e_{\mathrm{L}}}. \tag{1.25}$$

Application of charge conjugation to this yields

$$\pi^- \rightarrow e^- (\nu_{e_{\mathrm{L}}})^C. \tag{1.26}$$

This implies a left-handed anti-neutrino. A decay such as that described by equation (1.26) has not been observed so far. Only after the application of the parity operation does equation (1.26) become

$$\pi^- \rightarrow e^- \bar{\nu}_{e_{\mathrm{R}}}, \tag{1.27}$$

which now represents an observed decay mode. All interactions conserve CP exactly except for the weak interaction, for which CP non-conservation has been observed, although to date only in the neutral K-meson system.

1.3.3.2 CP violation

The neutral K-meson system consists of a K^0 (quark content $d\bar{s}$) and its anti-particle, $\bar{K}^0$ ($s\bar{d}$). The K^0 and $\bar{K}^0$ can be produced by the strong interaction as two clearly distinguishable states, as S (the flavour quantum number associated with the s-quark, (see section 1.3.5) is conserved. A K^0 is generated via a process such as

$$\pi^- + p \rightarrow \Lambda + K^0 \qquad (1.28)$$
$$S = 0 + 0 \rightarrow -1 + 1$$

whereas a $\bar{K}^0$ is produced via

$$\pi^- + p \rightarrow \bar{\Lambda} + \bar{K}^0 + 2n \qquad (1.29)$$
$$S = 0 + 0 \rightarrow 1 + -1 + 0.$$

There are therefore two clearly distinguishable neutral kaons with strangeness $+1$ and -1. Kaons propagating freely in space can decay via the weak interaction with $\Delta S = 1$ into 2 or 3 pions. At the same time however they can also convert into one another through virtual pion states so that:

$$
\begin{array}{ccc}
 & 2\pi & \\
\nearrow & & \searrow \\
K^0 & & \bar{K}^0 \qquad (1.30) \\
\searrow & & \nearrow \\
 & 3\pi &
\end{array}
$$

In this process strangeness changes by 2 units. Such strangeness oscillations are permitted as a second-order weak interaction. Neither K^0 nor $\bar{K}^0$ are eigenstates under CP transformations, but are related via

$$CP|K^0\rangle \rightarrow \eta|\bar{K}^0\rangle \qquad (1.31)$$
$$CP|\bar{K}^0\rangle \rightarrow \eta'|K^0\rangle \qquad (1.32)$$

by a phase factor η, η'. Through linear combinations of these states it is, however, possible to generate two CP eigenstates (K_1 and K_2) with well-defined CP eigenvalues,

$$|K_1\rangle = \frac{1}{\sqrt{2}}(|K^0\rangle + |\bar{K}^0\rangle) \quad CP = +1 \qquad (1.33)$$

$$|K_2\rangle = \frac{1}{\sqrt{2}}(|K^0\rangle - |\bar{K}^0\rangle) \quad CP = -1. \qquad (1.34)$$

The $CP = +1$ state is associated with the decay into 2 pions, since they also have $CP = +1$, whereas the 3 pion state usually has $CP = -1$. (The 3 pion state with $CP = -1$ is kinematically very strongly favoured.) Due to the larger phase-space for the 2π-decay, K_1 has a lifetime of 0.9×10^{-10} s and K_2 of 0.5×10^{-7} s. In 1964 it was verified experimentally that the K_2 can also decay into 2 pions, which is only possible via CP violation [Chr64]. Therefore the experimentally observed particles are only approximately identical to the CP eigenstates, so that it is necessary to define the observed states K_L ($\simeq K_2$) and K_S ($\simeq K_1$) as (see e.g. [Com83]):

$$|K_S\rangle = (1 + |\epsilon|^2)^{-1/2}(|K_1\rangle - \epsilon|K_2\rangle) \tag{1.35}$$

$$|K_L\rangle = (1 + |\epsilon|^2)^{-1/2}(|K_2\rangle + \epsilon|K_1\rangle). \tag{1.36}$$

CP violation caused by this mixing can be characterized by the parameter ϵ. The ratio of the amplitudes for the decay into charged pions may be used as a measure of the CP violation [Per87, PDG96]:

$$|\eta_{+-}| = \frac{A(K_L \to \pi^+\pi^-)}{A(K_S \to \pi^+\pi^-)} = (2.285 \pm 0.019) \times 10^{-3}. \tag{1.37}$$

For a more recent measurement see [Adl95].

A similar relation is obtained for the K^0-decay into 2 neutral pions, in analogy characterized as η_{00}. The complex amplitudes are written more usefully as $\eta_{+-} = |\eta_{+-}|e^{i\Phi_{+-}}$ and $\eta_{00} = |\eta_{00}|e^{i\Phi_{00}}$ respectively. Experimentally, measurements by the E731 experiment at Fermilab [Woo88a] and the NA31 experiment at CERN [Car90, Bar93b] have shown that

$$\left|\frac{\eta_{00}}{\eta_{+-}}\right| = (0.9931 \pm 0.0020) \times 10^{-3} \quad \Phi_{+-} = 46.0° \pm 2.2° \quad \text{(NA31)} \tag{1.38}$$

$$\left|\frac{\eta_{00}}{\eta_{+-}}\right| = (0.9904 \pm 0.0120) \times 10^{-3} \quad \Phi_{+-} = 42.8° \pm 1.2° \quad \text{(E731)} \tag{1.39}$$

Information about both experiments can be found in [Woo88, Bur88]. The ϵ appearing in equations (1.35) and (1.36), together with a further parameter ϵ', defined below, can be connected with η via the relation

$$\eta_{+-} = \epsilon + \epsilon' \tag{1.40}$$

$$\eta_{00} = \epsilon - 2\epsilon' \tag{1.41}$$

from which can be deduced (see e.g. [Com83])

$$\left|\frac{\eta_{00}}{\eta_{+-}}\right| \approx 1 - 3\text{Re}\left(\frac{\epsilon}{\epsilon'}\right). \tag{1.42}$$

All data are consistent with $|\epsilon| = 2.26 \pm 0.02 \times 10^{-3}$, while the value of ϵ' is somewhat more uncertain [Bar93b, Win92]

$$\epsilon'/\epsilon = (2.3 \pm 0.7) \times 10^{-3} \quad \text{NA31} \tag{1.43}$$

$$\epsilon'/\epsilon = (0.74 \pm 0.81) \times 10^{-3} \quad \text{E731.} \tag{1.44}$$

Evidence for a non-zero ϵ' would show that CP could be violated *directly* in the decay, i.e. in processes with $\Delta S = 1$, and does not only depend upon the existence of mixing [Com83]. Future experiments, such as KLOE at a ϕ-factory, as well as NA48 (CERN) or E832 (Fermilab), should increase the sensitivity of the measurement by an order of magnitude [For95].

Is it possible that there are oscillations in other meson systems containing heavy quarks, such as $D^0-\bar{D}^0$? It can be shown that (see e.g. [Nac86, For95]), in this system the expected effect is very much smaller than in the K^0 system. However, in the $B^0-\bar{B}^0$ system (B-mesons contain a b-quark), analogous flavour oscillation effects have been discovered [Alb87], which implies an oscillation of the bottom quantum number by two units. Two distinct neutral mesons exist, B_d^0 ($b\bar{d}$) and B_s^0 ($b\bar{s}$). Considering firstly B_d^0 and the ratio χ_d of the amplitudes

$$\chi_d = \frac{\Gamma(B^0 \to l^-X) \text{ (via } \bar{B}^0)}{\Gamma(B^0 \to l^{\pm}X)} = \frac{\Gamma(\bar{B}^0 \to l^+X) \text{ (via } B^0)}{\Gamma(\bar{B}^0 \to l^{\pm}X)} \qquad (1.45)$$

where l means a lepton and X the hadronic final state, the experimental results for the ratio r are [Bar93c, Alb94]

$$r = \frac{\chi_d}{1 - \chi_d} = 0.16 \pm 0.08 \quad \text{ARGUS} \qquad (1.46)$$

$$= 0.149 \pm 0.045 \quad \text{CLEO}.$$

The data were taken at the energy of the Υ (4S)-resonance (a bound state of the $b\bar{b}$ system at about 10.6 GeV). The ratio is in fact rather large [Sch89]. Figure 1.7 shows a completely reconstructed event, in which 2 B^0-mesons were found rather than a $B^0-\bar{B}^0$ pair. New results which measure the sum of the two neutral B-mesons, B_d^0 and B_s^0, and which enable statements to be made about the standard model and the CKM matrix elements (see section 1.5.2) have been obtained from LEP [For95].

The theoretical understanding of CP violation does present difficulties. One possible source of CP violation in the standard model is via the complex phase of the Cabibbo–Kobayashi–Maskawa matrix (CKM matrix) (see section 1.5.2 and [Jar89]). The unitarity of this matrix leads to relations between its matrix elements, such as

$$V_{ub}^* V_{ud} + V_{cb}^* V_{cd} + V_{tb}^* V_{td} = 0. \qquad (1.47)$$

The unitarity triangle of equation (1.47) (see figure 1.8) is a geometric illustration of this relation in the complex plane. The examination of various B decays allows the determination of individual matrix elements, as well as conclusions on possible physics beyond the standard model (see section 1.5.2 and chapter 2). In the standard model the predictions for the CP asymmetries in neutral B decay are clearly determined by the three angles α, β and γ. A definite proposal for the examination of the suitable channel $B \to J/\Psi + K_S$ and hence the determination of β exists in the form of the HERA-B experiment [Her94] at the ep-storage ring HERA in Hamburg. In the near future it is hoped that by building

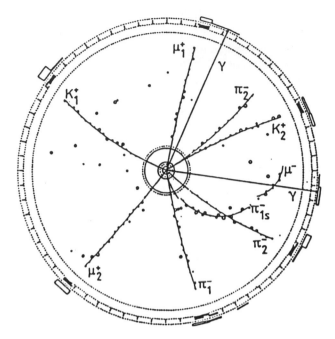

Figure 1.7. View of a fully reconstructed event from the decay of the $b\bar{b}$-system, $\Upsilon(4S) \rightarrow B^0B^0$, taken in the ARGUS detector at DESY. This event results from an oscillation from the original $B^0\bar{B}^0$ final state (from [Alb87]).

B factories at SLAC at Stanford in California (BaBar experiment [Bab95]) and at KEK in Japan (Belle experiment [Bel95b]) which can produce B mesons in very large numbers, the B system will produce more valuable information both on flavour oscillations and possible CP violation (see for example [Nir92]).

1.3.4 Time reversal T and the CPT-theorem

A further symmetry operation is time reversal, T. The result of this operation on a wavefunction is:

$$T\psi(\boldsymbol{x}, t) = \psi(\boldsymbol{x}, -t). \tag{1.48}$$

It can be compared to running a film backwards, which results in the reversal for example of all momenta. One consequence is the *principle of detailed balance*, which means that under certain very general conditions the matrix elements for a reaction and the time-reversed reaction are identical (see e.g. [Mui65, Heu76]).

One of the most important and most general theorems of modern quantum field theory is that of invariance under the three combined symmetry operations C, P and T (CPT-theorem) (see e.g. [Lüd54, Lüd57, Str64, Fon70, Lan75, Itz85]). The conditions for the validity of CPT invariance are so universal

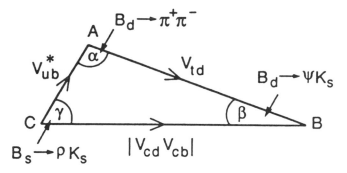

Figure 1.8. The unitarity triangle is a representation in the complex plane of the CKM matrix elements. The two poorly determined elements V_{ub} and V_{td} are shown. The first angle to be determined experimentally will be β (from [Nir92]).

that until now no theory has been conceived which does not obey it. *CPT* invariance results among other things in the identity of the mass and lifetime of particle and anti-particle, as well as equal but opposite magnetic moments (see table 1.4). The most precise test of *CPT* invariance to date stems from the limit on the mass difference between K^0 and $\bar{K}^0$ [PDG96]. Recently the charge to mass ratios of protons and anti-protons have been confirmed at LEAR to be equal within 10^{-9} [Gab95].

Table 1.4. The behaviour of some important physical quantities under *C*, *P* and *T* transformations.

Quantity	*P*	*C*	*T*
space vector r	$-r$	r	r
time t	t	t	$-t$
momentum p	$-p$	p	$-p$
spin σ	σ	σ	$-\sigma$
electric field E	$-E$	$-E$	E
magnetic field B	B	$-B$	$-B$

There is no direct evidence for T violation as yet; it merely follows from CP violation in K decay. Direct evidence for both T and P violation would be the demonstration of a non-zero electric dipole moment of the neutron. This would be caused by an asymmetric distribution of positive and negative charge within the neutron. An isolated neutron at rest has as its only preferred direction its spin s which would also be the axis for a possible dipole moment d_N. A dipole interaction with an external field would result in an interaction energy

H_{int}

$$H_{\text{int}} = d_N \cdot E \sim s \cdot E. \tag{1.49}$$

A P transformation would result in a change of $s, E \rightarrow s, -E$ and a T transformation in $s, E \rightarrow -s, E$ and therefore change the interaction energy. Such a dipole moment however has not yet been observed and its experimental upper limit is given by (see also chapter 11)

$$d_N < 1.2 \times 10^{-25} \text{ ecm.} \tag{1.50}$$

The CP violation established in the K^0 system, and its implementation in the standard model (which will be discussed later), also permits a dipole moment. However its size is [Wol86, He89]

$$d_N \approx 10^{-31} - 10^{-33} \text{ ecm.} \tag{1.51}$$

In summary, those quantum numbers conserved in the individual interactions are shown in table 1.5.

Table 1.5. Summary of conservation laws.

Conservation law	Strong	Electromagnetic	Weak
Energy	yes	yes	yes
Momentum	yes	yes	yes
Angular momentum	yes	yes	yes
B, L	yes	yes	yes
P	yes	yes	no
C	yes	yes	no
CP	yes	yes	no[a]
T	yes	yes	no[b]
CPT	yes	yes	yes

[a] Currently only in the K^0-system.
[b] Follows indirectly from CP violation and CPT invariance.

1.3.5 Baryon number B

This additive quantum number has not as yet been associated with a fundamental symmetry, and therefore it is quite possible that it is not in fact conserved. Every baryon is assigned a baryon number $+1$, every anti-baryon -1, and mesons and leptons baryon number 0. This inevitably implies that the baryon number of quarks is 1/3.

For each individual quark flavour there are separate flavour quantum numbers, i.e. the strangeness S (for s quark $S = -1$, for $\bar{s}$ $S = 1$, for all other quarks $S = 0$), charm C (for c quark $C = 1$, for $\bar{c}$ $C = -1$), bottom B (for b quark $B = -1$, for $\bar{b}$ $B = 1$) and top T (for t quark $T = 1$, for $\bar{t}$ $T = -1$).

Experiments specifically searching for baryon number violation are neutron–anti-neutron oscillation experiments ($\Delta B = 2$) and the search for proton decay ($\Delta B = 1$). Neither the oscillation experiments with their present limit on the oscillation period of $\tau_{n\bar{n}} > 10^8$ s [Bal94a], nor the proton decay experiments with their limits of $\tau_p > 9 \times 10^{32}$ years [PDG94] have so far given an indication for baryon number violation (see also chapter 2).

1.3.6 Lepton number L

The lepton number is also an additive quantum number and cannot as yet be associated with any fundamental symmetry. Each flavour has its own lepton number, L_e, L_μ, L_τ, and a total lepton number exists such that $L = L_e + L_\mu + L_\tau$. Each lepton has quantum number $+1$, each anti-lepton -1. There is as yet no evidence for the violation of either the total or any of the individual lepton numbers. A classic test for the conservation of individual lepton numbers is the capture reaction

$$
\begin{array}{lcccc}
 & \mu^- & + {}^A_Z X & \rightarrow {}^A_Z X & + e^- \\
L_e & 0 & + 0 & \rightarrow 0 & + 1 \\
L_\mu & 1 & + 0 & \rightarrow 0 & + 0
\end{array}
$$

This would violate both L_e and L_μ conservation, but would leave the total lepton number unchanged. A further test of individual lepton number violation would be neutrino oscillations, i.e. the transformation of one flavour of neutrino into another. Such experiments today play an important role in neutrino physics, as such an oscillation represents, among others, a possible solution to the so-called solar neutrino problem (see chapter 12). As mentioned previously, further processes which violate lepton number L would be proton decay

$$p \rightarrow \pi^0 + e^+ \Rightarrow \Delta L = -1 \tag{1.52}$$

and neutrinoless double beta-decay

$${}^A_Z X \rightarrow {}^A_{Z+2} X + 2e^- \Rightarrow \Delta L = 2 \tag{1.53}$$

Such processes have also not been observed to date (see chapter 2). In some unified theories (see chapter 2) L and B should however not be conserved separately, but rather $B - L$ [Moh88]. In this case, proton decay would be possible, but neutrinoless double beta-decay would not, since it would violate $B - L$ conservation. Other theories predict the conservation of other combinations of lepton flavour numbers rather than the individual lepton numbers, such as for example $L_e - L_\tau$ [Lan88]. Such predictions can also be tested via oscillation experiments.

1.4 Gauge theories

All modern theories of elementary particles are gauge theories. We will therefore attempt to indicate the fundamental characteristics of such theories without going into the details of a complete presentation. Theoretical aspects such as renormalization, the derivation of Feynman graphs or the triangle anomalies will not be discussed here and we refer to standard textbooks such as [Qui83, Hal84, Ait89, Don92]. However, it is important to realise that such topics do form part of the fundamentals of any such theory. One absolutely necessary requirement for such a theory is its *renormalizability*. Renormalization of the fundamental parameters is necessary to produce a relation between calculable and experimentally measurable quantities. Non-renormalizable theories which after all attempts of renormalization still contain divergent terms are not useful. The fact that it can be shown that gauge theories are *always* renormalizable, as long as the gauge bosons are massless, is of fundamental importance [t'Ho72, Lee72]. Only after this proof did gauge theories become candidates to model interactions. One well known non-renormalizable theory is the general theory of relativity. This behaviour makes the construction of a quantum theory of gravity very difficult; a solution to this may be beginning to emerge in terms of superstring theories (see chapter 2). A further aspect of the theory is its *freedom from anomalies*. The meaning of anomaly in this context is that a classical invariance of the equations of motion, or equivalently the Lagrangian, no longer exists in quantum field theoretical perturbation theory. The reason for this arises from the fact that in such a case a consistent renormalization procedure cannot be found.

1.4.1 The gauge principle

The gauge principle can be explained by the example of classical electrodynamics. The basis for this is the Maxwell equations and the electric and magnetic fields, measurable quantities which can be represented as the components of the field-strength tensor $F_{\mu\nu} = \partial_\mu A_\nu - \partial_\nu A_\mu$. Here the four-potential A is given by $A = (\phi, A)$, and the field-strengths are derived from it as $E = -\nabla\phi - \partial_t A$ and $B = \nabla \times A$. If $\rho(t, x)$ is a well-behaved, differentiable real function it can be seen that under a transformation of the potential such as

$$\phi'(t, x) = \phi(t, x) + \partial_t \rho(t, x) \tag{1.54}$$

$$A'(t, x) = A(t, x) + \nabla\rho(t, x) \tag{1.55}$$

all observable quantities remain invariant. The fixing of ϕ and A to particular values, in order to, for example, simplify the equations of motion, is called *fixing the gauge*.

In gauge theories this gauge freedom of certain quantities is raised to a fundamental principle. The existence and structure of interactions is determined by the demand for such gauge-fixable but physically undetermined quantities.

The inner structure of the gauge transformation is specified through a symmetry group.

So far success is the greatest justification for gauge theories, although that of course does not exclude justification from future, more fundamental, principles. For example, Kaluza–Klein theories attempt to trace back interactions to principles of differential geometry, [Kal21, Kle26, App87], which are also the foundations of general relativity. However, higher dimension geometric spaces are necessary for this (see also [Col89, Kla95]).

1.4.2 Global internal symmetries

Internal symmetries can be subdivided into discrete and continuous symmetries. We will concentrate on continuous symmetries. In quantum mechanics a physical state is described by a wavefunction $\psi(x, t)$. However, only the modulus squared appears as a measurable quantity. This means that as well as $\psi(x, t)$ the functions

$$\psi'(x, t) = e^{-i\alpha}\psi(x, t) \tag{1.56}$$

are also solutions of the Schrödinger equation, where α is a real (space and time independent) number. This is called a *global symmetry* and relates to the space and time independence of α. Consider again the wavefunction of a charged particle such as the electron. The relativistic equation of motion is the Dirac equation:

$$i\gamma^\mu \partial_\mu \psi_e(x, t) - m\psi_e(x, t) = 0. \tag{1.57}$$

The invariance under the global transformation

$$\psi_e'(x, t) = e^{ie\alpha}\psi_e(x, t) \tag{1.58}$$

is clear:

$$e^{ie\alpha}i\gamma^\mu \partial_\mu \psi_e(x, t) = e^{ie\alpha}m\psi_e(x, t) \tag{1.59}$$

$$\Rightarrow i\gamma^\mu \partial_\mu e^{ie\alpha}\psi_e(x, t) = me^{ie\alpha}\psi_e(x, t) \tag{1.60}$$

$$i\gamma^\mu \partial_\mu \psi_e'(x, t) = m\psi_e'(x, t). \tag{1.61}$$

Instead of discussing symmetries of the equations of motion, the Lagrange density $\mathcal{L}$ is often used. The equations of motion of a theory can be derived from the Lagrange density $\mathcal{L}(\phi, \partial_\mu \phi)$ with the help of the principle of least action (see e.g. [Gol85]). For example consider a real scalar field $\phi(x)$. Its free Lagrangian density is

$$\mathcal{L}(\phi, \partial_\mu \phi) = \tfrac{1}{2}\left(\partial_\mu \phi \partial^\mu \phi - m^2\phi^2\right). \tag{1.62}$$

From the requirement that the action integral S is stationary

$$\delta S[x] = 0 \quad \text{with} \quad S[x] = \int \mathcal{L}(\phi, \partial_\mu \phi)\mathrm{d}x \tag{1.63}$$

the equations of motion can be obtained:

$$\partial_\alpha \frac{\partial \mathcal{L}}{\partial(\partial_\alpha \phi)} - \frac{\partial \mathcal{L}}{\partial \phi} = 0. \tag{1.64}$$

The Lagrange density displays certain symmetries of the theory relatively clearly. In general it can be shown that the invariance of the field $\phi(x)$ under certain symmetry transformations, results in the conservation of a four-current, given by

$$\partial_\alpha \left(\frac{\partial \mathcal{L}}{\partial(\partial_\alpha \phi)} \delta\phi \right) = 0. \tag{1.65}$$

This is generally known as *Noether's Theorem* [Noe18]. Using this, time-, translation- and rotation-invariance imply the conservation of energy, momentum and angular momentum respectively. We now proceed to consider the differences introduced by local symmetries, in which α in equation (1.56) is no longer a constant, but a function of space and time.

1.4.3　Local (= gauge) symmetries

If the requirement for space and time independence of α is dropped, the symmetry becomes a local symmetry. It is obvious that under transformations such as

$$\psi'_e(x) = e^{ie\alpha(x)} \psi_e(x) \tag{1.66}$$

the Dirac equation (equation (1.57)) does not remain invariant:

$$(i\gamma^\mu \partial_\mu - m)\psi'_e(x) = e^{ie\alpha(x)}[(i\gamma^\mu \partial_\mu - m)\psi_e(x) + e(\partial_\mu \alpha(x))\gamma^\mu \psi_e(x)]$$
$$= e(\partial_\mu \alpha(x))\gamma^\mu \psi'_e(x) \neq 0. \tag{1.67}$$

The field $\psi'_e(x)$ is therefore not a solution of the free Dirac equation. If it were possible to compensate the additional term, the original invariance could be restored. This can be achieved by introducing a gauge field A_μ, which transforms itself in such a way that it compensates for the extra term. In order to achieve this it is necessary to introduce a covariant derivative D_μ, where

$$D_\mu = \partial_\mu - ieA_\mu \tag{1.68}$$

The invariance can be restored if all partial derivatives ∂_μ are replaced by the covariant derivative D_μ. The Dirac equation then becomes

$$i\gamma^\mu D_\mu \psi_e(x) = i\gamma^\mu (\partial_\mu - ieA_\mu)\psi_e(x) = m\psi_e(x). \tag{1.69}$$

If one now uses the transformed field $\psi'_e(x)$, it is easy to see that the original invariance of the Dirac equation can be restored if the gauge field transforms itself according to

$$A_\mu \to A_\mu + \partial_\mu \alpha(x). \tag{1.70}$$

The equations (1.66) and (1.70) describe the transformation of the wavefunction and the gauge field. They are therefore called *gauge transformations*. The whole of electrodynamics can be described in this way as a consequence of the invariance of the Lagrange density $\mathcal{L}$ or equivalently the equations of motion, under phase transformations. The resulting conserved quantity is the electric charge, e. The corresponding theory is called quantum electrodynamics, (QED), and as a result of its enormous success it has become a paradigm of a gauge theory. In the transition to classical physics, A_μ becomes the classical vector potential of electrodynamics. The gauge field can be associated with the photon, which takes over the role of an exchange particle. Further, it is found that generally in *all* gauge theories the gauge fields have to be *massless*. Any required masses have to be built in subsequently, via a phenomenon known as spontaneous symmetry breaking with which we will later become familiar. The case discussed here corresponds to the gauge theoretical treatment of electrodynamics. Group-theoretically the multiplication with a phase factor group can be described by a unitary transformation, in this case the U(1) group. It has the unity operator as generator. The gauge principle can easily be generalized for Abelian gauge groups, i.e. groups whose generators commute with each other. It becomes somewhat more complex in the case of non-Abelian groups and the resulting non-Abelian gauge theories (Yang–Mills theories) [Yan54].

1.4.4 Non-Abelian gauge theories (= Yang–Mills theories)

Non-Abelian means that the generators of the group do not commute, but are subject to certain commutator relations. One example are the commutation relations of the Pauli spin matrices σ_i,

$$[\sigma_i, \sigma_j] = i\hbar\sigma_k \tag{1.71}$$

which act as generators for the SU(2) group. Generally SU(N) groups possess $N^2 - 1$ generators. A representation of the SU(2) group is all unitary 2×2 matrices with determinant $+1$. Consider the electron and neutrino as an example. Apart from their electric charge and their mass these two particles are identical with respect to the weak interaction, and one can imagine transformations such as

$$\begin{pmatrix} \psi_e(x) \\ \psi_\nu(x) \end{pmatrix}' = U(x) \begin{pmatrix} \psi_e(x) \\ \psi_\nu(x) \end{pmatrix} \tag{1.72}$$

where the transformation can be written as

$$U(a_1, a_2, a_3) = e^{i\frac{1}{2}(a_1\sigma_1 + a_2\sigma_2 + a_3\sigma_3)} = e^{i\frac{1}{2}a\sigma} \tag{1.73}$$

The particles are generally arranged in multiplets; in this case a doublet. Considering the Dirac equation and substituting a covariant derivative for the normal derivative by introducing a gauge field $W_\mu(x)$ and a quantum number g

$$D_\mu = \partial_\mu + \frac{ig}{2}W_\mu(x) \cdot \sigma \tag{1.74}$$

does *not* lead to gauge invariance! Rather, because of the non-commutation of the generators, an additional term results, an effect which did not appear in the electromagnetic interaction. Only transformations of the gauge fields such as

$$W_\mu' = W_\mu + \frac{1}{g}\partial_\mu a(x) - W_\mu \times a(x) \qquad (1.75)$$

bring the desired invariance. (Note the difference compared with equation (1.70).) The non-commutation of the generators causes the exchange particles to carry 'charge' themselves (contrary to the case of the photon, which does not carry electric charge) because of this additional term. Among other consequences, this results in a self-coupling of the exchange fields. We now proceed to discuss in more detail the non-Abelian gauge theories of the electroweak and strong interaction, which are united in the *standard model of elementary particle physics*.

1.5   The standard model of elementary particle physics

We now consider a treatment of interactions in the framework of gauge theories. Gravitation will be excluded from this since so far no gauge theories exist which can describe it. The exposition will also be restricted to an outline, for a more detailed discussion we refer to standard textbooks, for example [Bec83, Hal84, Nac86, Ait89, Gre89, Don92, Mar92]. Theoretically the standard model group corresponds to an SU(3) ⊗ SU(2) ⊗ U(1)-group, as discussed below.

1.5.1   Quantum chromodynamics (QCD)

1.5.1.1   The properties of the strong interaction

We first consider the strong interaction. Previously the nuclear force was described as the exchange of mesons between the proton and the neutron. Today one generally describes the strong force through the exchange of gluons between quarks, and the inter-nuclear forces result from this as a kind of van der Waals force. In the 1950s, when the number of known 'elementary' particles grew larger and larger, Gell–Mann and Zweig constructed a new model [Gel64, Zwe64]. They explained all particles participating in the strong interaction as constructed from elementary building blocks called quarks. In this model, baryons consist of 3 quarks and mesons of a quark–anti-quark pair. This model has stood the test of time well. As the proton is composed of three so-called valence quarks, each quark has an electric charge which is a multiple of 1/3. The u, c and t quarks have charge $q = 2/3e$ and d, s and b quarks charge $q = -1/3e$. However, another new quantum number was necessary for a complete description of the Ω^- particle [Bar64]. In the quark model this consists of three s quarks with parallel spins. However, these have identical quantum numbers, and since quarks are also fermions, this leads to a violation of the Pauli exclusion principle. This was avoided by introducing a new quantum number to distinguish the quarks, the *colour* quantum number. Further evidence for colour was given by experiments at e^+e^- colliders. From these experiments the number of distinct colours can be established. Assuming that following the annihilation of the e^+ and e^- the virtual photon produced once again turns into a fermion–anti-fermion pair, the ratio R of the $e^+e^- \to \mu^+\mu^-$ and $e^+e^- \to \bar{q}q$ reactions (see e.g. [Per87, Pic95]) is given by:

$$R = \frac{\sigma(e^+e^- \to q\bar{q} \to \text{hadrons})}{\sigma(e^+e^- \to \mu^+\mu^-)} = \sum_q Q_q^2. \qquad (1.76)$$

Q_q denotes the quark charge as a fraction of the elementary charge e. If the u, d, s, c and b quarks contribute to R (which is the case above about 10 GeV) one expects for colourless quarks

$$R = (\tfrac{1}{3})^2 + (\tfrac{1}{3})^2 + (\tfrac{1}{3})^2 + (\tfrac{2}{3})^2 + (\tfrac{2}{3})^2 = \tfrac{11}{9}. \qquad (1.77)$$

If however there is a possibility of several colours existing, this value must be multiplied by the number of distinct colours (because of the higher number of decay channels into $q\bar{q}$-pairs). In the case of three colours the value of R is given by

$$R = 3 \cdot \tfrac{11}{9} = \tfrac{11}{3}. \qquad (1.78)$$

The experimental situation is shown in figure 1.9. For energies above 10 GeV, R is about 4 and thus in good agreement with the hypothesis of three colours. These colours are known as red, green and blue.

A 'free' colour charge has never been observed (all particles are colourless), and neither has a free quark (confinement). The experimental search for free quarks is specifically focusing on detecting fractional (1/3) electric charges. The examination of meteorites, ocean sediments and many other samples gives a limit for the number of free quarks of less than 5×10^{-27} per nucleon [Smi89, Hom92]. All baryons consist of three quarks of different colour. The sum of all three colours results in a 'colourless' particle (in analogy with the colours of the spectrum, which add up to give white). Mesons consist of quark–anti-quark pairs (colour and anti-colour result in 'colourless' particles). The exchange particles, the gluons, have for this reason to carry two colour charges (colour and anti-colour). With six quarks and three colours it is possible to describe all hadrons. Today, quarks are accepted as elementary, since they act as point-like particles even at the highest energies so far achieved in accelerators. Information about their distribution and the internal structure of the proton and neutron can be deduced from deep inelastic scattering of leptons on protons and neutrons (see e.g. [Hal84]). Figure 1.10 shows the behaviour of the total cross section for deep inelastic neutrino–nucleon scattering, which clearly supports the assumption of point-like quarks, as only in this case does the total cross section depend linearly

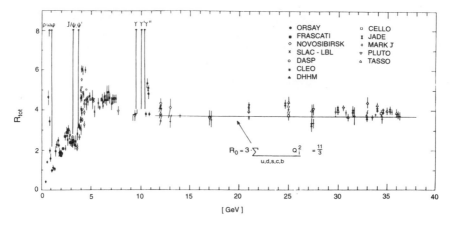

Figure 1.9. The ratio of the annihilation cross sections $R = \sigma\,(e^+e^- \to$ hadrons) $/\sigma(e^+e^- \to \mu^+\mu^-)$ as a function of centre of mass energy W. The number of active quark flavours can be determined from the value of the plateaux. The expected value for three colours is shown. The step-like increase in R at $W = 4$ GeV corresponds to crossing the threshold for c quark production. The points marked $\rho, \omega, \phi, J/\psi, \psi', \Upsilon, \Upsilon'$ and Υ'' show where the indicated vector mesons are produced. The experiments which produced the data used are listed in the figure (after [Loh83]).

on the energy [Per87]. The structure of the nucleon itself, expressed by the so-called structure functions, is however anything but simple and is currently being intensively examined at the HERA ep storage ring [Aid96, Der95, Der96]. Figure 1.11 shows one of the detectors involved in these investigations.

The gauge theory which describes the strong interaction is quantum chromodynamics (QCD) (*chromos* is the Greek for colour). It is based on invariance under rotation in colour space, which can be described by an SU(3) group. In this case the quarks are arranged in triplets and transform as

$$
\begin{pmatrix} \psi_1(x) \\ \psi_2(x) \\ \psi_3(x) \end{pmatrix}' = U(x) \begin{pmatrix} \psi_1(x) \\ \psi_2(x) \\ \psi_3(x) \end{pmatrix}. \tag{1.79}
$$

The matrix $U(x)$ is defined as

$$
U(x) = \mathrm{e}^{-\mathrm{i}\sum_l \alpha_l \lambda_l / 2} \tag{1.80}
$$

where a suitable choice of generators λ_l in the matrix representation is given by the so-called *Gell-Mann matrices*:

$$
\begin{pmatrix} 0 & 1 & 0 \\ 1 & 0 & 0 \\ 0 & 0 & 0 \end{pmatrix} \quad \begin{pmatrix} 0 & -\mathrm{i} & 0 \\ \mathrm{i} & 0 & 0 \\ 0 & 0 & 0 \end{pmatrix} \quad \begin{pmatrix} 1 & 0 & 0 \\ 0 & -1 & 0 \\ 0 & 0 & 0 \end{pmatrix}
$$

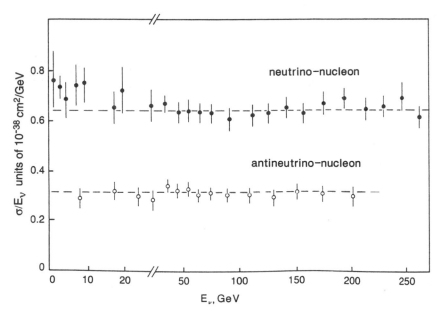

Figure 1.10. Total cross section for neutrino and anti-neutrino scattering from nucleons as a function of the neutrino energy. The constancy of the quantity σ/E over almost two orders of magnitude is a direct demonstration of point-like constituents inside the nucleon (from [Per87]).

$$\begin{pmatrix} 0 & 0 & 1 \\ 0 & 0 & 0 \\ 1 & 0 & 0 \end{pmatrix} \quad \begin{pmatrix} 0 & 0 & -i \\ 0 & 0 & 0 \\ i & 0 & 0 \end{pmatrix} \quad \begin{pmatrix} 0 & 0 & 0 \\ 0 & 0 & 1 \\ 0 & 1 & 0 \end{pmatrix} \quad (1.81)$$

$$\begin{pmatrix} 0 & 0 & 0 \\ 0 & 0 & -i \\ 0 & i & 0 \end{pmatrix} \quad \frac{1}{\sqrt{3}} \begin{pmatrix} 1 & 0 & 0 \\ 0 & 1 & 0 \\ 0 & 0 & -2 \end{pmatrix}.$$

Two characteristics of QCD are *asymptotic freedom* and *confinement*. Contrary to all other interactions, the force between two quarks continues to increase as their separation increases, or vice-versa, they are effectively free when they are close together. This is because the gluons are themselves carriers of colour (see chapter 2). Hence it is not possible to separate two quarks from one another. As more energy is supplied, further quark–anti-quark pairs form. The behaviour of the quarks is phenomenologically described by a potential of the form (see e.g. [Per87])

$$V(r) = -\frac{4}{3}\frac{\alpha_s}{r} + kr. \quad (1.82)$$

This potential can be tested by the spectroscopy of bound quark–anti-quark systems (quarkonium), for example the J/Ψ ($c\bar{c}$) or Υ resonance ($b\bar{b}$). There are also excited states of these systems, similar to those of positronium (the

Figure 1.11. View of the H1 detector at the *ep* storage ring HERA at DESY. The detector is able to recognise the scattered electron and quark, the latter manifesting itself in the form of a jet. The cryostat of the liquid argon calorimeter can be seen. The calorimeter measures the energy of the particles produced and surrounds the inner tracking chambers, which are required to measure their momenta (by kind permission of DESY, Hamburg).

e^+e^- bound state) which is extremely well described by QED. In quarkonium the spectrum of excited states can be described by the above potential (equation (1.82)). This leads to a value for the so-called *string constants*, k, of about 1 GeV fm^{-1} and for the *strong coupling constant*, α_s, at a few GeV to a value of $\alpha_s \approx 0.3$. This coupling constant has a strong energy dependence (see chapter 2). Because the theory is normally formulated in powers of the coupling constant and terms of higher order are ignored, perturbation theory fails for QCD at low energies, since α_s at such energies becomes of order unity, and therefore higher order terms have considerable influence. An attempt to overcome the difficulties of a non-perturbative theoretical approach to QCD and its connected divergences is made by constructing a discrete version of the space-time continuum and then studying the mathematical limit in which this lattice is evolved back to the continuum (lattice gauge theories, see e.g. [Cre83, Hua92]). At high energies (for instance 90 GeV), perturbative QCD does however give a good description, as at these energies the coupling constant is relatively small (about 0.1 at 90 GeV).

A further interesting aspect of QCD arises if nucleons can be compressed

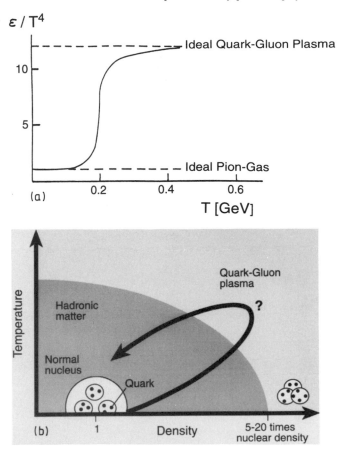

Figure 1.12. (*a*) Behaviour of the energy density ϵ as a function of the temperature at the phase transition to the quark–gluon plasma, as predicted by lattice calculations. The jump occurs because of the many new degrees of freedom of the quark–gluon plasma. Instead of the three degrees of freedom of a pion gas, there are 27 degrees of freedom in the quark-gluon plasma (from [Sat90]). (*b*) Have nuclear collisions at CERN already squeezed matter to create a quark–gluon plasma? The question is still open. The arrow indicates the way of a high energy experiment (from [Tau96a]).

far above the normal nuclear density. If the energy density within the nucleus is increased, at some point a state should be reached in which quarks will no longer be bound in the nucleons. Such a state is known as the *quark–gluon plasma*. Such a state, in which the nucleus consists effectively of free quarks and gluons, is predicted by lattice gauge theories (see e.g. [Mül85, Hwa90]). From this theoretically predicted transition follows a dramatic increase in the number of degrees of freedom, which implies a phase-transition between confinement and deconfinement (see figure 1.12), characterized by the quantity Λ_{QCD} (see chapter

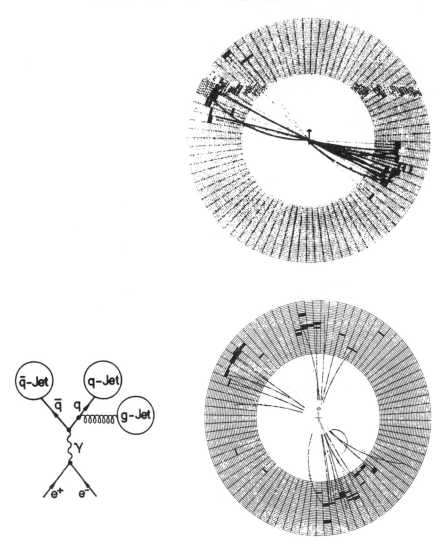

Figure 1.13. A 2- (*a*) and 3-jet (*b*) event taken with the JADE detector at the e^+e^- storage ring PETRA at DESY at $E_{CM} \approx 30$ GeV. The two quarks produced move away from each other in opposite directions, and further quark–anti-quark pairs are produced from the colour field. The hadrons which evolve from this process are bunched around the direction of the primary quarks (jets). The radiation of a hard gluon and its hadronization result via $e^+e^- \rightarrow q\bar{q}g$ in the production of 3-jet events (from [Qui83, Gro89, Gro90]).

2). The critical temperature for this phase transition is at about 200 MeV. It is expected that this state should appear typically at energy densities of about 2.5 GeV fm^{-3}. In normal $p\bar{p}$ collisions in accelerators 0.3 GeV fm^{-3} can

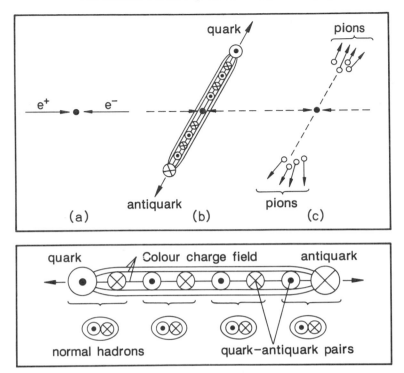

Figure 1.14. Schematic illustration of the process of hadronization in e^+e^- annihilation in a quark–anti-quark pair. More quark–anti-quark pairs are produced in the QCD colour field between the primary quark and anti-quark. They combine to form hadrons (mostly pions). The radiation of gluons can result in the production of multi-jet events (see figure 1.13), which can be described in QCD perturbation theory (from [Loh83]).

be attained. However, it is hoped to reach this state in ultra-relativistic heavy ion collisions (using S, Au or Pb nuclei as projectiles), since the achievable energy densities increase as $A^{\frac{1}{3}}$ [Sin93, Won94]. Due to the complexity of the reactions it is however currently still difficult to make predictions for experimental signatures [Sin93]. However, direct photons, the production of direct lepton-pairs, the increase of the Drell–Yan background or the suppression of J/Ψ production might be used as evidence for the quark–gluon plasma (see e.g. [Sat85, Sat90, Sin93, Won94, Mül95a, Har96a]). Some indications of the latter seem to occur in ultrarelativistic Pb + Pb collisions at CERN [Gon96].

The reaction $e^+e^- \rightarrow q\bar{q}$ in accelerators produces reaction products in the form of hadronic jets, which are illustrated schematically in figure 1.13. These are interpreted as emanating from the hadronization of quark–anti-quark pairs, which were produced in annihilation. The exact mechanism for this fragmentation is however still unclear (see figure 1.14). As only two particles were originally produced, they emerge back-to-back and hence the hadrons are also collimated. Through the emission of hard gluons it is also possible to produce multi-jet events, as the gluon also fragments and produces its own associated jet of hadrons. Recently the triple-gluon vertex has been experimentally verified for the first time at LEP [Ade90, Akr91, Dec92a]. This was achieved by studying the angular correlations of 4-jet events. This is the first direct confirmation that gluons really carry colour, and is evidence for the direct coupling of gauge bosons with each other.

1.5.2 The electroweak interaction

The weak interaction at low energies corresponds to the classic four-fermion point interaction of Fermi [Fer34]. An example is the beta-decay of the neutron $n \rightarrow p + e + \bar{\nu}_e$, which can be depicted as a current–current interaction between a hadronic (j_B^μ) and a leptonic $j_{\mu L}$ current:

$$H = \frac{G_F}{\sqrt{2}} j_B^\mu j_{\mu L}. \tag{1.83}$$

Assuming a $V - A$ (where V stands for a vector and A for an axial vector coupling) structure for the currents, they can be written as

$$j_B^\mu = p\gamma^\mu(1 - \gamma_5)n \tag{1.84}$$

and

$$j_{\mu L} = \nu_e \gamma_\mu(1 - \gamma_5)e. \tag{1.85}$$

In the weak interaction there are only left-handed currents (this is true only for processes which are produced by couplings of *charged* currents, see e.g. [Gro89, Gro90]). All effects of the weak interaction at low energies can be explained with the help of such current–current couplings. The approximation of a four-fermion point interaction becomes invalid at energies approaching the W and Z masses, where the effects of the exchange of these bosons become important. A description at these energies, first achieved by Glashow, Weinberg and Salam, unites the weak and electromagnetic force into a single, electroweak, force [Gla61, Wei67, Sal68]. The SU(2) $\otimes$ U(1) group acts here as the gauge group†. The left-handed quarks and leptons are arranged in doublets while the right-handed particles are singlets:

$$\begin{pmatrix} u \\ d' \end{pmatrix}_L, \quad \begin{pmatrix} c \\ s' \end{pmatrix}_L, \quad \begin{pmatrix} t \\ b' \end{pmatrix}_L, \quad \begin{pmatrix} e \\ \nu_e \end{pmatrix}_L, \quad \begin{pmatrix} \mu \\ \nu_\mu \end{pmatrix}_L, \quad \begin{pmatrix} \tau \\ \nu_\tau \end{pmatrix}_L;$$

$$u_R, \quad d_R, \quad s_R, \quad c_R, \quad b_R, \quad t_R, \quad e_R, \quad \mu_R, \quad \tau_R. \tag{1.86}$$

Here d', s' and b' are the Cabibbo mixed states (see equation (1.94)). The requirement of local gauge invariance results in four gauge bosons. The

---

† More precisely, the theory is made such that spontaneous symmetry breaking of SU(2)$_L \otimes$ U(1)$_Y$ yields an unbroken U(1)$_{EM}$ group describing electromagnetism.

Lagrange density contains expressions of the form (see e.g. [Per87])

$$\mathcal{L} = \frac{g}{\sqrt{2}}(j^{-\mu}W_{\mu}^{+} + j^{+\mu}W_{\mu}^{-}) + \frac{g}{\cos\theta_W}(j_{\mu}^{3} - \sin^2\theta_W j_{\mu}^{em})Z^{\mu} + g\sin\theta_W j_{\mu}^{em}A_{\mu}.$$

(1.87)

The first term describes charged weak currents via the exchange of the $W^{\pm}$ bosons. An example of such a process would be neutron decay. The second term describes neutral weak currents via the exchange of a Z^0 boson. Finally, the third term describes the electromagnetic interaction. The field A_{μ} corresponds to the photon. The coupling to the weak gauge bosons is described by two coupling constants g and g', where in equation (1.87) we have already exploited the relation

$$\tan\theta_W = \frac{g'}{g}$$

(1.88)

The quantity θ_W is known as the *Weinberg angle*. Since furthermore it is also true that

$$\sin\theta_W = \frac{e}{g}$$

(1.89)

it is possible, by precision measurement of the Weinberg angle, to determine the coupling constants and with that two of the fundamental parameters of the theory.

In the formulation of the theory all particles are assumed to be massless. The particles receive mass subsequently through the so-called *Higgs mechanism* [Hig64, Kib67], based on the phenomenon of spontaneous symmetry breaking.

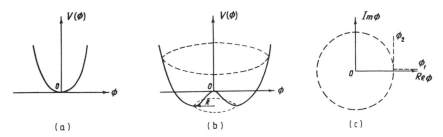

Figure 1.15. The Higgs potential for the description of spontaneous symmetry breaking. While in the case of (*a*), the symmetry is maintained and there is no degeneracy of the vacuum, in the situation shown in (*b*) the vacuum is infinitely degenerate. The symmetry of the system is broken by the choice of a particular ground state. Case (*c*) shows the top plan view in the complex plane.

1.5.2.1 Spontaneous symmetry breaking and the Higgs mechanism

Spontaneous symmetry breaking results in the ground state of a system no longer having the full symmetry corresponding to the underlying Lagrangian. As an example we can consider the ferromagnet. Above a certain temperature

(Curie temperature) all the spins are randomly oriented. Below this temperature, however, an ordering of spins occurs, which leads to ferromagnetism. Although the underlying description of the problem is completely symmetrical, the spins arrange themselves in a certain direction and the symmetry is broken. The Higgs mechanism functions analogously. Consider the potential of a scalar, complex field Φ of the form (see figure 1.15)

$$V(\Phi) = -\mu^2 \Phi^\dagger \Phi + \lambda(\Phi^\dagger \Phi)^2 \tag{1.90}$$

This potential is certainly symmetric under the interchange $\Phi \leftrightarrow -\Phi$. If $\mu^2 > 0$ and $\lambda > 0$, a minimum or equilibrium configuration occurs at $v = \sqrt{\mu^2/\lambda}$. The stable equilibrium positions are situated either at $\Phi = -v$ or $\Phi = v$. However, neither of these two ground states shows the full symmetry of the potential any longer. The symmetry is broken spontaneously. Generally it can be shown that spontaneous symmetry breaking is connected with degeneracy of the ground state. In the simplest way spontaneous symmetry breaking in the electroweak model is achieved via two complex scalar fields ϕ_1, ϕ_2, which are arranged in a doublet

$$\phi(x) = \begin{pmatrix} \phi_1(x) \\ \phi_2(x) \end{pmatrix} \tag{1.91}$$

leading to the following expression for the Lagrangian:

$$\mathcal{L} = (\partial_\mu \phi^\dagger)(\partial^\mu \phi) - (-\mu^2 \phi^\dagger \phi + \lambda(\phi^\dagger \phi)^2) \tag{1.92}$$

Here the minima, corresponding to the vacuum expectation values of ϕ lie at $\langle \Phi \rangle = v = \frac{1}{\sqrt{2}}\sqrt{\mu^2/\lambda}$. The orientation of this ground state in two-dimensional isospin space is however not defined. The Higgs field chooses from the infinite number of possible values one particular value, which results in the symmetry being broken, despite the problem being completely symmetrical. Perturbation theory is then developed around this new vacuum expectation value. Substituting the covariant derivative for the normal derivative leads directly to the coupling of the Higgs field to the gauge fields. Terms follow from the covariant derivative of the Higgs field which can be interpreted as mass terms for the gauge bosons and which lead to equations (1.98) and (1.99). The fermions also get their masses through coupling to the vacuum expectation value of the Higgs field. These are called Yukawa couplings and have the typical form

$$\mathcal{L} = -c_e \bar{e}_R \phi^\dagger \begin{pmatrix} v_{e_L} \\ e_L \end{pmatrix} + \text{h.c.} \tag{1.93}$$

In the spontaneous breaking of a global symmetry a massless, scalar particle develops which is called a *Goldstone boson*. These massless degrees of freedom do not appear in the spontaneous breaking of a local symmetry as they are 'eaten' by the gauge bosons, thereby giving them mass (see e.g. [Nac86]).

1.5.2.2　The CKM mass matrix

It has been shown experimentally that mass eigenstates do not have to be identical with flavour eigenstates. Thus strangeness changing weak currents show that the mass eigenstates of the d and s quark are not identical with the flavour eigenstates. The mass eigenstates s, d and the flavour eigenstates s', d' which take part in the weak interaction are connected by

$$\begin{pmatrix} d \\ s \end{pmatrix} = \begin{pmatrix} \cos\theta_C & \sin\theta_C \\ -\sin\theta_C & \cos\theta_C \end{pmatrix} \begin{pmatrix} d' \\ s' \end{pmatrix}. \tag{1.94}$$

The *Cabibbo angle* θ_C is about $13°$. Considering the general case of three generations leads to the so-called *Cabibbo–Kobayashi–Maskawa matrix* (CKM) [Kob73]. It can be parametrized by:

$$\begin{pmatrix} V_{ud} & V_{us} & V_{ub} \\ V_{cd} & V_{cs} & V_{cb} \\ V_{td} & V_{ts} & V_{tb} \end{pmatrix} = \begin{pmatrix} c_1 & s_1 c_3 & s_1 s_3 \\ -s_1 c_2 & c_1 c_2 c_3 - s_2 s_3 e^{i\delta} & c_1 c_2 s_3 + s_2 c_3 e^{i\delta} \\ -s_1 s_2 & c_1 s_2 c_3 + c_2 s_3 e^{i\delta} & c_1 s_2 s_3 - c_2 c_3 e^{i\delta} \end{pmatrix} \tag{1.95}$$

where $s_i = \sin\theta_i$, $c_i = \cos\theta_i$ $(i = 1, 2, 3)$. The individual matrix elements describe transitions between the quarks and have to be experimentally determined. The present experimental results and the constraints of unitarity give the values (90% confidence) [PDG96]:

$$\begin{pmatrix} 0.9745 - 0.9757 & 0.219 - 0.224 & 0.002 - 0.005 \\ 0.218 - 0.224 & 0.9736 - 0.9750 & 0.036 - 0.046 \\ 0.004 - 0.014 & 0.034 - 0.046 & 0.9989 - 0.9993 \end{pmatrix}. \tag{1.96}$$

The phase $e^{i\delta}$ can be linked with CP violation. The necessary condition for CP invariance of the Lagrangian is that the Cabibbo–Kobayashi–Maskawa matrix and its complex conjugate are identical, i.e. its elements are real. While this is always the case for two families, for three families in the above parametrization it is only true if $\delta = 0$ or $\delta = \pi$. This means that if δ does not equal one of those values then the CKM matrix is a source of CP violation (see e.g. [Nac86]). Another source of CP violation will be discussed in chapter 11. In the leptonic sector the issue seems somewhat simpler: as long as neutrinos are assumed to be massless there will be no mixing of weak eigenstates. However, if neutrinos should be massive, this will lead to a number of new processes, for example the effect of neutrino oscillations (see chapter 2). Furthermore, if neutrinos are Majorana particles, there would already be the possibility of CP violation with two families [Wol81].

1.5.2.3　Experimental tests

We now consider predictions of the theory which can be experimentally tested. The existence of the neutral gauge boson Z^0 in the theory implies the existence

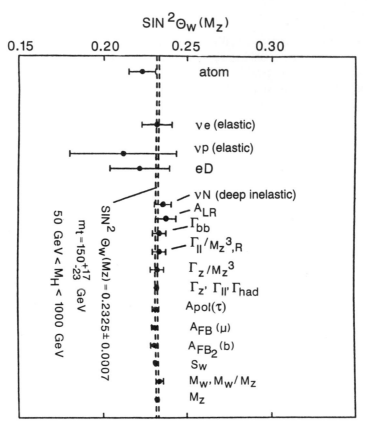

Figure 1.16. The measurement of the Weinberg angle at different energy scales, corrected to the value at the Z^0 mass. The measurements are consistent across the energy range, from parity violation in atoms to the Z^0 mass (from [Lan93a]).

of weak neutral currents. They were discovered at CERN in 1973, six years after the development of the theory, in the reactions $v_\mu N \rightarrow v_\mu X$ [Has73]. In the limit of low energies (four-momentum transfer $Q^2 \ll m_{W,Z}^2$) the electroweak model is equivalent to the Fermi point interaction and the parameters of both theories can be connected with one another, for example (see e.g. [Nac86])

$$\frac{G_F}{\sqrt{2}} = \frac{g^2}{8m_W^2}.$$ (1.97)

From equations (1.97) and (1.89) the masses of the vector bosons $W^\pm$ and Z^0 can be predicted:

$$m_W^2 = \frac{g^2}{4\sqrt{2}G_F} \rightarrow m_W = \frac{37.4}{\sin\theta_W} \text{ GeV}$$ (1.98)

Figure 1.17. View of the ALEPH detector as an example of one of the four detectors (together with OPAL, DELPHI and L3) at the e^+e^- storage ring LEP at CERN in Geneva. Starting from the centre and working outwards, the two time-projection chambers (TPC) can be seen, followed by the electromagnetic calorimeter, the magnet and the hadronic calorimeter (with kind permission of the ALEPH collaboration).

$$m_Z = \frac{m_W}{\cos\theta_W} = \frac{75}{\sin 2\theta_W} \text{ GeV.} \qquad (1.99)$$

The discovery of the $W^\pm$ and Z^0 bosons at CERN in 1983 was the final breakthrough for the standard model [Arn83]. The accelerators LEP (Geneva) and SLC (Stanford) have produced via e^+e^- annihilation several million Z^0 particles

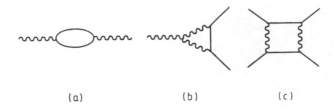

(a)   (b)   (c)

Figure 1.18. Typical examples of higher order corrections, which have to be taken into account in precision experiments: (*a*) vacuum polarization, (*b*) vertex correction, (*c*) box diagram.

since 1989. From these experiments it has been possible to determine the mass of the Z^0 very accurately as $m_Z = 91.187 \pm 0.007$ GeV [PDG96]. From the accurate determination of the width of the Z^0 resonance, important statements can be made about the number of neutrinos or the existence of certain exotic particles (see chapter 4). Z^0 decays also allow the possibility of determining the Weinberg angle, a central parameter of the theory. This can, however, also be extracted among other processes, from neutrino–electron or neutrino–nucleon scattering experiments. The best value obtained from all these experiments is [PDG96] (see figure 1.16)

$$\sin^2 \theta_W(m_Z) = 0.2315 \pm 0.0002 \pm 0.0003. \tag{1.100}$$

The measurement of the W mass at Fermilab (Chicago) and by the UA2 collaboration (CERN) result in $m_W = 80.41 \pm 0.18$ GeV [Abe95b] and $m_W/m_Z = 0.8813 \pm 0.0041$ [Ali92] which are in perfect agreement with the predicted values. So far the standard model has been able to explain all the available data to a remarkable degree of accuracy (see table 1.6).

1.5.2.4 *Precision tests at* LEP *and open questions*

Through the measurements at the e^+e^- collider LEP, a number of the parameters of the electroweak theory have been determined. These include in particular all parameters of the Z^0 resonance (see chapter 4). Figure 1.17 shows as an example one of the LEP detectors. In addition LEP supplies a precise determination of the strong coupling constant (see chapter 2) and also has made important contributions to the physics of B-mesons (see e.g. [Ste91]). From all these data, statements about possible missing ingredients of the standard model can be made, as these can contribute at higher orders. Figure 1.18 shows some of these typical loop diagrams. For example, the relation

$$\sin^2 \theta_W = 1 - \frac{m_W^2}{m_Z^2} \tag{1.101}$$

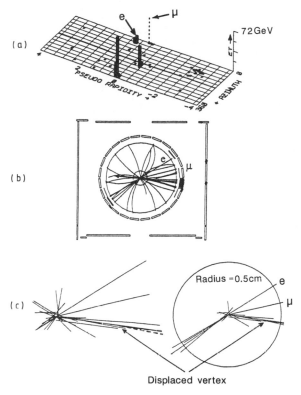

Figure 1.19. A candidate event from a top quark decay. The top quark was produced by the reaction $p\bar{p} \to t\bar{t}$ at $E_{CM} = 1.8$ TeV. It decays into a b quark and a W boson. In this case an electron and also a muon from the decay of the two W bosons are found. (*a*) Shows the event as an angular plot in which the isolated high energy electron as well as the high energy jets of the b quarks can be seen. (*b*) Shows in section the response of the muon detector as well as the evidence for the electron in the calorimeter. (*c*) The b decay is well identified by the displacement of the decay vertex. Position resolutions sufficiently good to allow the identification of such vertices have only been possible with the advent of silicon vertex detectors (from [Abe94]).

is corrected through higher order contributions (so-called radiative corrections) to [Ell91b]

$$m_W^2 \sin^2 \theta_W = m_Z^2 \cos^2 \theta_W \sin^2 \theta_W = \frac{(37.28 \,\text{GeV})^2}{1 - \Delta r}. \tag{1.102}$$

The size of the correction factor Δr depends, for example, on the masses of the top-quark and the Higgs boson. The scalar Higgs boson, arising from spontaneous symmetry breaking, should have a mass of $m_H = \sqrt{2\mu^2}$. Unfortunately the theoretical predictions are sufficiently uncertain that a range

Figure 1.20. The central part of the CDF detector at the Tevatron $p\bar{p}$ accelerator at Fermilab in Chicago. The two half-shells on the left and right edges are the hadronic calorimeter. In the centre is the magnet surrounding the tracking chambers. This experiment found the first evidence for the top quark (with kind permission of the CDF collaboration, Fermilab).

of between 7 GeV and 1.4 TeV is allowed (see e.g. [Nac86]). LEP results can exclude a Higgs boson lighter than about 60 GeV [Mor93a], from the reaction

$$e^+e^- \to Z^0 \to H\,e^+e^-   \text{ with }   H \to e^+e^-. \tag{1.103}$$

Energy and momentum conservation allow the topology of the decay mode to be distinguished from the normal e^+e^- modes. A major part of the current effort in particle physics is devoted to the search for the Higgs particle [Gun90]. A further important result of the LEP experiments is the measurement of the strong coupling constant at the Z^0 mass (see chapter 2).

The long-sought top quark has at last been discovered at the $p\bar{p}$ collider at Fermilab (Chicago) [Abe95a, Aba95]. Since it is heavier than the weak gauge

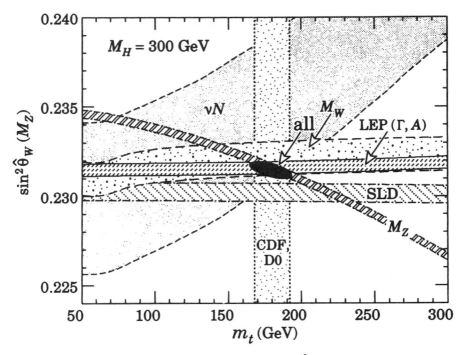

Figure 1.21. One-standard-deviation uncertainties in $\sin^2 \theta_W$ as a function of the top quark mass m_t, the direct CDF and D0 range 180 ± 12 GeV, and the 90% CL region in $\sin^2 \theta - m_t$ allowed by all the data, assuming $M_H = 300$ GeV (from [PDG96]).

bosons, its dominant decay channel is

$$t \rightarrow b + W \qquad (1.104)$$

By examining W-decays with simultaneous evidence for a b quark, several events have been extracted which are compatible with the signature of a top decay (see figure 1.19). Figure 1.20 shows as an example one of the detectors used in this discovery, the CDF detector. The top quark mass is determined to be 176.8 ± 4.4(stat.) ± 4.8(sys.) GeV (CDF experiment [Abe95a, Ger96a]) and 173.3 ± 5.6(stat.) ± 6.2(sys.) GeV (D0-Experiment [Aba95, Aba97]). This is much greater than that of other quarks. The results are in agreement with the LEP data, (see figure 1.21) which, from the contribution to the radiative corrections discussed above, predicted that the mass would be $m_t = 150^{+17}_{-23} \pm 17$ (the last error is due to the uncertainty in m_H) GeV [Ell91b].

With the higher energies which are attained at the 'LEP-II', it is possible to examine $W^+ W^-$ pair-production and hence the direct coupling of the gauge bosons with each other. In any case it should be possible to produce the Higgs particle at the planned LHC accelerator at CERN, as it will reach an energy

Table 1.6. Comparison of measured electroweak quantities with expectations from the standard model (from [Lan93a, Lan95]).

Quantity	Measurement	Standard model
M_Z (GeV)	91.187 ± 0.007	input
Γ_Z (GeV)	2.492 ± 0.007	$2.493 \pm 0.001 \pm 0.005$
$\Gamma_{l\bar{l}}$ (MeV)	83.33 ± 0.007	$83.74 \pm 0.03 \pm 0.13$
Γ_{had} (MeV)	1737.1 ± 6.7	$1741 \pm 1 \pm 4$
$\Gamma_{b\bar{b}}$ (MeV)	373 ± 9	$376.4 \pm 0.2 \pm 0.3$
Γ_{inv} (MeV)	504.6 ± 5.8	$500.8 \pm 0.1 \pm 0.9$
N_ν	3.04 ± 0.04	3
M_W	80.41 ± 0.18	$80.23 \pm 0.02 \pm 0.13$
M_W/M_Z	0.8813 ± 0.0041	$0.8798 \pm 0.0002 \pm 0.0014$
$g_A^e(\nu e \to \nu e)$	-0.503 ± 0.017	$-0.506 \pm 0.0 \pm 0.0014$
$g_V^e(\nu e \to \nu e)$	-0.025 ± 0.02	$-0.037 \pm 0.001 \pm 0.001$
$\sin^2\Theta_W$	0.2242 ± 0.0042	$0.2259 \pm 0.0003 \pm 0.0025$

region of greater than 10 TeV. Should the Higgs not be found, a revision of the standard model will be necessary.

Another important field in the future will lie in the investigation of CP violation. The search for direct CP violation (ϵ' non-zero) and a possible CP violation in the B system will receive special attention. The latter is particularly interesting, as a cosmic baryon asymmetry during the GUT phase transition cannot be explained by the size of the CP violation in the K system alone, and therefore additional mechanisms have to be invoked (see chapter 3). Furthermore the question of the mass of neutrinos has become a central plank in current research. The above form only a small part of the remaining open questions.

In spite of the successes of the standard model, it is generally believed that it cannot be the ultimate theory. The standard model has 18 free parameters which can only be determined by experiment. These are, for example, three coupling constants, six quark masses, three lepton masses, four free parameters of the CKM matrix and the W and Higgs mass. Furthermore it does not give an explanation for the mass hierarchies, the quantization of electric charge, etc. However, what has undoubtedly succeeded has been the unification of two of the fundamental forces at higher energies. The question now arises as to whether a further unification of forces at still higher energies can be achieved, and if in the end everything can be reduced to a single 'original force'. We turn to this subject in the next chapter.

Chapter 2

Grand unified theories (GUTs)

Having discussed in the previous chapter the successful unification of electromagnetic and weak interactions, we now consider an even wider unification. The aim is to derive *all* interactions from the gauge transformations of *one* group G. Such theories are known as grand unified theories (GUTs). The grand unified group must contain the $SU(3) \otimes SU(2) \otimes U(1)$ group as a subgroup, i.e.

$$G \supset SU(3) \otimes SU(2) \otimes U(1). \tag{2.1}$$

In addition the gauge group must be *simple* so that the unified interaction can be described by a single coupling constant (we will refrain here from discussing other, very specific solutions). The gauge transformations of a simple group, which act on the particle multiplets characteristic for this group, result in an interaction between the elements within a multiplet which is mediated by a similarly characteristic number of gauge bosons. From *one* simple group only *one* individual interaction with *one* typical coupling constant can be derived. The three well known and completely different coupling constants can be derived in the end from a single one only if the symmetry associated with the group G is broken in nature.

2.1 Coupling constants

By comparing the two coupling constants α_{em} and G_F it can be seen that their relative strengths are energy dependent. For lower energies the masses of the W and Z bosons, which appear as an effect of spontaneous symmetry breaking, play an important role and result in the point-like behaviour of the weak interaction at low energies. At higher energies, $(E \gtrsim m_{W,Z})$, a description using exchange particles is necessary. At such energies both forces have similar strength. In quantum field theory, for an unbroken symmetry, the self-interaction of the fields can give rise to energy dependent coupling constants. One example of such an effect is the vacuum polarization, which shows itself for example in the Lamb shift of atomic spectra. This results from the interaction of photons with

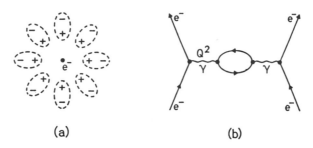

Figure 2.1. (*a*) An electric charge embedded into a dielectric produces polarization charges in its neighbourhood, which weakens the original field. In QED the vacuum can be understood to be the dielectric, and the virtual electron–positron pairs represent the polarization charges. This is called vacuum polarization. (*b*) The figure shows the modification of the e^+e^- interaction in the lowest order due to the vacuum polarization.

virtual e^+e^--pairs in the vacuum. Electric charges polarize the virtual e^+e^--pairs existing in the vacuum, thus resulting in a shielding of the true charge (see figure 2.1). The vacuum reacts to a charge in a similar manner as a dielectric. At larger distances, corresponding to lower energies, one sees the completely shielded charge, while at shorter distances, corresponding to higher energies, ever more of the real charge becomes visible (figure 2.3). Moreover, since the effects of vacuum polarization and other higher order effects cannot be switched off, the bare charge e_0 is not observable. In contrast to this behaviour, gluons produce an anti-shielding effect, as they themselves carry colour charge and thus contribute to a 'smearing out' of the colour charge of the quarks. From this the phenomena of confinement and asymptotic freedom can be deduced (see figure 2.2). A 'true' charge *in principle* cannot be measured. This is taken into account via an energy dependent, so-called *running* or *effective* coupling constant. Figure 2.2(*b*) shows the results of recent measurements of this effect for the strong coupling constant α_S.

Theoretically a consistent description is only possible in *renormalizable* theories. In these the assumption is made that the observed charges do not correspond to the bare parameters of the theory, but to the result of a corresponding perturbation theory (see e.g. [Ait89]). Perturbation theory is a series in powers of the coupling constant. A theory in which, at each order in the perturbation series, new types of divergence appear whose elimination requires an infinite number of parameters, is known as 'non-renormalizable'. If, on the other hand, only a finite number of divergences appear, which can be eliminated with a limited number of experimentally determined parameters to all orders, the theory is renormalizable. Therefore, the theoretical proof that all spontaneously broken or unbroken gauge theories are renormalizable was of fundamental importance [t'Ho71, t'Ho72]. Many of these higher order renormalization effects have already been taken into account in the energy dependent coupling constant.

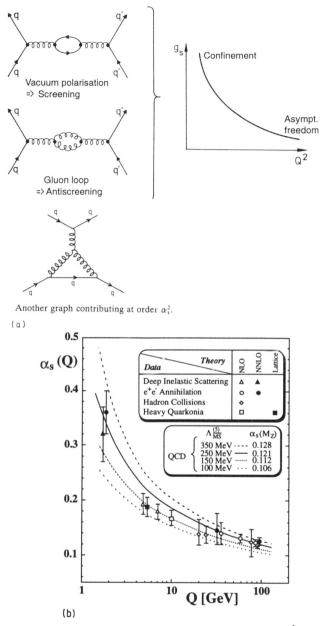

Figure 2.2. (*a*) Q^2-dependence of the strong coupling constant $\alpha_s = \frac{g_s^2}{4\pi}$. In addition to the screening effect (top left) analogous to figure 2.1, there is also an anti-screening effect (the two graphs bottom left) due to the gluon self-interaction; the anti-screening dominates and leads to the effects of asymptotic freedom and confinement. (*b*) Compilation of α_s measurements as a function of the energy scale (from [Pic95]).

Part of this renormalization effect is incorporated in the vacuum polarization. As these effects are dependent on the square of the four-momentum transfer, q^2, the electric charge, as the coupling constant of electromagnetism, is also dependent on q^2, i.e. $e_{\text{eff}} = e_{\text{eff}}(q^2)$, or $\alpha_{\text{em}} = \alpha_{\text{em}}(q^2)(\alpha_{\text{em}} = \frac{e^2}{4\pi})$. Because of energy and momentum conservation the exchanged virtual photon always has a negative q^2, so that the positive quantity $Q^2 = -q^2$ is normally introduced.

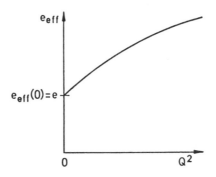

Figure 2.3. Modification of $e_{\text{eff}}(Q^2)$ by means of vacuum polarization (schematic).

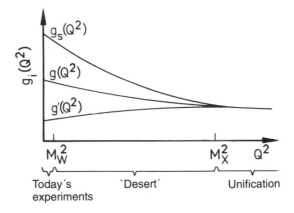

Figure 2.4. Schematic illustration of the Q^2 dependence of the effective coupling constants g and g' of the electroweak interaction as well as the strong coupling constant g_s. Above $Q^2 = M_X^2$ they all merge into a single coupling constant (from [Gro89, Gro90]).

According to Heisenberg's uncertainty principle, larger four-momentum transfers Q^2 correspond to smaller distances between the interacting charges, so that a smaller shielding effect is seen. This implies a difference between the observed charge e_{eff} in high-energy scattering experiments ($Q^2 \gg 0$) in comparison to the electrostatic case ($Q^2 = 0$). In the case of QED, the effective charge therefore grows with decreasing distance (see figure 2.3), leading to the following behaviour for the coupling constant [Ait89, Ber94a]:

$$\alpha_{\rm em}(Q^2) = \frac{\alpha(m_e^2)}{1 - \frac{\alpha(m_e^2)}{3\pi} \ln \frac{Q^2}{4m_e^2}}. \tag{2.2}$$

In contrast, an additional anti-screening effect is produced by the colour carrying gluons in QCD. The behaviour of the strong coupling constant can be described by [Gro73, Pol73]

$$\alpha_s(Q^2) = \frac{12\pi}{33 - 2n_{\rm f}} \frac{1}{\ln(Q^2/\Lambda_{\rm QCD}^2)}. \tag{2.3}$$

Provided the number of quark flavours $n_{\rm f} < 17$, the coupling constant decreases with increasing energy. The parameter $\Lambda_{\rm QCD}$ (which we have already met in the context of the quark–gluon plasma, see chapter 1) describes a scale factor, below which confinement becomes essential. It has a value of approximately 200 MeV. The energy dependence of the strong coupling constant reveals itself dramatically at lower energies. Precision measurements at the Z^0 resonance have made it possible to measure $\alpha_s(Q^2 = m_{Z^0}^2)$ in several different ways. One of the methods relies on the fact that in perturbation theory the hadronic width of the Z^0 is proportional to a power series in α_s. Another method relies on the measurement of three jet events, which originate from the emission of a hard gluon, the rate of which is therefore also proportional to the strong coupling constant. These investigations result in a value of (see e.g. [PDG96])

$$\alpha_s(Q^2 = m_{Z^0}^2) = 0.123 \pm 0.004 \pm 0.002. \tag{2.4}$$

In general there are equations in gauge theories which describe the behaviour of coupling constants α_i as a function of Q^2. These so-called 'renormalization group equations' have the general form up to second order perturbation theory (see e.g. [Lan93b, Lop94, deB94]):

$$\frac{d\alpha_i^{-1}}{d\ln\mu} = -\frac{b_i}{2\pi} - \sum_{j=1}^{3} \frac{b_{ij}\alpha_j}{8\pi^2} \tag{2.5}$$

where $\alpha_i(Q^2) = g_i^2(Q^2)/4\pi$, and g_i is the gauge coupling of the ith group. μ represents the Q^2 scale from which the extrapolation begins. This is usually taken to be the mass of the Z^0 boson. The b_i, b_{ij} are coefficients depending on the group and the particle content of the theory. Assuming that second order corrections can be neglected (corresponding to the second term in equation 2.5), the equation has the following solution:

$$\frac{1}{[\alpha(\mu)]} = \frac{1}{[\alpha(M_X)]} + \frac{b_i}{2\pi} \ln\left(\frac{\mu}{M_X}\right) \tag{2.6}$$

where M_X is a typical value for the unification scale. It is therefore possible with this equation to extrapolate to energies which are not accessible in accelerators. The coefficients b_i determine the behaviour of the coupling constants. Assuming also that the only particles which exist above the Z^0 mass are those known to be in the standard model, then the coefficients become [Lan93b]

$$b_i = \begin{pmatrix} b_1 \\ b_2 \\ b_3 \end{pmatrix} = \begin{pmatrix} 0 \\ -22/3 \\ -11 \end{pmatrix} + N_{\text{fam}} \begin{pmatrix} 4/3 \\ 4/3 \\ 4/3 \end{pmatrix} + N_{\text{Higgs}} \begin{pmatrix} 1/10 \\ 1/6 \\ 0 \end{pmatrix} \quad (2.7)$$

where N_{fam} represents the number of fermion families and N_{Higgs} the number of Higgs doublets. Figure 2.4 illustrates this behaviour schematically. Because of the precision measurements at LEP, the values of the coupling constants at the Z^0 mass are normally chosen as a reference point from which to extrapolate, (see e.g. [Lan93b]).

The electromagnetic coupling is given by:

$$\alpha_{em}^{-1}(Q^2 = m_{Z^0}^2) = 127.9 \pm 0.1 \quad (2.8)$$

Further we have equations (2.4) and (1.100). After a sufficiently large extrapolation is carried out, all three coupling constants should meet at a point roughly at a scale of 10^{15} GeV and from that point on an unbroken symmetry with a single coupling constant should exist (but see section 2.5). This would mean that, in this simple model, no new physics would be expected over 12 orders of magnitude in energy.

We now proceed to consider which groups are suitable as unification groups. The simplest group for our purposes is SU(5). We will therefore first discuss the minimal SU(5) model (Georgi–Glashow model) [Geo74].

2.2 The minimal SU(5) model

For massless fermions the gauge transformations fall into two independent classes, for left- and right-handed fields, respectively. Let us assume the left-handed fields are the elementary fields (the right-handed transformations are equivalent and act on the corresponding charge conjugated fields). We simplify matters by considering only the first family, consisting of u, d, e and ν_e, giving 15 elementary fields:

$$\begin{matrix} u_r, u_g, u_b, \nu_e \\ u_r^c, u_g^c, u_b^c, d_r^c, d_g^c, d_b^c \qquad\qquad e^+ \\ d_r, d_g, d_b, e^- \end{matrix} \quad (2.9)$$

By the notation d we mean in this context, and later, the mixed state d' (see section 1.5.2.2). In this arrangement the $SU(3)_C$-triplets are arranged horizontally and the $SU(2)_L$-doublets vertically. If one wished to classify these fields into 3 fundamental multiplets, each multiplet would contain 5 fields. However, through the combination of the weak $SU(2)_L$- and the strong $SU(3)_C$-transformations the 6 fields $u_r, u_g, u_b, d_r, d_g, d_b$ can be transformed into each other. Therefore they have to belong to one multiplet, as in principle only the fields of a multiplet can transform into each other. This implies that a higher representation is necessary, the next higher representation after the 5-dimensional being a decuplet (10-dimensional). In this way the fields can be classified into a 10- and a 5-dimensional representation (the complementary representation to the fundamental representation 5, although this is not significant for our current purposes). The following arrangement of fields results from the condition that the sum of the charges in every multiplet has to be zero:

$$\bar{5} = \begin{pmatrix} d_g^C \\ d_r^C \\ d_b^C \\ e^- \\ -\nu_e \end{pmatrix} \quad 10 = \frac{1}{\sqrt{2}} \begin{pmatrix} 0 & -u_b^C & +u_r^C & +u_g & +d_g \\ +u_b^C & 0 & -u_g^C & +u_r & +d_r \\ -u_r^C & +u_g^C & 0 & +u_b & +d_b \\ -u_g & -u_r & -u_b & 0 & +e^+ \\ -d_g & -d_r & -d_b & -e^+ & 0 \end{pmatrix}. \quad (2.10)$$

The minus signs in these representations are conventional. The SU(5)-gauge transformations of these multiplets can be written as

$$\bar{5}' = e^{i(\sum_{j=1}^{24} \alpha_j(x)\tilde{T}_j)}\bar{5} \tag{2.11}$$

and

$$10' = e^{i(\sum_{j=1}^{24} \alpha_j(x)\tilde{T}_j)} 10 e^{-i(\sum_{k=1}^{24} \alpha_k(x)\tilde{T}_k)}. \tag{2.12}$$

SU(5) has 24 generators T_j (it is well known that an SU(N)-group has $N^2 - 1$ generators), to which correspond 24 gauge fields B_j, which can be written in matrix form as:

$$24 = \sqrt{2} \sum_j T_j B_j =$$

$$\begin{pmatrix} G_{11} - \frac{2B}{\sqrt{30}} & G_{12} & G_{13} & X_1^C & Y_1^C \\ G_{21} & G_{22} - \frac{2B}{\sqrt{30}} & G_{23} & X_2^C & Y_2^C \\ G_{31} & G_{32} & G_{33} - \frac{2B}{\sqrt{30}} & X_3^C & Y_3^C \\ X_1 & X_2 & X_3 & \frac{W^3}{\sqrt{2}} + \frac{3B}{\sqrt{30}} & W^+ \\ Y_1 & Y_2 & Y_3 & W^- & -\frac{W^3}{\sqrt{2}} + \frac{3B}{\sqrt{30}} \end{pmatrix}.$$

$$\tag{2.13}$$

Here the 3×3 sub-matrix G characterizes the gluon fields of QCD, and the 2×2 submatrix W contains the gauge fields of the electroweak theory. In addition to the gauge bosons known to us there are, however, a further twelve gauge bosons X, Y, which mediate transitions between baryons and leptons.

The SU(5)-symmetry has, however, to be broken in order to result in the standard model. Here also the breaking occurs through the coupling to

Higgs fields. The breaking scale lies at about 10^{15} GeV. The breaking can be produced through a 24-dimensional Higgs field with a vacuum expectation value of about 10^{15}–10^{16} GeV. This means that all particles receiving mass via this breaking (e.g. the X, Y-bosons) have a mass which is of the order of magnitude of the unification energy. Suitable SU(5) transformations can result in only the X and Y bosons coupling to this vacuum expectation value, while the other gauge bosons remain massless. A SU(5)-invariant interaction of the Higgs fields with the fermions is also not possible, so that the latter also remain massless. For the breaking of the SU(2) at about 100 GeV a further, independent 5-dimensional Higgs field is necessary, which gives the W, Z-bosons and the fermions their mass. We can now leave this simplest unifying theory and consider its predictions. For a more detailed description see e.g. [Lan81].

Two predictions can be seen immediately from (2.10):

(i)   Since the sum of charges has to vanish in a multiplet, the quarks have to have 1/3 multiples of the electric charge. For the first time the appearance of non-integer charges is required.

(ii) From this also follows immediately the equality of the absolute value of the electron and proton charge.

A further result is:

(iii) The relation between the couplings of the $\mathcal{B}$-field to a SU(2) doublet (see equation (1.86)) compared to that of the W_3-field is, according to equation (2.13), given by $(3/\sqrt{15}) : 1$. This gives a prediction for the value of the Weinberg angle θ_W [Lan81]:

$$\sin^2 \theta_W = \frac{g'^2}{g'^2 + g^2} = \frac{3}{8}. \tag{2.14}$$

This value is only valid for energies above the symmetry breaking. If renormalization effects are taken into consideration, a slightly lower value of

$$\sin^2 \theta_W = (0.218 \pm 0.006) \ln \left(\frac{100 \, \text{MeV}}{\Lambda_{\text{QCD}}} \right) \tag{2.15}$$

results. This value is in agreement with the experimentally determined value (see chapter 1).

(iv) The theory also makes a prediction for the ratio of the b quark to τ lepton masses, which are in the same multiplet, i.e. $m_b/m_\tau \approx 3$, in good agreement with the experimental value of 2.4 [PDG94].

(v)   Probably the most dramatic prediction is the transformation of baryons into leptons due to X, Y exchange. This would, among other things, permit the decay of the proton and with it ultimately the instability of all matter.

Because of the importance of this last process we will deal with it in more detail.

2.2.1 Proton decay

As baryons and leptons are in the same multiplet it is possible that protons and bound neutrons can decay. The main decay channels in accordance with the SU(5) model are [Lan81]:

$$p \rightarrow e^+ + \pi^0 \qquad (2.16)$$

and

$$n \rightarrow \nu + \omega. \qquad (2.17)$$

Here the baryon number is violated by one unit. We specifically consider proton decay. The process $p \rightarrow e^+ + \pi^0$ should amount to about 30–50% of all decays (table 2.1). Because of the large mass of the exchange bosons, the force should be extremely short range. Assuming an effective coupling constant g_X such that

$$g_X = \frac{\alpha_5}{M_X^2} \qquad (2.18)$$

with the SU(5) coupling constant α_5 given by

$$\alpha_5 = \frac{g_5^2}{4\pi} \qquad (2.19)$$

(this is to be seen as formally analogous to the connection between the Fermi constant and W boson mass

$$\frac{G_F}{\sqrt{2}} = \frac{g^2}{8m_W^2} \Bigg), \qquad (2.20)$$

the proton decay can be calculated analogously to the muon decay, resulting in a lifetime:

$$\tau_p \approx \frac{M_X^4}{\alpha_5^2 m_p^5}. \qquad (2.21)$$

Using the renormalization group equations, the two quantities M_X and α_5 can be estimated as [Lan81, Lan86]

$$M_X \approx 1.3 \times 10^{14} \, \text{GeV} \frac{\Lambda_{\text{QCD}}}{100 \, \text{MeV}} \pm (50\%) \qquad (2.22)$$

$$\alpha_5(M_X^2) = 0.0244 \pm 0.0002.$$

The minimal SU(5) model thus leads to the following prediction for the dominant decay channel [Lan86]:

$$\tau_p(p \rightarrow e^+\pi^0) = 6.6 \times 10^{28\pm0.7} \left[\frac{M_X}{1.3 \times 10^{14}\text{GeV}} \right]^4 \text{years or} \qquad (2.23)$$

$$\tau_p(p \rightarrow e^+\pi^0) = 6.6 \times 10^{28\pm1.4} \left[\frac{\Lambda_{\text{QCD}}}{100\text{MeV}} \right]^4 \text{years}. \qquad (2.24)$$

With $\Lambda_{QCD} = 200$ MeV the lifetime becomes $\tau_p = 1.0 \times 10^{30\pm1.4}$ years. For reasonable assumptions on the value of Λ_{QCD}, the lifetime should therefore be smaller than 10^{32} years. Besides the uncertainty in Λ_{QCD}, additional sources of error in the form of the quark wavefunctions in the proton must be considered. These are contained in the error on the exponent.

Table 2.1. Expected branching ratios for proton decay in the SU(5) model (from [Luc86]).

Decay mode	Branching ratio (%)
$p \to e^+\pi^0$	31–46
$p \to e^+\eta$	0–8
$p \to e^+\rho^0$	2–18
$p \to e^+\omega$	15–29
$p \to \bar{\nu}_e\pi^+$	11–17
$p \to \bar{\nu}_e\rho^+$	1–7
$p \to \mu^+K^0$	1–20
$p \to \bar{\nu}_\mu K^+$	0–1

Table 2.2. μ and n fluxes (without shielding) in some underground laboratories (from [Kla95a]).

Laboratory	Depth (mwe)	μ flux ($m^{-2} \times d^{-1}$)	n flux ($cm^{-2} \times s^{-1}$)
Mont Blanc	5000	0.7	2.2×10^{-5}
Gran Sasso	3500	16	5.3×10^{-6}
Frejus	4000	8	$< 3 \times 10^{-5}$
Broken Hill Silver mine	3300	(20)	——
Solotvina Salt mine	1000	1.5×10^3	$< 2.7 \times 10^{-6}$
Baksan	660	7×10^3	3×10^{-5}
Windsor Salt mine	650 (350 m)	(7×10^3)	——

We now review the current experimental situation. Proton decay is, according to the above estimates, an extremely rare process. Even in one ton of matter (about 6×10^{29} nucleons), only about one decay per decade can be expected. It is therefore necessary to develop detectors of the order of magnitude of 100 tons and more, and also to reach a very low level of background. For this reason mines or tunnels are used in order to shield against cosmic radiation

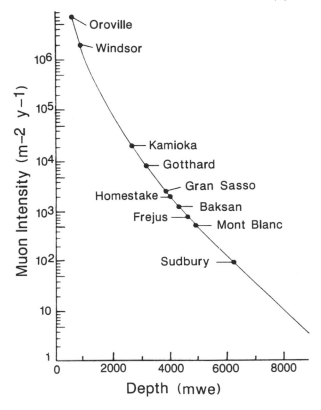

Figure 2.5. Muon intensities as a function of the depth of some underground laboratories. The depth is usually given in mwe (meter water equivalent) and corresponds to the height of a water column producing the same screening effect as the rock (from [Lon92, 94]).

(see figure 2.5 and table 2.2). The remaining major source of background is reactions induced by atmospheric neutrinos. In principle there are two different detection strategies: one using the Cerenkov effect and the other the calorimetric method (see e.g. [Per84, Gro89, Kla95]) (table 2.3). Figure 2.6 shows the current state of proton decay experiments as well as some future possibilities. Some of the detectors built using these methods will be discussed later in more detail in connection with neutrinos. The Frejus detector (see figure 2.7) (in operation from 1984–1988) consisted of 900 tons of iron and functioned as a calorimeter. The very thin (3 mm) iron plates were interspersed by orthogonally arranged spark chambers and Geiger counters, which enabled three-dimensional track reconstruction. The main signature for the decay $p \rightarrow e^+ + \pi^0$ would be two diametrically opposite electromagnetic showers. These originate from the annihilation of the positron and the decay of the pion ($\pi^0 \rightarrow 2\gamma$). In Cerenkov detectors the signature would be two oppositely positioned Cerenkov cones.

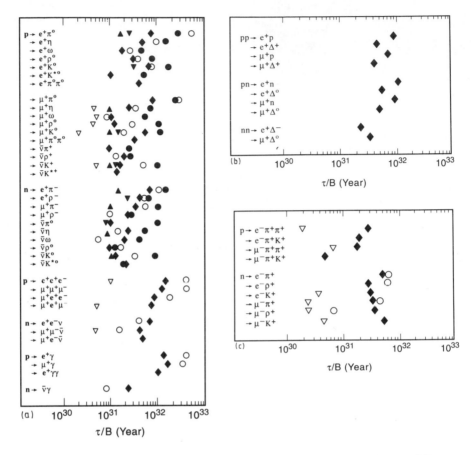

Figure 2.6. Present experimental half-life limits (90% confidence level) for the different decay channels of the nucleon of the form: (a) $\Delta B = 0$, $\Delta(B - L) = 0$ where $N \to \bar{l} +$ meson(s), $N \to \bar{l}ll$ and $N \to \bar{l}\gamma$, (b) $\Delta B = 1$, $\Delta(B - L) = 0$ where $NN \to \bar{l}N$ and $NN \to \bar{l} + \Delta$, (c) $\Delta B = 1$, $\Delta(B - L) = 2$ where $N \to l^-$ and meson(s). The symbols correspond to: ∇ = HPW, $\bullet$ = Kamioka, O = IMB, ∇ = Kolar, $\triangle$ = NUSEX and $\diamond$ = Frejus (from [Bar92a]). (d) Future possibilities with the ICARUS detector (with kind permission of C Rubbia). (e) Future possibilities with Super Kamiokande, see section 12.2.2 (with kind permission of Y Totsuka): $\bullet$ = Kamiokande limit, O = IMB, + = Frejus, /// = sensitivity of SuperK.

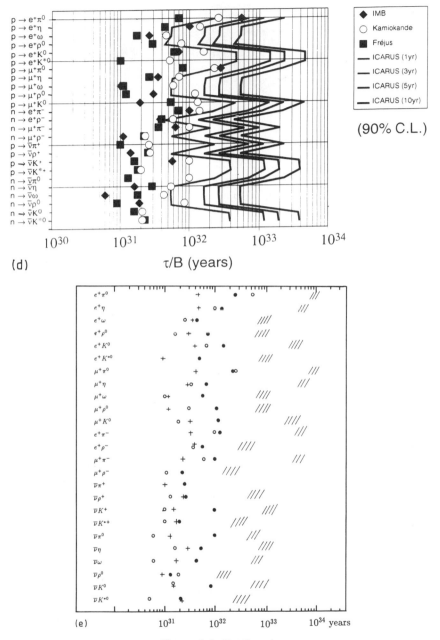

Figure 2.6. Continued.

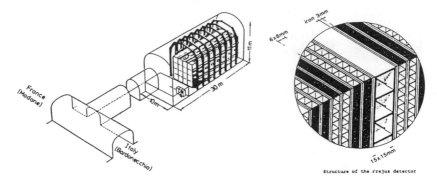

Figure 2.7. Illustration of the Frejus detector for the detection of proton decay. It consisted of a sandwich-like arrangement of spark chambers, Geiger tubes and iron plates (from [Mey86]).

However, no significant number of events have been detected above the expected background. The IMB experiment (see figure 2.8), also now no longer running, has produced for this decay channel an upper limit of $\tau_p > 5.5 \times 10^{32}$ years [Gaj92], which together with the limits from Kamiokande ($\tau_p > 1.3 \times 10^{32}$ years) [Hir89] and Frejus ($\tau_p > 0.7 \times 10^{32}$ years) [Ber91] can be combined to give a world limit of $\tau_p > 9 \times 10^{32}$ years [PDG94] (often the lifetime, with regard to the unknown branching ratios B, is given as τ_p/B). For an overview see [Bar92a]. The experimental lower limit clearly lies above the theoretically predicted value, which has led to doubts about the minimal SU(5) model. A modified version of the SU(5) model is the so-called 'flipped SU(5)', based upon a SU(5)$\times$ U(1) group [Lop94a]. This is, strictly speaking, not a simple gauge group, except when contained within a higher group. In this group an exchange of the fields $u^c \leftrightarrow d^c$ and $e \leftrightarrow \nu$ takes place. Its great advantage is that all spontaneous symmetry breakings are caused by low-dimensional Higgs representations (contrary to SU(5), which requires a 24-dimensional representation). It is motivated from superstring theories, which do not contain high-dimensional Higgs representations. The existing superstring GUTs thus lead to the flipped SU(5) theory [Ant88, Ant89]. In such theories the lifetime of the proton lies in the range of $\tau_p \approx 10^{33}$–10^{35} years, and is thus again compatible with the experimental observations. Supersymmetric SU(5) models also lead to larger lifetimes of about 10^{35} years (see [Lan86]), which therefore avoid conflict with the current experimental limits (see section 2.5.2).

A new generation of experiments (SuperKamiokande [Suz94, Suz96], ICARUS [Ben94a, Rub96] (figures 2.9 and 12.14) will be able to probe this range, at least in part. The limits attainable with these detectors are estimated at 10^{34} years [Bar92a] (see figure 2.6(d) and (e)).

Figure 2.8. View of the interior of the IMB detector. The IMB water Cerenkov counter contained 8000 tons of water viewed with 2048 photo-multipliers and was constructed in the Morton-Thiokol salt mine near Cleveland at a depth of about 700 m (1580 m water equivalent) (Photograph by Joe Stancampiano and Karl Luttrell, © National Geographic Society).

2.2.2 Successes and failures of SU(5)

In our examination of the first grand unified theory we have discussed the basic ideas of these theories. There were some successes:

- The success of unifying three forces, equivalent to a description using only one coupling constant.
- The necessity of multiples of 1/3 charge follows from the assignment to multiplets.
- From this also follows the equality of the absolute value of the proton and electron charge.
- The Weinberg angle, as well as the relation of the masses of the b quark to the τ lepton can be predicted from considerations more than twelve orders of magnitude higher in energy.

In addition to these successes there are, however, several unsatisfactory points, whose solution would have been expected in a unified theory:

- The arrangement into two multiplets seems arbitrary and not unified.

Table 2.3. (*a*) Characteristics of proton decay experiments (iron calorimeters) (from [Kla95a]). (*b*) Characteristics of proton decay experiments (water Cerenkov counters) (from [Kla95a]).

(*a*)	KGF	NUSEX	Fréjus	Soudan II
M_{tot} (t)	140	150	912	1000
M_{eff} (t)	60	113	550	600
Depth (m)	2300	1850	1780	760
Water equivalent (m)	7600	5000	4850	1800
Vertex resolution (cm)	10	1	0.5	~ 0.5
Place	Kolar goldmine	Mont-Blanc tunnel	Fréjus tunnel	Soudan mine

(*b*)	Kam I (II)	IMB I, III	HPW	SuperKam
M_{tot} (t)	3000	8000	680	50000
M_{eff} (t)	880 (1040)	3300	420	22000
Depth (m)	825	600	525	825
Water equivalent (m)	2400	1600	1500	2400
Vertex resolution (cm)	100 (20)	100		10
Place	Kamioka mine	Thiokol salt mine	King silver mine	Kamioka mine

- The built-in left–right asymmetry is unsatisfactory. The gauge group does not contain a right-handed $SU(2)_R$ as a counterpart to $SU(2)_L$. It is, however, unclear why nature should in principle prefer left-handedness.

- The 23 free parameters of the model are more than in the standard model.

- The model does not allow any statement about the number of particle generations.

- Equally, the mass spectrum of the elementary particles remains unsolved.

- It remains unclear why so many orders of magnitude lie between the electroweak and the GUT scale.

- Gravitation remains excluded from the unification.

- Neutrinos also remain massless in this theory.

- The lifetime of the proton is predicted to be too short.

Many of these criticisms are, however, characteristic of all GUT theories. A solution of the first two and last two of the above points can be found by using the SO(10) group as the gauge group.

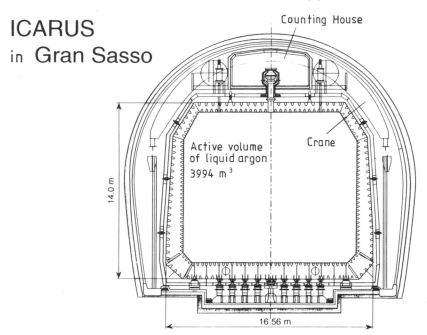

Figure 2.9. Cross section through the planned ICARUS detector in the Gran Sasso laboratory. This is a huge time projection chamber (TPC) which will be filled with 4000 tons of liquid argon (see section 12.4.5) (with kind permission of C Rubbia).

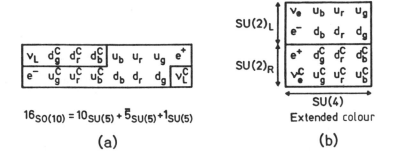

Figure 2.10. (*a*) All fermions of one family can be accommodated in *one* SO(10) multiplet. The 16th element is the as yet unseen right-handed neutrino ν_R, or equivalently its CP conjugate ν_L^C. The illustrations correspond to different SO(10) breaking schemes. (*b*) The breaking of the SO(10) multiplet according to the $SU(4)_{EC} \otimes SU(2)_L \otimes SU(2)_R$ structure is shown (from [Gro89, Gro90]).

2.3   The SO(10) model

Since the minimal SU(5) model has difficulties in correctly predicting the lifetime
of the proton, alternatives were sought. One such alternative is the SO(10) model
[Fri75, Geo75], which contains the SU(5) group as a subgroup. At the same
time the SO(10) model does represent the simplest left–right symmetrical theory.
The spinor representation is in this case 16-dimensional (see figure 2.10):

$$16_{SO(10)} = 10_{SU(5)} \oplus \bar{5}_{SU(5)} \oplus 1_{SU(5)}. \tag{2.25}$$

The SU(5) singlet cannot take part in any known SU(5) interaction. In principle
it could still participate in a U(1) interaction. However, as the hypercharge
within a multiplet has to disappear even this possibility is ruled out. This new
particle is therefore interpreted as the right-handed partner ν_R of the normal
neutrino (more accurately, the field ν_L^C is incorporated into the multiplet, see
e.g. [Gro89, Gro90]). ν_R does not take part in any SU(5) interaction, and in
particular does not participate in the normal weak interaction of the GWS model.
However ν_R does participate in a superweak interaction mediated by the new
SO(10) gauge bosons, which is the right-handed counterpart of the normal weak
interaction. This justifies the use of the term right-handed neutrino.

  Since the SO(10) symmetry contains the SU(5) symmetry, the possibility
now exists that somewhere above M_X the SO(10) symmetry is broken to the
SU(5) symmetry and that it then breaks down further as already discussed:

$$SO(10) \to SU(5) \to SU(3) \otimes SU(2)_L \otimes U(1). \tag{2.26}$$

Other breaking schemes for SO(10) do, however, exist. For example, it can
be broken down without any SU(5) phase, and even below the breaking scale
left–right symmetry remains. In this Pati–Salam model [Pat74] the symmetry
breaking happens as follows:

$$SO(10) \to SU(4)_{EC} \otimes SU(2)_L \otimes SU(2)_R \tag{2.27}$$

where the index EC stands for *extended colour*, an extension of the strong
interaction with the leptons as the fourth colour charge. The $SU(2)_R$ factor can
be seen as the right-handed equivalent of the left-handed $SU(2)_L$. It describes a
completely analogous right-handed weak interaction mediated by right-handed
W bosons. Figure 2.10 shows the splitting of the multiplet according to the two
symmetry breaking schemes.

  Contrary to the SU(5) model, which does not conserve B and L, but does
conserve $(B - L)$, $(B - L)$ does not necessarily have to be conserved in the
SO(10) model. A baryon number as well as a lepton number violation of two
units is possible and with that the possibility of not only neutrinoless double
beta decay but also of neutron–anti-neutron oscillations opens up. In the first
case

$$\Delta L = 2 \quad \Delta B = 0 \tag{2.28}$$

and in the second

$$\Delta B = 2 \quad \Delta L = 0. \tag{2.29}$$

Nonsupersymmetric SO(10) models can also solve the problem of SU(5), regarding the predictions as to the lifetime of the proton. Their predictions lie in the region of 10^{32} to 10^{38} years [Lee95]. Supersymmetric SO(10) models with automatic conservation of R-parity give far less clear predictions [Lee95a].

2.3.1 Neutron–anti-neutron oscillations

Theoretically the neutron–anti-neutron system can be described as a two-state system, like the $K^0 \bar{K}^0$ system or the neutrino oscillations which will be discussed later. If $|n_1\rangle$ and $|n_2\rangle$ are eigenstates of the mass matrix M, which is given by

$$M = \begin{pmatrix} m & \delta m \\ \delta m & m \end{pmatrix} \tag{2.30}$$

the resulting mass eigenvalues are

$$m_{1,2} = m \pm \delta m. \tag{2.31}$$

The off diagonal elements of M represent the transition energy between neutrons and anti-neutrons. The diagonal elements must be equal because of the CPT-theorem: $m_n = m_{\bar{n}}$. The time behaviour of the eigenstates follows from the Schrödinger equation and can be written as

$$n_i(t) = n_i(0)e^{-im_i t}e^{-\Gamma t} \quad \text{with} \quad \Gamma = \frac{\hbar}{\tau_n}. \tag{2.32}$$

The second exponential term describes the decay of the free neutrons, τ_n is the neutron lifetime. The physical eigenstates of the baryon number operator B, $|n\rangle$ and $|\bar{n}\rangle$ can be written as linear combinations such that

$$|n\rangle = \frac{1}{\sqrt{2}}(|n_1\rangle + |n_2\rangle) \tag{2.33}$$

$$|\bar{n}\rangle = \frac{1}{\sqrt{2}}(|n_1\rangle - |n_2\rangle). \tag{2.34}$$

The description of a neutron beam $n(t)$, from the above equations, takes the form

$$n(t) = \frac{1}{\sqrt{2}}e^{-\Gamma t}(n_1(0)e^{-im_1 t} + n_2(0)e^{-im_2 t}) \tag{2.35}$$

This corresponds to a superposition of two states and therefore to an oscillatory phenomenon. The probability $P_{n\bar{n}}(t)$, to find anti-neutrons after a time t in a pure neutron beam is given by

$$P_{n\bar{n}}(t) = e^{-\Gamma t} \sin^2 \delta m t \tag{2.36}$$

or, neglecting neutron decay, simply by

$$P_{n\bar{n}}(t) = \sin^2 \delta m t. \tag{2.37}$$

In order to characterize the phenomenon, the so-called oscillation period $\tau_{n\bar{n}}$ is defined as

$$\tau_{n\bar{n}} = \frac{\hbar}{\delta m} \tag{2.38}$$

so that

$$P_{n\bar{n}}(t) = \left(\frac{t}{\tau_{n\bar{n}}}\right)^2 \quad \text{for} \quad t \ll \tau_{n\bar{n}}. \tag{2.39}$$

This condition is in practice always fulfilled, as the oscillation period is considerably longer than the lifetime of the neutron (about ten minutes). Predictions for this period are given in table 2.4. The number of anti-neutrons $\bar{n}$ in an originally pure beam of n neutrons can easily be determined from this as

$$\bar{n}(t) = n \left(\frac{t}{\tau_{n\bar{n}}}\right)^2. \tag{2.40}$$

It is important to mention that all equations containing the probability $P_{n\bar{n}}(t)$ and the resulting conclusions are only valid for free neutrons. Neutrons bound in the nucleus, or neutrons in magnetic fields are affected by additional terms, e.g. from interactions with the magnetic moment (see e.g. [Kla95]).

One neutron–anti-neutron oscillation experiment is being carried out at the reactor in Grenoble (ILL). A neutron beam is transported over 35 m to a target of carbon foil. In order to realize as far as possible the condition that the neutrons are free, the entire path length is shielded from the Earth's magnetic field. Any anti-neutrons produced in the 0.1 s flight time most probably annihilate within the foil, which is practically transparent to neutrons. In every annihilation around 5 pions are produced, which form the experimental signature. From this and other similar experiments a lower limit for the oscillation period of [Bal94a]

$$\tau_{n\bar{n}} > 0.86 \times 10^8 \, \text{s} \quad (90\% \, \text{confidence level}) \tag{2.41}$$

can be derived. For a detailed description of the phenomenon of $n\bar{n}$-oscillations we refer to [Kla95a, Moh96a].

Symmetry groups even larger than SO(10), such as SU(15) [Fra90] or the exceptional group E_6 [Gur76, Ach77] have also been discussed as unifying groups. They contain the SO(10) group as a sub-group, and all fermions are contained in 27-dimensional multiplets. The E_6 group is particularly interesting owing to superstring theories ([Can85] and see section 2.7). Unfortunately, with larger groups the number of generators and free parameters increases rapidly, which makes clear experimental predictions extremely difficult. We therefore now leave this field and turn to a further experimental possibility to test GUTs.

2.4 Massive neutrinos

Probably the most promising candidate for giving further information about unified theories is the neutrino. In the standard model this particle is massless and takes part only in the weak interaction, which makes its detection extremely difficult. However, for many topics we wish to discuss in this book, this particle plays a central role. Moreover, many attempts to explain modern experiments and observations require massive neutrinos (see e.g. [Gro89, Gro90, Kay89, Moh91, Win91, Boe92, Kim93, Kla95a]). Even the particle type of the neutrino, i.e. either Dirac or Majorana particle, is today still unknown. While four different states exist for a Dirac particle, only two exist for a Majorana particle. In the latter case particles and anti-particles cannot be distinguished. If there is no right-handed interaction this distinction is irrelevant for massless particles. (For a detailed discussion see e.g. [Gro89, Gro90, Kay89, Kla95a].) We now consider models for the production of neutrino masses, and introduce some relevant experiments for detecting them (see also chapters 12 and 13).

We first consider the minimal SU(5) model. As only a left-handed neutrino exists in this model, Dirac mass terms are excluded, as they typically have the form $m\nu_{L,R}\bar{\nu}_{R,L}$. Such a particle can therefore only be a Majorana particle. As the SU(5) symmetry should be conserved, the mass term produced by a coupling to a Higgs field has also to be SU(5) invariant. A Majorana mass term would take the general form of (a detailed discussion can be found in [Lan81])

$$-m \sum_{\alpha} (\bar{\chi}_L \otimes (\Psi_L)^{CP})_\alpha \phi^\alpha + \text{h.c.} \qquad (2.42)$$

where χ and Ψ represent two fermions from the multiplets. As the SU(5) symmetry has to remain unbroken, the dimensions of the thus developed fermion combination have to be identical to those of the Higgs field. For a Majorana neutrino mass this implies $\chi_L = \Psi_L = \bar{5}$. The combination of two fermions results in a 10-dimensional and a 15-dimensional multiplet:

$$\bar{5} \otimes \bar{5} = 10 \oplus 15. \qquad (2.43)$$

The Higgs field thus must form either a 10- or a 15-dimensional multiplet. But the minimal SU(5) model contains only a 24- and a 5-dimensional Higgs field. So it is also not possible to generate Majorana masses, which is why in the minimal SU(5) model the neutrinos remain massless.

In the SO(10) model we have seen that the free singlet can be identified with a right-handed neutrino (see figure 2.10). It is therefore possible to produce Dirac mass terms. However, as the neutrinos belong to the same multiplet as the remaining fermions, their mass generation is not independent from that of the other fermions and one finds

$$m_{\nu_e} \simeq m_u \approx 5 \text{ MeV} \qquad (2.44)$$

in blatant contradiction to experiments where limits are given of order of magnitude of eV. With a 126-dimensional representation of the Higgs field ρ_{126}, it is possible to give all fermions Dirac masses, and neutrinos a Majorana mass. The component $\rho_{126}(1)$, which supplies the mass term $\langle\rho_{126}(1)\rangle\bar{v}_R(v_R)^{CP}$, is a singlet under SU(5)-transformations and is responsible for the breaking of the SO(10) to the SU(5). This is why $\langle\rho_{126}(1)\rangle$ can take on very large values up to M_X. The $\rho_{126}(1)$-singlet only couples to right-handed neutrinos. Thus under certain assumptions about ρ, it is possible to obtain no Majorana mass term for v_L and a very large term for v_R. In this case the mass matrix has the following form:

$$M = \begin{pmatrix} 0 & m^D \\ m^D & m^M \end{pmatrix} \tag{2.45}$$

where m^D is of the order of MeV–GeV, while $m^M \gg m^D$. If this matrix were diagonalized, two mass states are obtained such that

$$m_1 \approx \frac{(m^D)^2}{m^M} \tag{2.46}$$

and

$$m_2 \approx m^M. \tag{2.47}$$

This means that it is possible for a suitably large Majorana mass m^M in equation (2.45) to reduce the observable masses so far that they are compatible with experiment. This is the *see-saw* mechanism for the production of small neutrino masses [Gel78, Yan78]. If this is taken seriously a quadratic scaling behaviour of the neutrino masses with the quark masses or charged lepton masses follows, i.e.

$$m_{v_e} : m_{v_\mu} : m_{v_\tau} \sim m_u^2 : m_c^2 : m_t^2 \quad \text{or} \quad \sim m_e^2 : m_\mu^2 : m_\tau^2. \tag{2.48}$$

The vacuum expectation value of the Higgs singlet breaks the SO(10) symmetry and is thereby responsible for the non-conservation of the $(B - L)$-symmetry. In the SU(5) group B and L are not conserved separately, but the difference $B - L$ is. This is interesting, since $B - L$ is the only anomaly-free combination of these quantum numbers. The Goldstone boson produced in the spontaneous breaking of this global $(B - L)$ symmetry, is called the Majoron. The Majoron can appear as a triplet, a doublet, as well as a singlet state [Chi80, Gel81, San88]. The triplet Majoron does, however, contribute the equivalent of 2 neutrino flavours to the width of the Z^0 resonance [Nus81] and is therefore excluded by the LEP results. The same is also true for the doublet, which contributes half a neutrino width [San88]. A significant mixture of doublet and singlet, or a pure singlet Majoron, would, however, be compatible with the Z^0 width, as they possess an insignificantly small gauge coupling [Moh91]. A further possibility for its detection exists in the $\beta\beta$-decay (see section 2.4.2). Recent Majoron models and their testing via double beta-decay are discussed in [Bur93, Car93, Bur94, Bam95, Hir96a].

Table 2.4. GUT predictions for the period $\tau_{n\bar{n}}$ for $n\bar{n}$-oscillations (from [Moh89, Moh96a]).

GUT model	$\tau_{n\bar{n}} = 10^6 - 10^{10}$ s
standard model	no
SU(5)	no
$SU(2)_L \otimes SU(2)_R \otimes SU(4)_{ec}$	yes
SO(10)	no
E_6	no
SUSY $SU(3)_c \otimes SU(2)_L \otimes U(1)_Y$	yes[a]
SUSY left–right symmetric with E_6-type spectrum	yes

[a] But too fast without fine tuning of parameters.

Besides the hierarchy of neutrino masses for various flavours predicted in the see-saw mechanism, newer GUT models, such as SO(10) models with horizontal S_4 symmetry, predict almost *degenerate* masses for the different neutrino flavours, namely in the region of around 1 eV [Lee94, Pet94, Moh94, Ion94, Moh96c]. This could be of special significance for the analysis of solar neutrino experiments and the problem of dark matter. In the next section the experimental consequences of massive neutrinos and the experimental limits are discussed.

2.4.1 β-decay: the mass of the electron neutrino

It was nuclear beta decay, that is basically the process

$$n \rightarrow p + e^- + \bar{\nu}_e \tag{2.49}$$

that led Pauli in 1930 to postulate the neutrino, since otherwise the continuous energy distribution of the emitted electrons implied a violation of energy conservation. The distribution of the emitted electrons can be described, assuming massless neutrinos, by

$$N(E)\mathrm{d}E \sim p_e^2 F(Z, E)(E_0 - E)^2 \mathrm{d}E \tag{2.50}$$

where p_e is the electron momentum, $F(Z, E)$ is a function characteristic for every nucleus (Fermi function), and E_0 is the Q value of the nuclear transition. This energy is also the maximum energy of the emitted electrons. A non-zero rest mass of the neutrino influences the spectrum of the electrons:

$$N(E)\mathrm{d}E \sim p_e^2 F(Z, E)(E_0 - E)((E_0 - E)^2 - m_\nu^2)^{\frac{1}{2}} \mathrm{d}E. \tag{2.51}$$

A massive neutrino would therefore manifest itself by a reduction of the endpoint energy, since an energy equivalent to the rest mass of the neutrino is not available to the electrons. Displayed in a suitable form (the Kurie plot), this effect shows itself as a spectrum ending vertically at $E_0 - m_\nu c^2$. An especially suitable isotope for the search for a neutrino mass is tritium, as its endpoint energy lies relatively low at 18.59 keV, and the wavefunctions can be calculated relatively well. In this way relatively small limits on neutrino masses can be established. Several experiments of this type have been and are still being performed [Hol92a, Kün92], and at present the experimental upper limit is [Bel95, Lob96]

$$m_{\bar{\nu}_e} < 3.5 \text{ eV} \quad (95\% \text{ confidence level}) \tag{2.52}$$

(see however the discussion of this limit in [PDG96]). An illustration of various experimental limits from the tritium decay is shown in table 2.5. The fact that

Table 2.5. Results from experiments on tritium beta-decay.

m_ν (eV/c^2)	Confidence level (%)	Reference
< 250	——	[Lan52]
< 86	90	[Röd72]
< 60	90	[Ber72]
14–46	99	[Lub80]
< 65	95	[Sim81]
< 50	90	[Der83]
20–45	——	[Bor85]
< 18	95	[Fri86]
17–40	——	[Bor87]
< 32	95	[Kaw87]
< 27	95	[Wil87]
< 29	95	[Kaw88a]
< 15.4	95	[Fri91]
< 13	95	[Kaw91]
< 9.3	95	[Rob91]
< 11.6	95	[Hol92b]
< 5.6	95	[Bon96]
< 3.5	95	[Lob96]

most experiments yield a negative value for the neutrino mass squared, has led—besides explanations by possible experimental (e.g. molecular [Ott95]) effects—to speculations about long range anomalous neutrino interactions [Moh96b]. These could lead to clouds of neutrinos, with densities of $\sim 10^{15}$–10^{16} cm^{-3}, clustering on a range of the baryonic protocloud of our solar system. Absorption from this background of electron neutrinos via $\nu_e + {}^3\text{H} \rightarrow e^- \rightarrow {}^3\text{He}$ could lead to electrons in the anomalous endpoint region of tritium decay (for details see [Rob91, Moh96b, Ste96]).

Further information can be obtained from the shape of the spectrum, which has in recent times caused great excitement. Possible heavy neutrinos would lead to a kink in the spectrum, as the whole spectrum results from a superposition of all decay channels

$$N(E) = \sum_i |U_{ei}|^2 dN_i \tag{2.53}$$

where U_{ei} describes the admixture of a heavy neutrino to the electron neutrino (see also section 2.4.5) and dN_i is the spectrum for the emission of the mass eigenstate with mass m_i. Several groups have claimed to have found evidence in this way for a 17 keV neutrino at a level of about 1%. This would mean exciting new physics.

2.4.1.1 The 17 keV neutrino

The evidence for a 17.1 keV neutrino, which is mixed into the normal decay of tritium at about 3%, was already published as early as 1985 [Sim85]. In the following years, however, the negative results regarding this 17.1 keV neutrino increased. For example, the internal bremsstrahlung spectrum of ^{125}I was investigated at CERN and a mixture of such a neutrino with more than 2% was excluded with a 98% confidence [Bor86]. Other beta decay experiments with magnetic spectrometers, using for example ^{63}Ni, excluded an admixture of 0.25% with 90% confidence [Het87]. More recent experiments again gave evidence of 17 keV neutrinos in the β-decay of ^{3}H and ^{35}S with a mixing of less than 1% [Him89]. There seem to be, however, far more groups with negative evidence (for a more detailed summary see e.g. [Mor92, Him93]).

Nevertheless we wish to consider for a moment what the consequences of such a neutrino would be. Clearly it would mean an extension of the standard model, as here all neutrinos are massless. However, astrophysical observations also make certain assumptions about such a neutrino necessary. Such a neutrino would have to be unstable, since otherwise the number produced in the big bang model (see chapter 3) would result in too large a contribution to the matter density. Various decay modes, such as into axions, Majorons or the three light neutrinos have been discussed, but all of them have great uncertainties. So the supernova neutrino results (see chapter 13) and the explanation of the solar neutrino deficit (see chapter 12) get into difficulties with the existence of a 17 keV neutrino. Should it actually be a Majorana particle, it would necessitate the existence of a further heavy neutrino, or equally CP violation in the leptonic sector. For further discussions about the consequences of a 17 keV neutrino see [Kol91, Kra91, Gel91]. Meanwhile, the experimental counter arguments have increased to such a level (see e.g. [Him93, Abe93]), that hardly anyone continues to believe in the existence of the 17 keV neutrino (see e.g. [Zub93]).

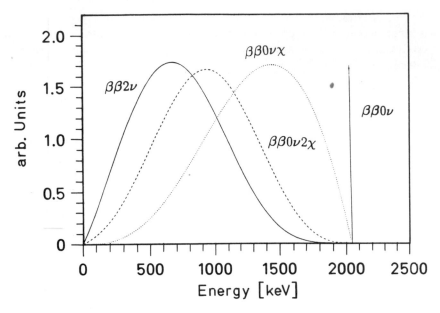

Figure 2.11. Spectrum of the summed energies of the two electrons emitted in double beta-decay. The neutrinoless decay shows itself as a sharp line, while the neutrino-accompanied or Majoron-accompanied decay results in a continuous spectrum. The end point energy corresponds to that of ^{76}Ge.

2.4.2 The $\beta\beta$-decay: effective mass of the electron neutrino

A larger sensitivity for a Marjorana neutrino mass than in single β-decay discussed above is found in one of the modes of the so-called double beta decay. There are two main reactions being studied:

$$2n \rightarrow 2p + 2e^- + 2\bar{\nu}_e \qquad (2.54)$$

and

$$2n \rightarrow 2p + 2e^-. \qquad (2.55)$$

There are 35 potential double beta emitters [Gro86b]. They are all even–even nuclei. The spectral form of the energy of the electrons for both decay modes is shown in figure 2.11. The first of the two decay modes (equation (2.54)) is a second order process in perturbation theory in Fermi theory (fourth order in the electroweak standard model). This process is therefore very rare with a half-life of order 10^{20} years or more. The process has been observed in geochemical experiments as an isotope anomaly (e.g. [Kir68]) and was seen for the first time in a detector experiment investigating the decay of ^{82}Se [Ell87a], later also for ^{100}Mo [Lal94], ^{76}Ge [Bal94c] and several other isotopes (see e.g. [Kla95]). The evidence for the neutrinoless $\beta\beta$-decay of equation (2.55) would have far greater consequences. This process is not permitted in the standard model, since it violates lepton number by two units and also violates $(B - L)$. It is only possible if the neutrino is a Majorana particle, as an anti-neutrino emitted at the first vertex has to be absorbed as a neutrino at the second vertex. This would throw light on the fundamental character of the neutrino. Furthermore the process requires that either the neutrinos are massive, or that right-handed currents have to contribute to the weak interaction. The anti-neutrino at the first vertex is right-handed and has, if there is no right-handed interaction, to be absorbed as a left-handed neutrino. This helicity change is only possible for massive neutrinos. It can be shown (see e.g. [Kay89]) that in *gauge theories* positive evidence for double beta decay *in any case* implies the existence of massive Majorana neutrinos. As in this case neutrino mixing can also not be excluded, the size of the measured effect is equivalent to the so-called *effective* Majorana neutrino mass $\langle m_\nu \rangle$. The decay rate is (neglecting right-handed currents)

$$\omega^{0\nu} = F^{0\nu} |M^{0\nu}|^2 \frac{\langle m_\nu \rangle^2}{m_e^2} \qquad (2.56)$$

where $F^{0\nu}$ is a phase space factor and $M^{0\nu}$ is the nuclear matrix element describing the transition. In order to extract the neutrino mass from the measured spectrum, a good knowledge of the matrix elements is necessary [Mut88, Sta90a, Kla94, Kla95, Sim97] (for an overview see [Gro89, Gro90]). The effective neutrino mass is given by

$$\langle m_\nu \rangle = \sum_i |U_{ei}^2 m_i| \qquad (2.57)$$

where m_i represents the mass eigenstates. In the case of a mixture of only two neutrino flavours, $\langle m_\nu \rangle$ is given by

$$\langle m_\nu \rangle = |m_1 \cos^2 \theta + e^{2i\beta} m_2 \sin^2 \theta|. \qquad (2.58)$$

A CP violating phase $e^{i\beta}$ already occurs for two families, and therefore the possibility of destructive interference exists. The above formula applies exactly only for neutrinos lighter than 10 MeV (for a more detailed discussion see [Gro86b]). The effects of heavy and superheavy neutrinos in the $0\nu\beta\beta$ decay are discussed e.g. in [Gro89, Gro90, Zub96]. In nearly all the GUTs which produce neutrino masses via the see-saw mechanism, this effective Majorana neutrino mass is equivalent to the actual electron neutrino mass [Lan88], which means that it is identical to the mass eigenstate.

To date the best limit from all $\beta\beta$-experiments is obtained from the Heidelberg–Moscow experiment using enriched ^{76}Ge (figure 2.12) [Kla87, Bal93, Bec93a, Kla94, Kla95a, Bal95b, Bal94c, Kla96, Kla96a, Gun97]. It operates 11.5 kg HP-Ge detectors enriched to 86% of ^{76}Ge (natural abundance about 7.8%). From the measurements so far (figure 2.13) for the neutrinoless decay mode a lower limit for the half life of [Kla97b]

$$T_{1/2} > 1.1 \times 10^{25} \text{ years} \quad (90\% \text{ confidence level}) \tag{2.59}$$

has been obtained, which using the matrix elements of [Sta90a] corresponds to a mass limit of

$$\langle m_\nu \rangle < 0.5 \text{ eV} \quad (90\% \text{ confidence level}) \tag{2.60}$$

(see table 2.6). This limit is already more precise by one order of magnitude than that from single β-decay, and is therefore the most precise limit for the mass of the electron neutrino overall. Alternative interpretations can provide *additional* information about supersymmetric theories or right-handed W bosons (see e.g. [Moh91, Hir95, Hir96b, Hir96b, Hir96d, Kla96, Kla96a] and section 2.5), leptoquarks [Hir96e, Hir96f, Kla97b], etc, which amoungst others proved to be important in the interpretation of the recently discovered 'high Q^2 events' at HERA (see papers in [Kla97e]). Another possibility for neutrinoless decay is the emission of two electrons and a Majoron. From this a model-independent Majorana-neutrino coupling can be obtained, which is given by

$$\langle g_{\nu\chi} \rangle = \sum_{ij} g_{\nu\chi} U_{ei} U_{ej}. \tag{2.61}$$

The present limit for $\langle g_{\nu\chi} \rangle$ is of order 10^{-4} [Bec93a]. Experimental half-life limits for different Majoron models are given in [Hir96a]. They lie in the region of 7×10^{21} years.

Table 2.6. $\beta\beta$ half-lives from the at that time most sensitive experiments and the upper limits calculated from them for $\langle m_\nu \rangle$ (matrix elements from [Sta90a] without regard for a right-handed weak interaction).

Decay	$T_{1/2}^{0\nu}$ (y) c.l.	$\langle m_\nu \rangle$ (eV) c.l.	Reference
$^{48}_{20}\text{Ca} \rightarrow ^{48}_{22}\text{Ti}$	$> 9.5 \times 10^{21} (76\%)$	$< 12.8^a (76\%)$	[Key91]
$^{76}_{32}\text{Ge} \rightarrow ^{76}_{34}\text{Se}$	$> 1.1 \times 10^{25} (90\%)$	$< 0.5 \quad (90\%)$	[Kla96a, Kla97b]
$^{82}_{34}\text{Se} \rightarrow ^{82}_{36}\text{Kr}$	$> 2.7 \times 10^{22} (68\%)$	$< 5.0 \quad (68\%)$	[Ell92]
$^{100}_{42}\text{Mo} \rightarrow ^{100}_{44}\text{Ru}$	$> 4.4 \times 10^{22} (68\%)$	$< 5.4 \quad (68\%)$	[Als93]
$^{116}_{48}\text{Cd} \rightarrow ^{116}_{50}\text{Sn}$	$> 2.9 \times 10^{22} (90\%)$	$< 4.1 \quad (90\%)$	[Dan95]
$^{128}_{52}\text{Te} \rightarrow ^{128}_{54}\text{Xe}$	$> 7.7 \times 10^{24} (68\%)$	$< 1.1 \quad (68\%)$	[Dek92]
$^{130}_{52}\text{Te} \rightarrow ^{130}_{54}\text{Xe}$	$> 1.8 \times 10^{22} (90\%)$	$< 5.2 \quad (90\%)$	[Ale94]
$^{136}_{54}\text{Xe} \rightarrow ^{136}_{56}\text{Ba}$	$> 4.2 \times 10^{23} (90\%)$	$< 2.3 \quad (90\%)$	[Ger96]
$^{150}_{60}\text{Nd} \rightarrow ^{150}_{62}\text{Sm}$	$> 2.1 \times 10^{21} (90\%)$	$< 4.1 \quad (90\%)$	[Moe95]

^{a} Calculated using [Mut91].

Figure 2.12. (*a*) The laboratory containing the Heidelberg–Moscow experiment at the Gran Sasso underground laboratory (Italy). Top left: one of the authors (Professor H V Klapdor-Kleingrothaus, spokesman of the Heidelberg–Moscow collaboration) with Professor M Morita. (*b*) Mounting of the first three of five detectors of enriched 'high purity' ^{76}Ge from this experiment, the first of their kind in the world, in their shielding (extremely low activity lead) in the Gran Sasso. In order to keep the background at an acceptable level, special extreme precautions must be taken.

For a detailed discussion of current experimental results and theoretical considerations for double beta decay see [Doi85, Mut88, Moh91, Moh86, 92, Lee95, Kla94, Hir95, Hir96a, Hir96b, Hir96c, Hir96d, Hir96e, Hir96f, Kla95, Kla95a, Kla96, Kla96a, Kla97b, Kla97e, Kla97f].

Double beta-decay could play a key role in the solution of the solar neutrino problem and the problem of dark matter. Recent GUT models predicting almost degenerate neutrino masses in the range of 1 eV claim that in this way they could solve the problems of solar neutrinos (see chapter 12), the atmospheric neutrino deficit (see chapter 8) and dark matter (see chapter 9) (see [Pet94, Lee94, Moh94, Ion94, Raf95, Cal95, Smi97]. This mass range is becoming accessible in the present second generation of double beta experiments [Kla96, Kla96a]. The planned new Heidelberg double–beta project GENIUS (germanium nitrogen underground set-up) [Kla97b] will even probe the ^{76}Ge halflife up to 10^{28} years and the Majorana neutrino mass down to 10^{-2} eV by operating 1000 kg of 'naked' enriched ^{76}Ge detectors in a liquid nitrogen tank

Figure 2.12. Continued.

of about 7–10 m in height and diameter (figure 2.14, see also figures 9.16 and 12.26).

2.4.3 The muon neutrino

The mass of the muon neutrino is determined from pion decay. The two-body decay

$$\pi^+ \rightarrow \mu^+ + \nu_\mu \tag{2.62}$$

allows the measurement of ν_μ from simple kinematic considerations:

$$m_{\nu_\mu}^2 = m_{\pi^+}^2 + m_{\mu^+}^2 - 2m_{\pi^+}(p_{\mu^+}^2 + m_{\mu^+}^2)^{1/2}. \tag{2.63}$$

With knowledge of the muon and pion mass only a precise measurement of the muon momentum [Abe84] is needed, which produces a mass limit of

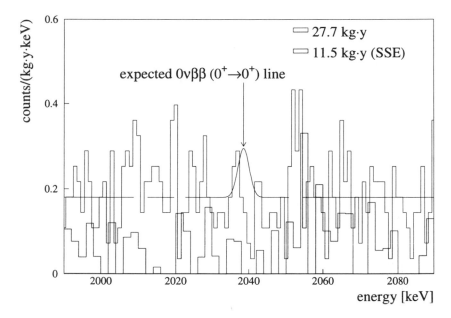

Figure 2.13. Region of the spectrum around the expected $0\nu\beta\beta$ line for ^{76}Ge decay in the Heidelberg–Moscow experiment. The spectrum observed in 27.7 kg years is shown together with the peak ruled out (solid curve) at 90% confidence level. It implies a lower limit for the $0\nu\beta\beta$ half-life of $T_{1/2}^{0\nu} > 1.1 \times 10^{25}$ years at 90% confidence level. The darkened histogram corresponds to data accumulated using a new pulse shape analysis for further background reduction [Hel96] in a measuring time of 11.5 kg y, (from [Kla97b]).

[Ass94, Ass96]

$$m_{\nu_\mu} < 170 \text{ keV} \quad (90\% \text{ confidence level}) \qquad (2.64)$$

2.4.4 The τ neutrino

The mass of the τ neutrino is even more uncertain. A mass limit for it can be obtained from the decay of the τ lepton, which can be produced in e^+e^- accelerators. The best limit so far has been reached at the LEP accelerator at CERN. The decay channel

$$\tau \rightarrow \pi^\pm + \pi^+ + \pi^+ + \pi^- + \pi^- + \nu_\tau \qquad (2.65)$$

is especially suitable. By examining the energy and momentum balance of all the produced pions it is possible, similarly to β-decay, to derive a mass limit for the τ neutrino of [Bus95, Pas96]

$$m_{\nu_\tau} < 18.2 \quad \text{MeV} \quad (95\% \quad \text{confidence level}). \qquad (2.66)$$

This limit is about 15 MeV lower than those of ARGUS [Alb92] and CLEO [Cin93], which give upper limits of 31 MeV and 32.6 MeV respectively (90% confidence limits).

Limits for possible neutrino masses from cosmological considerations will be discussed in chapter 3.

2.4.5 Neutrino oscillations

2.4.5.1 General

Under the assumption of massive neutrinos, the mass eigenstates ν_i are generally not identical with the flavour eigenstates ν_α of the weak interaction (see e.g. [Bil87]). This is, for example, also not the case in the quark sector (Cabibbo mixing, Kobayashi–Maskawa matrix). We discuss firstly the case of two neutrinos in a vacuum. The quantum mechanical description is identical to that of the previously discussed $(n\bar{n})$ and $(K^0\bar{K}^0)$ systems. The flavour and mass eigenstates are connected by

$$\begin{pmatrix} \nu_e \\ \nu_\mu \end{pmatrix} = \begin{pmatrix} \cos\theta_V & \sin\theta_V \\ -\sin\theta_V & \cos\theta_V \end{pmatrix} \begin{pmatrix} \nu_1 \\ \nu_2 \end{pmatrix} \qquad (2.67)$$

where θ_V is the vacuum mixing angle. The time development of an electron neutrino is described by

$$|\nu_e(t)\rangle = \cos\theta_V \exp(-iE_1 t)|\nu_1\rangle + \sin\theta_V \exp(-iE_2 t)|\nu_2\rangle. \qquad (2.68)$$

The probability of finding an original ν_e still in the same state after a time t is given by

$$|\langle\nu_e(t)|\nu_e(0)\rangle|^2 = 1 - \sin^2 2\theta_V \sin^2\left(\tfrac{1}{2}(E_1 - E_2)t\right). \qquad (2.69)$$

If one now also makes the simplifying assumption that both eigenstates have the same momentum, the energy difference of relativistic neutrinos is given by

$$E_2 - E_1 = \frac{m_2^2 - m_1^2}{2E} = \frac{\Delta m^2}{2E} \qquad (2.70)$$

where it is assumed that $m_2 > m_1$, otherwise the sign changes. With this it is possible to rewrite the probability (equation (2.69)) as

$$|\langle\nu_e(t)|\nu_e(0)\rangle|^2 = 1 - \sin^2 2\theta_V \sin^2\left(\frac{\pi R}{L_V}\right) \qquad (2.71)$$

where R is the distance travelled and L_V is the *oscillation length*. It represents a full transition cycle $\nu_e \rightarrow \nu_x \rightarrow \nu_e$ and is given by

$$L_V = \frac{4\pi E\hbar}{\Delta m^2 c^3} = 2.48\left(\frac{E}{\text{MeV}}\right)\left(\frac{eV^2}{\Delta m^2}\right) m. \qquad (2.72)$$

The oscillation length thus depends on the difference of the squares of the two mass eigenstates, and the oscillation amplitude depends on the mixing angle. Therefore oscillations do not occur for equal masses. It can also be seen that, to be sensitive to small mass differences, very large distances between source and detector are needed. Solar neutrino experiments are outstanding in this regime (see table 2.7).

Table 2.7. Neutrino sources and typical energies, together with the resulting oscillation lengths corresponding to the given mass parameters.

Source	Energy	L		
		$\Delta m^2 = 1\,\mathrm{eV}^2$	$\Delta m^2 = 10^{-6}\,\mathrm{eV}^2$	$\Delta m^2 = 10^{-11}\,\mathrm{eV}^2$
CERN SPS	100 GeV	250 km	2.5×10^8 km	2.5×10^{13} km
CERN PS,	5 GeV	12.5 km	1.25×10^7 km	1.25×10^{12} km
BNL AGS				
LAMPF	30 MeV	75 m	75000 km	7.5×10^9 km
Reactor	4 MeV	10 m	10000 km	1×10^9 km
Sun	0.2–10 MeV			1.5×10^8 km

Considering the general case of N neutrinos, a flavour eigenstate can be written as

$$|\nu_\alpha\rangle = \sum_{i=1}^{N} U_{\alpha i} |\nu_i\rangle. \qquad (2.73)$$

The unitary matrix U represents the mixture between the various states, similar to the Cabibbo–Kobayashi–Maskawa matrix in the quark sector. For two flavours it reduces to the above 2×2 matrix (equation (2.67). In an analogous manner the probability to find another flavour ν_β in an originally pure beam of ν_α is given by:

$$P(\nu_\alpha \to \nu_\beta) = \left| \sum U_{\alpha i} e^{-i\frac{m_i^2}{2E}t} U_{\beta i}^* \right|^2 \qquad (2.74)$$

For a detailed discussion see [Bil87, Kay89, Gro89, Gro90, Kla95a].

2.4.5.2 Experiments

Neutrino oscillations can be observed in 'appearance' or 'disappearance' experiments. Appearance experiments are based on the appearance of flavours that were not present in the original beam, while disappearance experiments try to prove that less neutrinos of the same flavour reach the detector than would be expected from the source. Accelerators are particularly suitable for the first type of experiments, while nuclear reactors are more useful for the second. Both

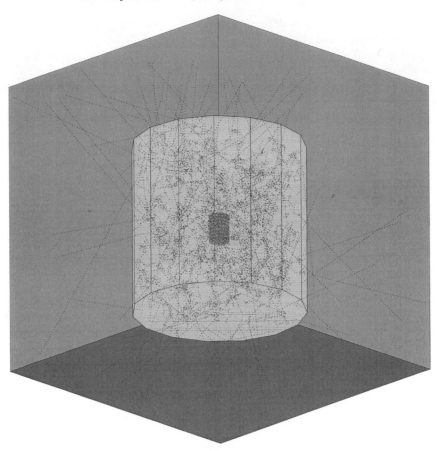

Figure 2.14. An Artist's view of the new Heidelberg project GENIUS. It is planned that it will use 1 ton of enriched ^{76}Ge ($\sim$ 300 detectors, dark shaded, inner region) in a liquid nitrogen tank of 7–10 m height and diameter for probing $0\nu\beta\beta$ half lives up to 10^{28} years and neutrino masses down to 10^{-2} eV. An upgraded 10 ton version could even probe the large angle solution of the solar neutrino problem directly, see figure 12.26. Tracks correspond to Monte Carlo simulated background events (from [Kla97b]).

methods have their advantages and disadvantages. Appearance experiments are very sensitive to small mixing angles, as in an ideal case one would have a pure neutrino beam without background, and the appearance of a *single* neutrino of another flavour would already count as a signal. In accelerators the conversion $\nu_\mu \to \nu_e$ and $\nu_\mu \to \nu_\tau$ is usually used for this. Disappearance experiments have the disadvantage that a precise knowledge of the neutrino flux is necessary. Their advantage, however, is that they measure all channels, e.g. $\nu_e \to \nu_\mu, \nu_\tau$ etc. Mainly reactor experiments are used here (table 2.8).

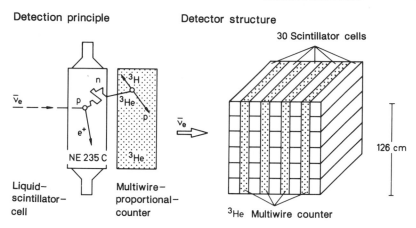

Detection principle   Detector structure

Figure 2.15. Construction of the Gösgen detector. In this sandwich-like construction of liquid scintillators and multi-wire proportional chambers, the positrons from the reaction $\bar{\nu}_e + p \rightarrow n + e^+$ are detected by means of their annihilation radiation in the scintillator, in coincidence with the neutron in the proportional chamber (from [Zac86]).

Table 2.8. A list of completed reactor experiments on neutrino oscillations.

Reactor	Thermal power (MW)	Distance (m)	Reference
ILL–Grenoble (France)	57	8.75	[Kwo81]
Bugey (France)	2800	13.6, 18.3	[Cav84]
Rovno (Ukraine)	1400	18.0, 25.0	[Afo85]
Savannah River (USA)	2300	18.5, 23.8	[Bau86]
Gösgen (Switzerland)	2800	37.9, 45.9, 64.7	[Zac86]
Krasnojarsk (Russia)	?	57.0, 57.6, 231.4	[Vid94]
Bugey III (France)	2800	15.0, 40.0, 95.0	[Ach95]

Table 2.9. Energy thresholds for reactions of the type $\bar{\nu}_l + p \rightarrow n + l^+$.

Reaction	Threshold energy (MeV)
$\bar{\nu}_e + p \rightarrow n + e^+$	1.804
$\bar{\nu}_\mu + p \rightarrow n + \mu^+$	100
$\bar{\nu}_\tau + p \rightarrow n + \tau^+$	3600

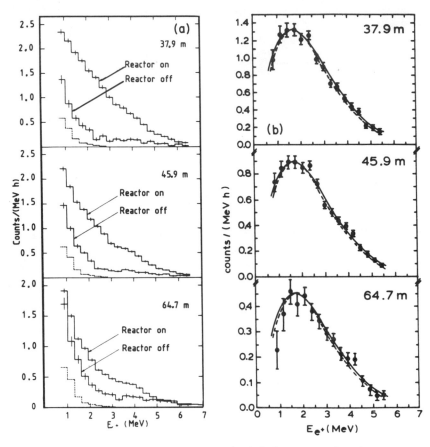

Figure 2.16. The Gösgen experiment: on the left the measured positron spectra at various distances and operational state of the reactor are shown. On the right are shown the positron spectra corrected for background. The solid and dotted lines have been calculated under the assumption of *no* oscillation. There is no sign of any oscillation phenomena to be seen (from [Zac86]).

Reactors

In the detector only the $\bar{\nu}_e$ flux can be detected, due to the high energy threshold for the reactions of the other flavours (table 2.9). In this way both methods complement each other; accelerator experiments are sensitive to the smallest mixing angles, while reactors to the smaller mass differences. The Gösgen experiment [Zac86] can be taken as a typical example of a reactor experiment. Common reactors in the GW region produce about 6 $\bar{\nu}_e$ per fission, and therefore a neutrino flux of about

$$N_{\bar{\nu}} \simeq 1.6 \times 10^{20} \bar{\nu}/(\text{s GW}). \tag{2.75}$$

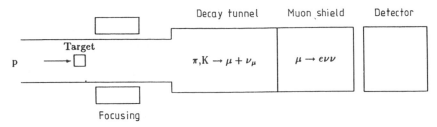

Figure 2.17. The principle of an oscillation experiment with neutrino beams in a fixed target reaction in accelerators. A secondary beam of pions and kaons is produced. These decay into muons and ν_μ. After shielding of the muons finally only ν_μ remain. Signatures of ν_e and ν_τ are looked for.

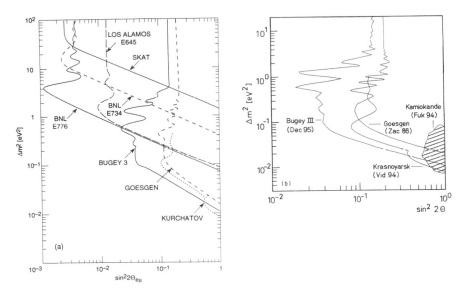

Figure 2.18. Exclusion curves of the oscillation parameters for ν_e–ν_μ oscillations from accelerator and reactor experiments. The areas to the right of the curves are not compatible with experiment: (*a*) (from [Gel95]) and (*b*) (from [Ach95]). The hatched area shows *allowed* regions for ν_e–ν_μ oscillations from the Kamiokande experiment [Fuk94] looking at atmospheric neutrinos.

The resulting neutrino spectrum stems from the β-decay of the fission products, mainly from ^{235}U, ^{238}U, ^{239}Pu, ^{241}Pu and extends to about 8 MeV. The expected neutrino spectrum is, however, subject to some uncertainty, due to the complicated processes within the reactor [Kla82a, Sch85a]. Furthermore, the burn-up of the fuel rods has to be taken into consideration, as this is a long term experiment. In this experiment the spectrum was measured at three different distances from the reactor, and examined for a spectral change. Cosmic

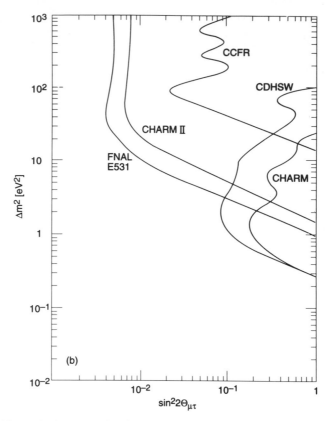

Figure 2.19. Exclusion curve for the oscillation parameters for ν_μ–ν_τ oscillations from accelerator experiments. The areas to the right of the curve are incompatible with experiment (from [Gel95]).

rays proved to be the main background component which could be suppressed to a great extent using active veto counters and coincidences. Furthermore, background spectra were taken during the fuel exchange of the reactors. The signature reaction for neutrinos is

$$\bar{\nu}_e + p \rightarrow n + e^+. \tag{2.76}$$

The positrons were detected in the Gösgen experiment in a scintillation counter based on mineral oil, which simultaneously acted as a target. The neutrons were detected in a ^{3}He filled position-sensitive multi-wire proportional chamber (see figure 2.15). By varying the detector distance from the reactor, it is possible to test various parameter regions. Figure 2.16 shows some spectra at different distances. Subsequent more sensitive experiments have been carried out at the reactor complex (3 reactors) in Krasnojarsk, as well as in Bugey (Bugey III experiment), in which measurements have been made at 57.0, 57.6 and 231.4 m

[Vid94] and 15, 40 and 95 m [Ach95], respectively. CHOOZ [Der93a] and Palo Verde (the former San Onofre) [Boe92a, Gra96] are new reactor experiments probing smaller Δm^2 than previous experiments.

Accelerators

Analogous experiments can be performed at accelerators where an almost pure neutrino beam (mainly ν_μ from kaon and pion decay) can be produced (table 2.10). The principle is illustrated in figure 2.17. Reactions which can only be produced by other neutrino flavours are searched for; these are therefore typical appearance experiments. The evidence rests on the reactions

$$\nu_i + N \rightarrow l_i^- + X \quad i = e, \mu, \tau. \tag{2.77}$$

As an example the detectors consist of a sandwich construction of spark chambers and lead plates. In the case of the muon neutrino, two tracks are produced in the detector, while the electron associated with the electron neutrino emits bremsstrahlung due to its small mass, and produces an electromagnetic shower. The difference between two tracks and one track plus electromagnetic shower is a clean signature. Current limits from the reactor and accelerator experiments for the two free parameters θ_V and Δm^2 for the three known flavours are shown in figures 2.18 and 2.19. A new experiment in Los Alamos (LSND) has recently claimed evidence for neutrino oscillations [Lou95, Ath95, Ath96, Ath97] (but see also [Hil95]). For an interpretation of these results and for an illustration of the connection to the solar neutrino problem and double beta-decay we refer to [Raf95, Cal96]. This result is only marginally compatible with the results of other experiments, e.g., the German–English KARMEN collaboration [Arm95, Kle96].

There are several new accelerator experiments either under construction or already in operation. These are partly motivated by the atmospheric neutrino deficit (see chapter 8) and the discussion of solar neutrino results (see chapter 12), as well as the possibility that ν_τ could be a candidate for dark matter (see chapter 9).

The most prominent new accelerator experiments in operation are CHORUS and NOMAD (both at CERN) [Pan92, Win95, Rub96a]. Their main objective is to search for ν_μ–ν_τ oscillations, and they could lead to a possible order of magnitude improvement in the current limits. CHORUS (see figure 2.20) uses emulsion plates in order to detect the associated τ lepton, whereas NOMAD detects the τ lepton from the kinematics of its decay. First limits can be found in [deS97, Mig97]. Directing an accelerator-produced beam of neutrinos to an underground detector a large distance away (so-called *'long-baseline' experiments*) is under discussion. As an example, there are proposals to point the Fermilab neutrino beam towards the Soudan mine (about 720 km away) by the beginning of the next millennium, and the CERN neutrino beam towards the Gran Sasso underground laboratory (see for example [Bal94b, Egg95, Rub96]

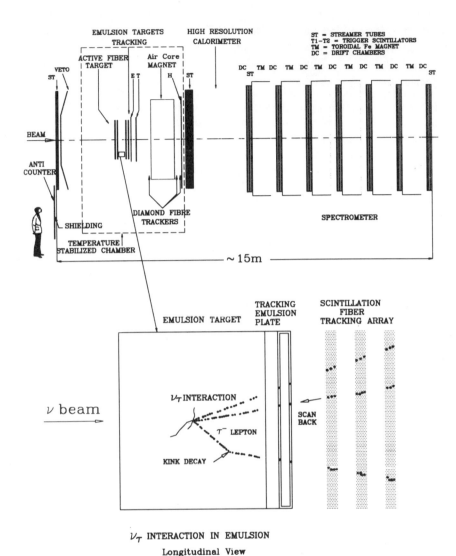

ν_T INTERACTION IN EMULSION

Longitudinal View

Figure 2.20. Schematic diagram of the CHORUS detector for neutrino oscillations at CERN (above). The neutrino beam which consists predominantly of ν_μ enters from the left. A possible ν_τ coming from oscillations can be detected by its transformation to a τ lepton. The possible decay of any τ produced can be detected using emulsion plates (below); the remaining detector elements are used to measure the momentum and energy of the secondary particles (with kind permission of K Winter).

Table 2.10. Results from neutrino oscillation experiments at accelerators.

Channel	Experiment	$(\Delta m^2)^a (eV^2)$	$(\sin^2 2\theta)^b$
$\nu_\mu \to \nu_e$	COL-BNL [Bak84]	< 0.6	$< 6 \times 10^{-3}$
	BNL-E734 [Ahr85]	< 0.43	$< 3.4 \times 10^{-3}$
	BEBC/PS [Ang86]	< 0.09	$< 1.3 \times 10^{-2}$
	LAMPF-E764 [Dom87]	< 0.67	$< 8 \times 10^{-3}$
	BNL-E776 [Bor92]	< 0.075	$< 3 \times 10^{-3}$
$\bar{\nu}_\mu \to \bar{\nu}_e$	FNAL [Tay83]	< 2.4	$< 1.3 \times 10^{-2}$
	LAMPF-E645 [Fre93]	< 0.14	$< 2.4 \times 10^{-2}$
$\nu_\mu \to \nu_\tau$	FNAL[Ush81]	< 3.0	$< 1.3 \times 10^{-2}$
	FNAL-E531 [Gau86]	< 0.9	$< 4 \times 10^{-3}$
	CERN SPS [Gru93]	< 1.5	$< 8.0 \times 10^{-3}$
$\bar{\nu}_\mu \to \bar{\nu}_{\tau_e}$	FNAL [Asr81]	< 2.2	$< 4.4 \times 10^{-2}$
$\nu_e \to \nu_\tau$	COL-BNL [Bak84]	< 8	< 0.6
	FNAL-E531 [Gau86]	< 9	< 0.12

[a] For maximal mixing ($\sin^2 2\theta = 1$).
[b] For large Δm^2.

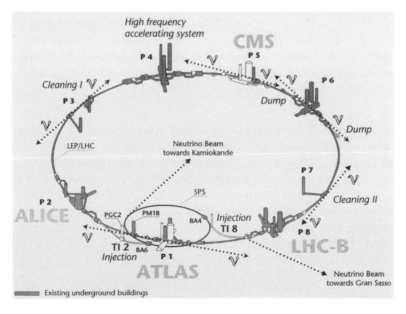

Figure 2.21. The possible creation of a neutrino beam from the planned large hadron collider (LHC) at CERN and its orientation towards various underground laboratories such as Gran Sasso and Kamioka (from [Egg95]).

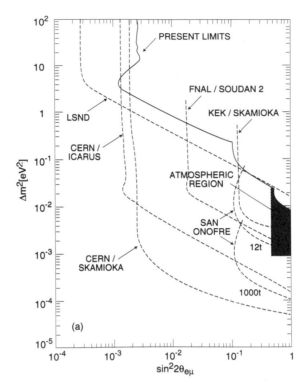

Figure 2.22. Expected exclusion curves for the oscillation parameters for (a) ν_e–ν_μ and (b) ν_μ–ν_τ oscillations from the next generation of accelerator experiments (from [Gel95]). The shaded region corresponds to an *allowed* region from Kamiokande results for atmospheric neutrinos (from [Fuk94]).

and figure 2.21). An experiment (KEK-E362) using a neutrino beam from KEK in Japan directed towards SuperKamiokande (see section 12.4.2.2) should give the first results in 1999 [Suz96]. Such experiments are sensitive to intermediate (around 1000 km) oscillation length scales. Figure 2.22 shows possible limits for neutrino oscillations which could be attained in future experiments. For a recent review see [Zub97]. Because of the large distance between the source and the detector, solar neutrinos are particularly suitable to look for small Δm^2. The possibility offered by solar and also atmospheric neutrinos in neutrino oscillation searches will be discussed in chapters 8 and 12.

2.4.6 Neutrino decay

If neutrinos have a finite mass, and if the flavour eigenstates are not equal to the mass eigenstates, there also exists the possibility of neutrino decay in addition to the previously discussed effects. Various decay possibilities exist, depending

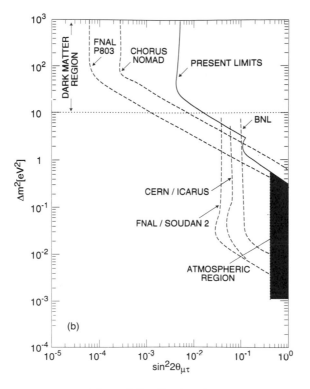

Figure 2.22. Continued.

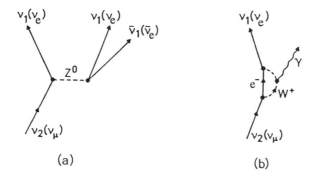

Figure 2.23. Two diagrams for the neutrino decay: (*a*) into three neutrinos (*b*) the radiative decay $\nu_2 \rightarrow \nu_1 + \gamma$. There are additional exotic decay channels, such as into Majorons.

on the mass difference, such as

$$\nu_H \rightarrow \nu_L + \gamma \qquad (2.78)$$

$$\nu_H \rightarrow \nu_L + l_i^+ + l_j^- \qquad l_{i,j} = e, \mu \dots \qquad (2.79)$$

$$\nu_H \rightarrow \nu_L + \bar{\nu}_L + \nu_L \qquad (2.80)$$

$$\nu_H \rightarrow \nu_L + \chi \qquad (2.81)$$

The radiative decay and the decay in Majorons χ are possible if $m_{\nu_H} > m_{\nu_L}$, while for the invisible decay into three neutrinos the condition $m_{\nu_H} > 3m_{\nu_L}$ is required. For the decay into charged leptons $m_{\nu_H} - m_{\nu_L} > m_{l_i} + m_{l_j}$ has to apply. Some graphs of these decay channels are illustrated in figure 2.23. Experimental limits from reactor experiments can be found in [Obe92]. Further limits exist from astrophysical considerations. For example, neutrinos with masses of between 100 eV and about 2 GeV have to be unstable, in order not to contradict the experimentally observed matter density (see chapter 3). In this case the decay channels of equations (2.79)–(2.81) apply. The observations of the supernova 1987a yields the strongest limits on radiative decay. Observations made by the Solar Maximum Satellite resulted in a limit on the lifetime of [Kol89]

$$\frac{\tau_{\nu_i}}{m_{\nu_i}} > 3, 3 \times 10^{14} \quad \frac{s}{eV}. \qquad (2.82)$$

The previously discussed 17 keV neutrino would also necessarily have implied a neutrino decay, which, taking into account the limits from all other observations, would most likely have been via the emission of Majorons (see e.g. [Kla92]). Neutrino decay has also been considered as an explanation of the solar neutrino problem [Fri88], but the latest results (see chapter 12) seem to make this rather unlikely.

2.5 Supersymmetry

We now consider an extension of the standard model by the introduction of a completely new kind of symmetry. The main motivation for this extension is the solution of the mass or *hierarchy problem*. This problem lies in the fact that some of the infinities within the field theoretical treatment are more difficult to control than others. In perturbation theory higher order corrections and renormalization only result in small corrections to the gauge couplings and fermion masses. A different situation arises for scalar particles such as the Higgs boson. Only scalar fields can have a non-zero vacuum expectation value without destroying the Lorentz invariance of the theory. In the description of all the previous GUT theories, all particles were initially massless and only gained mass by spontaneous symmetry breaking (Higgs mechanism). We have seen that Higgs particles are necessary at several completely different mass scales. On one hand they have to give vacuum expectation values of order M_X, and on the other hand, in the low energy region of order M_W. The problem is that these scales remain independent only at the tree level and mix via loop diagrams in higher order perturbation theory. The size of this correction depends on the momenta k which are taken into account in the loops. A small change in the cut-off scale, Λ, implies only a small change in the observed fermion mass, since this depends only logarithmically on the cut-off parameter. However, the situation is different for scalar particles. The Higgs mass receives a correction δm_H via this mechanism, where [Ell91b, Nil95]

$$\delta m_H^2 \sim g^2 \int^\Lambda \frac{d^4k}{(2\pi)^4 k^2} \sim g^2 \Lambda^2 \qquad (2.83)$$

If Λ is set at the GUT scale, the lighter Higgs particle would certainly undergo corrections of order M_X. In order to achieve a well defined theory it is necessary to fine tune all parameters to all orders. Two solutions have been suggested to avoid this problem, namely *technicolour* and *supersymmetry*. By introducing supersymmetry, compensation of the loop contributions can be achieved to all orders, since the contributions of fermions and bosons have opposite signs. This is the expression of a general principle of these models, namely that the non-renormalization theorems ensure that the higher order corrections in the renormalization of many quantities disappear. It is then sufficient to calculate them at the tree level, i.e. diagrams without closed loops. In order to produce acceptable corrections to the Higgs mass of up to 100 GeV, the difference of the boson and fermion masses has to lie at scales of less than about 1 TeV. If such energies can be reached in accelerators, supersymmetric particles should be discovered.

Supersymmetry implies a complete symmetry between fermions and bosons [Wes74]. This is a new symmetry, as fundamental as that between particles and anti-particles. We will first examine *global* supersymmetry. This expands the normal Poincaré algebra for the description of space-time with an extra generator, which changes fermions into bosons and vice versa.

Let Q be the generator of the supersymmetry such that

$$Q|(\text{Fermion})\rangle = |\text{Boson}\rangle$$
$$Q|(\text{Boson})\rangle = |\text{Fermion}\rangle.$$

In order to achieve this Q itself has to have half-integer spin. The algebra of the supersymmetry is determined by the following relationships:

$$\{Q_\alpha, Q_\beta\} = 2\gamma^\mu_{\alpha\beta} p_\mu \qquad (2.84)$$

$$[Q_\alpha, p_\mu] = 0 \qquad (2.85)$$

$$[p_\mu, p_\nu] = 0. \qquad (2.86)$$

Here p_μ is the four momentum operator. Note *that due to the anti-commutator relation equation (2.84) internal particle degrees of freedom are connected to the external space-time degrees of freedom.* This has the consequence that a *local* supersymmetry has to contain gravitation. Some general characteristics in relation to gauge theories can be derived immediately. A global supersymmetry always commutes with gauge symmetries, otherwise it would be a local supersymmetry. This has the consequence that all internal quantum numbers of particles of one supermultiplet are identical.

Because every fermion is now associated with a supersymmetric boson, and vice versa, the theory has twice as many particles as before. Particles and their super-partners are combined in superfields which are described in a superspace. Like the connection between four-momentum and a translation in our four-dimensional space-time (x^μ), Q_α is linked to a translation in a space whose coordinates (θ^α) anti-commute. Both spaces together combine to form a *superspace*, characterized by a set of co-ordinates (x^μ, θ^α).

Table 2.11. SUSY partners for some particles.

Normal particles	SUSY–partner	Symbol	Spin
quark	squark	$(\tilde{q})$	0
lepton	slepton	$(\tilde{l})$	0
gluon	gluino	$(\tilde{g})$	1/2
W-boson	Wino	$(\tilde{w})$	1/2
photon	photino	$(\tilde{\gamma})$	1/2
Higgs	higgsino	$(\tilde{h})$	1/2
graviton	gravitino	$(\tilde{G})$	3/2

The nomenclature of the supersymmetric partners is as follows: the scalar partners of normal fermions are designated with a preceding 's', so that for example the supersymmetric partner of the quark becomes the squark $\tilde{q}$. The super-partners of normal bosons receive the ending '-ino'. The partner of the photon therefore becomes the photino $\tilde{\gamma}$. The supersymmetric partners of the gauge bosons are usually known as *gauginos*. The partner of the graviton is a particle with spin 3/2 with the name gravitino. Table 2.11 shows some of the resulting particle pairs. To date no known particle could be identified as the supersymmetric partner of any other, so that supersymmetry must also be broken. If the supersymmetry is turned into a *local* (= gauge) theory, gravity will, as already mentioned, be included in the theories which are therefore known as *supergravity* theories (SUGRAs).

In most supersymmetric models the conservation of the so-called R-parity is assumed. (For the limits on R-parity violating interactions see section 2.5.2). R-parity is assigned as follows:

$$R_P = 1 \quad \text{for normal particles}$$
$$R_P = -1 \quad \text{for supersymmetric particles.}$$

R_P is a multiplicative quantum number and is connected to the baryon number

B, the lepton number L and the spin S by

$$R_P = (-1)^{3B+L+2S}. \tag{2.87}$$

Conservation of R-parity has three consequences:

(i)   Supersymmetric particles can only be produced in pairs in reactions such as $e^+e^- \rightarrow \tilde{e}^+\tilde{e}^-$.

(ii) Heavy supersymmetric particles can decay into lighter ones, for example $\tilde{e} \rightarrow e\tilde{\gamma}$.

(iii) The lightest supersymmetric particle (LSP) has to be stable due to R_P conservation.

We now consider the latter particle in greater depth. The LSP probably does not take part in the strong and electromagnetic interactions, as it otherwise would have formed bound states with normal matter, and would be condensed together with normal matter in stars, planets etc. Experimental limits on the abundances of very heavy isotopes exclude this possibility [Nor87]. Therefore only sneutrino, gravitino and the neutral gauge particles remain as candidates for the LSP. However, the experiments covered in the next section rule out the sneutrino. The gravitino also requires a very specific choice of parameters, so we will concentrate on the third possibility.

The state is then composed of a superposition of four neutral particles, the photino, zino (more precisely the neutral Wino $\tilde{W}^3$ and the Bino $\tilde{B}^0$), and two neutral Higgsino particles. These states are generally known as *neutralinos*. The lightest of them is of particular interest as a candidate for dark matter (see chapter 9). The mixing of the neutralinos is [Moh91, Moh86]:

$$\begin{pmatrix} \tilde{W}^3 & \tilde{B}^0 & \tilde{H}^0_1 & \tilde{H}^0_2 \end{pmatrix} \mathcal{M} \begin{pmatrix} \tilde{W}^3 \\ \tilde{B}^0 \\ \tilde{H}^0_1 \\ \tilde{H}^0_2 \end{pmatrix} \tag{2.88}$$

where the mixing matrix is

$$\mathcal{M} = \begin{pmatrix} M_2 & 0 & -\sqrt{\frac{1}{2}}g_2 v_1 & \sqrt{\frac{1}{2}}g_2 v_2 \\ 0 & \frac{5}{3}\frac{\alpha_1}{\alpha_2}M_2 & \sqrt{\frac{1}{2}}g_1 v_1 & -\sqrt{\frac{1}{2}}g_1 v_2 \\ -\sqrt{\frac{1}{2}}g_2 v_1 & \sqrt{\frac{1}{2}}g_1 v_1 & 0 & \mu \\ \sqrt{\frac{1}{2}}g_2 v_2 & -\sqrt{\frac{1}{2}}g_1 v_2 & \mu & 0 \end{pmatrix}. \tag{2.89}$$

Here g_1, g_2 are the gauge couplings of the SU(2) and U(1) groups, v_1, v_2 are the vacuum expectation values of the two Higgsino fields and M_2, μ are mass parameters. In the case of small μ and for $M_2 \rightarrow 0$ the photino is the lightest state with

$$m_{\tilde{\gamma}} = \frac{g_1 \tilde{W}^3 + g_2 \tilde{B}^0}{\sqrt{g_1^2 + g_2^2}}. \tag{2.90}$$

Due to the free parameters it is, however, also possible for the Higgsino state to be the LSP. The experimental results do not yet allow a definite conclusion (see e.g. [Bed94a]). The minimal extension of the standard model, in which five additional parameters are required, is known as the *minimal supersymmetric standard model* (MSSM). More detailed expositions of supersymmetry can be found in [Dra87, Wes86, 90, Moh91, Moh86, 92, Hab93, Kan93, Kan94, Mur94, Tat95, Nil95, Bar95, Lop96, Tat97].

2.5.1   The search for supersymmetry with accelerators

In the search for supersymmetric particles, accelerators are of course the most effective instruments. Assuming conservation of R parity supersymmetric particles are produced in pairs and decay eventually into the stable LSP. Strategies to detect these particles are based on two criteria. The first is to detect evidence for the LSP by measuring missing transverse momentum due to the escape of the lightest supersymmetric particle. As an example, the signature of the decay of a squark $\tilde{q} \to q + \tilde{\gamma}$ would be detected via missing transverse momentum from the escaping photino. This signature of missing particles can be distinguished from neutrinos since neutrinos are produced in conjunction with a charged lepton. The second criterion uses the different event topology (for example the angular distribution) compared to standard model processes. For a detailed discussion of the observable signatures see [Hab85, Kan94, Bae95]. The best limits for several supersymmetric particles come from LEP results, which thus far show no evidence for supersymmetic particles. A lower limit of $m_{\text{LSP}} > 18.4$ GeV is usually given for the LSP [PDG94]. This is based on the assumption of universal gaugino masses at the GUT scale. If this assumption is relaxed, the lower limit is reduced to around 3–5 GeV [Bed97b]. For sleptons lower limits of 40 GeV are obtained [PDG94] since decay chains like

$$e^+ e^- \to Z^0 \to \tilde{l}\tilde{l} \quad \text{and} \quad \tilde{l} \to l\tilde{\chi} \quad (l = \text{lepton}, \tilde{\chi} = \text{LSP}) \tag{2.91}$$

have not been observed. A lower limit for sneutrinos of 41.8 GeV is obtained from the width of the Z^0 (see chapter 4). Sneutrinos as well as neutrinos would contribute to invisible matter in the universe (see chapter 9). Lower limits for the squark mass of more than 224 GeV have been obtained from experiments at Fermilab [Abe96]. Further limits on SUSY particles can be found in [Dec92b, Bus95, Aba95a, Aid96a] and elsewhere. Another method of detection is the observation of the decay of quarkonia $(q\overline{q})$ into supersymmetric particles.

The high precision measurements of the coupling constants at the Z mass from LEP seem to indicate that when extrapolating in the standard model the

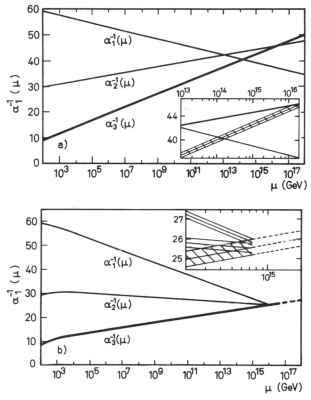

Figure 2.24. Evolution of the coupling constants determined by the precision measurements at LEP and extrapolated according to the standard model (above) and in the minimal supersymmetric standard model (below). A supersymmetric expansion to the standard model allows the unification of the coupling constants at one point (from [Ama91]).

coupling constants do not intersect at one point. By assuming supersymmetry it is, however, possible to unite the coupling constants at one point (see figure 2.24) [Ama91]. This is because the coefficients b_i we discussed in section 2.1 change in the transition to supersymmetry, and now have the form of

$$b_i = \begin{pmatrix} b_1 \\ b_2 \\ b_3 \end{pmatrix} = \begin{pmatrix} 0 \\ -6 \\ -9 \end{pmatrix} + N_{\text{Fam}} \begin{pmatrix} 2 \\ 2 \\ 2 \end{pmatrix} + N_{\text{Higgs}} \begin{pmatrix} 3/10 \\ 1/2 \\ 0 \end{pmatrix} \tag{2.92}$$

for the MSSM. This changes the behaviour of the coupling constants as a function of energy. The possibility of unifying the coupling constants with the help of supersymmetry may be counted as indirect evidence for the existence of supersymmetry (see, however, e.g. [Sha92]).

Similarly, the prediction of the Weinberg angle in supersymmetric models

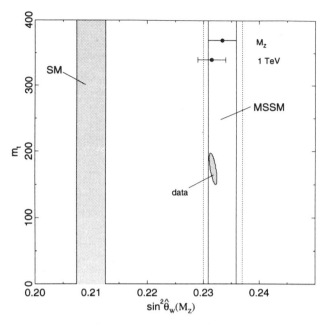

Figure 2.25. Prediction of $\sin^2 \theta_W$ in the standard model and the minimal supersymmetric standard model (MSSM). α and α_S have been used as input parameters. All new particles were assumed to be degenerate in mass, either at the Z mass or at 1 TeV (from [Lan93b]).

corresponds better to the experimentally observed value than that of GUT theories without supersymmetry (see figure 2.25). The predictions of these theories are [Lan93a, Lan93b]

$$\sin^2 \theta_W(m_Z) = 0.2334 \pm 0.0050 \quad \text{(MSSM)} \tag{2.93}$$

$$\sin^2 \theta_W(m_Z) = 0.2100 \pm 0.0032 \quad \text{(SM)}. \tag{2.94}$$

The experimental value is (see equation (1.100))

$$\sin^2 \theta_W(m_Z) = 0.2315 \pm 0.0002 \pm 0.0003. \tag{2.95}$$

2.5.2   The search for supersymmetry in non-accelerator experiments

In this section two experiments which are interesting in this connection are described. A further possibility exists in the search for dark matter (see chapter 9).

2.5.2.1   Supersymmetry and proton decay

By introducing supersymmetry we have greatly increased the number of particles. This has an effect on the unification scale since, due to equation (2.92) the number of particles enters into the renormalization group equations. As a consequence the unification point moves to about 10^{16} GeV. The increased unification energy results in a bigger M_X mass, which is in the minimal SU(5) SUSY GUT model [Lan86]

$$M_X \approx 4.8 \times 10^{15} \, \text{GeV} \left[\frac{\Lambda_{\text{QCD}}}{100 \, \text{MeV}} \right]. \tag{2.96}$$

This results in a substantially increased lifetime of the proton of about 10^{35} years, which is compatible with experiment. However, the dominant decay channel (see e.g. [Moh91, Moh86, 92]) changes in such models, such that the decays $p \rightarrow K^+ + \bar{\nu}_\mu$ and $n \rightarrow K^0 + \bar{\nu}_\mu$ should dominate. The experimentally determined lower limit [Hir89] of the proton lifetime of 1×10^{32} years for this channel is even less restrictive than the $p \rightarrow \pi^0 + e^+$ mode. Supersymmetric SO(10) models yield less clear predictions [Lee95a].

2.5.2.2 *Supersymmetry and neutrinoless double beta-decay*

The double beta-decay discussed in section 2.4.2 also provides information on the supersymmetric sector. Supersymmetric $0\nu\beta\beta$ decay is possible even when R-parity is conserved [Har95, Hir97a]. If R-parity conservation is no longer assumed, the Feynman graphs shown in figure 2.26(a) would provide a contribution to the decay amplitude [Moh91, Hir95, Hir96b, Hir96c].

    R-parity breaking can be introduced into supersymmetric models by adding an additional R-parity breaking term into the superpotential

$$W = W_{\text{MSSM}} + W_{R_p} \tag{2.97}$$

with

$$W_{R_p} = \lambda_{ijk} L_i L_j \bar{E}_k + \lambda'_{ijk} L_i Q_j \bar{D}_k + \lambda''_{ijk} U_i \bar{D}_j \bar{D}_k. \tag{2.98}$$

Here the indices i, j and k denote generations. L, Q denote lepton and quark doublet superfields and $\bar{E}$, $\bar{U}$ and $\bar{D}$ denote lepton and up, down quark singlet superfields respectively. Terms proportional to λ, λ' violate lepton number, those proportional to λ'' violate the baryon number. From proton decay limits it is clear that both types of terms cannot be present at the same time in the superpotential [Zwi83, Hir96b]. On the other hand, once the λ'' terms are assumed to be zero, the λ and λ' terms are not limited, and $0\nu\beta\beta$ decay can occur within the R_p-violating MSSM through the Feynman graphs shown in figure 2.26.

    Hence double beta-decay provides, amoungst others, limits on the strength of a R-parity violating interaction λ'_{111}. These are better than the best limits of the high energy accelerators, the Tevatron [Roy92] and HERA [Aid96b] (see figure 2.27). The limits for λ'_{ijk} from $\beta\beta$-decay are compared to those from other sources in [Kol97]. The limit for λ'_{111} from $0\nu\beta\beta$ decay excludes the possibility of squarks of first generation (of R-parity violating SUSY) being produced in high-Q^2 HERA events discussed recently (see [Cho97, Alt97, Kal97, Kla97b, Fra97]).

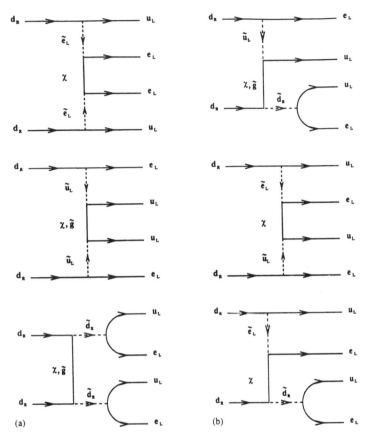

Figure 2.26. Feynman diagrams of supersymmetric contributions to the neutrinoless double beta decay. For the processes shown it is necessary for R-parity to be violated (from [Hir95, Hir96b]).

In R-parity conserving SUSY models $0\nu\beta\beta$ decay contributions can arise at the level of box diagrams via the $(B - L)$-violating sneutrino mass term [Hir97a, Hir97b, Hir97c] (see figure 2.28). From the data of the Heidelberg–Moscow double beta decay experiment then constraints on the Majorana-like sneutrino mass can be derived [Hir97a]. The $(B - L)$-violating sneutrino mass leads to a mass splitting in the sneutrino–anti-sneutrino ($\tilde{\nu} - \tilde{\nu}^C$) system and to the effect of lepton-number violating $\tilde{\nu} - \tilde{\nu}^C$ oscillations [Hir97a, Hir97b, Hir97c]. The constraints from double-beta decay leave room for searches for some of the manifestations of these effects at future colliders (e.g. the next linear collider (NLC)), only for the second and third generation $(B - L)$-violating sneutrinos mass term.

For limits on R-parity violation from studies of B meson decay see [Car95].

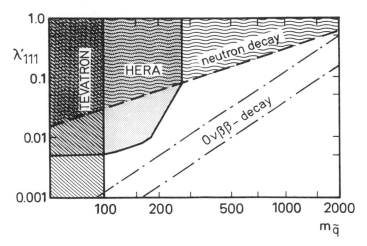

Figure 2.27. The Yukawa coupling λ'_{111} as a function of the squark mass in *R*-parity violating supersymmetric theories. Here λ'_{111} represents additional *R*-parity violating terms in the Lagrangian. The experimentally excluded regions from low energy experiments, high energy colliders and from double beta decay experiments are shown (excluded regions lie beyond the lines). The vertical line marks the lower limit for squark masses from the Tevatron [Roy92], the solid line corresponds to the limit reachable at HERA [But93, Dre94]. The dashed line is the best present limit from neutron decay [Bar89]. The dashed-dotted lines represent the limits from double beta decay (the Heidelberg–Moscow experiment), the upper (lower) line corresponding to a mass of the gluino of 1 TeV (100 GeV) respectively (from [Hir95, Hir96b, Kla96a]).

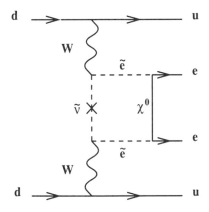

Figure 2.28. An example of one of the dominant contributions of the sneutrino to the $0\nu\beta\beta$ decay amplitude in *R*-parity conserving SUSY models (from [Hir97a]).

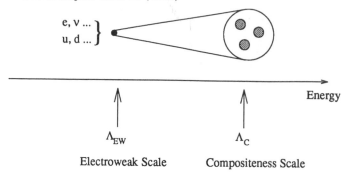

Energy

Λ_{EW} Λ_{C}

Electroweak Scale   Compositeness Scale

Figure 2.29. The idea of compositeness: at an energy scale Λ_C (the as yet unknown compositeness scale), normal quarks and leptons may reveal a substructure (from [Pan96]).

2.6 Compositeness

The question arises as to whether quarks and leptons are the most elementary building blocks of Nature. Up to now, every closer examination has revealed new substructures. If this were to continue, all quarks, leptons and Higgs bosons (maybe even the W and Z gauge bosons) would have to be constructed of even smaller particles, called preons (see figure 2.29). These would have to be bound by a new form of superstrong interaction, which would ensure that the preons combine to form quarks, leptons, etc. This hypothetical interaction goes under many names, such as hypercolour, metacolour, technicolour, etc. It is characterized by a compositeness scale Λ. We refer to [Sch85b, Mar92, Moh91, Moh86, 92, Sou92] for theoretical details. One possible experimental signature for compositeness is the search for excited leptons and quarks. Existing limits for this can be found in [Der95, Aid96c]. An interaction at high energy can appear as a point interaction at low energies (similar to the Fermi point interaction being a low-energy approximation for the exchange of heavy W or Z bosons). Such a contact interaction would be able to modify the expected cross section for many accelerator reactions. The non-observation of such effects leads to limits on the scale Λ of order 1–2 TeV. If, for example, the results from HERA were to be converted into a quark radius, it would be smaller than 2.7×10^{-18} m [Aid95]. In a similar way, an electron radius smaller than 10^{-19} m can be derived from the LEP data and the $(g-2)$ experiments [Kin90]. No evidence for any substructure has been found up to the above scales. The recent evidence for excess events with very high Q^2 at both HERA detectors H1 and ZEUS [Adl97, Bre97] has also been discussed, amoungst others, in connection with an indication of contact interactions and compositeness [Alt97, Fra97, Adl97a, Aka97]. Double beta-decay also provides stringent limits on compositeness [Pan94, Pan96, Tak96, Pan97].

2.7 Superstring theories

The unification of gravitation with the other interactions, despite all the successes of the theories discussed above, still presents considerable difficulties. This is because every theory with point-like objects becomes divergent for energies above the Planck scale. The Planck scale, or Planck mass, is the energy at which a quantum theory of gravitation is needed, which means the point at which the Schwarzschild radius

$$R = \frac{2Gm}{c^2} \tag{2.99}$$

and Compton wavelength

$$\lambda = \frac{\hbar}{mc} \tag{2.100}$$

of an object become of the same order of magnitude. These are the characteristic scales at which the general theory of relativity and quantum theory, respectively, have to be applied in order to give a sensible description. The Planck mass lies at

$$M_{Pl} = \left(\frac{\hbar c}{G}\right)^{\frac{1}{2}} \simeq 1.2 \times 10^{19} \text{ GeV} \tag{2.101}$$

corresponding to a Planck length and Planck time of

$$L_{Pl} = \left(\frac{\hbar G}{c^3}\right)^{\frac{1}{2}} \simeq 1.6 \times 10^{-33} \text{ cm} \tag{2.102}$$

$$t_{Pl} = \left(\frac{\hbar G}{c^5}\right)^{\frac{1}{2}} \simeq 5.4 \times 10^{-44} \text{ s.} \tag{2.103}$$

In the description of the early universe using the big bang model, a quantum theoretical description of gravitation is required above M_{Pl} and below L_{Pl}, t_{Pl}. One way to bypass the difficulties described above is to assume that the elementary objects are strings, whose character only appears above M_{Pl}. The typical extent of the strings is characterized by the Planck scale. String theories have already been used in the 1960s as an explanation for hadron physics, but with the continuing success of the standard model they had largely been forgotten. The use of strings experienced a renaissance when Green and Schwarz showed that a string theory with space-time supersymmetry which is both gauge and gravitation anomaly free, can only be described in ten dimensions and with the help of the internal symmetry groups SO(32) or $E_8 \otimes E_8$ [Gre86]. It was already known from the old string theories that both unitarity and Lorentz invariance for string theories could only be attained in higher dimensional spaces. The idea to use higher dimensional spaces to describe forces had either been already used by Kaluza and Klein in the 1920s, in order to describe electromagnetism and gravitation on a purely geometrical basis (Kaluza–Klein theories). Because of the in-built supersymmetry these new theories are called *superstring* theories. In the framework of these theories some of the quantum mechanical excitations of the strings (normal modes) are interpreted as the experimentally observed elementary particles. Such excitations can be either rotations, vibrations, or excitations of the internal degrees of freedom. Hence the entire spectrum of elementary particles results from a single fundamental string. The number of states with masses smaller than the Planck mass are finite and correspond to the observed particles. There is however also an infinite number of excitations with masses larger than the Planck mass. In general these modes are unstable and decay into lighter modes; however, stable solutions which can have exotic characteristics (such as magnetic charge, exotic values of electric charge) could also exist. It is also remarkable that in all particle spectra which correspond to the classic solutions of string theories, exactly *one* massless spin-2 graviton appears.

Superstring theories with fermions are formulated in 10-dimensional spaces. It is therefore necessary for our universe that 6 dimensions have become non-observable (*compactification*). This division into 4 space-time and 6 compactified dimensions does not seem to be stringent, but the sum of the compactified and non-compactified dimensions has to be 10. Hence universes with a different number of dimensions would seem to be possible. Strings appear in two different topologies, in the form of open strings with free ends or in the form of closed loops. They can in addition have an intrinsic orientation. The quantum numbers of open strings are found at their ends, while in closed loops the quantum numbers are smeared out along the string.

At present three consistent superstring theories exist. Superstring theories of type I consist of unoriented strings of either topology. The two other theories, type II and heterotic string theories, are based on oriented closed strings with different internal symmetries. They can be described by both the SO(32) and $E_8 \otimes E_8$ symmetry groups. The E_8 group is the largest finite exceptional group, and the product of the two groups has, like SO(32), 496 generators. These symmetry groups have now to be broken in order to be compatible with the previously discussed unified theories below 10^{15} GeV. It turns out that all this physics is contained in *one* E_8 group. The other E_8 group is only noticeable in our universe from the gravitational interaction, and it therefore leads a shadowy existence (*shadow matter*). The particles connected with this shadow matter are therefore very suitable as candidates for dark matter (see chapter 9).

Interactions are described within superstring theories by the breaking of a string or by the combination of two strings (see figure 2.30). The interactions are not described by Feynman diagrams, but by two dimensional 'world sheets', where in type II and heterotic strings the genus, the number of incoming strings, determines the order (see figure 2.31). As the interaction no longer takes place at a specifically defined space-time point (see figure 2.30) many of the problems with point-like particles vanish. For further information we refer to [Gre87, Kak88, Din90, Wit96]. Experimentally verifiable predictions from

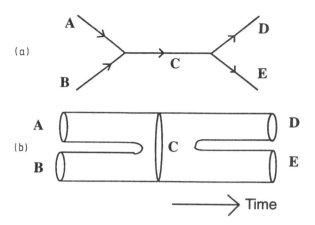

Figure 2.30. Space-time diagram of a string interaction with two strings, (A and B) (*b*), in comparison with a particle interaction (*a*). The space-time point of the string interaction (C) is ambiguous and depends on the Lorentz frame; the end products are D and E (from [Gro93]).

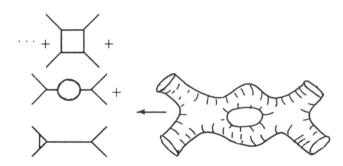

Figure 2.31. First order corrections to the Feynman diagrams reveal themselves in string theories by holes in world sheets. The number of holes (also called the genus) describes the order of perturbation theory (see e.g. [Wit96]).

superstring theories prove to be extremely difficult due to the many unknown parameters which they contain. A glimpse into superstring phenomenology can be obtained from [Ell93a].

We have now ventured into the most speculative area of unified theories. Even if a quantitative understanding still lies in the distant future, perhaps for the first time a genuinely unified description of all particles and interactions within the framework of superstring theories seems to have been achieved. Having gained the necessary insight into elementary particle physics, we now turn to the other important standard model, namely that of cosmology, the big bang theory.

Chapter 3

Cosmology

It is a reasonable assumption that, on the scales that are relevant to a description of the development of the present universe, of all the interactions only gravity plays a role. All other interactions are neutralized by the existence of opposite charges in the neighbourhood, and have an influence only on the detailed course of the initial phase of the development of the universe. The currently accepted theory of gravitation is Einstein's *general theory of relativity*. This is *not* a gauge theory; gravitation is interpreted purely geometrically as the curvature of four-dimensional space-time. As discussed in chapter 2, this therefore may not be the ultimate theory of gravitation. For a more detailed introduction to general relativity see [Wei72, Mis73, Sex87]. While general relativity was being developed (1917) the accepted model was that of a stationary universe. Friedman in 1922 examined *non-stationary* solutions of Einstein's field equations. All models based on expansion contain an initial singularity of infinitely high density. From this the universe developed via an explosion (the big bang). This type of model of the universe was experimentally confirmed when Hubble discovered galactic red-shifts in 1929 [Hub29] and interpreted their velocity of recession as a consequence of this explosion. With the discovery of the cosmic microwave background in 1964 [Pen65], which is interpreted as the echo of the big bang, the big bang model was finally established in preference to competing models, such as the steady state model. The proportions of the light elements could also be predicted correctly over 10 orders of magnitude within this model (see chapter 4). All this has resulted in the big bang model being today known as the *standard model of cosmology*. For further literature we refer to [Boe88, Gut89, Kol90, Kol93, Nar93, Pee93, Pee95, Goe94].

3.1 Cosmological models

Our present conception of the universe is that of a homogeneous, isotropic and expanding universe. Even though the observable spatial distribution of galaxies seems decidedly lumpy (see chapter 6), it is generally assumed that,

100

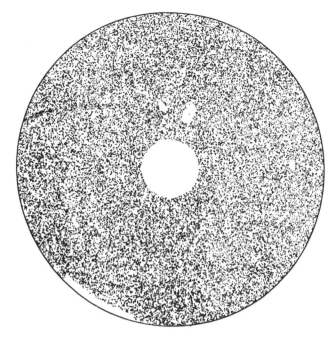

Figure 3.1. The distribution of radio sources in the Green Bank survey at 6 cm. The picture contains 33 000 radio sources. The galactic north pole is situated in the centre of the picture (surrounded by an area that has not been observed), and the galactic equator is the outer boundary. The homogeneous distribution of the radio sources with a significant fraction consisting of far away radio galaxies points to a homogeneity over very large distances (from [Pee93], see also [Lon94]).

at large enough distances, these inhomogeneities will average out and an even distribution will exist (see figure 3.1). At least this seems to be a reasonable approximation today. The isotropy of the microwave background radiation (chapter 7) also testifies to a very high isotropy of the universe. Whereas from isotropy inevitably follows homogeneity, the opposite is not necessarily true. These observations are embodied in the so-called *cosmological principle*, which states that there is no preferred observer, which means that the universe looks the same from any point in the cosmos. The space-time structure is described with the help of the underlying metric. In three-dimensional space the distance is given by the line element ds^2 with

$$ds^2 = dx_1^2 + dx_2^2 + dx_3^2, \tag{3.1}$$

whereas in the four-dimensional space-time of the *special theory of relativity* a line element is given by

$$ds^2 = dt^2 - (dx_1^2 + dx_2^2 + dx_3^2) \tag{3.2}$$

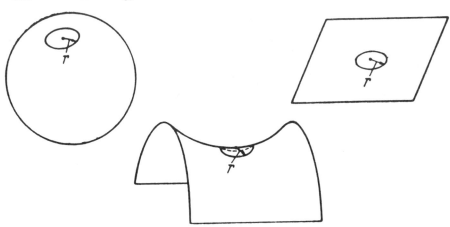

Figure 3.2. Surfaces, equivalent to two-dimensional spaces, with constant curvature $k > 0$ (case a), $k = 0$ (case b), $k < 0$ (case c), as models for curved spaces.

which in the general case of also non-inertial systems can be written as

$$ds^2 = \sum_{\mu\nu=1}^{4} g_{\mu\nu} dx^\mu dx^\nu. \tag{3.3}$$

Here $g_{\mu\nu}$ is the metric tensor, which in the case of the special theory of relativity takes on the simple diagonal form of

$$g_{\mu\nu} = (1, -1, -1, -1). \tag{3.4}$$

The simplest metric with which to describe a homogeneous isotropic universe in the form of spaces of constant curvature is the *Robertson–Walker metric* [Wei72], in which a line element can be described by

$$ds^2 = dt^2 - R^2(t) \left[\frac{dr^2}{1 - kr^2} + r^2 d\theta^2 + r^2 \sin^2\theta \, d\phi^2 \right]. \tag{3.5}$$

Here r, θ and ϕ are the three co-moving spatial coordinates, $R(t)$ is the scale-factor and k characterizes the curvature. A closed universe has $k = +1$, a flat Euclidean universe has $k = 0$ and an open hyperbolic one has $k = -1$ (see figure 3.2). In the case of a closed universe R can be interpreted as the 'radius' of the universe. The complete dynamics is embodied in this time dependent scale-factor $R(t)$[1], which is described by Einstein's field equations

$$R_{\mu\nu} - \tfrac{1}{2} R g_{\mu\nu} - \Lambda g_{\mu\nu} = 8\pi G T_{\mu\nu}. \tag{3.6}$$

† This name implies that the spatial separation of two adjacent 'fixed' space points (with constant r, ϕ, θ coordinates) is scaled in time by $R(t)$.

In this equation $R_{\mu\nu}$ is the Ricci tensor, $T_{\mu\nu}$ corresponds to the energy–momentum tensor and Λ is the cosmological constant [Wei72, Mis73, Sex87, Ber90e]. We first set the cosmological constant to zero. Its effects will be examined in a separate chapter (chapter 5). If we look at space only locally, we can assume to the first approximation a flat space, which means the metric is given by the Minkowski metric of the special theory of relativity. As $g_{\mu\nu}$ is diagonal here the energy–momentum tensor also has to be diagonal. Its spatial components are equal due to isotropy. The dynamics can be described in analogy to the model of a perfect liquid with density $\rho(t)$ and pressure $p(t)$ when averaging over galaxies and super-clusters. The energy–momentum tensor then has the form

$$T_{\mu\nu} = \text{diag}(\rho, -p, -p, -p). \tag{3.7}$$

From the zeroth component of Einstein's equations it follows that

$$\frac{\dot{R}^2}{R^2} + \frac{k}{R^2} = \frac{8\pi G}{3}\rho \tag{3.8}$$

while the spatial components give

$$2\frac{\ddot{R}}{R} + \frac{\dot{R}^2}{R^2} + \frac{k}{R^2} = -8\pi G p. \tag{3.9}$$

These equations (3.8) and (3.9) are called *Einstein–Friedmann–Lemaitre equations*. From these equations it is easy to show that:

$$\frac{\ddot{R}}{R} = -\frac{4\pi G}{3}(\rho + 3p) \tag{3.10}$$

Since currently $\dot{R} \geq 0$ (i.e. the universe is expanding), and on the assumption that the expression in brackets has always been positive, i.e. $\ddot{R} \leq 0$, it inevitably follows that R was once 0. This singularity at $R = 0$ can be seen as the 'beginning' of the development of the universe. Evidence for such an expanding universe came from the red-shift of far away galaxies [Hub29]. The further galaxies are from us, the more red-shifted are their spectral lines, which can be interpreted as a consequence of the velocity of recession v. The measurements resulted in the Hubble relation

$$v = H_0 r. \tag{3.11}$$

The proportionality constant H_0 is called the *Hubble constant* (see figure 3.3). The index 0 represents the current value both here and in what follow·

The relationship between equation (3.11) and the expanding universe can be seen by considering a light source emitting waves at a point r_1, at the time t_1, which are observed by us at a time t_0 at a position $r = 0$. As light travels along

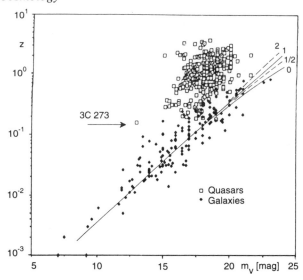

Figure 3.3. Hubble diagram for radio galaxies and quasars. The red-shift is plotted against the apparent visual brightness. The Hubble relation is adjusted for small z, and the different behaviour for different deceleration parameters q_0 is illustrated. In contrast to radio galaxies, quasars show a large variation in apparent brightness, due to their variability and the wide diversity in their activity. This results in them all lying above the Hubble line, the cosmological interpretation is supported by the fact that no quasar lies *under* the line for the radio galaxies (from [Uns92]).

geodesics ($ds^2 = 0$), the Robertson–Walker metric results in (without restricting the generality it can be assumed that $d\phi = d\theta = 0$)

$$\int_{t_1}^{t_0} \frac{dt}{R(t)} = \int_0^{r_1} \frac{dr}{(1 - kr^2)^{1/2}}. \tag{3.12}$$

The following equality holds for a second wave crest sent out from the source, stationary in the co-moving reference frame, a short time δt later:

$$\int_{t_1}^{t_0} \frac{dt}{R(t)} = \int_{t_1 + \delta t_1}^{t_0 + \delta t_0} \frac{dt}{R(t)}. \tag{3.13}$$

If δt is sufficiently small, R can be assumed to be constant over the integration period, so that:

$$\frac{\delta t_1}{R(t_1)} = \frac{\delta t_0}{R(t_0)}. \tag{3.14}$$

However, since δt is the time between two consecutive wave crests and therefore corresponds to the wavelength in both emission and absorption, it follows that:

$$\frac{\lambda_1}{\lambda_0} = \frac{R(t_1)}{R(t_0)}. \tag{3.15}$$

The red-shift z is therefore defined as

$$1 + z = \frac{\lambda_0}{\lambda_1} = \frac{R(t_0)}{R(t_1)} \qquad (3.16)$$

Thus it can be seen that the red-shift is related to the 'size' of the universe at a particular time. By observing z it is possible to obtain direct information about the development of $R(t)$. This can be demonstrated by expanding $R(t)$ as a Taylor series around the value it has today, giving

$$\frac{R(t)}{R(t_0)} = 1 + H_0(t - t_0) - \frac{1}{2} q_0 H_0^2 (t - t_0)^2 + \dots \qquad (3.17)$$

The Hubble constant H_0 is therefore

$$H_0 = \frac{\dot{R}(t_0)}{R(t_0)}, \qquad (3.18)$$

and the deceleration parameter q_0 is given by

$$q_0 = \frac{-\ddot{R}(t_0)}{\dot{R}^2(t_0)} R(t_0). \qquad (3.19)$$

Considering only the first term of the Taylor expansion, equation (3.16) can be rewritten using the equality $r = c(t - t_0)$ as

$$cz = H_0 \cdot r. \qquad (3.20)$$

This can be compared to equation (3.11) and provided cz is interpreted as the velocity, the expression (3.20) corresponds to the empirically observed Hubble relation. The effect manifests itself as a wavelength shift similar to the classic Doppler effect. The cosmological red-shift, however, differs from the Doppler effect in that it is not caused by the relative movement of source and observer, but rather by the stretching out of the wavelength due to the expansion of the universe. Due to the special importance of the Hubble constant for all cosmological distance determination, we discuss several methods for its determination in the next section.

3.1.1 Determination of the Hubble constant H_0

Models of the universe can be expressed in terms of H_0 and q_0, which are experimentally easily accessible, rather than the more abstract k and R. We first consider H_0 and discuss measurements of q_0 in chapter 5. The determination of H_0 is of extraordinary importance to cosmology. The first method of determination is straightforward and uses equations (3.11) and (3.20) to determine H_0 by measuring z and r. However, this requires an independent measurement of the distance. Thus, we come across the whole problematic nature of the cosmological distance scale.

3.1.1.1   Distance determination in space

The distance scale in the universe is built up in a pyramid-like shape. For a detailed discussion of the distance scale see [Row85b, Ber95]. While trigonometric methods ('parallaxes', i.e. the shift of star positions due to the annual movement of the Earth) are adequate for exploring our nearest surroundings, spectroscopic methods are needed for larger distances. The *parsec* (pc = short for parallax second) has become established as the unit of distance. This is the distance from which the radius of the Earth's orbit subtends an angle of 1 arc second. This corresponds to a distance of about 3.2 light years. An important connection between geometric and spectroscopic methods is the Hyades star cluster at a distance of about 50 pc, since its distance can be determined by both methods. All spectroscopic methods in the end depend upon the determination of an *absolute brightness M*, which, subtracted from the observed apparent brightness *m* produces the *distance modulus*

$$m - M = 5\log\left(\frac{r}{10\text{ pc}}\right) \tag{3.21}$$

from which the distance can be determined (interstellar absorption effects have been neglected here). Following the historical assignment of star brightness into magnitudes, the *apparent brightness* is defined via the radiation flux S (see e.g. [Uns92]) as

$$m = C - 2.5\log S \tag{3.22}$$

where C depends on the wavelength of the observation. This is usually indicated by a subscript (for example, m_V is the apparent brightness in the V-band at about 550 nm). The apparent brightness of the brightest star Sirius is, for example, $m_V = -1^m.5$. The smaller the number, the brighter the star. A difference of 5 magnitudes corresponds to a factor 100 in the brightness. The *bolometric brightness* is obtained through an integration over all wavelengths. As can be seen from the distance modulus, the absolute brightness of a star is given by the apparent brightness that a star has at a distance of 10 pc. If all stars were brought to this distance they would be distinguished by their intrinsic brightness, which is called *luminosity*. It is given by

$$L = 4\pi r^2 \sigma T^4 \tag{3.23}$$

where r is the radius of the star, T is its surface temperature and σ is the Stefan–Boltzmann constant.

A reliable knowledge of M is needed in order to determine the distance. Luckily M is connected with other characteristics of stars or galaxies. An example are variable stars. The evolution of stars is appropriately illustrated in a luminosity versus temperature diagram, which is also called the *Hertzsprung–Russel diagram* (see figure 3.4). If the observed stars are entered into this diagram it can be seen that the distribution of the stars is clearly correlated.

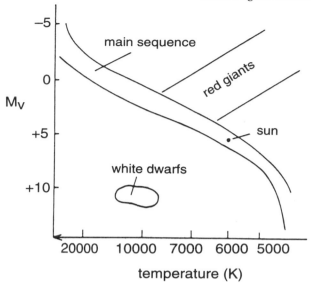

Figure 3.4. Schematic illustration of a Hertzsprung–Russel diagram, i.e. the absolute visual brightness M_V (approximately corresponding to the luminosity) as a function of temperature. The smaller M_V, the brighter the star. It can be seen that the stars are not distributed uniformly in this figure, but rather there are certain accumulation areas that mirror different stellar evolutionary phases. Apart from the main sequence on which the hydrogen burning stars are situated, there are, for example, regions of giant stars (to the right of the main sequence) and white dwarfs, a final stage of stellar evolution (bottom left) (from [Boe88]).

Normal hydrogen burning stars arrange themselves along the so-called *main sequence*. When the star has burned off about 10% of its hydrogen it starts to develop further and distances itself from the main sequence. Massive stars reach a state in which they become pulsation unstable. One class of these pulsation variables are the cepheids, named after the first known example of this kind, δ Cephei. As they are massive stars and therefore relatively short-lived, they are found preferentially in accumulations of young stars such as open clusters or OB associations. Typical pulsation periods are between 2 and 150 days. As early as 1907, Leavitt established a correlation between the pulsation period P and the absolute brightness in the V-band (550 nm) for cepheids. Today it is usually written:

$$M_V = a - b \log P + c(B - V). \tag{3.24}$$

Here a, b, c are constants to be determined, and $(B - V)$ represents the so-called colour-excess (B corresponds to an observation at 440 nm). Typical values would be $a = -3, b = 3.8$ and $c = 2.7$ [Row85b]. This relation can easily be understood. Considering the pulsation as standing sound waves in a star, a relation between the pulsation period P and the average density $\bar{\rho}$ can be established:

$$P \simeq \frac{1}{\sqrt{G\bar{\rho}}} \qquad (3.25)$$

Bright stars, i.e. having M_V strongly negative, according to equation (3.23) should also have a large radius. Such stars have a low surface density and therefore also a long pulsation period. From observations in our Milky Way we know the dependence of the absolute brightness on the pulsation period of the cepheid variables fairly well and can therefore determine the distance of other galaxies by observing these variables within them.

A further group of variables are the RR-Lyrae stars. These have typical periods of roughly 0.4–1 day. Investigation of these variables showed that they all have the same absolute brightness of $M_V \simeq 0^m.6$ but that the value is slightly dependent on the 'metal' content Z (all elements heavier than He). The above value applies to RR-Lyrae stars which have very little metal content. Distance determinations of up to 100 kpc and up to a few Mpc are possible for RR-Lyrae and cepheid variables respectively.

An alternative *standard candle* to these variable stars are supernovae (see chapter 13). Because of their extremely large luminosity they can be observed from far-away galaxies. Due to their identical development mechanism supernovae of the type Ia seem to be especially suitable. Their absolute brightness in the B-band at the time of the maximal emission t_{max} seems to be almost constant

$$M_B(t_{max}) = -19^m.12 \pm 0^m.3. \qquad (3.26)$$

The variables and supernovae are primary distance indicators. With the variables it is possible to determine the distance to the nearest objects, such as the Magellanic clouds and Andromeda. Recently they have even been extended to galaxy M87 [Pie94]. The supernova method is adequate for distances of up to 100 Mpc. An examination of 13 supernova type Ia light curves in galaxies in which the distance could also be determined via cepheids resulted in a value for the Hubble constant of $H_0 = (67 \pm 7)$ km s^{-1} Mpc^{-1} [Rie95]. An observation programme for evidence of supernovae in galaxies with large red-shifts has meanwhile discovered 7 supernovae at about $z \approx 0.4$; in future this will extend the importance of this method of distance determination via supernovae [Per95]. However, because of the scarcity of supernovae there are also other, so-called secondary distance indicators in use, which are mainly calibrated using the primary standard candles mentioned above. Here it is apparent how errors accumulate at greater distances. One example of such a secondary indicator seems to be the correlation between the number of globular clusters and the luminosity of a galaxy: the brighter a galaxy is, the more globular clusters it contains. The number n_H of clusters is given by [Row85b]

$$n_H \sim L^{1.0 \pm 0.3} \qquad (3.27)$$

and with some assumptions a correlation to M_B can be found [Row85b]:

$$\log n_H \simeq -0.3[M_B(\text{Galaxy}) + 11.0] \qquad (3.28)$$

There are, however, also drastic exceptions to this relation, for example the galaxy M87. Other methods use, for instance, the brightest blue or red stars of a galaxy, the brightest galaxies of a cluster etc in the distance determination, in order to venture thereby to ever larger distances. Two other methods should also be mentioned. For spiral galaxies the Tully–Fisher relation [Tul77] is often used, a dependence of the luminosity on the velocity dispersion σ (given in km s^{-1}), measured with the help of the 21 cm line of neutral hydrogen (the hyperfine transition of the hydrogen ground state, caused by the alignment of electron and proton spin). The movement of the hydrogen gas in a galaxy along the line of sight leads to a Doppler broadening of the line, which can be converted into a velocity dispersion. This results in

$$M_{pg} = -a \log \left(\frac{\sigma}{\sin i}\right) - b \qquad (3.29)$$

where M_{pg} is the photographic brightness and the constants a and b are dependent on the calibration and the observers, and have values of approximately $a = 6.5$ and $b = 3.5$. Furthermore, a dependence on the type of galaxy seems to exist (see chapter 6). The inclination i describes the angle of the galaxy with respect to the line of sight. For $i = 0°$ ('face-on'), there is no contribution from the rotation of the galaxy, in contrast to a galaxy at $90°$ ('edge-on') inclination, and the dispersion depends on the random movement of the gas. The inclination is one of the main problems, as the internal absorption of the emitting galaxy has to be corrected for. Therefore the velocity dispersion is measured at the 21 cm line, whereas the luminosity is measured in the infra-red region, for example at about 2 μm, where this effect is very small.

The Faber–Jackson method [Fab76] is very suitable for elliptical galaxies. A correlation between the velocity dispersion and the luminosity also exists here:

$$L \propto \sigma^4. \qquad (3.30)$$

Another method consists of a correlation between the velocity dispersion σ and the diameter of the galaxy for a particular surface brightness ($D_n - \sigma$ relation). All of these multi-parameter correlations for elliptical galaxies are different projections of a single *fundamental plane* within the parameter space [Kor89]. A deeper understanding of the latter will also further improve the other existing relations.

The problem at very large distances lies less in the relative distance measurement (the error here is typically 10%), but more in the calibration, i.e. the connection of these quantities to absolutely determined distances. This also causes the measurement of the Hubble constant to be uncertain. While initially values of around 500 km s^{-1} Mpc^{-1} were obtained, currently values of between

40 and 100 km s^{-1} Mpc^{-1} are preferred [Ton93, Nug95, Fre96]. Unfortunately, therefore, H_0 is still uncertain by a factor of two, which leads to the introduction of another parameter, h, such that

$$h = \frac{H_0}{100 \text{ km s}^{-1} \text{ Mpc}^{-1}} \quad \rightarrow \quad 0.4 \leq h \leq 1. \tag{3.31}$$

One of the problems is, for example, how far the local region, i.e. our nearest neighbourhood (up to 100 Mpc) is influenced by the irregular distribution of galaxies (see chapter 6), and at what point one actually measures the real Hubble expansion. A more exact determination of the distance scale is expected from observations with the Hubble space telescope [San92, Kin93] (figure 3.5). For first results see [Tam96].

In future it might even be possible to obtain information about the value of the Hubble constant independent of the distance ladder. The study of gravitational lenses (see chapter 9), supernovae (see chapter 13) and the Sunyaev–Zeldovich effect in relation to X-ray emissions from galaxies (see chapter 7) seem to be particularly promising [Gun78, Rep95]. For first results from galactic clusters see [Bir91, Jon93].

3.1.2   The density in the universe

From equation (3.8) it is clear that a flat universe ($k = 0$) is only reached for a certain density, the so-called *critical density*. This is given as [Kol90]:

$$\rho_{c0} = \frac{3H_0^2}{8\pi G} \approx 18.8h^2 \times 10^{-27} \text{ kg m}^{-3} \approx 11h^2 \quad \text{H-atoms m}^{-3}. \tag{3.32}$$

It is convenient to normalize to this density, and therefore a *density parameter* Ω is introduced, given by

$$\Omega = \frac{\rho}{\rho_c}. \tag{3.33}$$

$\Omega = 1$ therefore means an Euclidean universe. This is predicted, or rather favoured, by inflationary models, which will be discussed later, and is one of the main reasons to postulate dark matter (see chapter 9). An $\Omega > 1$ implies a closed universe, which means that at some time the gravitational attraction will stop the expansion, and the universe will collapse again. Thus, the possibility exists for another 'big bang', so that oscillatory solutions are also possible. An $\Omega < 1$ on the other hand means a universe which expands forever. In the case of a vanishing cosmological constant the following connection between Ω and the deceleration parameter q_0 exists:

$$\Omega = 2q_0. \tag{3.34}$$

If the Friedmann equation (3.8) is solved for the $\mu = 0$ component, the first law of thermodynamics results:

$$d(\rho R^3) = -p \, d(R^3). \tag{3.35}$$

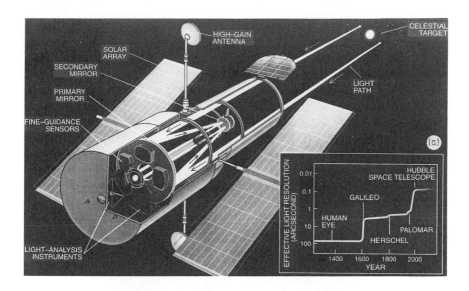

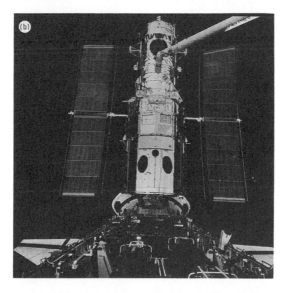

Figure 3.5. The Hubble space telescope HST. A more precise determination of the Hubble constant, should, among other things, be possible from its precise observations. (*a*) The HST is essentially constructed similarly to a terrestrial telescope. A primary mirror with a diameter of 2.4 m collects light and distributes it to five instruments. The HST improves the resolution, in comparison to terrestrial telescopes, by about the same factor by which Galileo's telescope improved on the human eye (from [Cha92]). (*b*) The HST shortly after recovery for repair on the service platform of a space shuttle in orbit (from [Alt94]).

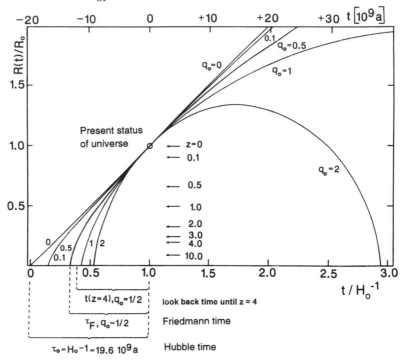

Figure 3.6. Behaviour of the scale factor $R(t)$ for different models of the universe. For all models $\Lambda = 0$ was assumed. Also shown are the various red-shifts, as well as the influence of various deceleration parameters q_0. We are currently in an era which allows no discrimination to be made about the model of the universe. A Hubble constant of 50 km s^{-1} Mpc^{-1} has been used (from [Uns92]).

This means simply that the change in energy in a co-moving volume element is given by the negative product of the pressure and the change in volume. Assuming a simple equation of state $p = k\rho$, where k is a time independent constant, the following relationships follow immediately:

$$\rho \sim R^{-3(1+k)} \qquad (3.36)$$

$$R \sim t^{\frac{2}{3}(1+k)}. \qquad (3.37)$$

The dependence of the density on R can hence be derived for different energy densities using the known thermodynamic equations of state. For the two limiting cases relativistic gas (the early radiation-dominated phase of the cosmos, particle masses negligible) and cold, pressure-free matter (the later, matter-dominated phase) we have:

Radiation $\rightarrow p = 1/3\rho$ $\rightarrow \rho \sim R^{-4}$
Matter $\rightarrow p = 0$ $\rightarrow \rho \sim R^{-3}$.

For the vacuum energy one has

$$\text{Vacuum energy} \quad \rightarrow p = -\rho \quad \rightarrow \rho \sim \text{constant.}$$

Hence in the considered Euclidean case a simple time dependence for the scale parameter (see figure 3.6) follows

$$R \sim t^{\frac{1}{2}} \quad \text{radiation dominated} \tag{3.38}$$

$$R \sim t^{\frac{2}{3}} \quad \text{matter dominated.} \tag{3.39}$$

From these different time dependences it follows that there was a point when the matter density and radiation density of the universe were equal. Using the present value for the matter density [Kol90]

$$\rho_{m0} = 1.88 \times 10^{-29} \Omega_0 h^2 \quad \text{g cm}^{-3} \tag{3.40}$$

and the radiation density of the 3K radiation (see chapter 7)

$$\rho_{s0} = 4.8 \times 10^{-34} \quad \text{g cm}^{-3} \tag{3.41}$$

it follows using the relationship

$$\frac{\rho_s}{\rho_m} \propto \frac{R_0}{R} = 1 + z \tag{3.42}$$

that

$$1 + z_{eq} = 2.32 \times 10^4 \Omega_0 h^2 \tag{3.43}$$

as well as [Kol90]

$$t_{eq} \simeq \frac{3}{2} H_0^{-1} \Omega_0^{-1/2} (1 + z_{eq})^{-3/2} \tag{3.44}$$

$$t_{eq} = 1.4 \times 10^3 (\Omega_0 h^2)^{-2} \quad \text{years.} \tag{3.45}$$

The transition from the radiation to the matter dominated universe thus took place for red-shifts of about $z = 1500$. This happened at almost exactly the same time as radiation decoupled from matter (see chapter 7).

3.1.3 The age of the universe

The question of the age of the universe is an interesting one. For a universe with no matter so that $\Omega_0 = 0$ one derives from the Hubble constant the *Hubble time*

$$\tau_0 = H_0^{-1} \Rightarrow 10 \times 10^9 < \tau_0 < 2.5 \times 10^{10} \quad \text{years.} \tag{3.46}$$

For an increasing Ω_0 the age decreases, so for example for a matter dominated universe with $\Omega_0 = 1$

$$\tau_0 = \frac{2}{3} H_0^{-1}. \tag{3.47}$$

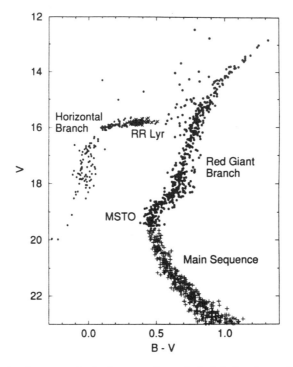

Figure 3.7. A colour-magnitude diagram of a typical globular cluster, M15. The vertical axis plots the magnitude (luminosity) of the star in the *V* wavelength region, with the brighter stars having smaller magnitudes. The horizontal axis plots the colour (surface temperature) of the stars, with cooler stars towards the right. For clarity, only about 10% of the stars on the main sequence have been plotted. By accurate determination of the main sequence turn over (MSTO) it is possible to determine the age of the globular cluster (from [Cha95]).

All this applies to the case of a vanishing cosmological constant. $\Lambda \neq 0$ results in models that are able to predict a significantly greater age for the universe (see chapter 5).

A first rough experimental estimate can be obtained from the age of our Sun. With the help of meteorites, the Sun can be dated relatively exactly to 4.5 billion years [Kir78]. Much higher limits for the age of the universe have been obtained, however, from the examination of globular star clusters. These objects surround our Milky Way almost uniformly and are some of the oldest objects in the universe. In the evolution of the globular clusters stars of very different masses, and therefore differing life-spans develop simultaneously. Massive stars, however, evolve away from the main sequence much faster than stars with smaller mass. If the stars of a globular cluster are entered into a Hertzsprung–Russel diagram, a bend can be seen in the main sequence (figures

3.4 and 3.7). The position of the bending point allows conclusions about the age of the star cluster. From such observations an age of order $(1.3–1.9) \times 10^{10}$ years can be obtained [Ber90a, Ren91, Ber95] with a conservative lower bound of 12 Gyr [Cha95]. Instead of looking at groups of stars, individual stars with small metal content can be analysed. These should be very old, since they have not accumulated the remains of dead stars, but contain only the primordial metal content. This should be small, as no heavy elements develop during the big bang (see chapter 4). These stars are called *population II stars*. From these the age of the universe can be estimated as approximately 1.5×10^{10} years [Ren91]. Similar ages can be obtained from the cooling of white dwarfs [Win87], so that generally today an age of between 15 and 20 billion years is seen as likely (see also [Gut89]). Another method of determining the age of the universe is via the 'nuclear clock', i.e. radioactive isotopes with half-lives of several billion years (nucleochronometers). A discussion of this follows in chapter 14, when we deal with the synthesis of heavy elements. With this method an age of around 20 billion years is predicted [Kla82b, Kla83, Thi83, Kla89].

3.2 The evolution of the universe

3.2.1 The standard model of cosmology

In this section we consider how the universe evolved from the big bang to what we see today. We start from the assumption of thermodynamic equilibrium for the early universe, which is a good approximation. A particle gas with g internal degrees of freedom, number density n, energy density ρ and pressure p obeys the following thermodynamic relations [Kol90]:

$$n = \frac{g}{(2\pi)^3} \int f(p) \mathrm{d}^3 p \tag{3.48}$$

$$\rho = \frac{g}{(2\pi)^3} \int E(p) f(p) \mathrm{d}^3 p \tag{3.49}$$

$$p = \frac{g}{(2\pi)^3} \int \frac{|p|^2}{3E} f(p) \mathrm{d}^3 p \tag{3.50}$$

where $E^2 = |p|^2 + m^2$. The phase space partition function $f(p)$ is given, depending on the particle type, by the Fermi–Dirac (+ sign in equation (3.51)) or Bose–Einstein (− sign in equation (3.51)) distribution

$$f(p) = [\exp((E - \mu)/kT) \pm 1]^{-1} \tag{3.51}$$

where μ is the chemical potentials of the corresponding type of particle. In the case of chemical equilibrium, the sum of the chemical potentials of the initial particles equals that of the end products. Consider a gas at temperature T. Since non-relativistic particles $(m \gg T)$ give an exponentially smaller contribution to the energy density than relativistic ($m \ll T$) particles, the former can be neglected, and thus for the radiation dominated phase we obtain:

$$\rho_R = \frac{\pi^2}{30} g_{\text{eff}} T^4 \tag{3.52}$$

$$p_R = \frac{\rho_R}{3} = \frac{\pi^2}{90} g_{\text{eff}} T^4 \tag{3.53}$$

where g_{eff} represents the sum of all effectively contributing massless degrees of freedom, and is given by [Kol90]

$$g_{\text{eff}} = \sum_{i=\text{Bosons}} g_i \left(\frac{T_i}{T}\right)^4 + \frac{7}{8} \sum_{i=\text{Fermions}} g_i \left(\frac{T_i}{T}\right)^4 \tag{3.54}$$

In this relation the equilibrium temperature T_i of the particles i is allowed to be different to the photon temperature T. At very low temperatures ($T \ll 1$ MeV) the three neutrino flavours are, for example, the only relativistic degrees of freedom (provided that no so far unknown particles need to be taken into consideration). Using the following expression (see chapter 7, equation (7.45))

$$T_\nu = \left(\frac{4}{11}\right)^{\frac{1}{3}} T_\gamma \tag{3.55}$$

it follows that $g_{\text{eff}} = 3.36$. For temperatures of above 300 GeV, all particles of the standard model of particle physics should contribute, which results in a g_{eff} of 106.75. Figure 3.8 illustrates the behaviour of g_{eff}. In addition to the temperature, the entropy also plays an important role. The entropy is given by:

$$S = \frac{R^3(\rho + p)}{T} \tag{3.56}$$

or in the specific case of relativistic particles by [Kol90]

$$S = \frac{2\pi^2}{45} g_s T^3 R^3 \tag{3.57}$$

where

$$g_s = \sum_{i=\text{Bosons}} g_i \left(\frac{T_i}{T}\right)^3 + \frac{7}{8} \sum_{i=\text{Fermions}} g_i \left(\frac{T_i}{T}\right)^3. \tag{3.58}$$

For the major part of the evolution of the universe, the two quantities g_{eff} and g_s were identical [Kol90]. The entropy per co-moving volume element is a conserved quantity in thermodynamic equilibrium, which together with constant g_s leads to the condition

$$T^3 R^3 = \text{constant} \Rightarrow R \sim T^{-1}. \tag{3.59}$$

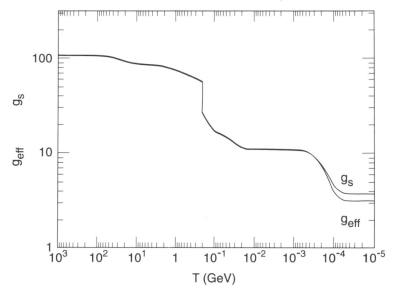

Figure 3.8. The cosmological standard model: behaviour of the summed effective degrees of freedom g_{eff} and g_S as a function of decreasing temperature. Only the particles of the standard model have been taken into consideration. One can see that both g_{eff} and g_S are identical over a wide range (from [Kol90]).

The adiabatic expansion of the universe is therefore clearly connected with cooling. In the radiation dominated phase it leads to a dependence of (see equations (3.38) and (3.59))

$$t \sim T^{-2}. \tag{3.60}$$

With the help of equation (3.60) the evolution can now be discussed in terms of either times or energies. During the course of the evolution at certain temperatures particles which were until then in thermodynamic equilibrium, ceased to be so. In order to understand this we consider the relation between the reaction rate per particle Γ and the expansion rate H. The former is

$$\Gamma = n\langle \sigma v \rangle \tag{3.61}$$

with a suitable averaging of relative speed v and cross section σ [Kol90]. The equilibrium can be maintained as long as $\Gamma > H$ for the most important reactions. For $\Gamma < H$ the corresponding particle is decoupled from equilibrium. This is known as *freezing out*. Let us assume a temperature dependence of the reaction rate of the form $\Gamma \sim T^n$. Consider two interactions mediated either by massless bosons such as the photon, or by massive bosons such as, for example, the Z^0 with a mass m_M. In the first case for the scattering of two particles a cross section of

$$\sigma \sim \frac{\alpha^2}{T^2} \quad \text{with} \quad g = \sqrt{4\pi\alpha} = \text{gauge coupling strength} \tag{3.62}$$

results. In the second case the same behaviour can be expected for $T \gg m_M$. For $T \leq m_M$

$$\sigma \sim G_M^2 T^2 \quad \text{with} \quad G_M = \frac{\alpha}{m_M^2} \tag{3.63}$$

holds. With a thermal number density, i.e $n \sim T^3$, for the case of massless exchange particles it follows that

$$\Gamma \sim \alpha^2 T. \tag{3.64}$$

In the radiation dominated phase there exists a relation between the Hubble constant and temperature [Kol90]. Using (2.101), (3.32) and (3.52) it follows:

$$H = \sqrt{\frac{8\pi\rho}{3m_{Pl}^2}} \approx 1.66 \times g_{\text{eff}}^{\frac{1}{2}} \frac{T^2}{m_{Pl}} \tag{3.65}$$

from which follows

$$\frac{\Gamma}{H} \sim \frac{\alpha^2 m_{Pl}}{T}. \tag{3.66}$$

This means that the reaction rates are sufficiently fast to produce a thermal equilibrium only for $T < \alpha^2 m_{Pl} \approx 10^{16}$ GeV. Thus we cannot necessarily assume that there has been an equilibrium between the Planck time and the time of the GUT symmetry breaking. For reactions involving the exchange of massive particles the following relationship holds:

$$\Gamma \sim G_m^2 T^5. \tag{3.67}$$

This means, that as long as

$$m_M \geq T \geq G_M^{-\frac{2}{3}} m_{Pl}^{-\frac{1}{3}} \approx \left(\frac{m_M}{100 \text{ GeV}}\right)^{\frac{4}{3}} \quad \text{MeV} \tag{3.68}$$

holds, such processes remain in equilibrium. We will return to this point in chapter 9.

   We will now discuss the evolution of the universe step by step (see figure 3.9). The earliest moment at which our present description can be applied is the already mentioned Planck time (see chapter 2). At this time, the Schwarzschild radius and Compton wavelength are of the same order. Before this point, a quantum mechanical description of gravity is necessary which does not exist currently. This time lies as about 10^{-43} s after the big bang, and at an energy of about 10^{19} GeV. All particles are highly relativistic, and the universe is radiation dominated. At the moment at which energies drop to around 10^{15} GeV the GUT symmetry breaking (discussed in the last chapter) took place, whereby the heavy gauge bosons X and Y froze out. At about 300 GeV a second symmetry breaking took place, which led to the interactions which can be observed in today's particle accelerators. At about 10^{-6} s the quarks and anti-quarks annihilated

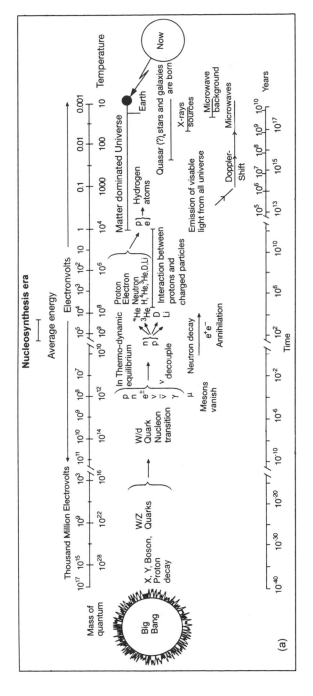

Figure 3.9 The chronological evolution of the universe since the big bang (*a*) from [Wil93] and (*b*) with kind permission of Microcosm (CERN).

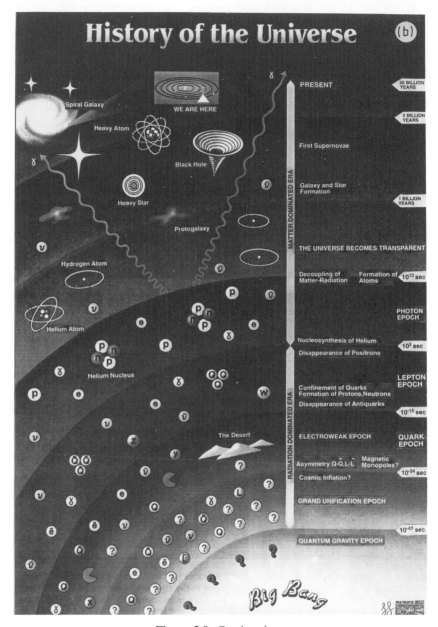

Figure 3.9. Continued.

and the surplus of quarks represents the whole of today's observable baryonic matter. The slight surplus of quarks is reflected in a baryon–photon ratio of about 10^{-9}. After about 10^{-5} s or at about 100 to 300 MeV, characterized by Λ_{QCD}, a further phase transition took place. This is connected with the breaking of the chiral symmetry of the strong interaction and the transition from free quarks in form of a quark–gluon plasma to quarks confined in baryons and mesons. At temperatures of about 1 MeV several things happened simultaneously, which will be discussed in more detail in chapter 4. During the period 10^{-2}–10^2 s the process of primordial nucleosynthesis took place. Therefore the observation of the lighter elements provides the furthest look back into the history of the universe. Around the same time, or more precisely a little before, the neutrinos decoupled and now developed further independently. In this way a cosmic neutrino background was produced, which has, however, not yet been observed. Also the almost total destruction of electrons and positrons took place at this time. The annihilation photons make up now part of the cosmic background radiation. A further crucial stage only takes place about 150 000 years later. By then the temperatures have sunk so far that nuclei can recombine with the electrons. As Thomson scattering (scattering of photons from free electrons) is strongly reduced, the universe suddenly becomes transparent, and the radiation decouples from the matter. This can still be detected today as 3 K background radiation (chapter 7). From this moment density fluctuations can increase, and therefore the creation of large scale structures as well as galaxies can begin. At roughly the same time the universe passes from a radiation dominated to a matter dominated state. This scenario, together with the discussed characteristics are called the *standard model of cosmology* (see table 3.1).

3.2.2 Baryon asymmetry in the universe

3.2.2.1 *General conditions and the GUT phase-transition*

Under the assumption of equal amounts of matter and anti-matter at the time of the big bang we observe today an enormous preponderance of matter compared with anti-matter. If we assume that anti-matter is not concentrated in regions which are beyond the reach of current observation, this asymmetry has to originate from the earliest phases of the universe. Here matter and anti-matter destroy themselves almost totally and from an existing excess of quarks *before* the destruction [Kol90] of

$$\frac{n_q - n_{\bar{q}}}{n_q} \approx 3 \times 10^{-8} \tag{3.69}$$

the baryonic matter developed. Generally it is believed that, in order to accomplish this imbalance, three conditions have to be fulfilled [Sac67b] (see also [Wei79]):

(i)   Both a C and a CP violation of one of the fundamental interactions

Table 3.1. GUT cosmology (from [Gro89, Gro90]).

	Time	Energy	Temperature	'Diameter' of the universe
	t (s)	$E = kT$ (GeV)	T (K)	R (cm)
Planck time t_{Pl}	10^{-44}	10^{19}	10^{32}	10^{-3}
GUT SU(5) breaking M_X	10^{-36}	10^{15}	10^{28}	10
$SU(2)_L \otimes U(1)$ breaking M_W	10^{-10}	10^2	10^{15}	10^{14}
Quark confinement $p\bar{p}$-annihilation	10^{-6}	1	10^{13}	10^{16}
ν decoupling, e^+e^--annihilation	1	10^{-3}	10^{10}	10^{19}
light nuclei form	10^2	10^{-4}	10^9	10^{20}
γ decoupling, transition from radiation-dominated to matter-dominated universe, atomic nuclei form, stars and galaxies form	10^{12} ($\approx 10^5$ y)	10^{-9}	10^4	10^{25}
Today, t_0	$\approx 5 \times 10^{17}$ ($\approx 2 \times 10^{10}$ y)	3×10^{-13}	3	10^{28}

(ii) Non-conservation of baryon number

(iii) Thermodynamic non-equilibrium

The production of the baryon asymmetry is usually associated with the GUT transition. The violation of baryon number is not unusual in GUT theories, since in these leptons and quarks are situated in the same multiplet, as we discussed in chapter 2. That a CP violation is necessary can be seen in the following illustrative set of reactions

$$X \xrightarrow{r} u + u \quad X \xrightarrow{1-r} \bar{d} + e^+ \tag{3.70}$$

$$\bar{X} \xrightarrow{\bar{r}} \bar{u} + \bar{u} \quad \bar{X} \xrightarrow{1-\bar{r}} d + e^- \tag{3.71}$$

In the case of CP violation $r \neq \bar{r}$. A surplus of u, d, e over $\bar{u}$, $\bar{d}$ and e^+ would follow therefore for $r \geq \bar{r}$. This is, however, only possible in a situation of thermodynamic non-equilibrium, as a higher production rate of baryons will otherwise also lead to a higher production rate of anti-baryons. In equilibrium the particle number is independent from the reaction dynamics.

One problem is to integrate CP violation into the formalism.   A spontaneously broken CP invariance cannot usually produce a sufficiently large baryon asymmetry. This is because the critical temperature between the CP invariant and CP broken phase lies typically of order 1 TeV. The relevant decays of the X, Y and Higgs bosons therefore all take place in the CP-invariant phase, and thus produce no significant baryon asymmetry. If, on the other hand, we have a *manifest* CP violation, i.e. it is explicitly built into the theory, for example through the QCD vacuum structure, the so-called θ-problem results (see chapter 11).

For a summary of the non-GUT baryon production, in which possibly *none* of the conditions (i)–(iii) mentioned above are really necessary for baryogenesis and the formation of a matter–anti-matter asymmetry, see [Dol92].

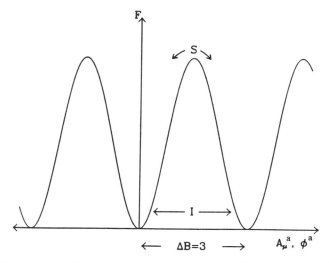

Figure 3.10. Schematic illustration of the potential with different vacua, which appear as different possible vacuum configurations of the fields A_μ, Φ in non-Abelian gauge theories. The possibility of the instanton tunnelling (I) *through* the barrier of height T_C, as well as the sphaleron (S) jumping *over* the barrier are indicated. In the transition $B + L$ changes by $2N_F$, and B by N_F, where N_F represents the number of families (in this case 3) (from [Kol90]).

3.2.2.2   The electroweak phase transition

In recent times another very different possibility to produce baryon number violation has emerged, namely the electroweak phase transition.   Its scale is characterized by the vacuum expectation value of the Higgs boson in the electroweak standard model, and therefore lies at around 200 GeV. In this case the three conditions for a baryon–anti-baryon asymmetry are fulfilled in the following way. It has been shown that non-Abelian gauge theories have non-trivial vacuum structures, and with a different number of left- and right-handed fermions can produce a baryon and lepton number violation [t'Ho76, Kli84]. Figure 3.10 shows such vacuum configurations, which are characterized by different topological winding numbers and are separated by energy barriers of height T_C. We will return to this in chapter 11, when we will discuss axions. In the case of $T = 0$ a transition through such a barrier can only take place by means of quantum mechanical tunnelling, (*instantons*), and is therefore suppressed by a factor $\exp(2\pi/\alpha_w) \approx 10^{-86}$ (α_w = weak coupling constant). However, this changes at high temperatures [Kuz85]. Now thermal transitions are possible, and for $T \gg T_C$ the transition is characterized by a Boltzmann factor $\exp(-E_{sph}(T)/T)$. Here E_{sph} represents the sphaleron energy. The *sphaleron* is a saddle point in configuration space which is classically unstable. This means that the transition takes place mainly via this configuration. The sphaleron energy E_{sph} is equivalent to the height of the barrier T_C and therefore is also temperature dependent. It lies at about m_w/α_w and therefore between about 8 and 14 TeV. Therefore effects with such vacuum transitions should be visible in the next generation of accelerators (e.g. LHC). A significant B-violation takes place particularly in the region of the phase transition. Thus there is a danger that all baryon–anti-baryon asymmetries generated earlier could be washed out, but at the same time a new one can be produced. This happens when the phase transition is at least a first order weak one. This means (see e.g. [Goe94]) that it is not a continuous transition, but that areas of the new phase (bubbles) develop, which expand. At the walls of the developing bubbles (see i.e. [Coh93]) a significant baryon number violation then takes place.

If one proceeds from one vacuum to the next the combination of $B + L$ changes by $2 \times N_F$, where N_F is the number of families, of which three are currently known. As, furthermore, $B - L$ is free of gauge-anomalies, which means that $B - L$ does not change here, the vacuum transition leads to $\Delta B = 3$. The condition of CP violation must also be fulfilled, but additional contributions besides the CP phase from the CKM-matrix are also needed here. Information about this can be obtained from the study of B-mesons (see i.e. [Nir92] and chapter 1).

In order that in addition the third condition, that of non-equilibrium, should be fulfilled, it is necessary that a phase-transition of at least weak first order should occur. It therefore seems possible to explain the baryon asymmetry also in the framework of the electroweak standard model, though many questions still remain to be answered [Din92]. A baryon number violation can also be produced through a lepton number violation (as $B - L = 0$). Majorana interactions are well suited for the latter, so that from this process limits can be placed on Majorana neutrino masses. Depending on the model the upper limits obtained range from 50 keV to 1 eV, or indeed lead to no limit at all [Fuk90a, Cam91, Gel92, Fla96]. For details of the electroweak phase transition see [Coh93, Bar95, Jan95b].

3.3 Problems of the standard model

As wonderful as the successes of the cosmological standard model are, there are still some points that require further consideration. There is, first of all, the flatness of the present universe. Further problems are the horizon and monopole problem. Furthermore it is seen as unsatisfactory that the evolution of the universe up to the present day is critically dependent on the initial parameters that existed at the Planck time. It would be much more desirable to have a model in which the subsequent development would be almost independent of the initial conditions at t_{Pl}. One possible solution of these problems can be found in the so-called inflationary models (see section 3.4).

3.3.1 The flatness problem

The metric is, as an initial condition, a quantity which is fixed once and for all, i.e. a spherical universe will always remain a spherical universe, and similarly for the two other possibilities. The question of the metric is therefore fundamental for the underlying model of the universe. The future evolution will be decisively influenced by it. A spherical universe will at some point collapse, while a hyperbolic universe (assuming a vanishing cosmological constant) will expand indefinitely.

In order to be able to discuss the curvature a scale is necessary. The only term in the Friedmann equations (equations (3.8) and (3.9)), which is not invariant against a change of R, is the curvature term $-k/R^2$. It disappears as $R \rightarrow \infty$. In the case of small curvature it follows that

$$\left| \frac{k}{R^2(t)} \right| \ll \left| \frac{8\pi G}{3} \rho(t) \right| \tag{3.72}$$

and the curvature term has hardly any influence on the development of $R(t)$. The opposite case of a strong curvature

$$\left| \frac{k}{R^2(t)} \right| \gg \left| \frac{8\pi G}{3} \rho(t) \right| \tag{3.73}$$

can only be realized in a hyperbolic universe. For a spherical universe, however,

$$\left| \frac{k}{R^2(t)} \right| \leq \left| \frac{8\pi G}{3} \rho(t) \right| \tag{3.74}$$

always applies. Therefore it makes more sense to define the degree of curvature as a deviation of the density $\rho(t)$ from the critical density ρ_c . Let us assume conservatively that today

$$0.1\rho_c \leq \rho_0 \leq 10\rho_c. \tag{3.75}$$

This allows solutions for all cases $k = -1, 0, +1$. Any deviation from $\rho_c(t)$ will grow in the course of time. From the dependences for a radiation-dominated universe ($\rho_s \sim R^{-4}$) and for a matter-dominated universe ($\rho_m \sim R^{-3}$) it is clear that the right side of equation (3.74) is growing faster than that of the curvature term (left side). In order that our universe does in fact lie so near to the critical density (equation (3.75)), the deviation at, for example, 10^{-36} s after the big bang would have had to be

$$\frac{|\rho(t) - \rho_c(t)|}{\rho_c(t)} \leq 10^{-50}. \tag{3.76}$$

Even 1 s after the big bang, corresponding to about 1 MeV, the deviation would have to have been less than 10^{-14}. The early universe would therefore have to have been extremely flat. To explain the flatness in a natural way, without very specially selected initial conditions, has to be the aim of any model.

3.3.2   The horizon problem

The distance up to which we can receive information today is called the *event-horizon*. It is defined by

$$d_H(t) = R(t) \int_0^t \frac{dt'}{R(t')}. \tag{3.77}$$

Two observers, separated from each other by $2d_H$, are therefore completely independent from each other, and were also at the time of the GUT-symmetry breaking. If we assume for the latter a characteristic temperature of about 10^{15} GeV and an inverse proportionality between R and T (equation (3.59)), the present value T of the background radiation results in an increase in R until today of a factor of 4×10^{26}. If the GUT time, 10^{-35} s after the big bang, is inserted into equation (3.77), a horizon-size of $2d_H \simeq 6 \times 10^{-25}$ cm is obtained. This would result today in a value of 2.40 m for d_H. The background radiation is, however, today isotropic at a distance scale of 10^{28} cm. This can only be possible if isotropy is demanded right from the start; otherwise it cannot be explained since such large areas were earlier not causally connected. Or, to put it in another way, as in all expanding universe models $R(t) \sim t^n$, and $n < 1$, it follows from equation (3.77) that

$$d_H = \frac{ct}{1-n}. \tag{3.78}$$

In the case of a vanishing curvature and with $p = k\rho$ a matter dominated universe leads to

$$R \sim t^{\frac{2}{3}} \rightarrow d_H = 3ct. \tag{3.79}$$

We should therefore observe several causally independent areas. However, there is then no explanation as to why the observed 3 K radiation is so highly isotropic.

3.3.3   The monopole problem

A further problem is the number of heavy particles left over from the GUT phase transition. One such example is magnetic monopoles, which we will discuss further in chapter 10. These are topological defects which develop at the edges of areas (domains) of space with different GUT phases. The predicted number is uncertain, but a simple estimate can be made [Kib76]. Here it is assumed that at least one monopole develops per domain. If the domain has a radius r a monopole density n_M of

$$n_M \gtrsim r^{-3} \qquad (3.80)$$

and a contribution to the matter density of

$$\rho_M = m_M n_M \gtrsim m_M r^{-3} \qquad (3.81)$$

result. Two facts now lead to a problem. The conservative assumption for the present matter density (equation (3.75)) of course also applies for monopoles. On the other hand, r cannot be larger than the event horizon at the time of the symmetry breaking as every area has to be causally connected. Hence r is associated with a temperature T_M, given by $kT_M = m_M$. From this condition and equations (3.75) and (3.81) it follows that

$$kT_M \leq 10^{10} \text{ GeV}. \qquad (3.82)$$

This means that a phase-transition for a typical SU(5)-breaking at temperature $kT \approx 10^{15}$ GeV (see chapter 2) and therefore a similar monopole mass would have led to an unacceptably high contribution by monopoles to the energy density of about $\Omega h^2 \approx 10^{10}$.

3.4   The inflationary phase

A solution to these problems can perhaps be found in the model of inflation [Gut81, Alb82, Lin82, Lin84]. For more recent outlines see [Oli90a, Kol90, Lin90, Nar91, Lid95, Lin96a, Tur97]. It should be mentioned that there is at present nothing which could be called a standard model of inflation. Up to now we have only considered the case of $\Lambda = 0$. In chapter 5 it will be shown that Λ has the meaning of a vacuum energy density ρ_V in modern quantum field theory. The equations (3.8) and (3.10) can then be written more generally ($\rho \to \rho + \rho_V$ and $p \to p + p_V$) as

$$\frac{\dot{R}^2}{R^2} + \frac{k}{R^2} = \frac{8\pi G}{3}(\rho + \rho_V) \qquad (3.83)$$

and

$$\frac{\ddot{R}}{R} = -\frac{4\pi G}{3}(\rho - 2\rho_V + 3p). \qquad (3.84)$$

As shown in chapter 5, a positive vacuum energy corresponds to a negative pressure $p_V = -\rho_V$. Should this vacuum energy at some time be the dominant contribution, that is dominate over all matter and curvature terms, new exponential solutions for the time behaviour of the scale-factor R result. The condition for inflation therefore means $\ddot{R} > 0$. So, in the case of $\rho_V > 0$

$$R(t) \approx R(0) \exp(Ht) \tag{3.85}$$

where

$$H^2 = \frac{8\pi G\rho_V}{3}. \tag{3.86}$$

Such exponentially expanding universes are called *de Sitter* universes. In the specific case in which the negative pressure of the vacuum is responsible for this, we talk about *inflationary* universes. Inflation is generally generated by scalar fields ϕ, sometimes called inflaton fields, which only couple weakly to other fields. An example of such a scalar field is the Higgs field (see section 1.5.2.1). In realistic models ϕ is usually related to other physical scales, like the GUT scale. In this case the exponential phase of inflation is restricted to the GUT phase transition. It remains to be explained how a phase can arise in which vacuum energy dominates. A fundamental condition is that the weakly interacting scalar Higgs field Φ is not in its ground state. How this is achieved depends on the model. One possibility to produce the initial shift of the field Φ from its ground state is via spontaneous symmetry breaking. In the discussion of spontaneous symmetry breaking we have seen (see chapter 1) that in the broken phase the vacuum expectation value of the Higgs field is different from zero, while in the case of unbroken symmetry $\langle \Phi \rangle = 0$. Above the critical temperature T_C, which characterizes the phase transition, the vacuum state lies at $\langle \Phi \rangle = 0$. For $T \simeq T_C$ a second minimum is developed at $\langle \Phi \rangle = \sigma$, which is lower lying for $T < T_C$ than that at $\langle \Phi \rangle = 0$. The minimum at $\langle \Phi \rangle = 0$ is called a *false vacuum*, since the vacuum is the state of lowest energy, and this now lies at $\langle \Phi \rangle = \sigma$ (see figure 3.11). In the original inflationary model of Guth [Gut81], as long as the field was situated in the false vacuum, it resulted in an exponential expansion. Locally, then, the transition to the right vacuum resulted either from quantum mechanical tunnelling or from thermal fluctuations. Thus bubbles developed in the right state, which merged into one another and stopped the inflation. Such a 'bubble-like' phase transition is called a phase transition of the first order. But in order to get a sufficiently large inflation to solve the problems we have discussed previously, the false vacuum has to remain relatively stable. This led to a very small amount of bubbles, and also the melting rate was ineffective. The problem of these models was therefore mainly that they led to either a runaway inflation or a very inhomogeneous universe. An improved step towards a solution of these problems was a temperature dependent Higgs potential, also called the Coleman–Weinberg potential $V_{CW}(\Phi, T)$ [Col73, Wei74] (see figure 3.12). For $T = 0$ this is given by

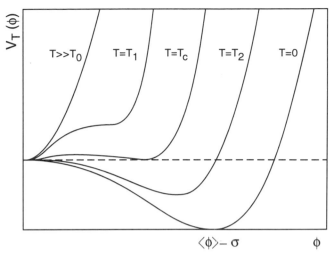

Figure 3.11. Change in the potential near a first order phase transition. From the parabolic shaped potential with vertex at zero a more complex potential develops, with a minimum at $\langle \Phi \rangle = \sigma$. This state is the new actual vacuum state, and the system tries to make a transition from the false vacuum (at $\langle \Phi \rangle = 0$) to the new vacuum (from [Kol90]).

$$V_{CW}(\Phi) = \frac{1}{2}B\sigma^4 + B\Phi^4 \left(\ln \frac{\Phi^2}{\sigma^2} - \frac{1}{2} \right). \tag{3.87}$$

B is related to the coupling constant of the particular GUT model (in this case SU(5)), and has a value of around 10^{-3}. For non-zero temperatures a barrier exists at $\Phi \approx T$ with a height of $\Phi \sim T^4$. The phase transition originates from the breaking of the GUT symmetry. This takes place at about 10^{15} GeV. Only at a temperature of about 10^9 GeV does the above barrier become unstable and the tunnelling probability becomes close to 100%. The subsequent rolling down on the flat part of the potential is what defines the time of the exponential expansion. If this time Δt is very much larger than the expansion scale H^{-1}, e.g.

$$\Delta t = 100 H^{-1} \tag{3.88}$$

an expansion by a factor $e^{100} = 3 \times 10^{43}$ follows! This rolling is described with the same equation of motion as the rolling of a ball on an inclined plane with friction, i.e.

$$\ddot{\Phi} + 3H\dot{\Phi} + V'(\Phi) = 0. \tag{3.89}$$

In the third step the field now carries out coherent oscillations around the new vacuum expectation value. These are damped by the expansion of the universe and the energy released is converted into the production of relativistic particles. The subsequent thermalization leads to an enormous heating up of the universe to temperatures which existed at the beginning of the inflation. At the same time an enormous entropy increase takes place (see e.g. [Kol90]). The whole process of inflation is limited to a time interval of about 10^{-35} s–10^{-30} s after the big bang. The great advance made by the Coleman–Weinberg potential was the decoupling of inflation from the barrier penetration. The duration of the inflation is now determined by the duration of the rolling on the flat part of the potential rather than by the duration of the stay in the false vacuum.

The original scenario of Coleman and Weinberg is in the meantime untenable, because it assumes a massless Higgs boson and requires a top quark lighter than 85 GeV, in contradiction to the experimental value, (see chapter 1). Also density perturbations (see chapter 6) contradict experiment in this scenario. But like the ruled out SU(5) model of grand unification theory (chapter 2), the Coleman–Weinberg potential is useful for studying the basic ideas.

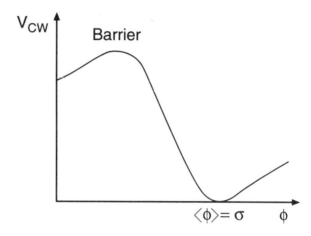

Figure 3.12. Behaviour of the Coleman–Weinberg potential, from which it is possible to produce an inflationary phase in the universe (from [Gro89, Gro90]).

The spontaneous symmetry breaking and the above example of the Coleman–Weinberg potential are not necessary conditions for inflation, but it is possible to produce such inflationary models using them. On the other hand for example, Linde's model of chaotic inflation [Lin82] assumes a potential of the form

$$V(\Phi) = \lambda \Phi^4 \tag{3.90}$$

where the initial values of the field Φ are distributed chaotically around space and the minimum of the potential lies at $\langle \Phi \rangle = 0$, which implies no spontaneous symmetry breaking. This potential has been introduced solely in order to lead to inflation. The decisive condition seems to be the existence of a flat part of the potential, on which the field can 'roll' sufficiently long and slow enough to result in an effective inflation.

We now discuss whether in fact such models can solve the problems we have encountered previously. Because of the exponential expansion some regions of space are enormously expanded. Our visible universe is now only a tiny part of a space domain. This gigantic expansion has the effect that any initial curvature is now so far flattened that the universe always seems approximately flat to us. From this follows the important statement that Ω_0 has to lie very close to 1. As we now see only a small part of the causally connected volume of space, no difficulties remain with respect to the isotropy of the background radiation. As only one monopole is expected per space domain, their density has been dramatically reduced and is no longer in contradiction with the experimental limits.

Support for inflationary models comes from the recently observed small anisotropies in the background radiation, as its spectrum can be well explained from the predictions of inflation (see chapter 7).

The extension of the big bang hypothesis through an inflationary phase 10^{-35} s after the big bang has proven to be very promising and successful. This is the reason why today the combination of the big bang model with inflation is often called the standard model of cosmology.

Having covered the basics of cosmology and particle physics in the first three chapters, we now turn to an examination of some of the points we have touched upon in more detail.

Chapter 4

Primordial nucleosynthesis

In this chapter we turn our attention to a very important foundation of the big bang model, namely the synthesis of the light elements in the early universe. They are, among others, H, D, ^{3}He, ^{4}He and ^{7}Li. Together with the synthesis of elements in stars and the production of heavy elements in supernova explosions (see chapter 14), this is the third important process in the formation of the elements. The fact that their relative abundances are predicted correctly over more than ten orders of magnitudes can be seen as one of the outstanding successes of the standard big bang model. Furthermore studying the abundance of ^{4}He allows statements to be made about the number of possible neutrino flavours. This prediction has lately been confirmed and measured more precisely at LEP. We turn firstly to the experimentally observable primordial relative abundances of the relevant isotopes. For a more detailed discussion we refer the reader to [Yan79, Yan84, Boe85, Boe88, Mal93, Wil94, Sar96].

4.1 Observed abundances of the elements

4.1.1 The ^{4}He abundance

The ^{4}He abundance is usually described by its relative mass fraction Y. As the primordial helium is enriched in the course of time via helium extracted in stars, the measured abundance is usually taken as an upper limit. In 1967 Faulkner could already show from examinations of sub-dwarfs that $Y > 0.20$, while others extracted values of about 0.30 from examinations of globular star clusters. These observations at least determined the order of magnitude. The next object to be investigated is our Sun. As the luminosity is, amongst other factors, dependent upon the average molecular weight μ ($L \sim \mu^{7.5}$), from the knowledge of the mass and radius, Y can be fixed to around 0.23. In 1978 the helium abundance of our nearest neighbour, α Centauri, was found to be different from that of the Sun by only about 0.01 . Other observations all point to a similar value. The problem is that the easily observable galactic H II-regions (these are hot areas of ionized hydrogen, which also contain He; the recombination lines, particularly the He^+ line, are used in order to detect the presence of helium) are contaminated with non-primordial helium, while the H II-regions in far-away metal depleted galaxies, which contain relatively little non-primordial helium, are difficult to observe. As only the recombination line is observable, models of the H II-regions have to be constructed in order to determine the amount of neutral helium. Observational techniques are discussed in e.g. [Dav89]. As helium is produced in stars, just like heavier elements, there is a correlation between Y and the mass fraction of metals, Z

$$Y = Y_{\text{prim}} + \frac{dY}{dZ} Z. \tag{4.1}$$

The primordial part then follows from an extrapolation to $Z = 0$. Since it is difficult to measure the total abundance of heavy elements (and thus the contribution of the stellar contamination), the fraction of the oxygen abundance to that of the hydrogen, O/H, is often used as an estimate. There is a correlation between Y and O/H of the form that in areas with little oxygen (O/H small), small helium abundances are also expected. Oxygen is mainly produced in the evolution of massive stars ($M > 8M_\odot$), while helium and e.g. nitrogen are more likely to be produced in intermediate mass stars ($M > 2M_\odot$). It is therefore preferable to utilize the information on the nitrogen abundance (see figure 4.1). Typical values for Y, including both correlations mentioned above, are [Wal91]:

$$Y = 0.226 \pm 0.005 \pm 160(\pm40)\text{O/H} \tag{4.2}$$

$$Y = 0.231 \pm 0.003 \pm 2800(\pm700)\text{N/H}. \tag{4.3}$$

Conservatively it can be said that the observations imply a value of

$$Y = 0.23 \pm 0.02. \tag{4.4}$$

For a detailed discussion of possible systematic errors in such measurements see [Sar96]. Latest observations with the Hopkins Ultraviolet telescope (HUT) during a Space Shuttle mission for the first time were able unambiguously to detect intergalactic helium [Sky95a]. The observations point to the fact that at least half of the helium produced in the big bang is contained in the intergalactic medium.

The situation is very different for other isotopes, which are very difficult to observe compared to helium.

4.1.2 Deuterium and ^{3}He

Let us first of all look at deuterium. Apart from primordial deuterium, it exists practically nowhere in the universe, except inside stars, where it is immediately converted into ^{3}He. For this reason its highest observed abundances have to be seen as a lower limit, as some deuterium has been lost, but no new deuterium

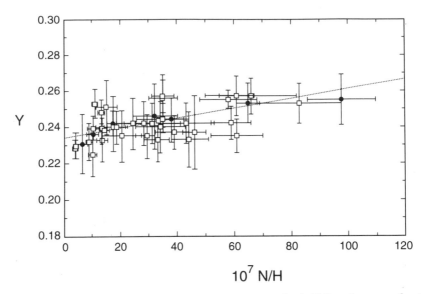

Figure 4.1. ^{4}He mass fraction Y, observed in 41 extra-galactic H II regions, as a function of the observed nitrogen abundance. Outliers in N/O versus O/H are shown as filled circles (from [Oli95a]).

enrichment of interstellar matter has taken place. At temperatures above 6×10^5 K it is easily destroyed via the reaction $p + D \to {}^3He + \gamma$, so that its observation is most suitable in the interstellar medium. Examinations of the DHO and CH$_3$D molecular lines in the large planets showed values for D/H of about $1 - 3 \times 10^{-5}$ [Kun82]. Possibilities for observation in the interstellar medium exist in the Lyman-series in the UV-region or through the hyperfine structure line at 92 cm wavelength, where the observation is made from the absorption lines in the spectra of hot stars, or even far away quasars [Rog73]. An observation of the interstellar matter in the direction of the star α Aurigae with the Hubble space telescope resulted for example in a value of [Lin92]

$$\frac{D}{H} = \left(1.65^{+0.07}_{-0.18}\right) \times 10^{-5}. \tag{4.5}$$

This is in good agreement with an analysis [McC92] of IUE and Copernicus satellite observations, which imply $D/H = (1.5 \pm 0.2) \times 10^{-5}$. A further possibility of obtaining such information will be via $Ly\,\alpha$ absorption systems (see section 6.5), [Son94a, Tyt94, Tyt96, Sch97a].

From all the observations we obtain [Kol90] limits for the abundance of

$$10^{-5} \le \frac{D}{H} \le 2 \times 10^{-4}. \tag{4.6}$$

Due to the great intensity differences of the D- and H-lines these are extremely difficult measurements.

^{3}He is best observed in H I-regions via the hyperfine structure line of the ^{3}He$^+$ at 3.46 cm. Due to its production in stars the lowest observed abundance is here to be interpreted as an upper limit. The limits are [Ban87]

$$1.2 \times 10^{-5} \le \frac{^3\text{He}}{\text{H}} \le 1.5 \times 10^{-4}. \tag{4.7}$$

A new analysis of observations performed on H II regions seems to yield ^{3}He/H$\approx (1-4) \times 10^{-5}$ [Bal94e].

Because of the further fast reaction from D to ^{3}He these two are often considered together (see e.g. [Den90]). In this context note the following: the observed D/H ratio only makes up a fraction f of the primordial value:

$$\frac{\text{D}}{\text{H}} = f \left(\frac{\text{D}}{\text{H}}\right)_{\text{prim}} \tag{4.8}$$

The destroyed D was, however, transformed into ^{3}He, so that we can say

$$\frac{^3\text{He}}{\text{H}} = (1-f)g \left(\frac{\text{D}}{\text{H}}\right)_{\text{prim}} + g \left(\frac{^3\text{He}}{\text{H}}\right)_{\text{prim}} + \left(\frac{^3\text{He}}{\text{H}}\right)_{\text{prod}} \tag{4.9}$$

where g is the fraction of the destroyed ^{3}He, and the last term takes account of the stellar production. This results in

$$g^{-1} \left(\frac{^3\text{He}}{\text{H}}\right) > (1-f) \left(\frac{\text{D}}{\text{H}}\right)_{\text{prim}} + \left(\frac{^3\text{He}}{\text{H}}\right)_{\text{prim}} \tag{4.10}$$

which therefore can be written as

$$\left(\frac{\text{D}+^3\text{He}}{\text{H}}\right)_{\text{prim}} < \frac{\text{D}}{\text{H}} + g^{-1}\frac{^3\text{He}}{\text{H}} \tag{4.11}$$

or

$$\left(\frac{\text{D}+^3\text{He}}{\text{H}}\right)_{\text{prim}} < \frac{\text{D}+^3\text{He}}{\text{H}} + (g^{-1}-1)\frac{^3\text{He}}{\text{H}}. \tag{4.12}$$

If we take the values, thought to be pre-solar, which are found in a special kind of meteorite, the carbonaceous chondrites, [Jef70]

$$\left(\frac{\text{D}+^3\text{He}}{\text{H}}\right)_{\text{prim}} < [4.3 + 1.9(g^{-1}-1)] \times 10^{-5} \tag{4.13}$$

and a value of $g > 0.25$ resulting from calculations of ^{3}He survival probabilities in stars [Dea86], abundances of

$$\left(\frac{\text{D}+^3\text{He}}{\text{H}}\right)_{\text{prim}} \le 10^{-4} \tag{4.14}$$

result. An outline of hydrogen observations in the interstellar medium can be found in [Boe85, Tur96a].

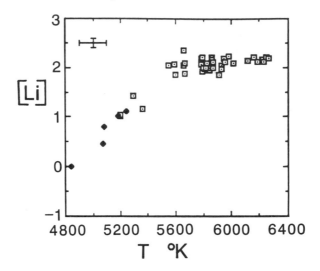

Figure 4.2. The ^{7}Li abundance for the most metal depleted population II stars as a function of temperature. The diamonds correspond to upper limits, the cross at the top left shows typical error bars. An approximately constant behaviour above about 5500 K can clearly be seen (from [Wal91]).

4.1.3 ^{7}Li, ^{9}Be, ^{11}B

^{7}Li is even more difficult to measure and really only stellar atmospheres, with all the resulting disadvantages, can be used. Several interesting observations do exist: if the ^{7}Li abundance is plotted against the effective temperature, a constant abundance is found above about 5500 K (see figure 4.2). Below this the abundance falls, which can be attributed to larger convection zones in the star. This leads to more ^{7}Li brought into the hotter inner regions and consequently being destroyed. For stars with an already low metal content the ^{7}Li abundance above 5500 K is independent of the former. One further finding is the fact that for three stars low in metal an upper limit for beryllium of ^{9}Be/H $< 2.5 \times 10^{-12}$ has been found. This excludes production of the observed ^{7}Li in this case, as otherwise significant ^{9}Be would have been produced, due to spallation processes in the cosmic radiation. For the observation very old, metal depleted, so-called population II halo stars are used, leading to abundances of order [Spi82, Mol95]

$$\frac{^7\text{Li}}{\text{H}} = (1.4 \pm 0.2) \times 10^{-10}. \qquad (4.15)$$

This is about a factor of ten less than the value of population I stars within the galactic disc. There are now controversial views as to which of the two abundances is actually the primordial one. For example, in younger stars there could have been material in the proto-star enriched with ^{7}Li from supernova explosions. On the other hand it could, of course, be possible that in the old population II stars a significant de-enrichment had occurred during their life, due to ^{7}Li$(p, \alpha)^4$He reactions. We assume the first possibility here, and a ^{7}Li/H-fraction of about 10^{-10}. This is supported by the latest discovery of ^{6}Li, which is even more susceptible to destruction than ^{7}Li, in an extremely metal depleted population II star [Smi93a].

The abundances of heavier elements such as ^{9}Be and ^{11}B are smaller by further orders of magnitude. There also exist observations of Be in population II stars [Rya90, Gil91]. Lately a group discovered a beryllium abundance corresponding to one thousand times the value expected from the standard model, and furthermore found lines of boron in the very old metal depleted star HD 140283 [Dun92]. The importance of this observation for nucleosynthesis is, however, disputed, as the abundances are correlated with metallicity, and therefore suggest a production via spallation instead of a primordial one. For a more detailed discussion of the experimentally determined abundances of light elements see [Wil94, Ree94, Oli95, Cop95, Sar96].

In summary, any model has to be able to explain a difference between the measured abundances of the elements of ten orders of magnitude.

4.2   The process of nucleosynthesis

The ability to predict the abundances of light elements over 10 orders of magnitudes is one of the big successes of the big bang model. As previously discussed in chapter 3, the synthesis of the light elements took place in the first three minutes after the big bang, which means at temperatures of about 0.1–10 MeV. Let us first of all look at the initial conditions ($T \gg 1$ MeV, $t \ll 1$ s). All protons and neutrons, together with any light nuclei, were in a state of thermal as well as chemical equilibrium. This means that the abundance of a nucleus $A(Z)$ is given by a Boltzmann distribution [Kol90]

$$ n_A = g_A A^{3/2} 2^{-A} \left(\frac{2\pi}{m_N T} \right)^{\frac{3(A-1)}{2}} n_p^Z n_N^{A-Z} \exp(B_A/T). \qquad (4.16) $$

The binding energies B_A and statistical factors g_A (see chapter 3) of some light nuclei are given in table 4.1. We consider the period of nucleosynthesis in three steps. The first begins at about 10 MeV, equivalent to $t = 10^{-2}$ s. Protons and neutrons are in thermal equilibrium through the weak interaction via the reactions

$$ p + e^- \longleftrightarrow n + \nu_e \qquad (4.17) $$

$$ p + \bar{\nu}_e \longleftrightarrow n + e^+ \qquad (4.18) $$

$$ n \longleftrightarrow p + e^- + \bar{\nu}_e \qquad (4.19) $$

Table 4.1. Binding energies and statistic factors of some light nuclei.

$^A Z$	B_A (MeV)	g_A
^{2}H	2.22	3
^{3}H	6.92	2
^{3}He	7.72	2
^{4}He	28.3	1
^{12}C	92.2	1

and the relative abundance is given in terms of their mass difference $\Delta m = m_n - m_p$ (neglecting chemical potentials) as:

$$\frac{n}{p} = \exp\left(-\frac{\Delta mc^2}{kT}\right). \tag{4.20}$$

The reaction rate $\Gamma(pe \to \nu n)$ is given by,

$$\Gamma(pe \to \nu n) \simeq G_F^2 T^5 \quad T \gg Q, m_e. \tag{4.21}$$

If this is compared to the expansion rate

$$\frac{\Gamma}{H} \approx \left(\frac{T}{0.8 \text{ MeV}}\right)^3 \tag{4.22}$$

we can see that from about 0.8 MeV the weak reaction rate becomes less than the expansion rate. The neutron–proton ratio begins to deviate from the equilibrium value. One would expect a significant production of light nuclei here, as the typical binding energies per nucleon lie in the region of 1–8 MeV. However, the large entropy, which manifests itself in the very small baryon–photon ratio η, prevents such production as far down as 0.1 MeV.

The second region begins at a temperature of about 1 MeV, or equivalently at 0.02 s. The neutrinos have just decoupled from matter, and at about 0.5 MeV the electrons and positrons annihilate. This is also the temperature region in which the above interaction rates become less than the expansion rate, which implies that the weak interaction freezes out, which leads to a ratio of

$$\frac{n}{p} = \exp\left(-\frac{\Delta mc^2}{kT_f}\right) \simeq \frac{1}{6}. \tag{4.23}$$

The third region begins at 0.3 to 0.1 MeV, corresponding to about 1 to 3 minutes after the big bang. Here practically all neutrons are converted into ^{4}He via the reactions

$$n + p \leftrightarrow D + \gamma \tag{4.24}$$

and e.g. (see figure 4.6)

$$D + D \leftrightarrow He + \gamma \tag{4.25}$$

$$D + p \leftrightarrow {}^{3}He + \gamma \tag{4.26}$$

$$D + n \leftrightarrow {}^{3}H + \gamma. \tag{4.27}$$

From this follows at once the amount of primordial helium

$$Y = \frac{2n_n}{n_n + n_p}. \tag{4.28}$$

Meanwhile the initial n/p-fraction has fallen to about $1/7$, due to the decay of the free neutrons. The equilibrium ratio, which follows from an evolution according to equation (4.20), would be $n/p = 1/74$ at 0.3 MeV. The behaviour of the neutron–proton ratio is shown in figure 4.3. The non-existence of stable nuclei of mass 5 and 8, as well as the now essential Coulomb barriers, very strongly inhibit the creation of ^{7}Li, and practically completely forbid that of even heavier isotopes (see figure 4.4). Because of the small nucleon density it is equally not possible to get over this bottleneck via 3α-reactions, as can be done in stars. This is the principal model of primordial element synthesis (see figure 4.5) [Yan84]. The underlying reaction chain (see figure 4.6) has hardly changed during the last 25 years (the current programmes are based mainly on the work of [Wag67, Kaw88b]); only the experimentally measurable parameters and, for example, reaction rates have been improved and have led to ever more precise predictions. For a detailed discussion of the results see [Kra90b, Wal91, Smi93b, Sar96].

4.2.1 Parameters controlling the ^{4}He abundance

The predicted abundances (especially ^{4}He) depend on a number of parameters. These are, in principle, three, namely the lifetime of the neutron τ_n, the fraction of baryons to photons $\eta = n_B/n_\gamma$ and the number of relativistic degrees of freedom g_{eff}. In the following these are investigated with particular reference to ^{4}He.

4.2.1.1 *The lifetime of the neutron τ_n*

The calculation of all weak reaction rates (equations (4.17)–(4.19)) depends on the nuclear matrix element for the β-decay of the neutron

$$|M|^2 \sim G_F^2(1 + 3g_A^2) \tag{4.29}$$

where g_A is the axial vector coupling constant of the nucleon. This can also be expressed in terms of the neutron lifetime

$$\tau_n^{-1} = \frac{G_F^2}{2\pi^3}(1 + 3g_A^2)m_e^5\epsilon \tag{4.30}$$

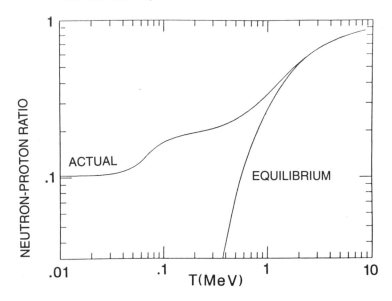

Figure 4.3. The real and equilibrium behaviour of the neutron/proton ratio over the relevant temperature region for primordial nucleosynthesis (from [Kol90]).

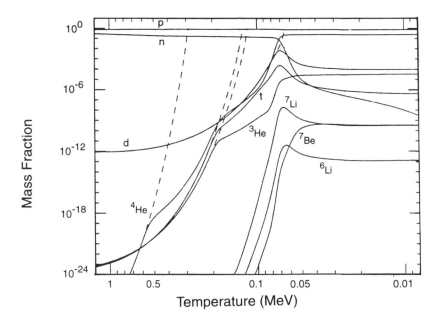

Figure 4.4. Development of the abundances of the light elements during primordial nucleosynthesis (from [Ree94]).

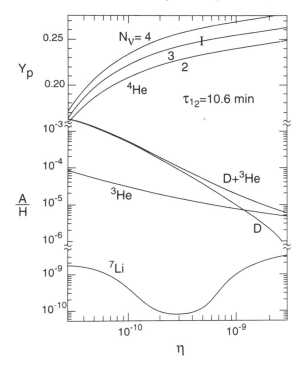

Figure 4.5. The primordial abundances of the light elements, as predicted by the standard model of cosmology, as a function of today's baryon density n_B or of $\eta = n_B/n_\gamma$. The ^{4}He fraction is shown under the assumptions $N_\nu = 2$, 3 and 4 where N_ν is the number of light neutrino flavours. A consistent prediction is possible over 10 orders of magnitude (from [Tur92a]).

where $\epsilon \simeq 1.6$ is a numerical factor. A longer lifetime thus provides a lower reaction rate Γ, and therefore a higher freeze-out temperature T_F, such that

$$\Gamma \sim \frac{T^5}{\tau_n} \quad \Rightarrow \quad T_F \sim \tau_n^{1/3}. \tag{4.31}$$

Consequently this leads to higher ^{4}He production. In addition, between the freezing-out and the helium synthesis fewer neutrons would decay, although this plays a rather subordinate role, while nevertheless also tending to increase the ^{4}He abundance. Figure 4.7 illustrates the influence of different lifetimes on the ^{4}He abundance. Table 4.2 gives a summary of measurements of the neutron lifetime. Laboratory experiments lead to [PDG94]

$$\tau_n = 887 \pm 2 \text{ s}. \tag{4.32}$$

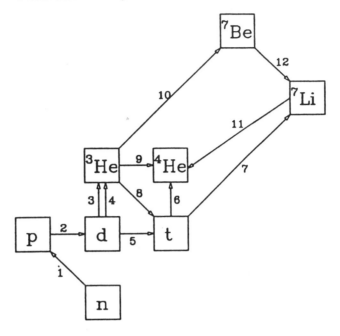

Figure 4.6. The 12 fundamental reactions in the chain of synthesis of the light elements, illustrating which elements can be built up in this way (from [Smi93b]). Labels indicate the following reactions: (1) $n \leftrightarrow p$, (2) $p(n, \gamma)d$, (3) $d(p, \gamma)^3\mathrm{He}$, (4) $d(d, n)^3\mathrm{He}$, (5) $d(d, p)t$, (6) $t(d, n)^4\mathrm{He}$, (7) $t(\alpha, \gamma)^7\mathrm{Li}$, (8) $^3\mathrm{He}(n, p)t$, (9) $^3\mathrm{He}(d, p)^4\mathrm{He}$, (10) $^3\mathrm{He}(\alpha, \gamma)^7\mathrm{Be}$, (11) $^7\mathrm{Li}(p, \alpha)^4\mathrm{He}$, (12) $^7\mathrm{Be}(n, p)^7\mathrm{Li}$.

Table 4.2. Measurements of the neutron lifetime.

Lifetime (s)	Reference
918±14	[Chr72]
903±13	[Kos86]
891±9	[Spi88]
876±21	[Las88]
877±10	[Pau89]
888±3	[Mam93]
878±30	[Kos89]
894±5	[Byr90]
888.4±4.2	[Nes92]
882.6±2.7	[Mam89]
887.0±2.0	[PDG94]

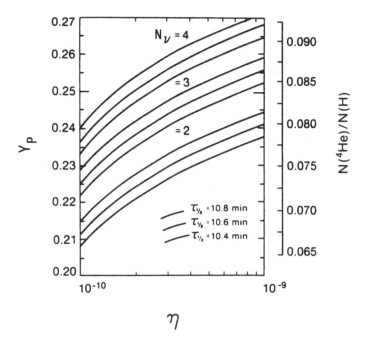

Figure 4.7. A more detailed illustration of the ^{4}He abundance as a function of the baryon/photon ratio $\eta = n_B/n_\gamma$. The influence of the number of neutrino flavours and the neutron lifetime on the predicted ^{4}He abundance can clearly be seen (from [Yan84]).

4.2.1.2 The baryon–photon ratio η

The abundance X_A of a nucleus $A(Z)$ in equilibrium depends on the baryon–photon ratio $\eta = n_B/n_\gamma$ according to $X_A \sim \eta^{A-1}$ [Kol90]. This means that higher values of η imply earlier production of D, ^{3}H, ^{3}He and therefore, due to the greater n/p-ratio, also more ^{4}He. A much more drastic effect is produced by η for ^{7}Li. ^{7}Li is produced in two different processes. For $\eta \leq 3 \times 10^{-10}$ the reaction ^{4}He(^{3}H, γ)^{7}Li dominates, while for higher values the reaction ^{4}He(^{3}He, γ)^{7}Be with subsequent electron-capture producing ^{7}Li is dominant. This creates the 'hole' at about 3×10^{-10} (see figure 4.5) and makes ^{7}Li an especially sensitive test for η and therefore for primordial nucleosynthesis. As the reaction rates for D, ^{3}H to ^{4}He are also dependent on η, their abundance for small η increases, as ever more material remains unburned. A conservative analysis of the observations yields [Sar96] $4.1 \times 10^{-11} < \eta < 9.2 \times 10^{-10}$ which can be tightened by some assumptions to $2.6 \times 10^{-10} < \eta < 3.3 \times 10^{-10}$. With knowledge of the photon density from the cosmic background radiation (see chapter 7) and η it follows [Kol90]

$$n_B = 1.13 \times 10^{-5}\Omega_B h^2 \text{ cm}^{-3} \tag{4.33}$$

which can be converted into a fraction of baryonic matter in the universe of

$$0.011 \leq \Omega_B h^2 \leq 0.037. \tag{4.34}$$

4.2.1.3 *The relativistic degrees of freedom g_{eff}, and the number of neutrino flavours.*

The expansion rate H is proportional to the number of relativistic degrees of freedom of the available particles (see equation (3.65)). According to the standard model, at about 1 MeV these are photons, electrons and three neutrino flavours. The dependence of the freeze-out temperature on the number of degrees of freedom then results from equations (3.65) and (4.22):

$$H \sim g_{\text{eff}}^{1/2}T^2 \quad \Rightarrow \quad T_F \sim g_{\text{eff}}^{1/6}. \tag{4.35}$$

Each additional relativistic degree of freedom (further neutrino flavours, axions, Majorons, right-handed neutrinos, etc) therefore means an increase of the expansion rate and therefore a freezing-out of the above reactions at higher temperatures. This again is reflected in a higher ^{4}He abundance. This dependence meant that it was possible to restrict the number of neutrino flavours N_ν already long before the accelerator experiments [Yan79, Yan84]. The maximum number of neutrino flavours could be determined to be at most four (see figure 4.7) [Oli90b]. The dependence of the primordial helium on the parameters discussed here can be parametrized as follows [Ber89b, Kra95]:

$$Y = 0.230 + 0.013(N_\nu - 3) + 0.014(\tau_n - 922) + 0.011\ln(\eta \times 10^{10}). \tag{4.36}$$

Using the latest results from LEP that only three light neutrino flavours are allowed, and the measured neutron lifetime, the following rather restrictive limits

$$0.236 \leq Y \leq 0.243. \tag{4.37}$$

can be derived. If the photon–baryon ratio permitted by all the observations (see figure 4.5), is converted into a baryonic density, then [Wal91]

$$0.02 < \Omega_B < 0.11 \quad \text{for} \quad 0.4 < h_0 < 0.7. \tag{4.38}$$

Therefore, according to primordial nucleosynthesis it is not possible to produce a closed universe from baryons alone! However, if Ω_B has a value close to the upper limit, a significant fraction can be present in dark form, as the luminous part is significantly less ($\Omega_B^L < 0.02$) than given by equation (4.38). It is therefore at least possible to use baryonic matter to explain the rotational curves of galaxies (see chapter 9).

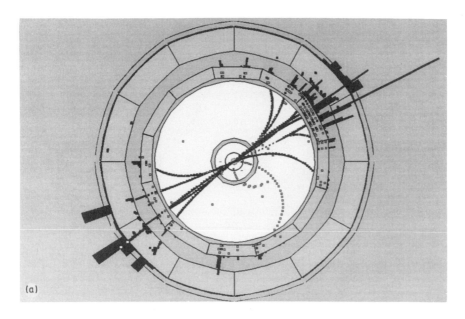

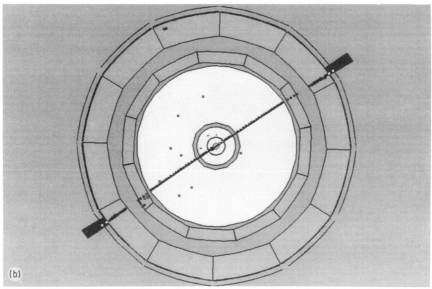

Figure 4.8. Typical decays of the Z^0 boson, as recorded by the ALEPH detector at LEP at CERN. (*a*) The decay $Z^0 \rightarrow q\bar{q}$ in the form of two jets. (*b*) The decay $Z^0 \rightarrow \mu^+\mu^-$ (with kind permission of the ALEPH Collaboration, CERN).

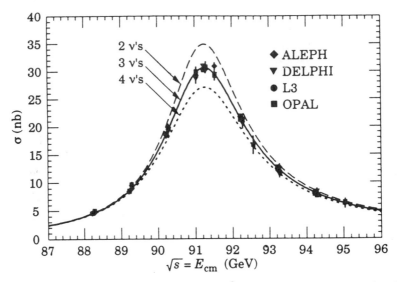

Figure 4.9. LEP results for the width of the Z^0 resonance. As every neutrino family contributes to the width of the resonance, the number can in this way be established very precisely as three. The curves correspond to 2, 3 and 4 neutrino families, the points to the combined measured values from the four LEP experiments (from [PDG94]).

4.3 Accelerators and the number of neutrino flavours

The results of the LEP storage ring in Geneva provide the most stringent limits from accelerator experiments on the number of families. Here the separate decay channels of the Z^0 can be closely examined (see figure 4.8). Table 4.3 and figure 4.9 show the measured decay widths of the Z^0 and the expectations from the standard model for various channels. The decay width into neutrinos ignoring radiative corrections is [Nac86]:

$$\Gamma(Z^0 \to \nu\bar{\nu}) = \frac{Gm_Z^3}{12\sqrt{2}\pi} = N_\nu(174 \pm 11) \text{ MeV}. \tag{4.39}$$

From the measured total width Γ_Z, the mass m_Z and the observed partial decay widths $\Gamma_{l^+l^-}$ for the decay of the Z^0 into lepton pairs $Z^0 \to l^+l^-$ and into quark–anti-quark pairs $Z^0 \to q\bar{q}$ (henceforth called the hadronic decay width Γ_{had}), the invisible width Γ_{inv} can be derived by subtracting these individual partial widths from the total width:

$$\Gamma_{inv} = \Gamma_Z - \Gamma_{had} - 3\Gamma_{l^+l^-}. \tag{4.40}$$

Table 4.3. Data on the Z^0 boson, from the four LEP- experiments; M_Z in GeV/c^2, decay width in MeV/c^2 (from [PDG96]).

	ALEPH	DELPHI	L3	OPAL	Average
M_Z	91.187	91.187	91.195	91.182	91.187
	±0.013	±0.013	±0.013	±0.013	±0.007 (LEP)
Γ	2501	2483	2494	2483	2490
	±56	±56	±56	±54	±52
Γ_e	84.61	83.31	83.43	83.63	83.83
	±0.49	±0.54	±0.52	±0.53	±0.3
Γ_μ	83.62	84.15	83.72	83.83	83.84
	±0.75	±0.77	±0.79	±0.65	±0.39
Γ_τ	84.18	83.55	84.04	82.90	83.68
	±0.79	±0.91	±0.94	±0.77	±0.44
Γ_{lepton}	84.40	83.56	83.49	83.55	83.84
	±0.43	±0.45	±0.46	±0.44	±0.27
Γ_{hadron}	1746	1723	1748	1741	1740.7
	±10	±10	±10	±10	±5.9
Γ_{invis}	450	509.4	540	539	517
	±68	±7	±120	±43	±22
N_ν	2.983	3.057	2.981	2.946	2.991
	±0.034	±0.040	±0.050	±0.045	±0.016

The factor 3 corresponds to the three existing lepton families. From this the number of neutrino flavours is given by [Ste91]

$$N_\nu = \frac{\Gamma_{\text{inv}}}{\Gamma_{\nu\bar\nu}} = \frac{\Gamma_{e^+e^-}}{\Gamma_{\nu\bar\nu}} \left[\sqrt{\frac{12\pi\Gamma_{\text{had}}}{m_Z^2 \sigma_{\text{had}} \Gamma_{l^+l^-}}} - \frac{\Gamma_{\text{had}}}{\Gamma_{l^+l^-}} - 3 \right] \tag{4.41}$$

where σ_{had} represents the peak of the hadronic cross section, given by

$$\sigma_{\text{had}} = \frac{12\pi \Gamma_{l^+l^-} \Gamma_{\text{had}}}{m_Z^2 \Gamma_Z^2}. \tag{4.42}$$

The measured shape of the Z^0 interpreted in this way results in a value from all four LEP experiments for the number of neutrino species of [PDG96]

$$N_\nu = 2.991 \pm 0.016. \tag{4.43}$$

This means that there are only three light neutrinos with masses smaller than 45 GeV, and hence only three lepton and quark families.

To be more precise, primordial nucleosynthesis and LEP actually measure somewhat different things. While from the LEP results the number of particles coupling to the Z^0 can be determined, the energy density during nucleosynthesis depends on *all* relativistic degrees of freedom around 1 MeV. These two quantities are not necessarily identical. For example, the singlet Majoron (see chapter 2) has a vanishing coupling to the Z^0 and would not contribute to its decay width. However, since it is very light, it would contribute to the relativistic degrees of freedom g_{eff}. Its contribution corresponds to 4/7 of the contribution of a neutrino flavour. From the good compatibility of the LEP data with three neutrinos and the bounds from primordial nucleosynthesis, limits on further light particles can therefore be deduced [Ste92, Sar96].

4.4   Inhomogeneous nucleosynthesis

The observation of an unexpectedly high ^{9}Be abundance in very old, metal depleted stars [Dun92], leads to the question of whether and how the production of such isotopes (^{9}Be and ^{11}B) can be increased in the framework of primordial nucleosynthesis. One possibility might be an inhomogeneous scenario. Inhomogeneous nucleosynthesis could take place during the quark–hadron phase transition [Wit84]. Here the free quarks of the quark–gluon plasma (see chapter 1) transform into hadrons. If this phase transition is a transition of the first order, i.e. both phases exist side by side, it can result in significant inhomogeneities in the baryon density [Alc87, Sch87a]. It is, however, not at all theoretically proven that such a first order phase transition actually exists. Hadronization works by freeing latent heat, which heats up the universe. This results in the growth of the hadronized areas at the expense of the quark–gluon plasma. However, at some point the heat becomes insufficient to sustain the growth, and the hadronized areas decouple from equilibrium. On further cooling the rest of the quark–gluon plasma hadronizes, but with a greater baryon density. The consequence of these inhomogeneities is that there now exist areas in the universe with high densities of protons, neutrons and electrons, and others with low densities. However, in the course of time the neutrons diffuse out of the areas of high density, which is not possible for the protons and electrons due to their electromagnetic attraction. When nucleosynthesis takes place, a relatively high concentration of neutrons will exist in areas of low density, which allows alternative formation processes. Instead of the reaction ^{7}Li$(p, \alpha)^4$He there now exists the possibility of

$$^7\text{Li} + n \rightarrow {^8\text{Li}}. \qquad (4.44)$$

^{8}Li is radioactive with a half-life of 0.8 s. However, if it were to encounter a He nucleus before decaying, the following reaction is possible:

$$^8\text{Li} + {^4\text{He}} \rightarrow {^{11}\text{B}} + n. \qquad (4.45)$$

^{11}B can, however, also be produced via ^{7}Li$(^3$H$, n)^9$Be$(^3$H$, n)^{11}$B. It can be seen that in this way significant amounts of beryllium and boron can be built up. From these carbon and nitrogen can also be produced. It seems even possible to synthesize elements up to ^{22}Ne in this way [App88]. For some time even a primordial r-process (see chapter 14) with the production of very heavy isotopes ($A > 60$) seemed not to be impossible to some authors [Cow91, Mal93].

The main advantage of inhomogeneous nucleosynthesis is that possibly the non-vanishing abundance of beryllium and boron, even in the oldest stars, could be explained in this way, assuming spallation plays no role. However, such theories received a setback when it was realized that the back-diffusion of the neutrons also had to be taken into account, and that this could easily lead to an over-production of ^{7}Li and ^{4}He [Kur91].

Chapter 5

The cosmological constant

Einstein's field equations, which have already been discussed in chapter 3, are unable to describe a *static* universe without the Λ term. Since in those days there was a strong belief in a static universe, Einstein introduced a free parameter, Λ, into his equations in 1917 [Ein17]. The meaning of Λ can be seen immediately if the field equations (equation (3.6)) are written in the form [Zel67]:

$$R_{\mu\nu} - \frac{1}{2} R g_{\mu\nu} = \frac{8\pi G}{c^4} T_{\mu\nu} + \Lambda g_{\mu\nu}. \tag{5.1}$$

Obviously $\Lambda c^4 / 8\pi G$ has the same dimension as the energy–momentum tensor. When evidence for an expanding universe began to accumulate, the cosmological constant lost some importance. A $\Lambda \neq 0$ would however also be necessary if the Hubble time H^{-1} (for $\Lambda = 0$) and astrophysically determined data led to different ages for the universe. Λ has experienced a revival through modern quantum field theories. In these the vacuum is not necessarily a state of zero energy, but the latter can have a finite expectation value. The vacuum is only defined as the state of lowest energy. Due to the Lorentz invariance of the ground state it follows that the energy–momentum tensor in every local inertial system has to be proportional to the Minkowski metric $g_{\mu\nu}$. This is the only 4×4 matrix which in special relativity theory is invariant under Lorentz 'boosts' (transformations along a spatial direction). According to the above the cosmological constant can be associated with the energy density ϵ_V of the vacuum to give:

$$\epsilon_V = \frac{c^4}{8\pi G} \Lambda = \rho_V c^2. \tag{5.2}$$

The equation of state for the vacuum follows at once from the analogy between the diagonal energy–momentum tensor and that of a perfect fluid

$$p_V = -\rho_V. \tag{5.3}$$

All terms contributing in some form to the vacuum energy density also provide a contribution to the cosmological constant. There exist, in principle, three different contributions:

- The static cosmological constant Λ_{geo}. It is identical to the free parameter introduced by Einstein.
- Quantum fluctuations Λ_{fluc}. According to Heisenberg's uncertainty principle, virtual particle–anti-particle pairs can be produced at any time even in a vacuum.
- Additional contributions of the same type as (ii) due to possible, currently unknown, particles and interactions Λ_{inv}.

The sum of all these terms is what can be experimentally observed

$$\Lambda_{tot} = \Lambda_{geo} + \Lambda_{fluc} + \Lambda_{inv}. \tag{5.4}$$

Neglecting the third type, it would *in principle* be possible to fix the free parameter Λ_{geo} from the observation of Λ_{tot} and the predictions for Λ_{fluc} from quantum field theories. That these quantum fluctuations really exist was proved clearly via the Casimir effect [Cas48, Lam97]. Two parallel metal plates were placed close to each other in a vacuum. Because of the small separation of the plates only a limited number of wavelengths can develop in the vacuum between the plates, in contrast to the situation in the surrounding vacuum. This difference, i.e. the fact that not just any quantum fluctuation can build up between the plates, contrary to the surrounding volume, manifests itself as a small attractive force between the plates (see also [Plu86]).

Before we discuss the contributions of quantum fluctuations in more detail, we first discuss the effects of Λ .

5.1   Cosmological models with $\Lambda \neq 0$

Under the assumption of an homogeneous isotropic universe, Einstein's field equations—as mentioned already in section 3.1—reduce themselves to the Einstein–Friedmann–Lemaitre equations, this time with an additional contribution due to the vacuum energy (see section 3.4):

$$\left(\frac{\dot{R}}{R}\right)^2 = \frac{8\pi G}{3}(\rho + \rho_V) - \frac{k}{R^2} \tag{5.5}$$

$$\frac{\ddot{R}}{R} = -\frac{4\pi G}{3}(\rho - 2\rho_V + 3p). \tag{5.6}$$

We consider first the *static solutions* ($\dot{R} = \ddot{R} = 0$). The equations are then written (for $p = 0$) as:

$$8\pi G\rho = 2\rho_V. \tag{5.7}$$

$$\frac{8\pi G\rho}{3R} + \frac{\rho_V R^2}{3} = k \tag{5.8}$$

From the first equation it follows that $\rho_V > 0$, and therefore the second equation has a solution only for $k = 1$:

$$R^2 = \frac{2}{8\pi G \rho}. \tag{5.9}$$

Equation (5.8) represents the equilibrium condition for the universe. The attractive force due to ρ has to exactly compensate the repulsive effect of a positive cosmological constant in order to produce a static universe. This closed static universe is, however, unstable, since if we increase R by a small amount, ρ decreases, while Λ remains constant. The repulsion then dominates and leads to a further increase of R, so that the solution moves away from the static case.

We now consider *non-static* solutions. As can easily be seen, a positive Λ always leads to an acceleration of the expansion, while a negative Λ acts as a brake. As can be seen in equation (5.5), Λ always dominates for large R, since ρ_V is constant. A negative Λ therefore always implies a contracting universe and the curvature parameter k does not play an important role. For positive Λ and $k = -1, 0$ the solutions are always positive, which therefore results in a continually expanding universe. For $k = 1$, there exists a critical value

$$\Lambda_c = 4 \left(\frac{8\pi G}{c^2} M \right)^{-2} \tag{5.10}$$

exactly the value of Einstein's static universe, dividing two regimes. For $\Lambda > \Lambda_c$ static, expanding and contracting solutions all exist. Solutions without an initial singularity also exist, if $0 < \Lambda < \Lambda_c$, as illustrated in figure 5.1. A very interesting case is that with $\Lambda = \Lambda_c(1 + \epsilon)$ with $\epsilon \ll 1$ (Lemaitre universe). Here there is a phase in which the universe is almost stationary, before continuing to expand again. In this case it is possible to derive a much greater age of the universe [Sex87].

As we live today in a matter dominated universe we can neglect the pressure term and solve the equations. However, the equation system is certainly over determined. The solutions can be expressed in terms of the following quantities:

- The present Hubble constant H_0
- The age of the universe t_0
- The present matter density ρ_0
- Λ or the vacuum energy density ρ_V
- The metric k

If three of these quantities are known, the two remaining ones can be calculated from them. Definite experimental values do exist for H_0 and t_0, and an estimate for ρ_0 can be obtained from primordial nucleosynthesis (see chapter 4), from which the baryonic density ρ_B in the universe can be determined. It then follows that $\rho_0 > \rho_B$. In a numerical analysis of the equations a positive Λ is found for nearly all values of the parameters [Kla86a]. A quick estimate for the value of

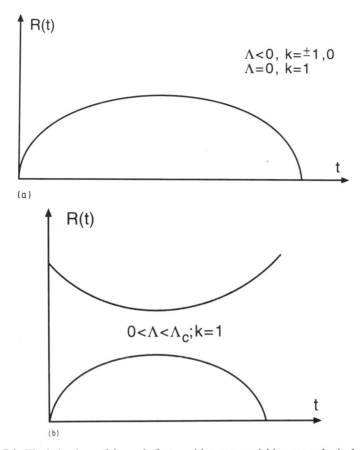

Figure 5.1. The behaviour of the scale factor with a non-vanishing cosmological constant: (*a*) Solutions of the Friedmann equation for $\Lambda < 0$ and $\Lambda = 0$, $k = 1$. (*b*) Solutions for $k = 1, 0 < \Lambda < \Lambda_c$, where Λ_c is the value for a static universe. There are now also models without an initial singularity. (*c*) Some further solutions of the Friedmann equation. (*d*) Lemaitre universe: a value of Λ only a little larger than Λ_c produces a phase in which the expansion of the universe is almost at a standstill, before a further expansion sets in (from [Sex87]).

Λ follows from the fact that the observed density ρ_0 is not very different from the critical value. This means

$$|\rho - \rho_c| \leq \frac{3H_0^2}{8\pi G} \Rightarrow |\Lambda| \leq H_0^2 \tag{5.11}$$

from which follows

$$|\rho_V| \leq 10^{-29} \text{ g cm}^{-3} \simeq 10^{-47}\hbar^{-3} \text{ GeV}^4. \tag{5.12}$$

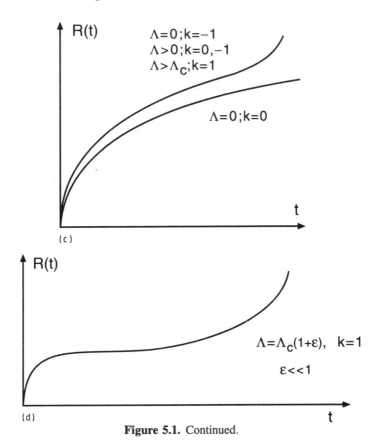

Figure 5.1. Continued.

From the assumption of an inflationary phase (i.e. also $k = 0$) and the experimental bounds for H_0, t_0 and ρ_0, a more accurate estimate for Λ can be obtained [Kla86a, Gro89, Gro90]:

$$3 \times 10^{-57} \text{ cm}^{-2} \leq \Lambda \leq 34 \times 10^{-57} \text{ cm}^{-2} \qquad (5.13)$$

and

$$1.6 \times 10^{-30} \text{g cm}^{-3} < \rho_V < 18 \times 10^{-30} \text{ g cm}^{-3}. \qquad (5.14)$$

These relationships are illustrated in figure 5.2. It can, for example, be seen that for $k = 0$, $\Lambda = 0$ and $H_0 = 75$ km s^{-1} Mpc^{-1}, the resultant age of the universe is only 8.7×10^9 years, in contradiction to observations. For $k = 0$, ρ_0 would have to take the value $\rho_0 = \rho_c \simeq 20\rho_B$. This can only be possible after the introduction of a considerable density of non-baryonic dark matter (see chapter 9). A model with $\Lambda > 0$, on the other hand, could provide solutions with $k = 0$ that lead to an age of the universe of $(15–20) \times 10^9$ years and only require a small density of dark matter [Gro89, Gro90].

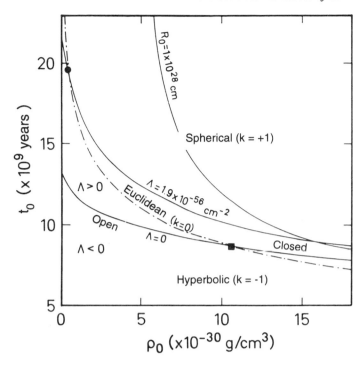

Figure 5.2. The present age t_0 of the universe, as a function of today's mass density ρ_0 for different cosmological models ($H_0 = 75$ km s^{-1} Mpc^{-1}). In the standard models with $\Lambda = 0$ only a relatively small t_0 is reached even for $\rho_0 = 0$. Greater values can be reached with a positive Λ. Also illustrated is the curve on which all Euclidean world models lie, and an example of a curve for models with fixed spatial curvature R_0. The value 10^{28} cm for the assumed R_0 corresponds roughly to the diameter of the presently observable universe (from [Gro89, Gro90]).

5.2   Direct determination of Λ

As already seen, Λ has the dimension [length^{-2}]. The quantity $1/\sqrt{\Lambda}$ is therefore a characteristic scale on which effects of the cosmological constant should be expected. The experimental determination of Λ is performed indirectly through the determination of the deceleration parameter q_0 (see equation (3.19)), which in the case of a non-zero vacuum energy is given by

$$q_0 = \frac{4\pi G}{3H_0^2}(\rho_0 - 2\rho_V) = \frac{1}{2}\Omega_0 - \Omega_V \tag{5.15}$$

where $\Omega_V = (\rho_V/\rho_c)$. It should therefore be possible, from an independent measurement of q_0 and Ω_0, to obtain information on Λ. Therefore we now discuss some methods for the determination of q_0; the various determinations of Ω_0 are dealt with in other chapters (see chapters 3, 4 and 9).

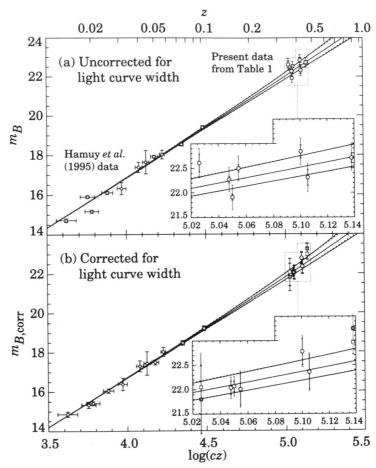

Figure 5.3. Hubble diagrams for the first seven high red-shift supernovae; (*a*) uncorrected m_B, with low red-shift supernovae of Hamuy *et al* (1995) for visual comparison; (*b*) $m_{B,\text{corr}}$ after 'correction' for the width–luminosity relation. Insets show the high red-shift supernovae on magnified scale. The solid curves in (*a*) and (*b*) are theoretical m_B for $(\Omega_M, \Omega_V) = (0, 0)$ on the top, (1, 0) in the middle, and (2, 0) on the bottom. The dotted curves in (*a*) and (*b*), which are practically indistinguishable from the solid curves, represent the flat universe case, with $(\Omega_M, \Omega_V) = (0.5, 0.5)$ on the top, (1, 0) in the middle and (1.5, −0.5) on the bottom (from [Per96a]).

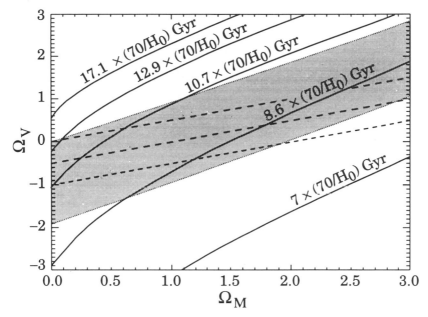

Figure 5.4. Contour plot of the 1σ (68%) confidence region in the Ω_V versus Ω_M plane, for the first seven high red-shift supernovae (shaded region). The solid lines show contours for the age of the universe (in Gyr), normalized to $H_0 = 70$ in units of km s^{-1} Mpc^{-1}. The dashed lines are the contours of the constant deceleration parameter $q_0 = 0.0$ (top), 0.5 (middle) and 1.0 (bottom) (from [Per96a]).

5.2.1 The determination of q_0

Three procedures to determine q_0 are discussed. All of these methods, however, have one uncertainty in common. Due to the large red-shifts, the time evolution of the objects used as 'standard candles' plays an important role. This is, however, not very well known and therefore limits the predictive power of the methods.

5.2.1.1 Luminosity distance–red-shift relation

Assuming that a galaxy has luminosity L (given as energy per unit time), one distance definition rests solely on the r^{-2}-dependence of the flux Φ (given as energy per unit time and area F). The distance d_l, defined by:

$$d_l^2 = \frac{L}{4\pi F} \tag{5.16}$$

is called *luminosity distance*. In a static universe it corresponds exactly to the physical distance. However, due to the expansion it is modified, as the observer subtends a different spatial angle at the time of emission than at the time of observation. d_l in this case is given by

$$d_l^2 = R^2(t_0)r^2(1+z)^2 \tag{5.17}$$

where r is the co-moving space coordinate and z the red shift. The factor $(1+z)^2$ corresponds to the energy decrease of single photons, which is connected to the red-shift, as well as to the time dilation for the observer in comparison to the emission (equation (3.14)). A Taylor expansion of $R(t)$ around R_0 (equation (3.17)) for small $H_0(t-t_0)$, i.e. not too long ago in the past, results in:

$$(t_0 - t) = H_0^{-1}\left[z - \left(1 + \frac{q_0}{2}\right)z^2 + \ldots\right]. \tag{5.18}$$

If, on the other hand, (see e.g. equation (3.12))

$$\int_{t_1}^{t_0} \frac{dt}{R(t)} = \int_0^{r_1} \frac{dr}{(1 - kr^2)^{\frac{1}{2}}} \tag{5.19}$$

is used in the Taylor expansion, then

$$r_1 = R(t_0)^{-1}[(t_0 - t_1) + \tfrac{1}{2}H_0(t_0 - t_1)^2 + \ldots]. \tag{5.20}$$

Inserting equation (5.20) in equation (5.18) and using (5.17) results in

$$H_0 d_l = z + \tfrac{1}{2}(1 - q_0)z^2 + \ldots. \tag{5.21}$$

It can be seen that the first correction term for the Hubble relation allows a determination of q_0. Such a Hubble diagram is shown in figure 3.3 for galaxies and quasars and in figure 5.3 for supernovae. The measurements are currently unable to distinguish whether q_0 is greater than, equal to or smaller than 0.5.

It is possible to use the 'standard candle' properties of type Ia supernovae rather than galaxies [Goo95a]. The equal absolute brightness of the maxima of such supernovae (see equation (3.26)) in galaxies with high red-shifts (typically $0.3 < z < 0.6$) has been used. A preliminary analysis of 7 such supernovae gives a value of $q_0 = 0.8 \pm 0.35 \pm 0.3$ [Per96] (figures 5.3 and 5.4).

5.2.1.2 *Angular diameter–red-shift relation*

Consider an object with diameter D emitting light at time $t = t_1$ which is observed by an observer at $r = 0$ at time $t = t_0$. The observer sees an object with angular diameter Θ [Kol90], where:

$$\Theta = \frac{D}{R(t_1)r_1}. \tag{5.22}$$

If the angular distance d_A is defined as

$$d_A = \frac{D}{\Theta} = R(t_1)r_1 \tag{5.23}$$

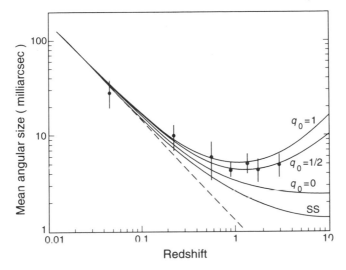

Figure 5.5. The average angular diameter Θ as a function of the red-shift z for 82 compact radio sources. Despite the large errors a value for the deceleration parameter of around $q_0 = 0.5$ may be deduced, equivalent to an Euclidean universe (from [Kel93]).

then using $d_A = d_l(1 + z)^{-2}$ (from equations (3.16) and (5.17)) with the approximation from equation (5.21) gives

$$H_0 d_A = z - \tfrac{1}{2}(3 + q_0)z^2 + \dots . \tag{5.24}$$

Data from compact radio sources resulted in evidence for q_0 of approximately 0.5, corresponding to $\Omega_0 = 1$ [Kel93]. Once again, the time evolution of the source is the decisive factor in the uncertainty (figure 5.5).

5.2.1.3 Galaxy number-count–red-shift relation

In this method a galaxy count is carried out as a function of the red-shift. Considering a co-moving volume element dV with dN galaxies, their number density $n(t)$ is given by (see for example [Kol90]):

$$dN = n(t)dV = n(t)\frac{r^2}{(1 - kr^2)^{\frac{1}{2}}}dr d\Omega. \tag{5.25}$$

Since r is not directly measurable, equation (5.25) can be rewritten using equations (5.18) and (5.20) to give

$$\frac{r^2}{(1 - kr^2)^{\frac{1}{2}}}dr = (H_0 R_0)^{-3}z^2 dz (1 - 2(q_0 + 1)z + \dots) \tag{5.26}$$

and inserting equation (5.26) into (5.25) gives:

$$\frac{1}{z^2}\frac{dN}{dz d\Omega} = (H_0 R_0)^{-3}n(z)(1 - 2(q_0 + 1)z + \dots). \tag{5.27}$$

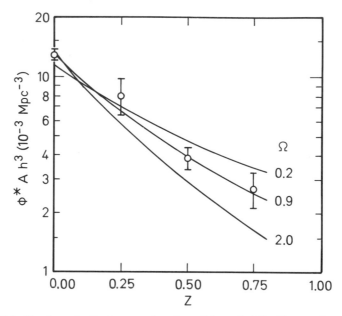

Figure 5.6. Number of galaxies as a function of the red-shift. The number of galaxies in a volume $dz d\Omega$ is proportional to $z^2 dz d\Omega h^3 A(z) \Phi^*$, where Φ^* is proportional to the galaxy density in a co-moving volume element, and $A(z)$ depends on the cosmological model. The observation is relatively well described by an Ω of 0.9, however, the statistics is limited and the observational data allow substantial freedom (from [Loh86]).

Under the assumption of a constant $n(z)$, i.e. there is no creation or destruction of galaxies (which is justified for small z), q_0 can be determined from this equation. However, two difficulties have to be overcome. Firstly, the evolution of galaxies with red-shift is still a subject of current research, but exactly such evolutionary effects have to be understood in order to apply this method sensibly. Secondly, a really complete set of very faint galaxies is also needed. Here errors in observation can lead to very large uncertainties. Loh and Spillar have carried out an examination with nearly 1000 infra-red galaxies, and obtained the result (see figure 5.6) [Loh86]:

$$\Omega_0 = 0.9^{+0.7}_{-0.5} \quad \rightarrow q_0 = 0.45^{+0.35}_{-0.25}. \tag{5.28}$$

This result is however not undisputed. The problem of the data analysis is such that the authors proceed from a luminosity distribution which in the past differed from today's only by a constant. Due to the great range of covered time periods, the evolutionary effects of galaxy development also have to be taken into account, which does not necessarily lead to a constant distribution of luminosity. An analysis taking into account the effects of evolution increases the error bars by a large amount [Bah88b].

5.2.2   Future alternatives to determine Λ

As well as these dynamic methods of determining q_0, new ideas on the observation of the effects of a cosmological constant are being developed, due to constantly improving observational data. One of the possibilities for this is the observation of quasar absorption lines (see also chapter 6). These are caused by hydrogen clouds along the line of sight. Theoretical considerations show that the number density of such lines per defined red-shift interval depends on the value of the cosmological constant [Tur92]. If a statistically significant number of observations for such systems can be obtained, particularly for quasars with small red-shifts, new information can be produced.

Gravitational lenses (see chapter 9) could also play an important role. A non-vanishing cosmological constant can lead to a significant change of the probability for such a lens effect [Fuk90b, Tur90a]. However, the difficulties of observing such effects and of obtaining theoretical calculations of clear signatures do not yet allow this method to produce interesting limits [Car92]. A cosmological constant also influences the growth of density fluctuations (see chapter 6); for the relevant investigations we refer to [Kof93].

At present, even at the greatest observed distances in the universe ($R \approx 10^{28}$ cm), there is no evidence for any effect of a cosmological constant, which, from the relation $R \simeq 1/\sqrt{\Lambda}$ leads to an experimental limit of $\Lambda < 10^{-56}$ cm^{-2} [Abb88].

5.3   The Λ problem

We now return to the contribution of quantum fluctuations to the vacuum energy density. First consider a quantum mechanical harmonic oscillator. Its energy eigenvalues are given by

$$E_n = \hbar\omega\left(n + \frac{1}{2}\right) \qquad n = 0, 1, \dots. \tag{5.29}$$

The vacuum ($n = 0$) therefore has a finite amount of energy (zero point energy). A relativistic field can be considered as a sum of harmonic oscillators of all possible frequencies ω. In the simple case of a scalar field with mass m, the vacuum energy is the sum of all contributions:

$$E_0 = \sum_j \frac{1}{2}\hbar\omega_j. \tag{5.30}$$

This summation can be carried out by putting the system in a box of size L^3 and then considering the limit as L tends to infinity (see e.g. [Car92]). Assuming periodic boundary conditions, equation (5.30) becomes

$$E_0 = \frac{1}{2}\hbar L^3 \int \frac{\mathrm{d}^3 k}{(2\pi)^3} \omega_k \tag{5.31}$$

where $k = 2\pi/\lambda$ corresponds to the wave vector. Now, using the relation

$$\hbar^2 \omega_k^2 = \hbar^2 k^2 + m^2 \qquad (5.32)$$

and a maximum cut-off frequency $k_{max} \gg m$, the integration can be carried out, resulting in

$$\rho_V = \lim_{L \to \infty} \frac{E_0}{L^3} = \int_0^{k_{max}} \frac{4\pi k^2}{(2\pi)^3} dk \frac{1}{2} \sqrt{k^2 + m^2} = \hbar \frac{k_{max}^4}{16\pi^2}. \qquad (5.33)$$

Assuming the validity of the general theory of relativity up to the Planck scale $(l_{Pl} \simeq (8\pi G)^{-\frac{1}{2}})$, $l_{Pl} = k_{max}$ results in a value which lies 121(!) orders of magnitude above the experimental value. This would lead to a vacuum energy of [Car92]

$$\rho_V = 10^{74} \hbar^{-3} \text{ GeV}^4 \approx 10^{92} \text{ g cm}^{-3}. \qquad (5.34)$$

It is rare for an estimate to be quite so incorrect. In addition, a value for k_{max} at approximately the electroweak scale of about 200 GeV, leads to a discrepancy of 54 orders of magnitude. Even with a k_{max} of order Λ_{QCD}, the prediction is still 42 orders of magnitude away from observation. In order to estimate the contribution of a single particle species, we assume that the virtual particles produced take up for a short time their Compton volume, L_c^3 (L_c stands for the Compton wavelength), where

$$L_c = \frac{\hbar}{mc} \to \rho_V = \frac{m}{L_c^3} = \frac{c^3 m^4}{\hbar^3}. \qquad (5.35)$$

Inserting for m the mass of, for example, the u, d quarks, their contribution to the vacuum energy density alone should produce effects on a scale of about 1/1 km^2. Contributions of the W and Z bosons would already be noticeable on scales of 1/20 cm^2. This means that effects of the curvature of space would have to appear on scales of metres to kilometres, in total contradiction to reality. A further significant contribution to the vacuum energy comes from the complex Higgs field Φ, which we discussed in chapter 1 in the framework of the electroweak theory to produce particle masses. The potential is given by

$$V(\Phi^\dagger \Phi) = -\mu^2 \Phi^\dagger \Phi + \lambda (\Phi^\dagger \Phi)^2 \quad \text{with} \quad \mu^2 > 0, \lambda > 0 \qquad (5.36)$$

which has a minimum at

$$\langle \Phi \rangle = \sqrt{\frac{\mu^2}{2\lambda}}. \qquad (5.37)$$

The contribution of the Higgs field to the cosmological constant can quickly be estimated as follows. With $M_{Pl} = \sqrt{G^{-1}}$ (equation (2.101)) and the vacuum energy density at the minimum of the potential

$$\Lambda = 8\pi M_{Pl}^{-2} V_{min} \qquad (5.38)$$

it follows immediately by substitution that the field in the broken phase provides a contribution of

$$\Lambda_\Phi = -4\pi G\mu^2 |\langle\Phi\rangle|^2. \tag{5.39}$$

Stability against higher order corrections in perturbation theory implies that μ must be greater than about 7 GeV [Lin76] and therefore

$$\Lambda_\Phi \leq -6 \times 10^{-32}\,\text{GeV}^2 = -1.55 \times 10^{-4}\,\text{cm}^{-2} \tag{5.40}$$

This corresponds to the contribution of the Higgs field at the time of the symmetry breaking, but it should still provide some contribution today. A very much smaller value of Λ would already dominate the current universe, due to the r dependence of the matter and the radiation density (see section 5.1). This value is also in complete contradiction with the observed value. It gives the wrong sign and is about 53 orders of magnitude off.

Instead of looking at the discrepancy in quantum field theoretical terms it can be viewed using the length scales involved. In the current observable universe there are no effects due to the curvature of space visible up to scales of 10^{28} cm, which can be converted to a limit of $\Lambda < 10^{-56}$ cm^{-2}. However, assuming the validity of the general theory of relativity up to the Planck scale, then according to equations (5.33) and (5.2) $\Lambda \approx l_{Pl}^{-2} \approx 10^{66}$ cm^{-2}. This also corresponds to a difference of 122 orders of magnitude.

It is interesting to speculate why the individual, completely independent contributions in the summation of equation (5.4) cancel almost completely. Perhaps they are not after all independent, and there could be a deeper, so far unknown relation. At the moment, this must remain a speculation.

5.3.1 Suggested solutions for the Λ problem

Many different possibilities, such as supersymmetry, fluctuations in the topology of the geometry of space-time, unimodular theories and many others are under consideration in order to explain the Λ problem. We now consider some of these is a little more detail. For a detailed discussion of these suggestions we refer to [Wei89, Car92, Wei96, Mar97].

A simple possible solution would be a shift of the zero point energy of the vacuum by the value of the vacuum energy. It does, however, appear to be a very unsatisfactory solution, as it is highly arbitrary. Another possibility would be supersymmetric theories. For an unbroken supersymmetry, with the operators discussed in chapter 2, the following relations hold:

$$Q_\alpha |0\rangle = Q_\alpha^\dagger |0\rangle = 0 \tag{5.41}$$

which implies

$$\langle 0|p^\mu|0\rangle = 0. \tag{5.42}$$

This implies a vacuum expectation value of zero, which is not modified by higher order corrections, as for every fermion contribution there is a corresponding boson contribution, which neutralize each other. This means that there is no contribution to the vacuum energy density and therefore a vanishing cosmological constant. The problem is that supersymmetry is clearly broken in nature, and then such arguments can no longer be applied easily. However, since this approach seems promising, efforts have been made, using supergravitation and superstring theories, to produce a vanishing cosmological constant even with a broken supersymmetry [Wit85, Din85, Wei89, Car92]. An additional mechanism, which would automatically provide a very small or even vanishing Λ, is hopefully expected from a quantum theoretical description of the whole universe [DeW67, Whe68, Kol90]. Indeed, this approach seems to promise the best chance of success. As we have no consistent theory of quantum gravitation, generally approximation methods are used, based on Feynman's path integral formalism of quantum mechanics [Fey65]. The wavefunction of a particle in a state ϕ with an initial state ϕ_0 is described through an integration of all paths which link the two points:

$$\Psi(\phi) \sim \int [\mathrm{d}\rho] e^{iS[\rho]/\hbar} \qquad (5.43)$$

where ρ represents the path from ϕ_0 to ϕ and $S[\rho]$ is the associated action. In the case of quantum gravity, the state has to be viewed as a three dimensional cut Σ in four dimensional space-time, and the wavefunction is replaced by the wavefunction of the universe $\Psi(\Sigma)$, interpreted as the probability amplitude that the universe contains Σ. Since oscillatory integrals such as equation (5.43) do not in general converge, the time parameter is analytically continued into the imaginary plane, $t \rightarrow i\tau$. As a result the component g_{00} of the metric changes sign and the paths to be considered lie in a Euclidean space. At the same time the action becomes imaginary, $S \rightarrow iS_E$. The integral is then damped by an exponential function and converges, provided that there is a lower limit for the Euclidean action S_E. This is discussed in [Har83, Vil88]. The wavefunction of the universe is then given by an integration over all manifolds $\mathcal{M}$:

$$\Psi(\Sigma) \sim \int [\mathrm{d}\mathcal{M}] e^{-S_E[\mathcal{M}]/\hbar}. \qquad (5.44)$$

$\mathcal{M}$ represents all four-dimensional Euclidean spaces with three-dimensional cuts Σ. Assuming that the universe is homogeneous on large distance scales and that gravity is the dominating force on these scales, the action can be approximated through the Euclidean action of the general theory of relativity [Col88]:

$$S_E \approx \frac{1}{16\pi G} \int \mathrm{d}^4x \sqrt{g}(2\Lambda - R) \qquad (5.45)$$

where g is the determinant of the metric and R is the Ricci scalar $R = g_{\mu\nu}R^{\mu\nu}$. From this the stationary points follow immediately from the principle of least action (see chapter 1) as a solution of Einstein's equations with a cosmological constant. For a Euclidean space this is a four-dimensional sphere. With the resulting $R = 4\Lambda$ and $\int d^4x \sqrt{g} = 24\pi^2/\Lambda^2$ it follows that

$$\Psi \approx e^{3\pi/\hbar G\Lambda}. \tag{5.46}$$

If we consider Λ as a free parameter, it can be seen that this expression has a prominent maximum for $\Lambda = 0$, which would solve the problem of the cosmological constant. To put it another way, universes with a vanishing cosmological constant dominate the path integral and make it extraordinarily probable that the cosmological constant of our universe will vanish. One possibility to turn Λ into such a free parameter exists using so-called *worm holes*, fluctuations in the topology of the geometry of space-time, which connect certain regions of Euclidean space in the form of strings [Col88] (figure 5.7). By this means an interaction between our universe and other universes is possible. One result of this is that all constants of nature known to us do not, in fact, have totally fixed values, but are the most probable values of distributions. In particular, the value of zero is favoured for Λ, since the action is stationary here. Moreover, in such calculations it turns out that indeed $\Psi \approx \exp(3\pi/\hbar G\Lambda)$ and therefore the maximum at zero becomes even sharper. Such models therefore give elegant solutions of the problem of the cosmological constant in the framework of quantum gravity. An interesting point of this exposition in [Col88] is to have achieved this without having invoked new physical laws or entirely new phenomena, but merely by the introduction of worm hole configurations in the path integral.

Due to the absence of a consistent theory, we should also discuss other possibilities. Einstein's field equations can be derived from the action (equation (5.45)) using the principle of least action and variation of the metric tensor. It is described by two parameters R and Λ. If, however, the determinant of the metric is a particular constant, the cosmological constant becomes irrelevant in the classical action (so-called *unimodular theories*) [Ng92]. Einstein's field equations can also be derived from such an action, leading again to a cosmological constant. The fundamental difference is, however, that it is now an arbitrary integration constant, completely independent of the parameters in the original action, contrary to the old Einstein–Hilbert action, where it is a fixed parameter of the theory. Purely classically, no value of Λ is now preferred. In a corresponding quantum theory we expect however that the state vector of the universe will be a superposition of all states with different values of Λ [Ng92]. Indeed, we again get a dependence of the form of equation (5.46), i.e. the configurations with $\Lambda = 0$ dominate.

Other theories prefer a time-dependent cosmological constant, not only during phase transitions such as inflation [Pee88, Wet94]. However, consistent formulations of such a theory seem to run into difficulties with primordial nucleosynthesis and the cosmic background radiation. The introduction of scalar fields whose only motivation would be to compensate the contribution of the

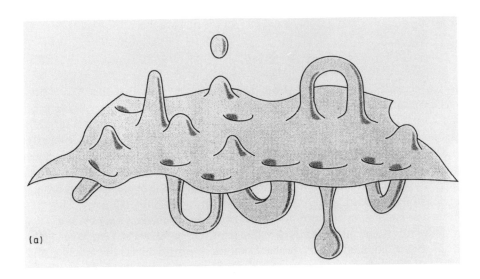

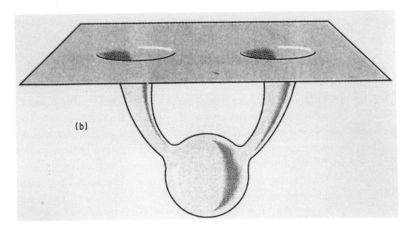

Figure 5.7. (*a*) Quantum gravity suggests that the well known flat four-dimensional space-time continuum should disappear on very small length scales, and instead be described by a 'foam-like' space-time with hills, valleys, bridges and worm holes. (*b*) Furthermore the possibility exists of many different parallel universes with some different dimensions. These would often be connected by worm holes (from [Dav92c]).

cosmological constant [Dol83], similar to the Peccei–Quinn mechanism for the solution of the strong CP problem (see chapter 11), would work as well.

Another consideration refers to the so called 'anthropic principle' [Wei89, Wei96]. Limitations on the cosmological constant are deduced from the requirement that the universe must survive long enough to allow for the evolution of life (for negative ρ_V) and that objects like galaxies can form before the universe begins its final exponential expansion (for positive ρ_V).

It must in conclusion be admitted that the problem of the cosmological constant remains unsolved, both theoretically and experimentally. This quantity is of fundamental importance, as it can be approached both from the viewpoint of elementary particle physics as well as from cosmology and currently leads to a discrepancy between theory and experiment of nearly 120 orders of magnitude. The solution of this problem will certainly lead to a great leap forward in our understanding.

Chapter 6

Large scale structures in the universe

As discussed in chapter 3, the Friedmann–Robertson–Walker metric describes a homogeneous and isotropic universe. Is this consistent with reality? At least at the level of stars a look at the night sky shows that it is not. In fact stars are arranged in galaxies. These inhomogeneities continue at larger scales. Galaxies group themselves into clusters, and the clusters into superclusters, separated by enormous regions with low galaxy density, the so-called voids. These empty spaces are not really empty, but contain a relatively low galaxy density. However, the distribution of radiogalaxies with a typical red shift of $z = 2$ (see figure 3.1), as well as the cosmic background radiation (chapter 7) with $z \approx 1000$, show that these inhomogeneities die out at large enough distances, and that we indeed seem to find an isotropic, homogeneous universe. We now turn our attention to the individual building blocks of the large scale structure. For further literature we refer to [Pee80, Pad93, Pee93, Bah97].

6.1 Galaxies

Galaxies appear in a multitude of forms, masses and sizes (figure 6.1). A simple classification is the Hubble classification scheme (also called the Hubble sequence), which distinguishes 4 main types (figure 6.2). For a detailed description of the Hubble sequence and its connection with physical quantities and the evolution of galaxies we refer to [Sil93].

Elliptical galaxies have almost no inner structures such as bars and spirals and consist mainly of old stars. Furthermore they contain very little or no dust and cold gas. The relative abundance of bright elliptical galaxies depends on their surroundings—in areas of low mass density they amount to about 10% of all galaxies, while for the central areas in dense galactic clusters their percentage grows to about 40%. Elliptical galaxies are referred to with the symbol En, which gives information about the relation of the major and minor axes a and b via $b/a = 1 - n/10$. The ellipticity ϵ is given by $\epsilon = 1 - b/a$. The most elliptical systems are of the type E7. Whether they are axially symmetric or

Figure 6.1. Different types of galaxies. (*a*) The spiral galaxy M31 (Andromeda) is the nearest large spiral galaxy with an appearance similar to our Milky Way. It is situated at a distance of about 700 kpc and belongs to the Hubble class Sb. M31 has two elliptical dwarf galaxies NGC 205 (type E5) and M32 (type E2) accompanying it (from [San94]). (*b*) The giant elliptical galaxy M87 in the centre of the Virgo cluster at a distance of about 15 Mpc. According to the Hubble classification it is of type E1. Most of the weak objects are globular clusters accompanying M87 (from [San94]). (*c*) The large Magellanic cloud (LMC) as an example of an irregular galaxy. It is our closest neighbour at a distance of about 50 kpc (from [San94]). (*d*) The gigantic spiral galaxy NGC 309 in the constellation Cetus (Whale). As a comparison for size the galaxy M81 in the constellation Ursa Major (Great Bear) is shown (small box), which is itself a huge spiral galaxy (from [Arp91]).

tri-axial systems can be seen from the isophotic (isophotes are contour lines of constant surface brightness) twist (a bending of the isophotic level as a function of intensity). The surface brightness profile of most elliptical galaxies can be described with de Vaucouleur's $R^{1/4}$-law [Vau48]

$$I(R) = I(0)\exp(-kR^{0.25}) = I_e \exp(-7.67((R/R_e)^{0.25} - 1)) \qquad (6.1)$$

where the effective radius R_e corresponds to the isophotal level within which half of the luminosity lies. I_e is the surface brightness at R_e. The luminosity of elliptical galaxies varies by about 7 orders of magnitude. The luminosity function $\Phi(L)$ ($\Phi(L)dL$ is the number of galaxies in a luminosity interval from

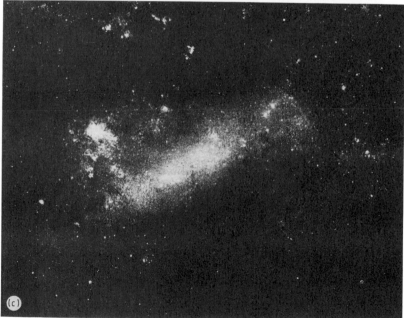

Figure 6.1. Continued.

Figure 6.1. Continued.

L to $L + dL$) to a good approximation is given by Schechter's law [Sch87a]

$$\Phi(L)dL = n_0 \left(\frac{L}{L_0}\right)^\alpha \exp(-L/L_0)\frac{dL}{L_0} \qquad (6.2)$$

where $n_0 = 1.2 \times 10^{-2}h^3$ Mpc^{-3}, $\alpha = -1.25$ and $L_0 = 1.0 \times 10^{10}h^{-2}L_\odot$ are good approximation values. $L_\odot$ is the luminosity of the Sun. Of course there are discrepancies from this law at low luminosity, where it becomes divergent.

Lens galaxies form a kind of intermediate stage between spiral and elliptical galaxies and are designated S0. They have a structureless disc without gas and dust, similar to ellipses, but they show an exponential behaviour of the surface brightness similar to spirals. Their abundance also changes with the surroundings, from less than 10 % in regions of low density up to a proportion of 50% in regions of high density.

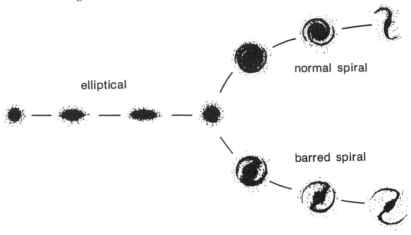

Figure 6.2. The classic classification scheme for galaxies. Proceeding from elliptical galaxies, which are classified according to their eccentricity, the spiral galaxies are divided into barred spirals and normal spiral galaxies. A further criterion corresponds to the openness of the spiral arms and increasing weakness of the nucleus. The content of gas and young stars grows from left to right. The irregular galaxies are grouped towards the right of the spiral galaxies (from [Rol88])

Spiral galaxies have a disc of young stars, together with gas and dust, as well as extended spiral arms of young stars in which active star formation is still taking place (figure 6.3). Their abundance also changes with the surroundings. They are the dominant part (about 80% of light galaxies) in areas of low density, while in core areas of galaxy clusters they are very rare (about 10%). Spiral and S0 galaxies are generally also called *disc galaxies*. The surface brightness of the disc galaxies obeys [Vau78]

$$I(R) = I_0 \exp(-R/R_d) \tag{6.3}$$

where $R_d \simeq 3h^{-1}$ kpc, and $I_0 \simeq 140 L_\odot$ pc^{-2}. Disc galaxies show Newtonian rotation curves, i.e. their masses can be derived from the measurement of the rotation speed as a function of distance. In elliptical galaxies, in contrast, the observed rotation curve is mainly caused by an anisotropic velocity field. Spiral galaxies have a bulge in the centre of their disc, and are surrounded by a spherical halo of old stars and globular clusters. The ratio between the luminosity of the bulge and the disc determines essentially the position in the Hubble sequence.

The last category is that of the *irregular galaxies*. These cannot be ascribed to any of the previous categories. They are mostly rich in gas and low in luminosity. The main bulk of their luminosity is produced by massive young stars and H II-regions.

An important role is also played by the *dwarf galaxies* . These are systems with little mass and little surface brightness. In models of cold dark matter

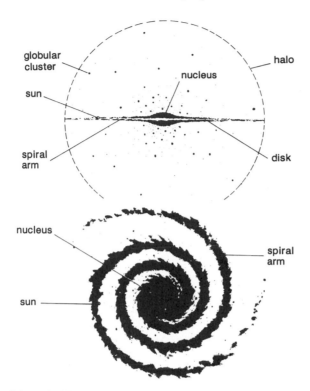

Figure 6.3. Schematic illustration of our Milky Way as an example of a spiral galaxy. Most of the constituents are situated in a disc-like object with a central dense nucleus (bulge). This structure is surrounded by spherically distributed globular star clusters within a dark halo. The distance from the Sun to the galactic centre is about 8.5 kpc (from [Rol88]).

(see section 6.8.1) these objects are the first to uncouple from the general expansion and can then form larger systems by merging with other dwarf galaxies [Blu84, Sil93]. The 'dwarf spheroidals' are the objects with the least mass known. They play an important role in the question of dark matter (see chapter 9).

6.2 Clusters, superclusters and voids

We now consider the distribution of galaxies in the sky. In the 1930s Shapley and Ames produced a catalogue of galaxies brighter than the 13th magnitude; after measurement of the red-shifts of these galaxies, it was already clear that the distribution of galaxies was far from homogeneous [San78]. Rather, they were grouped in clusters. Today there exist three large catalogues of galactic clusters, the Abell catalogue [Abe58], Zwicky's catalogue [Zwi68] and the catalogue of

Shectman [She85], based on the galaxy count of Shane and Wirtanen [Sha67]. Abell, while examining the photographs of the Mount Palomar Observatory Sky Survey (POSSI), was able to classify 2712 rich galactic clusters. A great number of further clusters were discovered in the X-rays by the ROSAT All Sky Survey [Trü93]. This catalogue contains more than 60 000 entries (see chapter 7). Galaxies group themselves in structures ranging from double to multiple systems up to rich clusters of up to several hundreds to a thousand members and radii of 5–10 Mpc. Our Milky Way, together with M31 (Andromeda) represent the biggest members of a cluster called the Local Group, an accumulation of about 30 galaxies. The distance to M31 (named after the Messier catalogue from the year 1784) is about 690 kpc. The distances between medium sized galaxies amount typically to about 3 Mpc. The Magellanic clouds also belong to the Local Group.

In principle there are two sorts of such clusters, regular and irregular ones. In regular clusters the galaxies are distributed mainly spherically, with a clear concentration in the centre. In the centre of such clusters elliptical galaxies are dominant, while in the outer parts spiral galaxies dominate. Irregular clusters, on the other hand, are more asymmetrical and can mostly be split into several smaller sub-clusters. One example of a regular cluster is the Coma cluster, which is at a distance of about $65h^{-1}$ Mpc. A typical example of an irregular cluster is the Virgo cluster. It is about $15h^{-1}$ Mpc away. In its centre is the enormous elliptical galaxy M87 (figure 6.1(b)), with an estimated mass of $10^{13} M_{\odot}$ ($1M_{\odot}$ corresponds to one solar mass, see chapter 12), one of the most massive galaxies known. Further known clusters in the immediate neighbourhood of our Milky Way are the Perseus cluster (about $50\ h^{-1}$ Mpc distant), and the Centaurus cluster (about $35\ h^{-1}$ Mpc distant). The more recent observations have now proved that these structures are also not independent, but are related on even larger scales (over $100\ h^{-1}$ Mpc). These features are called *Superclusters* [Oor83]. A galaxy within a supercluster has a typical velocity with respect to the cluster of 1000 km s^{-1} and therefore takes 3×10^{11} years to traverse it, a time much greater than the age of the supercluster. We are therefore dealing here with unrelaxed objects, which means that the virial theorem (see chapter 9) cannot be used to estimate the mass. These structures contain about 10^{15}–$10^{16} M_{\odot}$ and mostly have a filament-like appearance. The local group, together with the Virgo cluster and some other clusters form the *local supercluster*. This has a disc-like structure. Many examples of superclusters are now known, for example the Coma and the Pisces–Perseus superclusters. The major axis of such superclusters seems to lie along the direction of the neighbouring clusters, a tendency which becomes stronger the closer neighbours are [Boe88]. This seems to suggest a roughly simultaneous origin of the superclusters. In between these superclusters there are vast voids, where there are practically no galaxies [Kir81, deL86], see also [Bah88a, Roo88]. These also have sizes of 40–100 h^{-1} Mpc. The superclusters seem to arrange themselves as filaments around these voids, so that they produce a large scale structure rather reminiscent of a honeycomb.

In the following sections we discuss the methods by which we obtain information about these structures.

6.3   Red-shift surveys

In order to get a realistic picture of the three-dimensional distribution of galaxies it is necessary to measure three coordinates. Their position in the sky is given by the *right ascension* α (this corresponds to a longitude and is divided into 24 hours), the *declination* δ (which corresponds to a latitude) and the red-shift in the form of the quantity

$$cz = H_0 r + v_{pec} \tag{6.4}$$

which determines the depth. What is measured therefore is the sum of the Hubble velocity and the peculiar velocity (see section 6.4) of individual galaxies due to mass-concentrations. There are two ways to achieve this. One is to examine all possible galaxies with relatively small red-shifts, this means examine our near neighbourhood. Another possibility is to measure relatively narrow areas up to high red-shifts (pencil-beams). We begin with discussing the first method [Gio91]. In Zwicky's catalogue mentioned above the positions of about 30 000 galaxies with a brightness of up to $15^m.7$ have been measured. In this two-dimensional picture inhomogeneities were already visible. A first large-scale survey of our local supercluster took place in the 1970s. Tully and Fisher observed about 1800 galaxies in the region of the 21 cm line [Fis81]. This is the hyperfine structure line of neutral hydrogen, a favoured line in radio-astronomy. This observation resulted in a relatively exact picture of the local supercluster, but it did not contain any information about elliptical galaxies and galaxies with cz greater than 1000 km s^{-1}. However, an empirical relation between luminosity and the width of the 21 cm line discussed in chapter 3 (therefore also called the Tully–Fisher relation) was determined. This is generally used today to determine the distances of distant spiral galaxies. A more recent project to establish a truly three-dimensional catalogue, is being undertaken by the Harvard-Smithsonian Institute for Astrophysics (CfA) [Gel89, Huc90]. The red-shift and position of 14 383 galaxies down to an apparent brightness $m_B = 15^m.5$ with greater than $30°$ declination have been measured. Some of the already completed areas are shown in figure 6.4. The voids are clearly seen here, as well as the filament-like arrangement of the superclusters. A previously unknown feature, called the 'Great Wall', is only bounded by the size of the observation field. Its two-dimensional projection results in a size of at least $60\ h^{-1}$ Mpc $\times\ 170\ h^{-1}$ Mpc, and it contains about half of all the galaxies in these regions. In comparison, the current observable Hubble volume (sphere with radius H_0^{-1}) has a radius of about $H_0^{-1} \simeq 3000 h^{-1}$ Mpc. Mass estimates for this part of the Great Wall result in about $2 \times 10^{16} M_\odot$, which is ten times more than in the local supercluster. It is also important to note that this structure is only limited by the extent of observation and in this sense should be interpreted as a lower limit.

Information about the arrangement of galaxies is obtained via the 'counts in cells' method [Pee80]. In this method the three-dimensional space is rastered in cells. All cells with at least one galaxy per cell are taken, and for these cells an average number $\langle n \rangle$ of galaxies per cell is calculated. If the cells are enlarged, and $\langle n \rangle$ subsequently grows proportional to the diameter of the cell, the large scale distribution is one dimensional. If, however, $\langle n \rangle$ grows proportional to d^2, then the distribution is two dimensional. If we now compare the simulated with the measured data, a wall-like, rather than a rod-like distribution of galaxies does indeed seem to result. An observation of the area around the galactic south pole, the Southern Sky Red-shift Survey (SRSS2), gives a qualitatively similar picture of galaxy distribution [DaC94]. The inhomogeneous structure of the universe is also shown in a new, more comprehensive survey of redshifts, the Las Campanas Redshift Survey [She96]. Advances in technology should allow even more comprehensive surveys of this type to be carried out in future. A particularly ambitious project of this type is the Sloan Digital Sky Survey (SDSS), which plans to measure more than 1 million red-shifts [Lov96b]. A survey of infra-red galaxies at 60 μm was made possible with the help of the infra-red satellite IRAS. This wavelength has the advantage that absorption within our galaxy is smaller. On the other hand the absorption behaviour is less well known, and emission from dust and cool stars starts to have a disturbing effect. More than 17 000 galaxies, with a flux of more than 0.5 J y have been observed [Str88]. Here also structures become clearly visible, even on scales which exceed those of the optical survey, due to their greater depth. From the IRAS survey actually two surveys were made, the IRAS 1.2 J y survey [Fis95] and the QDOT survey [Sau91]. The QDOT survey looks at all galaxies with a flux of more than 0.6 J y, and contains 2184 galaxies. The QDOT survey contains a complete set of galaxies up to a distance of $140h^{-1}$ Mpc, and is therefore suitable to test the density structure of our immediate neighbourhood.

In order to derive the mass distribution in the universe from the above three-dimensional structures, we have to proceed from the disputed assumption that light is really a good indicator for matter. One of several possibilities (for a review see [Dek94]) to determine the distribution of velocities from the observations is provided by the POTENT method of Bertschinger and Dekel [Dek90, Dek92]. The problem with all the red-shift observations is that the velocity-component can only be measured in the direction of the line of sight. The basic assumption of the POTENT method is that, if gravitation really were the cause of the peculiar velocity (see below), the equations of motion for everything would be given by

$$\frac{dv}{dt} + Hv = g. \tag{6.5}$$

This is equivalent to the condition that the curl of the velocity field v is smaller than the gradient

$$|\nabla \times v| \ll |\nabla \cdot v|. \tag{6.6}$$

Right ascension

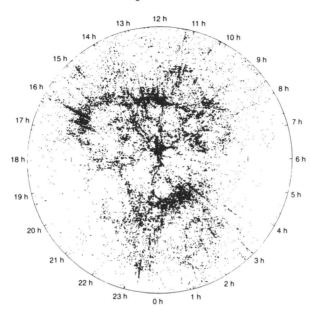

Figure 6.4. The distribution of galaxies in the nearby universe, as determined from the Centre for Astrophysics (CfA) red-shift survey. A total of over 14 000 galaxies is shown and forms a complete set of data in the region of declination from 8.5° to 44.5°. All galaxies have velocities of recession of less than 15 000 km s^{-1}. The Milky Way is situated in the centre. The filament-like structure, i.e. areas of very high density (superclusters), as well as voids, can clearly be seen (with kind permission of M Geller, J Huchra and the Harvard–Smithsonian Center for Astrophysics, see also [Lon92, 94]).

Then the velocity field can be expressed in terms of a velocity potential

$$v = -\nabla\phi. \tag{6.7}$$

POTENT now tries to reconstruct the velocity distribution and potential, e.g. from the observed IRAS red-shifts. From this the gravitational potential Φ and the density distribution δ can be determined and compared to the distribution of luminous objects in the sky (see figure 6.5). These are given by [Pee80]

$$\Phi \approx \tfrac{3}{2}\Omega^{0.4}\phi \tag{6.8}$$

$$\delta \approx -\Omega^{0.6}\nabla \cdot v. \tag{6.9}$$

In both quantities the density parameter Ω is included, which means that information about the structure of the universe can be obtained. The POTENT results are in agreement with the observations for values of Ω between 0.3 and 2.5 [Dek93]. This is significantly larger than the dynamic values which are

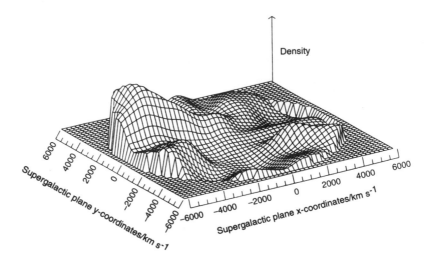

Figure 6.5. The distribution of the mass density in the plane of the local supercluster, characterized by X and Y, as obtained from the data using the POTENT method. The distances are given in km s^{-1}, corresponding to the intrinsic velocity relative to the general Hubble motion. The Milky Way and the local group are situated in the centre; the strong enhancement on the left hand edge corresponds to the Great Attractor (from [Ber90a]).

obtained from smaller scales. Furthermore it is also possible to make statements about the initial conditions for structure formation. The usefulness of this method will probably be proved further in the future [Kol95]. While inhomogeneities on the scale of 100 h^{-1} Mpc can be seen here also, the distribution of clusters seems to be relatively homogeneous on the largest scales, as evidenced by observations of radio galaxies [Gre91] (see figure 3.1).

The IRAS galaxies also seem to show a large scale movement and hence a dipole term [Lah88]. This lies in the direction of $l = 248° \pm 10°$ and $b = 46° \pm 10°$ and therefore only differs from the value obtained from the microwave background by $26° \pm 10°$ (see chapter 7). Here l and b are galactic coordinates, that means that the centre of the Milky Way lies at $l = 0°$ and $b = 0°$.

The alternative to these surveys is the previously mentioned 'pencil-beam' observations. In 1981 Kirshner, Oemler, Schechter and Schechtman discovered using this method the existence of a vast region devoid of galaxies at about 15 000 km s^{-1}, with a width of 6000 km s^{-1}, corresponding to 40 Mpc [Kir81]. This area is now known as the Bootes void after the constellation Bootes, and its discovery was a decisive breakthrough for the description of large scale structure in terms of superclusters and voids.

6.4   Peculiar velocities

In order to examine the gravitational field directly the large scale movement of many galaxies has to be measured directly, rather than by their red shift. If we assume as a basis a homogeneous expansion of the universe, mass concentrations will cause deviations of the expected speed, which is called the *peculiar velocity*. As well as the measurement of the red-shift, a z-independent distance-measurement, as discussed in chapter 3, is necessary for determining these velocities [Bur90]. The question arises of whether the additional gravitational attraction caused by the condensation of the local supercluster is sufficient to explain the speed of our Milky Way relative to the cosmic background radiation of about 600 km s^{-1} (see chapter 7). A new measurement of the peculiar velocities of about 400 elliptical galaxies shows a 'bulk motion' over a far greater area (see figure 6.6) [Dre91b]. All clusters and superclusters in the near neighbourhood seem to move towards a point with the coordinates $l = 312°\pm11°$ and $b = 6° \pm 10°$. The mass concentration seems to lie near $cz = 5000$ km s^{-1}, corresponding to a distance of 45 h^{-1} Mpc. Galaxies closer to this point have a greater peculiar speed than the local group, while galaxies further away move more slowly. This points to an even greater mass concentration, which is thus known as the *Great Attractor* [Dre87]. If we model the gravitationally attracting mass, we obtain an estimate for the Great Attractor of about $5 \times 10^{16} M_\odot$. In order to really present evidence for the effect of such a mass concentration, we would have to measure a smaller red-shift for galaxies further away from the centre of the concentration than is to be expected according to the Hubble expansion, since the concentration acts as a brake. Indeed, latest measurements, both with elliptical as well as with spiral galaxies, seem to confirm this behaviour (figure 6.7) [Dre91a]. In the above mentioned IRAS catalogue [Str88] two mass-concentrations are noticeable, one may be associated with the Great Attractor and the other with the Pisces–Perseus supercluster. First measurements of the peculiar speeds in the Pisces–Perseus cluster seem, however, to indicate no such mass concentration [Wil90]. Rather Pisces–Perseus as a whole also shows movement in the direction of the Great Attractor. Whether its attraction is sufficient to explain this effect, or whether greater and currently unknown concentrations are responsible for this, will only be known in future. The centre of the Great Attractor might be associated with the galaxy cluster Abell 3627 according to recent measurements [Kra96]. A new examination of the brightest galaxy clusters also shows a large scale movement [Lau94].

The method of peculiar velocities differs from those earlier mentioned surveys in that here the gravitational field is really being fully surveyed, and there is no dependence on the assumption that light is a good tracer of mass. The Great Attractor described above is the first structure surveyed in this way (although its existence is by no means universally accepted). Further examples should follow in future.

As we have seen, matter on scales of up to 100 Mpc is not equally

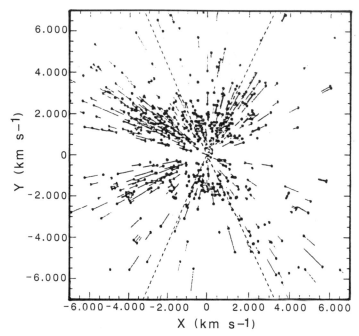

Figure 6.6. Peculiar velocities of galaxies in the direction of motion of the local group. Each point represents a galaxy, where the point marks the position in space and the line represents the absolute value of the velocity in the direction of the line. All galaxies seem to be moving towards one point, the Great Attractor (from [Dre91b]).

distributed, and even larger structures probably exist. Today the density in galaxies is about a factor of 10^5 and in galaxy clusters about 10^3 times larger than the average density. It is assumed that these matter concentrations have evolved from the density variations at the time the universe became matter-dominated. As this was the case around the time of the formation of the microwave background radiation, these perturbations should manifest themselves in anisotropies of the 3 K radiation. One of the most important questions is how the observed large scale structure can be described consistently with the incredible degree of isotropy of the background radiation (see chapter 7).

6.5 Quasars

Knowledge about the large scale structure of the universe is also obtained from the examination of the most distant objects known, the *quasars* (quasi stellar radio sources). Closely connected to this is the phenomenon of active galaxies. For a more detailed discussion of this subject see [Wee86, Bla90, Dus92, Rob96].

Today only about 1% of known galaxies are 'active', although this abundance is significantly higher at greater red-shifts. It was realized that this

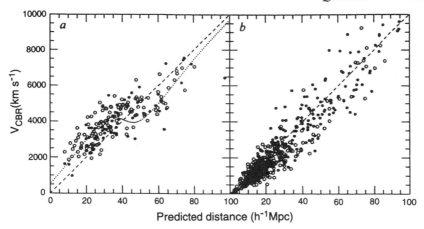

Figure 6.7. Hubble diagram of predicted distance against observed velocity with respect to the cosmic background radiation. (*a*) The dotted curve is the prediction of a model for the Great Attractor, while the broken curve shows the normal Hubble flow. It can clearly be seen that the points lying closer to us than to the Great Attractor are faster than expected from the normal Hubble flow, while further away galaxies appear to be slowed down. (*b*) Hubble diagram for galaxies far away from the area of the Great Attractor. Here the expected linear behaviour manifests itself (from [Dre91b]).

activity can show variations on very short time scales, such as days or months. Therefore the source causing these deviations should also have a corresponding spatial extent of light-days to -months. The nuclei of galaxies are thought to be the place of this activity, possibly powered by a super-heavy (several million solar masses) black hole [Beg84]. This activity often shows itself in the form of jets shooting out of the nucleus. Their examination is done mainly with the help of radio-astronomy, which by connecting several telescopes, such as the very long baseline array (VLBA), or the very large array (VLA), reaches a very high spatial resolution using interferometry (VLBI) (figure 6.8). The whole variety of observations of quasars and related phenomena such as, for example, radio and Seyfert galaxies and BL-Lac objects cannot be discussed in detail here (see e.g. [Uns92]), but nevertheless a rather unified picture of these objects has developed (see figure 6.9).

Due to their intense luminosity quasars are observable up to high red-shifts. While galaxies are typically observed only up to red shifts of $z \approx 0.5$, the furthest away observed quasar has a red-shift of $z = 4.89$ [Sch91b]. At such red-shifts the universe had only 7% of today's age, which means that the quasars make it possible to look far back into the past. The quasar with the smallest red-shift, 3C 273, lies at $z = 0.158$, which still corresponds to a distance of 450 Mpc. From the distribution of the radio-galaxies and quasars we can therefore learn about large scale structures [Koo87]. Their distance makes quasars interesting for other reasons as well. They provide a practical 'lamp' at large distances and

Figure 6.8. (*a*) Central part of the '*Y*' of the Very Large Array (VLA) in New Mexico, and its telescopes. These are constructed to be mobile and can be electronically connected, thereby attaining a very high spatial resolution (from [Zei91]). (*b*) With 6 telescopes, each with a diameter of 22 m, the Australian counterpart to the VLA is one of the best radio-interferometers in the southern hemisphere. In order to further increase the resolution, such arrays can also be connected inter-continentally (from [Sky93]).

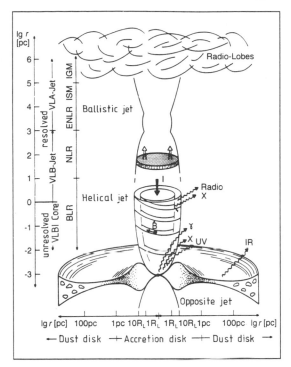

Figure 6.9. Model of the quasar phenomenon and related effects. Depending on the angle of observation, different aspects of this structure can be seen, and therefore the richness of the phenomenon becomes understandable (from [Qui93]).

therefore make it possible for us to discover information about the universe lying between us and the quasars. For example with the help of the gravitational lens effect (see chapter 9), where the light of a quasar is split up into several images by the mass of an intermediate galactic cluster. A second effect relies on the large red-shift. By the latter the strong Lyman-α-emission line of hydrogen (it is the first line of the Lyman-series in hydrogen and describes the transition of the L-shell to the ground state, laboratory wavelength: 121.6 nm) is partly pushed into the visible region. In the neighbouring continuum to short wavelengths an immense number of sharp absorption-lines can be seen (on average about 50 per quasar). This is called the *Ly-α-forest* (see figure 6.10). It has its origin in neutral hydrogen clouds which have a smaller red-shift than the quasar. It is therefore possible to investigate their distribution along the line of sight to the quasar [Sar80]. An estimate of the mass of these Ly α-clouds results in a value of 10^7–$10^8 M_\odot$ and temperatures of about 30 000 K. The investigation of these structures is extremely interesting for theories of galaxy evolution [Wol93].

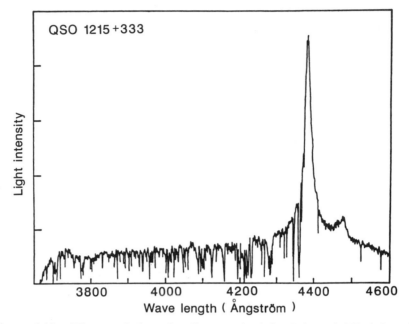

Figure 6.10. Pile up of absorption lines to the left of the red-shifted Ly-α-line ('Ly-α-forest') of a quasar. These are caused by hydrogen accumulations at smaller red-shifts, which lie along the line of sight (from [Sch87b]).

6.6 Description of structures

The simplest means to describe structure formation is the so-called mass-correlation function $\xi(r)$ [Pee80]. This describes the probability to observe two objects at a distance r. If these objects are galaxies we speak about a galaxy–galaxy correlation function ξ_{gg}, defined as

$$\xi_{gg}(r) = \langle \delta n(x + r)\delta n(x)\rangle \tag{6.10}$$

where $n(x)$ represents the number density of galaxies at position x, or, to put it another way, $\langle n \rangle \delta V[1 + \xi(r)]$ is the probability to find one galaxy at a distance r from another. For a random distribution it follows immediately that $\xi = 0$. From observations (figure 6.11) the following behaviour applies [Dav83]

$$\xi_{gg}(r) \simeq \left(\frac{r}{5h^{-1}\,\text{Mpc}}\right)^{-1.8} \tag{6.11}$$

on scales of about 10 kpc to $10h^{-1}$ Mpc, but the exponent might change with the observed region. Under the assumption that the number density is proportional to the mass density, equivalent to the claim that light is a good indicator of mass, it follows immediately that $\xi(r) = \xi_{gg}(r)$. As the underlying mass distribution for galaxies and clusters should be the same, one expects for a cluster–cluster correlation function $\xi_{cc} = \xi_{gg}$. Experimentally (figure 6.11) we find [Bah88a]

$$\xi_{cc}(r) \simeq \left(\frac{r}{25h^{-1}\,\text{Mpc}} \right)^{-1.8} \simeq 20\xi_{gg} \qquad (6.12)$$

and furthermore a dependence on the richness (a measure of the number of galaxies) of the cluster. So, as matter on the scale of galaxies tends to produce clusters, a kind of 'anti'-cluster formation should take place for greater scales due to mass conservation, which would show itself as a negative correlation function. For a recent discussion see [Sah95, Bah96a].

6.7 The development of fluctuations

The question now arises as to how the previously described structure formation occurred in the expanding universe. If we were to define a density contrast according to [Pee80]

$$\delta(\boldsymbol{x}) = \frac{\delta\rho(\boldsymbol{x})}{\langle\rho\rangle} = \frac{\rho(\boldsymbol{x}) - \langle\rho\rangle}{\langle\rho\rangle} \qquad (6.13)$$

where $\langle\rho\rangle$ is the mean density. Regions with excess density $\delta > 1$ can act as gravitation centres and collapse. The question is how to get a δ of the order one, from initial conditions which are particularly isotropic and homogeneous? Once one has managed to reach this condition the growth happens non-linearly and the evolution towards bound structures happens quite rapidly. The questions which arise are mainly

(i) How could a density contrast of about one develop in an expanding universe?

(ii) What were the initial conditions of the density inhomogeneities, i.e. what was the spectrum of the fluctuations from which structures developed?

The density contrast of galaxies : clusters : superclusters is in the ratio $10^6 : 10^3 : 1$–10. As the matter density in the universe is proportional to $(1 + z)^3$ and superclusters are not yet fully in dynamic equilibrium, it follows that galaxies as discrete objects cannot have formed at red-shifts greater than 100, clusters at red-shifts of about 10 and superclusters in the region of $z = 1$. The actual presently observable structures have therefore only separated from the expanding gas at red-shifts smaller than 100, and therefore in any case during the matter dominated phase of the universe.

The problem of the growth of small perturbations under the influence of gravitation was initially studied by Jeans in 1902 [Kol90]. Newton's approximation is used to describe the perturbation growth and a static model $(\dot{R} = 0)$ of an ideal liquid with density ρ, velocity v and pressure p is assumed.

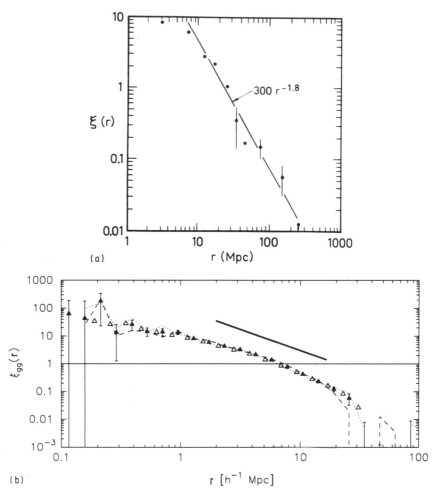

Figure 6.11. (*a*) Behaviour of the spatial correlation function for clusters of galaxies. The $r^{-1.8}$-dependence is clearly visible (from [Bah83, Bah88a]). (*b*) Galaxy–galaxy correlation function $\xi_{gg}(r)$ as extracted out of the Las Campanas red-shift survey. The filled triangles correspond to the north and south galactic cap sample, the dashed line to the northern cap and the dotted line to the southern cap alone. From 2.0 h^{-1} Mpc $< r < 16.4h^{-1}$ Mpc the data can be described by a power law with slope $\gamma = -(1.52 \pm 0.03)$ shown as a line with an artificial offset (from [Tuc96]).

The underlying equations are:

$$\frac{\partial \rho}{\partial t} + \nabla(\rho v) = 0 \quad \text{Continuity equation} \qquad (6.14)$$

$$\frac{\partial v}{\partial t} + (v \cdot \nabla)v + \frac{1}{\rho}\nabla\rho + \nabla\Phi = 0 \quad \text{Euler equation} \qquad (6.15)$$

with the gravitational potential Φ given by the Poisson equation

$$\nabla^2\Phi = 4\pi G\rho. \qquad (6.16)$$

Under the assumption of small deviations of all these quantities (such as $\rho = \rho_0 + \rho_1$ with $\rho_1 \ll \rho_0$), which allows the use of linearized equations, it follows that

$$\frac{\partial^2\rho_1}{\partial t^2} - v_s^2\nabla^2\rho_1 = 4\pi G\rho_0\rho_1. \qquad (6.17)$$

Here adiabatic deviations are assumed, which means that the speed of sound v_s corresponds to

$$v_s = \sqrt{\frac{\partial p}{\partial \rho}}. \qquad (6.18)$$

For the static case the solutions are plane waves $\exp(i(k \cdot r - \omega t))$ with a dispersion relation

$$\omega^2 = v_s^2 k^2 - 4\pi G\rho_0. \qquad (6.19)$$

Equation (6.17) can be adapted to the non-static case, which leads to the following expression for the density contrast

$$\frac{d^2\delta}{dt^2} + 2\left(\frac{\dot{R}}{R}\right)\frac{d\delta}{dt} = \delta(4\pi G\rho_0 - v_s^2 k^2). \qquad (6.20)$$

From the dispersion relation of the static case, a critical value ($\omega = 0$), called the *Jeans wavenumber* k_j, can be defined, given by

$$k_j = \left(\frac{4\pi G\rho_0}{v_s^2}\right)^{\frac{1}{2}}. \qquad (6.21)$$

Solutions with $k > k_j$ describe sound waves, where the internal pressure gradient is large enough to withstand gravity. For $k < k_j$, equivalent to an imaginary ω, the solutions describe exponentially growing or decaying modes. Hence the *Jeans mass* can be defined as the mass within a sphere of radius $\lambda_j/2 = \pi/k_j$:

$$M_j = \frac{4\pi}{3}\left(\frac{\pi}{k_j}\right)^3 \rho_0 = \frac{\pi^{\frac{5}{2}}}{6}\frac{v_s^3}{G^{\frac{3}{2}}\rho_0^{\frac{1}{2}}}. \qquad (6.22)$$

Masses greater than the Jeans mass are unstable with respect to gravitational collapse. Here the instability due to the self gravitation of the high density region becomes greater than the internal pressure gradient. Instead of the general solution of equation (6.20) we consider the case of large wavelengths ($k < k_j$ and the related neglect of the pressure term $v_s^2 k^2$) for universes with $\Omega = 1$ and $\Omega = 0$.

First case $\Omega = 1$:

In this case the following equation applies

$$4\pi G\rho = \frac{2}{3t^2} \quad \text{and} \quad \frac{\dot{R}}{R} = \frac{2}{3t}. \tag{6.23}$$

Hence equation (6.20) can be rewritten as

$$\frac{d^2\delta}{dt^2} + \left(\frac{4}{3t}\right)\frac{d\delta}{dt} - \frac{2}{3t^2}\delta = 0. \tag{6.24}$$

On the basis of the power dependence of t the solution must be of the form

$$\delta = At^{\frac{2}{3}} + Bt^{-1}. \tag{6.25}$$

As the second term describes damped modes, it can be ignored today. The first term, on the other hand, describes modes growing with a dependence of

$$\delta \sim t^{\frac{2}{3}} \sim R = (1+z)^{-1}. \tag{6.26}$$

The effect of the expansion is therefore to slow down the growth of deviations from an exponential behaviour to a power law! A relativistic treatment of the deviation also results in a power law, but modified to give [Kol90]

$$\delta \sim t \sim R^2 = (1+z)^{-2} \tag{6.27}$$

Second case $\Omega = 0$:

In this case

$$\rho = 0 \quad \text{and} \quad \frac{\dot{R}}{R} = \frac{1}{t} \tag{6.28}$$

from which follows

$$\frac{d^2\delta}{dt^2} + \left(\frac{2}{t}\right)\frac{d\delta}{dt} = 0. \tag{6.29}$$

This leads to a solution of the form

$$\delta = At^0 + Bt^{-1} \tag{6.30}$$

which implies one solution which decays away, together with one with constant amplitude. From these results the development of small deviations can be understood. At the beginning of the matter dominated phase the universe can be well described by an Einstein–de Sitter universe, and the amplitude of the density contrast grows linearly with R. For later phases, when the universe is rather similar to a $\Omega = 0$-model, the amplitude grows only very slowly, and in the limit $\Omega = 0$ does not grow at all.

We now consider the behaviour of the Jeans mass as a function of time. A detailed discussion can be found in [Kol90]. For reasons of simplicity we assume a universe consisting only of baryons and photons, i.e. $\rho = \rho_B + \rho_\gamma$. In the radiation-dominated phase the pressure is caused by the photons and therefore $v_s^2 = (1/3)c^2$. Hence the Jeans mass in the radiation-dominated phase can be written as

$$M_J = 2.8 \times 10^{30} z^{-3} \Omega_B h^2 M_\odot. \qquad (6.31)$$

Thus the Jeans mass grows proportional to R^3. The Jeans mass corresponding to a solar mass lies at a red-shift of about 10^{10} and grows to the typical mass of a large galaxy of about $M = 10^{11} M_\odot$ at a red-shift of 3×10^6. However, a dramatic change takes place at the time of the decoupling, at a red-shift of about 1200 (see chapter 7). This is caused by the abrupt decrease of the velocity of sound, as the pressure is now maintained only by non-relativistic hydrogen atoms, for which

$$v_s^2 = \frac{5}{3} \frac{kT}{m_H}. \qquad (6.32)$$

This results in a rapid decrease of the Jeans mass from $10^{16} M_\odot$ to $10^6 M_\odot$. Interestingly enough this corresponds to the typical mass scales of globular clusters, which are some of the oldest objects in the universe. All larger masses become gravitationally unstable, δ grows proportional to R, until $\Omega z \approx 1$. This tells us that density fluctuations have grown since the time of the recombination period (see chapter 3) by a factor $(t_0/t_R)^{\frac{2}{3}} = 1 + z \simeq 10^3$. Figure 6.12 shows the behaviour of the Jeans mass as a function of the age of the universe, or the temperature, in this simple model.

The simple model described above does, however, need at least two further important extensions. Weakly interacting particles, such as neutrinos, can escape without interaction from areas of high density to areas of low density, which can therefore lead to an equalization of the inhomogeneities. This process of free streaming, or collision-less damping, is important before the Jeans instability becomes effective. The typical scale λ_{FS} for this equalization is given by [Kol90]

$$\lambda_{FS} \simeq 30 \left(\Omega_X h^2\right)^{-1} \left(\frac{T_X}{T}\right)^4 \text{ Mpc}. \qquad (6.33)$$

X denotes the corresponding weakly interacting particle. If we consider neutrinos for instance, $T_\nu/T \approx 0.71$ and it follows that

$$\lambda_{FS} \simeq 20 \left(\frac{m_\nu}{30 \text{ eV}}\right)^{-1} \text{ Mpc} \qquad (6.34)$$

corresponding to a mass scale of

$$M_{FS} \simeq 4 \times 10^{14} \left(\frac{m_\nu}{30 \text{ eV}}\right)^{-2} M_\odot. \qquad (6.35)$$

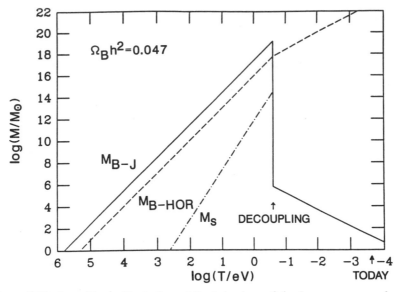

Figure 6.12. Logarithmic illustration of the behaviour of the Jeans mass as a function of temperature for a mixture of baryons and photons in a baryon dominated universe. Illustrated are the Jeans mass in baryons M_{B-J} (full line), the baryonic mass within the horizon $M_{B-\text{Hor}}$ (dashed line) and the behaviour of the Silk mass M_S (dashed-dotted line). A baryonic density of $\Omega h^2 = 0.047$ was assumed. The steep fall at the decoupling time results from the drastic decline of the velocity of sound (from [Kol90]).

Smaller mass scales are effectively washed out and no structure formation takes place.

A second process is mainly effective at the time of the recombination. Due to the suddenly increasing mean free path for photons, these can also effectively flow away from areas of high density. Their spreading, due to frequent collisions, does however correspond to diffusion rather than to a free flow. This kind of damping is called collisional damping, or *Silk-damping* [Efs83]. In a simple model considering only baryons and photons typical scales of [Kol90]

$$\lambda_S \simeq 3.5 \left(\frac{\Omega_0}{\Omega_B} \right)^{\frac{1}{2}} \left(\Omega_0 h^2 \right)^{-\frac{3}{4}} \text{ Mpc} \tag{6.36}$$

result, and the Silk mass is given by

$$M_S \simeq 6.2 \times 10^{12} (\frac{\Omega_0}{\Omega_B})^{\frac{3}{2}} \left(\Omega_0 h^2 \right)^{-\frac{5}{4}} M_\odot. \tag{6.37}$$

Here also smaller scales are effectively smeared out, as, due to the frequent interaction of the photons, inhomogeneities in the photon–baryon plasma are washed out.

For the theoretical description of the development of the fluctuations we have to now decompose the density contrast into its Fourier coefficients

$$\delta(r) = \frac{V}{(2\pi)^3} \int \delta_k e^{-ik\cdot r} d^3 k. \tag{6.38}$$

Due to the expansion of the universe the real wavenumber is, however, not k, but k/R. The growth of different modes δ_k at different times of the expansion can be described through the comparison of two scales. The first scale marks the growth of a mode $\delta_k(t)$, characterized by k, and happens either due to gravitational instabilities, or by the increase of the wavelength

$$\lambda = \frac{2\pi}{k} R(t) \tag{6.39}$$

due to the expansion of the universe. Each wavelength λ is connected to a characteristic scale λ_0, which develops according to

$$\lambda_0 \left[\frac{R(t)}{R_0} \right] \sim t^n \tag{6.40}$$

if, as in all models of the universe, $R \sim t^n$ (see chapter 3). The second scale for the expansion of the universe, on the other hand, is given by the Hubble radius $cH^{-1}(t) = cn^{-1}t$. Since in realistic models $n < 1$, an increasing ratio $\lambda(t)/cH^{-1}(t)$ follows for early times of the universe. At some time $\lambda(t)$ will be larger than the Hubble radius, which has the following consequences. Connected with any characteristic scale is a characteristic mass, according to

$$M(\lambda) = \frac{4\pi}{3} \langle \rho(t) \rangle \left(\frac{\lambda(t)}{2} \right)^3 = 1.5 \times 10^{11} M_\odot \Omega_0 h^2 \left(\frac{\lambda(t)}{1 \text{ Mpc}} \right)^3. \tag{6.41}$$

In this way, for example, with $\lambda = 2$ Mpc, masses of about $10^{12} M_\odot$ are connected, which are typically galactic masses. In case no non-linearities would occur due to gravitational instabilities, this would be the typical spatial size for such a mass. By non-linearity we understand a development of density inhomogeneities according to

$$\frac{\delta\rho}{\rho} \sim R^n \quad (n \geq 3), \qquad \frac{\delta\rho}{\rho} \geq 1 \tag{6.42}$$

whereas linear development is characterized by

$$\frac{\delta\rho}{\rho} \sim R, \qquad \frac{\delta\rho}{\rho} \leq 1. \tag{6.43}$$

However, in the current universe galaxies have a typical size of about 50 kpc, which means that these scales are already in the non-linear region. The ing line between linear and non-linear lies nowadays at $10\ h^{-1}$ Mpc. If xtrapolate this back to the point of transition from a radiation- to matter-dominated universe, we see that this scale already lies outside the Hubble radius. As even greater masses are associated with clusters and superclusters, their scales also lie outside. Since we have seen in chapter 3 that physical processes can only proceed within the Hubble volume, it is necessary to explain how structure could have happened at all. An exact treatment in the framework of the liquid-model has to take account of relativistic effects. For a detailed description see [Lon89, Kol90].

6.8 The evolution of structures

For the moment we continue to consider a universe of baryons and photons. As previously discussed, all small scales are already washed out at the time of recombination, and the fluctuations have masses of the order of $10^{15} M_\odot$. This is a typical mass scale of superclusters, and therefore these objects are the first to form after the drastic decrease of the Jeans mass after recombination. Since an exactly spherical mass distribution is unlikely, it is preferable to start from an ellipsoid in a collapse. This has the consequence that shrinkage occurs preferably along its shortest half-axis, which therefore results in a pancake-like structure. This is roughly in keeping with the original *pancake theory* of Zeldovich [Zel70], and seems to be confirmed by new computer simulations [Sha95]. From this clusters and single galaxies form through fragmentation and collapse of individual sub-regions. Because of the prior formation of the largest structures, leading down to galaxies and stars, such theories are called *top-down theories*. Also the collision of galaxies ('merging') plays an important role in the evolution of galaxies. However, there are difficulties with equation (6.50). As density fluctuations of the order of 1 are observed today, there must have been deviations of about 10^{-3} at the time of recombination. These would have to manifest themselves in the adiabatic case in corresponding temperature deviations of the microwave background radiation. This is, however, on the contrary highly isotropic and shows no signs of such large fluctuations (see chapter 7). One possible solution is the incorporation of dark matter (see chapter 9).

6.8.1 Dark matter and structure formation

There are two extreme candidates for dark matter: hot and cold dark matter. Hot dark matter consists of relativistic particles, such as neutrinos with masses of about 10 eV. Due to their good mobility these allow a relatively long time for the washing-out of all fluctuations. Structures only form when particles become non-relativistic. The cold dark matter, on the other hand, is different. These very heavy particles, with masses of at least in the GeV region, are non-relativistic at an early stage, and therefore provide very early seeds for mass concentrations. A medium stage is provided by warm dark matter, which would correspond roughly to keV-mass particles, e.g. neutrinos. The advantage of using cold dark matter is that their fluctuations can grow already at the time of the matter dominance, while the baryonic fluctuations only grow from the time of decoupling. The baryons can then already see a gravitational potential produced by the dark matter. As hot dark matter leads to an effective washing-out of small scales, it stimulates the formation of very massive large structures, such as those in the previously discussed pancake model. On the other hand, cold dark matter produces a gravitational potential, into which structures can fall directly after recombination. As here the Jeans mass is of the order of globular star clusters and small galaxies, such structures can now form initially. Large structures can subsequently evolve through the gravitational interaction. Such models are called *bottom-up models*. In such models the possibility of a first generation of stars presents itself (so-called population III stars), which provide for an enrichment of the universe with heavy elements. Latest experimental signs of the existence of such objects come from the Keck telescope on Hawaii, which has detected carbon in a cloud assumed to be primordial [Sky95]. However, carbon is mainly produced inside stars (see chapter 12), and not in the primordial nucleosynthesis (see chapter 4).

The initial spectrum of fluctuations in the presence of dark matter is shown in figure 6.13. With the help of computer simulations of N-body interactions the behaviour of these models can now be investigated (see e.g. [Boe88]). Both models have their advantages and disadvantages. In computer simulations it does not seem to be possible to produce, via cold dark matter, the large scale structures of the universe, such as superclusters, quickly enough, without getting into conflict with the age of the universe. On the other hand, in hot dark matter models galaxy formation does not proceed quickly enough. Experimental limits for the spectrum of fluctuations come mainly from the anisotropy measurements of cosmic 3 K background radiation via the COBE satellite (see chapter 7) and galaxy surveys like the 1.2 J y IRAS survey [Fis92] (see figure 6.13). The observations can be best explained by assuming a mixture of 70% cold dark matter and 30% hot dark matter [Dav92a, Tay92]; however, alternative explanations such as a non-vanishing cosmological constant or a threshold factor (bias parameter) between fluctuations in normal and in dark matter are possible. The determination of the power spectrum is also being attempted from other large scale surveys, such as the Las Campanas survey [Lin96].

6.9   The initial spectrum of density fluctuations

The general picture of structure formation rests upon the existence of initial regions of higher density, which after the recombination era can concentrate through gravity and thereby form the starting points for the formation of structure. Two appropriate scenarios for the *beginning* of the structure formation are discussed in more detail. These are density fluctuations within the baryons,

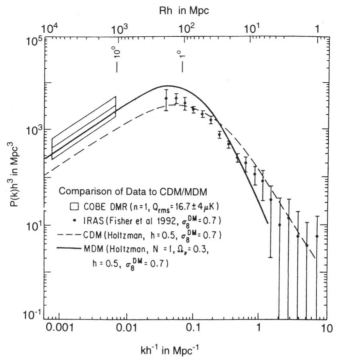

Figure 6.13. The power spectrum of the density fluctuations as a function of the wavenumber k or R, respectively. The experimental limits from the COBE results, as well as from the data obtained from the IRAS survey of galaxies are included. These two sets of data already pin down relatively well the shape of the spectrum. Also shown are two model predictions, one consisting only of cold dark matter (dotted line), which is not in agreement with the COBE data, and a model with a mixture of 70% cold and 30% hot dark matter (full line), which seems to describe both sets of data satisfactorily (from [Pec93]).

which have a Gaussian distribution, and cosmic strings, with an isothermal spectrum. How do these fluctuations occur, and what is the initial distribution? Returning briefly to the 2-point correlation function, it can be shown from equation (6.38) that

$$\xi(r) = \frac{V}{(2\pi)^3} \int |\delta_k|^2 e^{-ik \cdot r} d^3k. \qquad (6.44)$$

The quantity $|\delta_k|^2$ is known as the *power spectrum*. The power spectrum is therefore the Fourier transform of the 2-point correlation function and vice versa. If we assume an isotropic correlation function, integration over the angle coordinates gives

$$\xi(r) = \frac{V}{(2\pi)^3} \int |\delta_k|^2 \frac{\sin kr}{kr} 4\pi k^2 dk. \tag{6.45}$$

The aim is to theoretically predict this power spectrum, in order to describe the experimentally determined correlation function. The observation suggests that it is a wide spectrum without favoured scales, and we therefore assume a power law solution

$$|\delta_k|^2 \sim k^n. \tag{6.46}$$

For $kr > 1$ the integration can easily be carried out, leading to

$$\xi(r) \sim r^{-(n+3)} \sim M^{-\frac{(n+3)}{3}}. \tag{6.47}$$

The last relation follows from the fact that the mass of a fluctuation is proportional to r^3. As can easily be seen, all $n > -3$ imply a decrease of fluctuations to larger mass scales, which results in a homogeneous and isotropic universe for very large scales. A pure noise spectrum obeys a Poisson distribution and $n = 0$, leading to $\xi(r) \sim M^{-1}$. An interesting case results for $n = 1$. The spectrum has the characteristic that the mean square density contrast of fluctuations outside the horizon is equal for all scales. Such a scale invariant spectrum is, however, exactly what is predicted by inflationary models. The primordial fluctuations, presumed to be quantum fluctuations, developed either during the inflationary phase by the scalar field ϕ, or even earlier, more close to the Planck time. Equation (6.38) describes this. For Gaussian-like, and therefore uncorrelated, fluctuations, there is a connection between the mean square mass fluctuation and the power spectrum $|\delta_k|^2$, which contains all the information about the fluctuation [Kol90]

$$\langle \delta^2 \rangle_\lambda \simeq V^{-1} \left(k^3 |\delta_k|^2 / 2\pi^2 \right)_{k \approx 2\pi/\lambda}. \tag{6.48}$$

The spectrum of the density fluctuations is most easily discussed at the time when the fluctuations cross the horizon, i.e. when the wavlength of the fluctuation corresponds to the Hubble length, $\lambda \approx H^{-1}$. In standard cosmology without inflation all perturbations start at scales outside the horizon, i.e. $\lambda > H^{-1}$. Perturbations with scales of $\lambda \leq 13(\Omega_0 h^2)^{-1}$ Mpc enter during the radiation-dominated era into the horizon, while larger scales in the matter-dominated phase. Inflation now changes this picture. Instead of a single crossing of the horizon the behaviour now becomes more complex. All cosmologically interesting scales begin within the horizon, but are transported by inflation across the horizon. Roughly speaking, the quantum fluctuations form within the horizon, and after crossing the horizon they are effectively frozen, and return later as density perturbations. Larger scales which are the first to cross the horizon are also the last to re-enter it (see figure 6.14). For a detailed description see [Kol90]. The quantum fluctuations form during inflation due to quantum fluctuations of the scalar field which causes the inflation, e.g. the Higgs field. For a massless scalar field (a good approximation to the flat part of the Higgs potential, see chapter 3) in a de Sitter universe, the spectrum of the fluctuations is given by [Kol90]

$$(\Delta\Phi)_k^2 = V^{-1}k^3|\delta\phi_k|^2/2\pi^2 = \left(\frac{H}{2\pi}\right)^2 \tag{6.49}$$

where $\delta\phi_k$ represents the Fourier components of the scalar field. The evolution in the phase outside the horizon can be described by $\delta\phi_k = $ constant. As H only changes very slowly during the inflation, and all cosmological scales cross the horizon quite early, inflation predicts an almost scale invariant spectrum. Each fluctuation has, at the time of crossing, about the same physical size, approximately the inverse of the Hubble parameter, and the universe has the same expansion rate. This means that all scales have about the same amplitude. Such a spectrum is called a *Harrison–Zeldovich spectrum* [Har70, Zel72]. As furthermore the quantum fluctuations are uncorrelated, a Gaussian-like distribution is expected.

There are two kinds of initial fluctuation spectrum, namely the adiabatic fluctuations and those with constant curvature, which in the case $\rho_y \gg \rho_B$ are equivalent to isothermal fluctuations. Adiabatic fluctuations are characterized by

$$\frac{\delta T}{T} \approx \frac{1}{3}\frac{\delta\rho}{\rho}. \tag{6.50}$$

Fluctuations of constant curvature mean to a lesser extent variations in the energy density, but rather fluctuations in the local equation of state. They are characterized by $\delta\rho = 0$. This could be a spatial variation in the distribution of the different types of particles, as the local pressure is not only dependent on density, but also on composition. Both kinds of fluctuations produce in principle the same evolution, but they can be distinguished by their behaviour outside the horizon. In contrast to the situation for adiabatic fluctuations, isothermal fluctuations do not increase there.

We now consider the spectral form of the density fluctuations. It is best to define the form of the fluctuations at the time at which they cross the horizon. We consider fluctuations at different times, but due to the growth of the horizon the spectra can be recalculated easily for simultaneity. In the absence of an exact model of the original fluctuations, a general ansatz is made [Kol90]:

$$\left(\frac{\delta\rho}{\rho}\right)_H = \frac{k^{\frac{3}{2}}|\delta_k|}{\sqrt{2\pi}} \sim AM^{-\alpha} \tag{6.51}$$

where $\alpha = \frac{1}{2} + \frac{n}{6}$. For a Harrison–Zeldovich spectrum in an inflationary universe the condition $n = -1$ results. Restrictions on α result from the fact that galaxy formation suggests that

$$\left(\frac{\delta\rho}{\rho}\right)_H \approx 10^{-4\pm1} \tag{6.52}$$

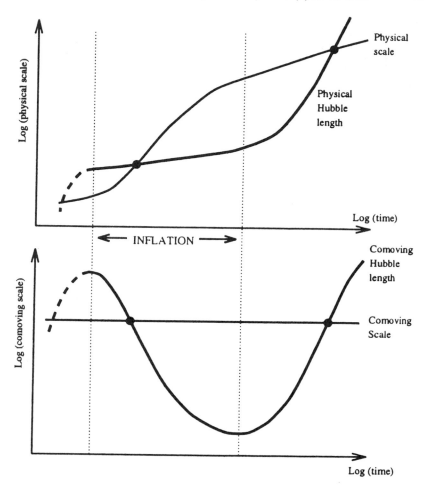

Figure 6.14. The behaviour of a given co-moving scale relative to the Hubble length, both during and after inflation, shown using (*a*) physical co-ordinates and (*b*) co-moving co-ordinates. The Hubble length H^{-1}, measured in co-moving co-ordinates, decreases during inflation. Because of the exponential expansion all physical scales are very rapidly increasing, the characteristic scale of the universe H^{-1} is becoming smaller, when measured relative to that expansion (from [Lid96]).

on scales of $M = 10^{12} M_\odot$ [Kol90]. Since

$$\left(\frac{\delta\rho}{\rho}\right)_H \leq 10^{-4} \tag{6.53}$$

follows from the isotropy of the cosmic microwave background radiation on present horizon scales of about $10^{22} M_\odot$ [Kol90], α should be < -0.1. On the other hand non-observations of primordial black holes imply a limit of $\alpha < 0.2$. This results from the fact that primordial black holes with masses above 10^{15} g evaporate via the Hawking process (particle–anti-particle production at the Schwarzschild radius) [Haw74, Pag76]. The above limit is produced by the requirement that the γ-radiation produced by such processes is compatible with experimentally observed values. Alternatively, the observed correlation function corresponds to an n of -1.2. This assumes, however, that $\xi(r)$ really reflects the initial fluctuation spectrum. If all the spectra are now corrected to a particular point in time, such as the time of recombination, a density spectrum as illustrated in figure 6.13 results. The determination of the spectrum from observations by COBE and the IRAS survey already provide the possibility to distinguish between different theoretical models. This again allows coclusions about the composition of dark matter (see chapter 9). In figure 6.13 two theoretical models (cold dark matter alone, and a mixture of hot and cold dark matter) are shown; it seems that mixed models can consistently describe both data sets [Pec93].

For an extensive and detailed description of the connections between density fluctuations and their spectra we refer to [Pee80, Lon89, Kol90]. In addition to these more or less conventional explanations there exists a completely different way to explain structure formation, namely via topological defects as condensation seeds for matter. These are discussed in the next section.

6.10   Cosmic strings

The GUT phase transition mentioned in chapter 3 is not perfect, any more than for instance the transition from water to ice. Disruptions can occur, which are called *topological defects*. During phase transitions the false vacuum is retained in the defects. The zero-dimensional defects, corresponding to vacancies in crystals, correspond to *magnetic monopoles*. These are discussed in a chapter in their own right (see chapter 10). One-dimensional defects, analogous to dislocations in crystals, are called *cosmic strings*, two-dimensional defects are called *domain walls*, and three-dimensional defects, which cannot be readily conceptualized, are called *textures*. Here we restrict ourselves to one-dimensional defects. For the cosmological production of topological effects see e.g. [Pre84, Vil85, Vil87, All90, Hin94]. We consider Abelian Higgs models with a spontaneously broken U(1)-symmetry. The Lagrange density with a gauge field A_μ and a complex Higgs field Φ is given by [Kol90]

$$\mathcal{L} = D_\mu \Phi D^\mu \Phi^\dagger - \frac{1}{4} F_{\mu\nu} F^{\mu\nu} - \lambda \left(\Phi^\dagger \Phi - \frac{\sigma^2}{2} \right)^2. \qquad (6.54)$$

In the broken phase the minimum of the field lies at the vacuum expectation value $\langle \Phi \rangle = \sigma e^{i\alpha}$ with $0 \le \alpha \le 2\pi$. Consider a closed loop, on whose edge the field is assumed to be everywhere at the minimum. From the fact that the solution must be single valued it follows that $\Delta\alpha = 2\pi n$. This does not change even if the surface is allowed to tend to zero. The uniqueness of the solutions now leads to a contradiction for $n \ne 0$, except there exists a point at which the phase is undefined. This is equivalent to a value of higher energy. Mathematically speaking this means that the surface cannot be simply connected, which in group theoretical language is expressed as

$$\Pi_1(M) \ne 1 \qquad (6.55)$$

where Π_1 represents the first homotopic group, applied to the manifold M of the vacuum states. If we now deform the loop, the same arguments continue to apply, and a chain of such points results which correspond precisely to our cosmic string. The requirement to maintain single valuedness implies that strings are either infinitely long, or exist in the form of loops. The formation of the strings is connected to a phase transition. Phase transitions are characterized by a critical temperature and a coherence length ξ. For spatial distances larger than ξ there is no connection between two points, and the phase α is arbitrary. Therefore during the phase transition a network of strings with a typical repetition-distance of ξ develops. Such a network of strings is shown schematically in figure 6.15. For the calculation of these networks see [Vil85, Alb89, All90]. From causality ξ has to be smaller than the Hubble radius. The evolution of the strings is non-trivial, so that for example its length is *not* given by

$$L(t) \sim R(t)L \Rightarrow \rho_{st} \sim R^{-2}. \qquad (6.56)$$

This is easily seen since, for the radiation dominated phase $\rho_r \sim R^{-4}$, so that

$$\frac{\rho_{st}}{\rho_r} \sim R^2 \sim T^{-2} \sim t. \qquad (6.57)$$

It can be seen that the contribution of strings to the energy density would already have been so dominant, that there would, in the case of a trivial evolution, be a contradiction with the age of the universe. In fact the strings lose energy through interactions with themselves or other strings. The loops emitted in this have a radius corresponding to the coherence length at the corresponding time. These oscillating loops now decay via emission of gravitational waves. The typical lifetime of a loop with radius r is given by [Vac85]

$$\tau(r) \sim (\gamma G \mu)^{-1} r. \qquad (6.58)$$

Here γ is a numerical factor of the order 100, G is the gravitational constant and μ is the mass-density per unit length. Values of $G\mu \approx 10^{-6}$ imply about 10^8 oscillations/lifetime.

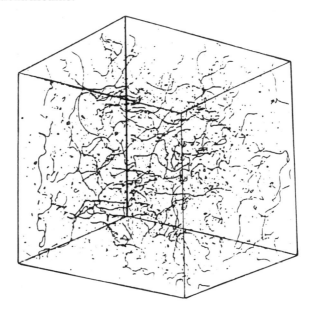

Figure 6.15. Numerical simulation of the production of large-scale spatial structures through cosmic strings. These act as condensation seeds and are either infinitely long or appear in the form of closed loops (after [Ben88, Pre89, All90, Kol90]).

Cosmic strings are extremely thin, with diameter of the order 10^{-30} cm. Due to the very high energy of the vacuum before the phase transition, their mass density per unit length, μ, is correspondingly very high. It corresponds to about 10^{22} g cm^{-1}. Using this fact it is possible to construct an alternative model of structure formation [Vac86]. The beauty of this theory is its relatively concrete predictions, and that contradictory experimental data would probably destroy the whole theory. Structures form by loops acting as condensation seeds for the matter accumulations after the transition from the radiation to the matter dominated universe. More precise estimates of how large masses are accreted, rely on only one free parameter μ. As discussed previously, loops of all sizes exist down to a lower cut-off radius r_l, given by [Kol90]

$$r_l = \gamma G \mu t_{eq}. \tag{6.59}$$

Large loops serve as seeds for structures from which superclusters subsequently develop, while smaller loops are responsible for galaxy formation. Objects on all mass scales between galaxies and superclusters should therefore occur. If one is only interested in the prediction of spatial connections, such as the correlation function, the dependence on μ also disappears. This predicts a universal form of the correlation function, which so far has been experimentally confirmed [Tur86].

Is there a way to prove that cosmic strings exist? It can be shown that they can cause fluctuations in the 3 K radiation of the order [Ste88a]

$$\frac{\delta T}{T} \simeq 10 G \mu v. \tag{6.60}$$

From the measured isotropy (see chapter 7) it follows therefore that $G\mu < 10^{-5}$. In scenarios with cosmic strings no secondary Doppler peaks appear in the multipole expansion of the 3K background radiation [Mag96] (see chapter 7 and in particular figure 7.4).

Due to their strong gravitational wave emission, strings would have a perturbing influence on the time behaviour of millisecond pulsars [Alb89]. The fact that this is not observed experimentally implies that $G\mu < 10^{-6}$. The most likely evidence, however, should come from the gravitational lens effect (see chapter 9). Due to its high mass density a string should lead to the splitting of the images of far away objects, such as quasars.

As exotic as these structures and ideas may sound, they are after all only the consequences of an application of the theories of elementary particles to cosmological questions.

Chapter 7

Cosmic background radiation

7.1 The 3 K background radiation

We now discuss one of the most important supports for the big bang theory, namely the cosmic microwave background. Gamov, Alpher and Herman already predicted in the 1940s that if the big bang model were correct, a remnant noise at a temperature of about 5 K should still be present [Gam46, Alp48]. Already in 1941 during the observation of interstellar CN-gas, Fraunhofer lines were found in the direction of the star ξ Ophiuchi, which indicates that rotational levels were being excited. However, no cosmological source was considered as an excitation mechanism, and the observation remained unsolved (see section 7.1.2). For further literature concerning the 3K radiation we refer to [Par95].

7.1.1 Spectrum and temperature

During the radiation dominated era, radiation and matter were in a state of thermodynamic equilibrium. Thompson scattering, particularly from free electrons, resulted in an opaque universe. As the temperature continued to fall, it became possible for more and more of the nucleons and electrons to recombine to form hydrogen. As most of the electrons were now bound, the mean free path of photons became much larger (of the order c/H), and they decoupled from matter. As the photons were in a state of thermodynamic equilibrium at the time of decoupling, their intensity distribution $I(\nu)\mathrm{d}\nu$ should correspond to a black-body spectrum

$$I(\nu)\mathrm{d}\nu = \frac{2h\nu^3}{c^2} \frac{1}{\exp\left(\frac{h\nu}{kT}\right) - 1}\mathrm{d}\nu. \tag{7.1}$$

It is easy to show that the black-body form in a homogeneous Friedmann universe remains unchanged despite expansion. Since in an adiabatic expansion the term $T_\gamma (R^3)^{\kappa-1}$ remains constant ($\kappa = 4/3$ for photons), it follows that $T_\gamma R$ also remains constant. Inserting the relations $T_\gamma = T_{\gamma 0}(1 + z)$ and $h\nu = h\nu_0(1 + z)$, the preservation of the black-body form follows immediately. The maximum of this distribution lies, according to Wien's law, at a wavelength of

$$\lambda_{max} T = 2.897 \times 10^{-3}\,\text{mK} \tag{7.2}$$

which for 5 K radiation lies at about 1.5 mm. Indeed, in 1964 Penzias and Wilson of the Bell laboratories discovered an isotropic radiation at 7.35 cm, with a temperature of (3.5 ± 1) K [Pen65]. The energy density of the radiation is found by integrating over the spectrum (Stefan–Boltzmann law)

$$\rho_\gamma = \frac{\pi^2 k^4}{15 h^3 c^3} T_\gamma^4 = a T_\gamma^4. \tag{7.3}$$

From equation (3.48) we obtain the following relationship

$$n_\gamma = \frac{30\zeta(3)a}{\pi^4 k} T_\gamma^3 \approx 20.3 T_\gamma^3\,\text{cm}^{-3} \tag{7.4}$$

for the number density of photons. Here $\zeta(3)$ is the Riemann ζ function of 3, which is $1.202\,06\ldots$.

We now consider this picture in a more quantitative way. As mentioned in chapter 3, a particle species decouples from equilibrium when the most important interaction rate becomes smaller than H, or equivalently, when the mean free path $\lambda \simeq \Gamma^{-1}$ becomes larger than the Hubble radius. For photons the most important interaction is Thompson scattering, and the rate is given by

$$\Gamma_\gamma = n_e \sigma_T. \tag{7.5}$$

Here n_e is the electron density, and $\sigma_T = 6.65 \times 10^{-25}\,\text{cm}^2$ is the cross section for Thompson scattering. The essential quantity of the recombination era is now the development of the free electron density. The number density of any particle participating in a reaction in thermodynamic equilibrium is given by

$$n_i = g_i \left(\frac{m_i T}{2\pi}\right)^{\frac{3}{2}} \exp\left(\frac{\mu_i - m_i}{kT}\right) \qquad i = e, p, H, B. \tag{7.6}$$

Here n_e, n_H, n_B are the electron, hydrogen and baryon densities, μ_i are the corresponding chemical potentials and g_i the statistical weights. The charge neutrality of the universe requires $n_e = n_p$, and the conservation of baryon number implies $n_B = n_p + n_H$. The reaction

$$p + e \rightarrow H + \gamma \tag{7.7}$$

ensures that $\mu_e + \mu_p = \mu_H$. Hence it immediately follows that

$$n_H = \frac{g_H}{g_p g_e} n_p n_e \left(\frac{m_e kT}{2\pi \hbar^2}\right)^{-3/2} \exp\left(-\frac{E_B}{kT}\right) \tag{7.8}$$

where E_B is the binding energy of hydrogen ($E_B = 13.59$ eV). This is the Saha equation. With $x = n_e/n_B$ it can be rewritten as

$$\frac{x^2}{1-x} = \frac{1}{n_B} \left(\frac{m_e kT}{2\pi\hbar^2} \right)^{3/2} \exp\left(-\frac{E_B}{kT} \right). \tag{7.9}$$

From this we can determine the ionization fraction $x(T)$ for any temperature. This quantity still depends, however, on n_B. For realistic assumptions on n_B we obtain a rapid decline for x from 1 to 0 at T between about 2500 and 5000 K, corresponding to red-shifts of $z = 1000$–1500 (figure 7.1). Particularly in this region the ionization fraction shows a strong dependence on the red-shift. A detailed analysis [Jon85] results in

$$x = 2.4 \times 10^{-3} \frac{(\Omega h^2)^{1/2}}{\Omega_B h^2} \left(\frac{z}{1000} \right)^{12.75} \tag{7.10}$$

where Ω represents the density (normalized to the critical density) and Ω_B the baryonic density (see chapter 3). As the surface of last scattering is determined as the region in which the optical depth τ is of order 1, it follows that [Lon89]

$$\tau = 0.37 \left(\frac{z}{1000} \right)^{14.25} \Rightarrow z \approx 1070. \tag{7.11}$$

The probability distribution of the last scattering events

$$\frac{dp}{d\tau} = e^{-\tau} \frac{d\tau}{dz} \tag{7.12}$$

is approximately Gaussian with a mean of $z = 1070$, and a standard deviation of 80. This means that roughly half of the last scattering events took place at red-shifts of between 990 to 1150. This interval in the red-shift today corresponds to a length scale of $\lambda \simeq 7(\Omega h^2)^{1/2}$ Mpc, and an angle of $\theta \simeq 4\Omega^{1/2}$ [arc minutes]. Structures on smaller angular scales are smeared out.

Besides the recombination, almost simultaneously, the transition from a radiation- to a matter-dominated universe took place. Due to the different dependences of the energy densities of the scalar factor R, equality between matter and radiation density is obtained at about $z = 1500$ (see chapter 3). It is therefore usually assumed that the recombination already took place in a matter dominated universe. After recombination the radiation decoupled from matter. The decoupling of matter and radiation did not happen abruptly, but Thompson and Compton scattering did ensure equality of the temperatures up to about $z = 100$. This picture can also be modified if a significant ionization of gases happened at some later time. The occurrence of these three events can be approximated as follows [Kol90]

$$1 + z_{eq} = 2.32 \times 10^4 \Omega_0 h^2 \quad \approx 1500 \quad \text{Equality of the densities} \tag{7.13}$$

$$1 + z_{rec} = 1380(\Omega_b h^2)^{0.023} \quad \approx 1240\text{–}1380 \quad \text{Recombination} \tag{7.14}$$

$$1 + z_{dec} = 1100(\Omega_0/\Omega_b)^{0.018} \approx 1100\text{–}1200 \quad \text{Decoupling.} \tag{7.15}$$

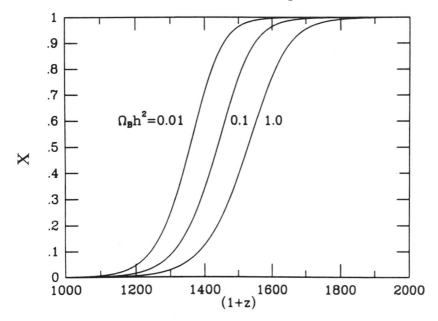

Figure 7.1. Ionization fraction x of hydrogen as a function of the red-shift. It can be interpreted as a measure of the optical opaqueness at that particular time. Coming from high red-shifts, a steep decline can be seen in the region around $z = 1400$, connected with an increasing transparency of the universe (from [Kol90]).

Any alternative supply of energy can now produce a deviation of the observed spectrum from the Planck form. Effects at $z > 10^7$ can be excluded immediately, since there is sufficient time to thermalize them completely. Within a time span of $10^7 > z > 10^4$ thermalization also still takes place, however the Planck form is modified by a chemical potential μ, i.e. $h\nu/kT \rightarrow (\mu + h\nu)/kT$. This only has an effect on the Rayleigh–Jeans region (low frequencies), where the temperature change is given by [Mel90]

$$\frac{\Delta T}{T} = \frac{\mu}{h\nu/kT}. \tag{7.16}$$

The chemical potential here is directly correlated to the energy injection. The measured maximum deviation of 0.03% from the black-body spectrum [Mat94] implies that a possible energy supply has to contribute less than 1% of the mean radiation energy. Recent measurements give an upper limit of $|\mu| < 9 \times 10^{-5}$ for the chemical potential [Fix96]. Energy supplies in the area of $z < 10^4$, on the other hand, are clearly noticeable in the spectrum. According to Sunyaev and Zeldovich [Sun80] the change can be described by a multiplication of the

Planck spectrum with the factor

$$1 + Yx\frac{e^x}{e^x - 1}\left(\frac{x}{\tanh(x/2)} - 4\right) \tag{7.17}$$

where $x = h\nu/kT$. Two extreme cases can result:

$$\frac{\Delta T}{T} = -2Y \quad \text{(Rayleigh–Jeans region)} \tag{7.18}$$

and

$$\frac{\Delta T}{T} = 5.4Y \quad \text{(Wien region)}. \tag{7.19}$$

The physical interpretation of Y depends on the unknown nature of the energy injection [Mel90]. In the simplest case Y measures the cross section for Compton scattering with the electrons of a hot gas, and depends on its density and temperature. This effect can also lead to perturbations in the post recombination time. The latest observations in the X-ray region have shown that many galaxies contain a great deal of hot ($T_e \approx 10^8$ K) gas. In these the Sunyaev–Zeldovich effect could take place. It manifests itself in temperature changes of about 1 per mille. This effect has already been observed in the direction of several galactic clusters [Bir84]. A very interesting side effect is that it could be possible to determine the Hubble constant from this method. The effect can be described by the following expression [Boe88]

$$\frac{\Delta T}{T} = -\frac{4kT_e}{m_e c^2}\sigma_T n_e R \tag{7.20}$$

where R is the radius of the gas cloud. Values for n_e and T_e can be extracted from the observed X-ray emission, and hence it is possible to extract R using equation (7.20). If the absolute radius R is known, it is possible, with the help of the observed angular diameter, to determine the Hubble constant directly without including all the uncertainties in the distance scales. The data from the FIRAS detector on COBE restrict the perturbations to $|Y| < 1.5 \times 10^{-5}$ [Fix96]. For a compilation of observations of the Sunyaev–Zeldovich effect see [Rep95].

A further source of perturbations in the post recombination time could be population III stars. These are assumed to be a kind of very massive stars of the first generation, with relatively short lifetimes, which so far have not yet been observed. The radiation of this very first class of stars could be thermalized by dust and could lead to significant fluctuations in the background.

In the next section we address the experimental question of whether the background radiation is really a type of black-body radiation, and what is its precise temperature [Par88].

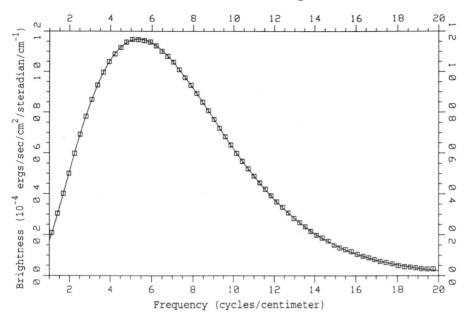

Figure 7.2. Spectrum of the cosmic background radiation, measured with the FIRAS detector on the COBE satellite. It shows a perfect black-body behaviour. The smooth curve is the best fit black-body spectrum (from [Mat90a]).

7.1.2   Measurement of the spectral form and temperature of the 3K radiation

Due to the importance of the microwave background radiation a great number of measurements have been made at the most varied wavelengths (see e.g. [Mel90, Par95]). The experimental difficulties in Earth-bound observations result mainly from the existence of the atmosphere. Firstly it produces temperature fluctuations, and secondly the gas molecules emit and absorb very strongly in the interesting area. Also, our Milky Way dominates the background above a certain frequency. Therefore, many observations have been performed at heights of 30–40 km using balloons. The satellite COBE (cosmic background explorer), built specifically to investigate the background radiation, was launched in 1989 [Smo90]. It surveyed the entire sky in different wavelengths. In previous measurements only a few wavelengths had ever been measured, and these were different in every experiment. COBE measured the region in wavenumbers k from 1–100 cm^{-1} very precisely. The FIRAS detector (far infra-red absolute spectrophotometer) of the COBE satellite is used to measure the Planck shape and any possible deviations. Using one antenna it measures the background radiation, and with another it measures the temperature of an internal reference radiator. Both are then compared with the help of a Michelson interferometer.

Furthermore it is possible to insert an external reference source into the actual beam, and thereby carry out an exact calibration. The measured spectrum shows a perfect black-body form at a temperature of (2.728 ± 0.004) K [Wri94a, Fix96] (figure 7.2). No deviations whatsoever are seen in the spectral form. In figure 7.3 the results from the first COBE measurement are shown together with other results. A submillimetre excess found in 1988 in a balloon flight could be excluded using these results. This illustrates the difficulties of Earth-bound observations and their corrections. The number density of photons can be determined from the COBE data as

$$n_\gamma = (412 \pm 2)\, \text{cm}^{-3} \tag{7.21}$$

from which it is possible to reduce the error in the number density of the former from about 20% to less than 0.5%. The number density is particularly interesting for the photon–baryon ratio (see chapter 4), whose main source of error now rests upon our lack of knowledge of the baryonic matter density in the universe.

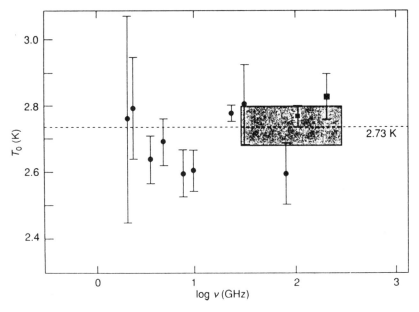

Figure 7.3. Comparison of the COBE measurements, shaded area, (the error bars are meanwhile smaller by a factor of 2.5) with other temperature measurements of the cosmic background radiation at different wavelengths (from [Par95]).

A further method for measuring the cosmic background radiation exists. This is based on the excitation of various rotational levels of interstellar CN molecules. From the occupation number of the different energy levels the temperature at 1.32 (0–2 transition) and 2.64 mm (0–1 transition) can be determined. The temperature values of 2.73 K deduced from this are in agreement with the COBE data [Cra86, Rot93]. Indeed this was the excitation which had been observed as early as 1941 (see section 7.1), but which had not been interpreted as having a cosmological origin. The molecular lines allow the $(1+z)$ dependence of the temperature to be tested (see chapter 3). Measurements at a red-shift of $z = 2.9092$ result in a temperature of the background radiation of less than 13.5 K. The value expected from COBE is 10.66 K [Son94a]. Another measurement at a red-shift of $z = 1.776$ resulted in an upper limit for the temperature of 16 K [Mey86]. A later measurement gave the value of 7.4 ± 0.8 K compared to an expectation of 7.58 K [Son94b]. This can also be expressed in the following way: should the temperature of the background radiation have a dependence on the red-shift in the form of $(1 + z)^{\alpha}$, the measurement at $z = 2.9$ implies $\alpha < 1.15$ (90% confidence level), and from the measurement at $z = 1.77$, $\alpha < 1.73$. The new measured value at $z = 1.77$ corresponds to $\alpha = 0.98 \pm 0.11$.

In addition to the spectral form and its distortions, the homogeneity and isotropy are also of extraordinary interest, as they allow conclusions as to the expansion of the universe and are an extremely important boundary condition for all models of structure formation (see chapter 6). In the next section we examine these anisotropies.

7.1.3   Anisotropies in the 3 K radiation

Anisotropies in the cosmic background radiation are of extraordinary interest, on one hand for our ideas about the formation of large scale structures and of galaxies in the universe, and on the other hand for our picture of the early universe. The former reveals itself through anisotropies on small angular scales (arc minutes up to a few degrees), while the latter is noticeable on larger scales (up to 180 degrees) (see below). We therefore consider those two possibilities separately [Whi94]. For an overview see also [Rea92, Hu95].

7.1.3.1   Measurement of the anisotropy

A measurement of the background radiation in a certain direction n results in

$$T(n) = T_0(1 + \Delta T(n)). \tag{7.22}$$

Here T_0 is the mean background temperature and $\Delta T(n)$ describes a possible deviation in the direction n. The data are now treated statistically with the help of a correlation function $C(\theta)$, given by

$$C(\theta) = \langle \Delta T(n) \Delta T(n') \rangle \tag{7.23}$$

with $n \cdot n' = \cos \theta$ (figure 7.4). The spherically symmetrical radiation field can be expanded in terms of the spherical harmonic functions Y_l^m and similarly the

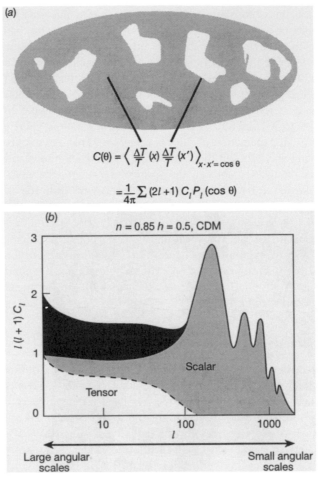

Figure 7.4. (*a*) The temperature autocorrelation function, $C(\theta)$, is obtained from a map of the sky (here represented by the oval) displaying the difference in the microwave background temperature from the average value, $\Delta T/T$. $C(\theta)$ is computed by taking the map average of the product $\Delta T/T$ measured from any two points in the sky separated by angle θ. If $C(\theta)$ is expanded as a sum of $(4\pi)^{-1}(2l+1)P_l(\cos\theta)$, where $P_l(x)$ are the Legendre polynomials, the co-efficients C_l are the *multipole moments* (from [Ste96a]). (*b*) A plot of $l(l+1)C_l$ versus the multipole moment number l is the cosmic microwave background power spectrum. For a given l, C_l is dominated by fluctuations on an angular scale $\theta \sim \pi/l$. In inflation, the power spectrum is the sum of two independent, scalar and tensor contributions (from [Ste96a]).

correlation function can be expressed in Legendre polynomials $P_l(x)$

$$T(n) = T_0 \sum_{l,m} a_l^m Y_l^m \qquad (7.24)$$

$$C_l(\theta) = \frac{1}{4\pi} \sum_l (2l+1) c_l P_l(\cos\theta). \qquad (7.25)$$

For a Gaussian distribution the following holds

$$(2l+1)c_l = \sum_{-l}^{+l} |a_l^m|^2. \qquad (7.26)$$

The monopole term $(l = 0)$ clearly drops out, and in addition the dipole term $(l = 1)$ has a special meaning (see next section), so that the analysis of anisotropies in the background radiation starts only with the quadrupole term $(l = 2)$. If we construct the rotationally invariant quantity

$$a_l^2 = \sum_m |a_l^m|^2 \qquad (7.27)$$

the quadrupole anisotropy can be defined as

$$Q = \left(\frac{a_2^2}{4\pi} \right)^{1/2}. \qquad (7.28)$$

A further anisotropy follows from the movement of the last scattering surface and the associated Doppler effect. This results mainly from thermo-acoustic oscillations of baryons and photons [Hu95, Smo95, Teg95]. This produces a series of peaks in the power spectrum which are predicted by most models of inflation. The typical order of magnitude lies between $100 < l < 1500$. The position and shape of the peaks are determined by several cosmological parameters, which can therefore by determined from the observations. For example, the position of the first peak is determined by $l \propto \Omega_0^{-1}$, the height of the peaks is dependent upon Ω_b and the Hubble constant h influences both the height and position of the peaks. In theories in which the structure formation is determined by cosmic strings (see chapter 6) no secondary Doppler peaks occur. The first Doppler peak is expected around $l = 200$ (see figure 7.4 and 7.5). It will be one of the main goals of the next generation satellite missions like MAP and PLANCK [Smo97b] to determine these cosmological quantities. The resolution function of the detector plays an important role in the corresponding measurements.

7.1.3.2   The dipole anisotropy

Due to its isotropy the background radiation is a preferred frame of reference for any co-moving observer. It is, however, not an absolute reference frame

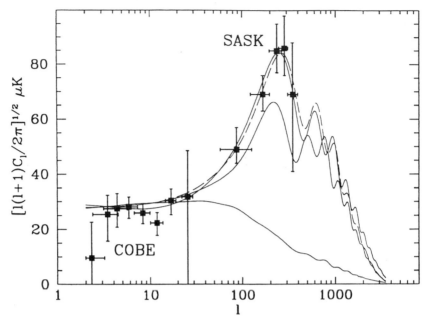

Figure 7.5. The anisotropy power spectrum C_l as a function of the multipole order l according to four variants of the standard model with cold dark matter of [Sug95] (see chapter 9), all with $n = 1$ and $\Omega_b = 0.05$, in comparison with data from COBE and the Saskatoon experiment [Nut96]. The first Doppler peak is clearly visible (from [Teg96]).

in a general relativistic sense, as it is in no way singled out by Einstein's field equations (see chapter 3). As already mentioned the dipole anisotropy plays a special role. It is the dominant anisotropy on large scales, and is interpreted as the Doppler effect, caused by our relative movement through the microwave background [Lin96b]. The motion is caused by the gravitational interaction. This means that in the case when all masses which contribute to the velocity of the local group are observed, the dipole deduced from this mass distribution should equal that observed in the 3 K radiation. The dipole anisotropy is given by

$$T(\theta) = T_0 \left(1 - \frac{v^2}{c^2}\right)^{1/2} \left(1 - \frac{v}{c}\cos\theta\right)^{-1} \simeq T_0 \left(1 + \frac{v}{c}\cos\theta + O\left(\frac{v^2}{c^2}\right)\right).$$
(7.29)

This dipole anisotropy is illustrated in figure 7.6. The value measured with COBE of $\Delta T/T = (3.372 \pm 0.007) \times 10^{-3}$ [Fix96, Kog96] implies a velocity of the Sun, compared to the background radiation, of (370 ± 1) km s^{-1}. If the movement of the Sun within the Milky Way and the movement of the Milky Way within the local group are added to this, a velocity of the latter can be derived. The result is a value of (627 ± 22) km s^{-1} in the direction of the galactic co-

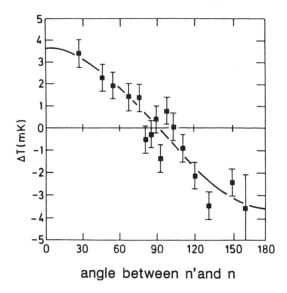

Figure 7.6. Temperature difference ΔT as a function of the angle between two observation directions n and n'. The dipole anisotropy of the cosmic background radiation is clearly visible. It is caused by the Doppler movement of the Earth relative to the frame of reference of the 3 K radiation (from [Boe88]).

ordinates $l = 264.3° \pm 0.3°$ and $b = 48.2° \pm 0.1°$ [Fix96]. This value does not as yet agree very well with the value for the local group, extracted from large scale movements (see chapter 6). This implies that as yet not all the important mass concentrations have been identified. The more distant IRAS galaxies also show a dipole effect, whose direction with respect to the microwave background can be derived as between $10°$ and $30°$ depending on the model [Str93].

7.1.3.3 Anisotropies on small scales

Anisotropies have to be divided into two types, depending on the horizon size at the time of decoupling. Fluctuations outside the event horizon are independent of the microphysics present during decoupling and so reflect the primordial perturbation spectrum (see chapter 6), while the sub-horizontal fluctuations depend on the details of the physical conditions for the time of decoupling. The event horizon at the time of decoupling today corresponds to an angular size of [Kol90]

$$\Theta_{\text{dec}} = 0.87 \Omega_0^{1/2} \left(\frac{z_{\text{dec}}}{1100} \right)^{-1/2} \quad \text{[degrees]}. \tag{7.30}$$

Below about $1°$ therefore the fluctuations mirror those which show up in galaxy formation. Starting from the Jeans mass, there is a correlation between the mass scale and the corresponding characteristic angular size of the anisotropies (see e.g. [Nar83])

$$(\Delta\theta) \simeq 23 \left(\frac{M}{10^{11} M_\odot} \right) (h_0 q_0^2)^{1/3} \quad \text{[arcseconds]}. \tag{7.31}$$

Typical density fluctuations that led to the formation of galaxies therefore correspond today to anisotropies on scales of $20''$. Assuming that density fluctuations develop adiabatically, the temperature contrast in the background radiation should be given by:

$$\left(\frac{\delta T}{T} \right)_R = \frac{1}{3} \left(\frac{\delta\rho}{\rho} \right)_R \tag{7.32}$$

where the subscript R stands for 'at the recombination time'. In order to produce the density currently observed in galaxies, observable temperature anisotropies of $\delta T/T \approx 10^{-3}$–$10^{-4}$ would be expected. This is, however, not observed. The density fluctuations in the baryon sector must therefore be smaller than expected and additional terms are needed in the density fluctuations. A significant amount of dark matter in the form of WIMPs could, for example, produce such an effect (see chapter 9). In this case

$$\left(\frac{\delta\rho}{\rho} \right)_{\text{WIMP}} > \left(\frac{\delta\rho}{\rho} \right)_B \tag{7.33}$$

since WIMP fluctuations can grow as soon as the universe is matter dominated. At decoupling time one therefore expects a relation between the density fluctuations of baryons and WIMPs of [Kol90]

$$\left(\frac{\delta\rho}{\rho} \right)_B \approx 0.05 (\Omega_0 h^2)^{-1} \left(\frac{\delta\rho}{\rho} \right)_{\text{WIMP}}. \tag{7.34}$$

So, while the expected temperature anisotropy in a baryon-dominated universe is given by equation (7.32), we expect in a WIMP-dominated universe anisotropies of [Kol90]

$$\left(\frac{\delta T}{T} \right)_\theta = \frac{(\Omega_0 h^2)^{-1}}{60} \left(\frac{\delta\rho}{\rho} \right)_\lambda \tag{7.35}$$

where the co-moving wavelength λ corresponds to an angular scale θ on the sky of $\theta = 34.4'' \, (\Omega_0 h)(\lambda/\text{Mpc})$.

By including the WIMPs it is therefore possible to allow for significantly smaller density variations within the baryons, and therefore to make the temperature anisotropies compatible with the observed values. From these limits it follows that any baryon dominated model without re-ionization in recent time is now excluded, independent of the exact size of Ω_B. For measurements of temperature anisotropies at these small scales see [Whi94, Ben95, Smo97a, Smo97b]. Several additional future experiments will provide further restrictive data.

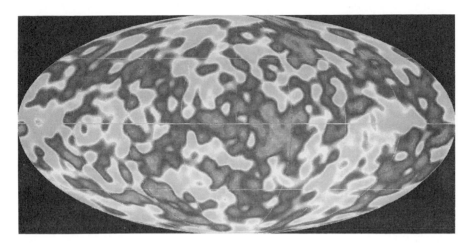

Figure 7.7. Sky map of the cosmic background radiation obtained with the COBE satellite. The anisotropy is reflected by the different brightness distributions (with kind permission of the COBE collaboration, see also [Kog96]).

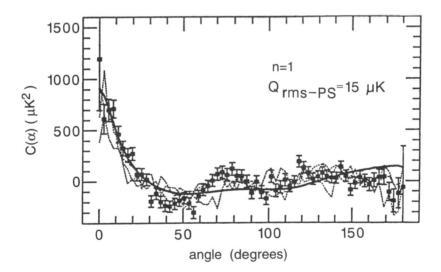

Figure 7.8. Correlation function $C(\alpha)$ of the COBE data, showing a clear structure. This is the first indication of anisotropies in the background radiation. The fluctuation spectrum ($n = 1$), corresponding to inflationary models, is also shown (solid curve). The quoted Q, corresponding to equation (7.28), results from a fit to the data (from [Ber93c], see also [Hin96a]).

7.1.3.4 *Anisotropies on large scales*

Another interesting possibility is that the universe was not initially homogeneous, but that it reached its homogeneity only later through energy and momentum transfer. In this case, the largest connected, homogeneous region is defined by the particle horizon (see chapter 3) at that particular time. Different values of the gravitational potential at the time of the recombination should therefore show up in the form of anisotropies at large scales (Sachs–Wolfe effect) [Sac67a]. Regions of high density appear as colder regions, as the photons experience an additional red shift, due to the gravitational potential, while the hotter areas reflect regions of lower density. To measure these anisotropies the COBE satellite has 6 radiometers, measuring in pairs at three different wavelengths (3.3 mm, 5.7 mm and 9.5 mm). As COBE has an angular resolution of 7°, evidence for anisotropies directly samples the initial fluctuation spectrum, as it is sensitive outside the event horizon. This apparatus, known as DMR (differential microwave radiometers), did indeed measure, after subtracting the Earth's motion and a correction for galactic emission, an anisotropy of the background radiation. In this historic measurement, anisotropies show up in all angular regions, from the lowest experimental resolution of 7°, up to 180° (see figure 7.7) [Smo92, Wri92, Ben95, Hin96, Kog96]. The entire measured correlation function (see figure 7.8) is dominated by the quadrupole term. The measured quadrupole anisotropy Q (see equation (7.28)) corresponds to a temperature of $(15.3^{+3.8}_{-2.8})\mu K$, corresponding to a anisotropy of

$$\frac{\Delta T}{T} \simeq 6 \times 10^{-6}. \tag{7.36}$$

Assuming a power law dependence for the density inhomogeneities (see chapter 6), implies temperature variations of the form

$$\Delta T \sim \Theta^{(1-n)/2}. \tag{7.37}$$

Analysis of the COBE data results in a value $n = 1.2 \pm 0.3$ [Hin96]. This is particularly interesting because an inflationary universe predicts a scale invariant fluctuation spectrum ($n = 1$) [Har70, Zel72] (see chapter 6). This implies a strong experimental indication for an inflationary phase, although no conclusive evidence. In order to obtain information about galaxy formation, it is necessary to extrapolate into the small angle region. For this purpose, it is assumed that indeed $n = 1$, and that a power law dependence is valid. Independent of the extrapolation, measurements have been made at small angles [Wat92, Tau94, Tau96]. As an example results from the Saskatoon experiment are shown in figure 7.9. The observations of the FIRS experiment (far infra-red survey) seem to be in good agreement with the extrapolated COBE DMR data [Gan93]. The anisotropies observed in this balloon experiment, on scales of 0.5°, result in a $Q = 19$ μK and an $n = 1$; furthermore they seem to have the same spatial origin.

The discrimination power of these results with respect to structure formation is, however, unfortunately rather limited, as models both with cold or hot dark matter, as well as those with a non-vanishing cosmological constant, can be brought into agreement with the observations. Nevertheless, limits on some models can be obtained from the data. It seems that the cold dark matter model works only with a certain normalization factor ('bias'). This indicates a threshold beyond which the density fluctuations increase. However, a model with cold dark matter and $\Omega = 1$ also appears to be in reasonable agreement with the observations [Sil93]. Models with a mixture of cold (70%) and hot (30%) dark matter also give good agreement [Row85a, Kly92, Tay92, Pri97].

Apart from density fluctuations there is a further source of anisotropies, in the form of gravitational waves. These develop as inevitably as the quantum fluctuations during inflation. For an analysis of the COBE anisotropies and the limits implied for gravitational waves see [Dav92b].

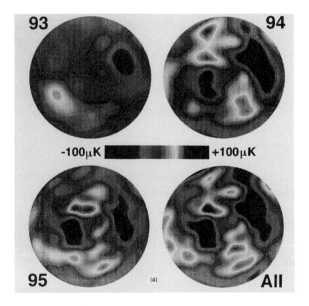

Figure 7.9. (*a*) Sky maps of the cosmic microwave background radiation (CMB) measured with the Saskatoon experiment. The CMB temperature is shown in coordinates where the north celestial pole is at the centre of a circle of 16° diameter, with RA being zero at the top and increasing clockwise. The first three panels show the maps using 1993, 1994 and 1995 data sets, respectively. The last panel (bottom right), shows the map based on all three years of data. Temperature fluctuations are clearly seen (from [Teg96a]). (*b*) Saskatoon in relation to COBE (with kind permission of M Tegmark, 1997).

Since the discovery of the anisotropies by COBE at least nine further

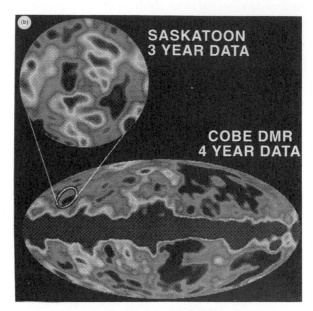

Figure 7.9. Continued.

measurements of anisotropies have been published (figures 7.9 and 7.10) [Sco95]. The development in this field is very fast and future experiments will provide further valuable data. For an overview see [Smo95].

Finally, for the sake of completeness, we mention the third telescope of the COBE mission, used for a measurement of the absolute brightness of the sky between 1–300 μm. It is used among other things in the search for a diffuse IR background which could have been produced by proto-galaxies from the early universe [Par67].

7.2   The cosmic X-ray background

Beside the microwave background, information about the early universe can also be obtained from other regions of the spectrum, such as the X-ray and gamma regions (see figure 7.11). In order to better understand the X-ray background, the ROSAT (Roentgen Satellite) satellite began measurements in 1990 (see figure 7.12). Apart from identifiable X-ray point sources there is also a diffuse component, whose origin especially above 3 keV is mainly extra-galactic. Generally it is assumed, mainly for reasons of energy, that, as long as the radiation is not red-shifted γ radiation, its origin has to be in the region of $0 < z < 10$. This is the area in which significant evolution of galaxies took place. The integrated emission of active galactic nuclei, quasars and Seyfert galaxies is generally believed to be the origin of the X-ray background,

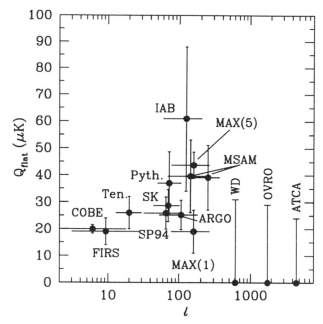

Figure 7.10. The present state of observations of anisotropies in the background radiation as a function of the multipole order l. The amplitudes illustrated correspond to the quadrupole moment for a flat anisotropy spectrum (i.e. a scale invariant spectrum of primordial perturbations, corresponding to a horizontal line) (from [Sco95]).

although this has not been fully established. As these cannot be resolved, they cause a diffuse background [Bol87, Has91, Fab92]. Close objects of this class have been observed, however, and show a softer spectrum than the diffuse X-ray background, so that strong evolutionary effects have to be involved. The ROSAT data provide at present the deepest insight into the X-ray background and suggest that at least 75% of it is caused by discrete sources [Trü93]. Further X-ray missions like ASCA [Cha95a] and the recently launched Beppo-SAX satellite [Pir97] also contribute significantly to our understanding of the X-ray sky.

The dipole anisotropy observed in the X-ray background does correspond within errors to that of the 3 K radiation, and thereby confirms the interpretation of the dipole as a Doppler effect caused by gravity [Bol87, Fab92].

Galactic clusters also show considerable X-ray emission. They are interpreted as thermal emission from a 10^7–10^8 K, (corresponding to 40 keV), hot gas [Pon93, Mul93, Böh95]. The X-ray emission shows a complex behaviour, which changes from cluster to cluster. Despite that, there seems to exist a general correlation: regular clusters with fewer spiral galaxies show smooth X-ray contours with a high luminosity, whereas irregular clusters with many spiral galaxies show irregular X-ray emissions with small luminosity. The

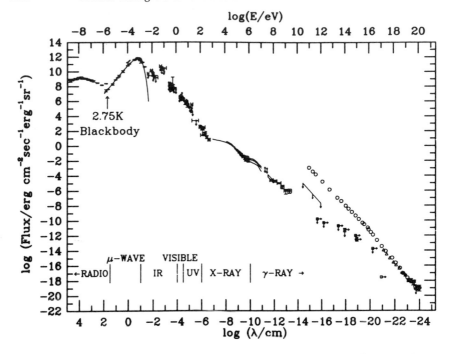

Figure 7.11. The complete observable diffuse photon background over the spectral region of 10^5 cm $\geq \lambda \geq 10^{-24}$ cm. The dominant part of the 3 K radiation can clearly be seen, as well as a decrease towards high energies. Vertical arrows mark upper limits, horizontal arrows correspond to integrated fluxes. The open circles denote the total flux of the cosmic radiation, from which an upper limit for the photon flux can be deduced (from [Kol90]).

gas in irregular clusters is also somewhat cooler than that in regular clusters. Furthermore, the emission from clusters with a dominant galaxy is more strongly concentrated towards the centre. This characteristic can be understood if the cluster is in dynamic equilibrium, and both the gas and the galaxies feel the same gravitational potential [Mul93]. X-ray observations are therefore important to trace the gravitational potential of galaxy clusters resulting in information on their dark matter contents.

7.3   The cosmic neutrino background

Analogous to the photon background there should also exist a cosmic neutrino background (see chapter 3). The evidence for this is now summarized. At temperatures above 1 MeV neutrinos, electrons and photons are in thermal equilibrium with each other via reactions such as $e^+e^- \leftrightarrow \gamma\gamma$ or $e^+e^- \leftrightarrow \nu\bar{\nu}$. In chapter 4 we showed that neutrinos decouple at about 1 MeV (see also chapter

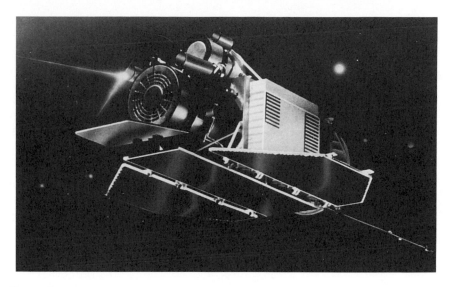

Figure 7.12. The ROSAT satellite, launched on June 1st, 1990 with a Delta II rocket, designed to explore the X-ray sky. It is situated in an Earth orbit of 580 km. Its first task was to establish a sky map of the low energy X-ray region. More than 60 000 X-ray sources (130 000 until 1997) could be detected (in comparison the catalogue resulting from the HEAO-1 satellite, launched in 1977, consists of only 840 sources). ROSAT has therefore opened up a new era of X-ray astrophysics (from [Hen91]).

3). As the temperature drops further to less than the rest mass of the electron, all energy is transferred to the photons via pair annihilation thereby increasing their temperature. The entropy of relativistic particles is given by (see equations (3.57) and (3.58)):

$$S = \frac{4}{3} k_B \frac{R^3}{T} \rho \tag{7.38}$$

where ρ represents the energy density. As $\rho \sim T^4$, according to the Stefan–Boltzmann law, it follows that, for constant entropy,

$$S = (TR)^3 = \text{constant.} \tag{7.39}$$

From the relations mentioned in chapter 3 it follows that

$$\rho_{\nu_i} = \frac{7}{16} \rho_\gamma \tag{7.40}$$

and, for $kT \gg m_e c^2$

$$\rho_{e^\pm} = \frac{7}{8} \rho_\gamma. \tag{7.41}$$

Using the appropriate degrees of freedom and from entropy conservation we obtain:

$$(T_\gamma R)_V^3 \left(1 + 2\frac{7}{8}\right) + (T_\nu R)_V^3 \sum_{i=1}^{6} \rho_{\nu_i} = (T_\gamma R)_N^3 + (T_\nu R)_N^3 \sum_{i=1}^{6} \rho_{\nu_i} \qquad (7.42)$$

where V represents times $kT > m_e c^2$ and N times $kT < m_e c^2$. Since the neutrinos had already decoupled, their temperature developed proportional to R^{-1}, and therefore the last terms on both sides cancel. Hence:

$$(T_\gamma R)_V^3 \frac{11}{4} = (T_\gamma R)_N^3 \qquad (7.43)$$

However, since prior to the annihilation phase of the $e^+ e^-$-pairs $T_\gamma = T_\nu$, this means

$$\left(\frac{T_\gamma}{T_\nu}\right)_N = \left(\frac{11}{4}\right)^{1/3} \simeq 1.4. \qquad (7.44)$$

On the assumption that no subsequent significant changes to these quantities have taken place, the following relation between the two temperatures exists today:

$$T_{\nu,0} = \left(\frac{4}{11}\right)^{1/3} T_{\gamma,0}. \qquad (7.45)$$

A temperature of $T_{\gamma,0} = 2.728$ K corresponds to a neutrino temperature of about 1.95 K. This only applies to massless neutrinos. For massive neutrinos the temperature is correspondingly lower. If the photon background consists of a number density of $n_{\gamma,0} \approx 410$ cm^{-3}, the particle density of the neutrino background is $n_{\nu,0} \approx 340$ cm^{-3}. This follows from the fact that for each light neutrino flavour (see [Gel88, Gro89, Gro90])

$$n_\nu = \frac{3}{11} n_\gamma. \qquad (7.46)$$

The very small cross section for such neutrinos has so far thwarted any experimental attempt to obtain evidence for their existence.

Chapter 8

Cosmic radiation

An extensive area of particle astrophysics is that of the cosmic radiation. Classical cosmic radiation consists of ionized nuclei that hit the Earth at a rate of about 1000 events cm^{-2} s^{-1}, and which were first discovered by Hess in 1912, using balloon flights. Since their energies span 15 orders of magnitude, different experimental strategies are necessary to examine all aspects of the phenomena (see figure 8.1). We first discuss the primary energy and particle spectrum observed above the Earth's atmosphere then turn our attention to information on the different secondary particles. Next we consider possible sources, acceleration mechanisms and the propagation of the cosmic radiation (see figure 8.2). New fields have recently been developed which give information on the sources of cosmic radiation, i.e. X-ray and gamma-ray astronomy, and also high energy neutrinos. This chapter therefore has three main sections:

- The classical cosmic radiation
- X-ray and gamma-ray astronomy
- High energy neutrinos

For more information on these subjects we refer to [Sok89, Ber90b, Gai90, Lon92, 94, Mur93, Gai95]. We first consider classical cosmic radiation.

8.1 Classical cosmic rays

8.1.1 The primary spectrum

Classical cosmic rays consist of about 98% nuclei and 2% electrons; the nuclei are divided into about 87% protons, 12% α-particles and 1% in the form of heavy elements. The element composition of cosmic rays has been directly experimentally determined in the region from a few MeV up to a few TeV [Mül91]. There is good agreement between the relative element abundances in the Sun and in cosmic rays (see figure 8.3) [Sim83]. This points to a similar creation mechanism, namely to stellar production. However, there are also some significant differences: in cosmic radiation, H and He are depleted compared

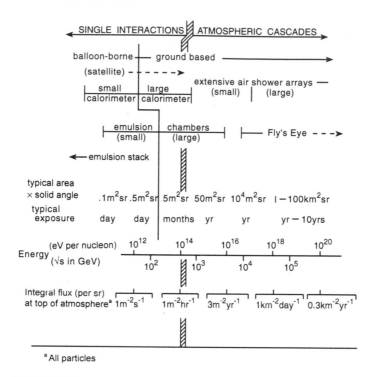

Figure 8.1. The different experimental techniques for the detection of cosmic radiation. Due to the enormous flux differences and the large energy range very different types of detectors are necessary (from [Ric87]).

to the heavy elements, which could on one hand be due to the high ionization potential of H, and the subsequently difficult acceleration, or to another source composition. On the other hand the elements Li, Be and B and the nuclei just below iron are enriched. These elements can be explained as spallation products of C, N and Fe, since Li, Be and B do not appear as end products of stellar nucleosynthesis. The ratio between primary (C, N and Fe) and secondary (Li, Be and B) particles allows the length of stay within the galaxy is estimated to be about 10^6 years [Gai90]. The relative abundance distribution turns out to be approximately (with the exception of Fe) energy independent. In the low energy region (below about 1 GeV/nucleon) the influence of the solar wind on the arriving plasma becomes apparent. In periods of increased solar activity the increased solar wind, due to its greater spatial extent, keeps the low energy cosmic radiation away from the Earth, and itself contributes more strongly to the observed particle flux in this region. This 11 year modulation due to the Sun's activity has been observed in the cosmic radiation [Web74]. The geomagnetic field also influences the observation of cosmic rays and produces a dependence on the geographical latitude.

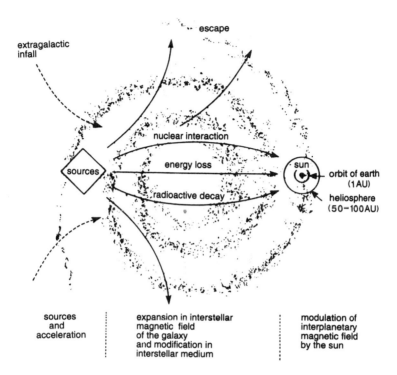

escape

extragalactic
infall

nuclear interaction

energy loss

sources

radioactive decay

sun

orbit of earth
(1 AU)

heliosphere
(50–100 AU)

| sources and acceleration | expansion in interstellar magnetic field of the galaxy and modification in interstellar medium | modulation of interplanetary magnetic field by the sun |

Figure 8.2. Schematic illustration of the processes influencing the observable spectrum of cosmic rays. Beside the sources of cosmic rays, it is the processes of acceleration, propagation and other modifications (such as interactions with the interstellar medium or decay) that modify the primary spectrum before its detection on Earth (from [Rol88]).

Above several TeV the element composition is unknown, and only the integrated energy spectrum, independent of the element composition, is known experimentally. This extends to at least 10^{20} eV, where the observed fluxes are extremely small (see figure 8.4). The spectrum is well described over a wide range by a power law of the form

$$\frac{dN}{dE} \sim E^{-\gamma}. \tag{8.1}$$

Up to values of around 10^{15} eV $\gamma \simeq 2.7$. From here on the spectrum becomes steeper with $\gamma \simeq 3$ ('knee'), which could point to a different origin for the two regions [Hil84]. From about 10^{18} eV the spectrum becomes less steep again ('ankle'). The behaviour at 10^{20} eV has been the subject of extensive discussion as to whether an energy maximum has been reached at 5×10^{19} eV

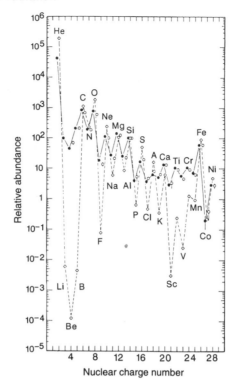

Figure 8.3. Comparison of the cosmic (closed circles) and solar (open circles) element abundances (normalized to silicon). The main observation is an excess of elements in the cosmic radiation just below iron and below carbon. This can be explained by spallation processes. In contrast, the elements H and He are depleted in the cosmic radiation (from [Sim83], see also [Uns92]).

(Greisen–Zatsepin–Kuzmin cut-off), because of the interaction of cosmic rays with the cosmic background radiation [Gre66, Zat66, Zat95, Ded95, Kuz97], whether a plateau is forming, or whether the flux simply becomes too small to be reliably measured (see e.g. [Sok92]). The observation of an event at an energy of $(3.2 \pm 0.9) \times 10^{20}$ eV and of other events at energies higher than 10^{20} eV [Efi88, Bir93, Bir95, Hay94, PDG96] has given rise to speculative ideas of the origin of these events from the decay of new superheavy dark matter particles [Chu97, Kuz97].

Electrons only contribute about 2% of the cosmic radiation. In the relativistic case they can mainly be detected in the radio region via their synchrotron radiation. It is possible from this to make statements about the galactic magnetic field. Electrons can either be produced directly from the sources of cosmic radiation, or as secondary products from nuclear reactions in

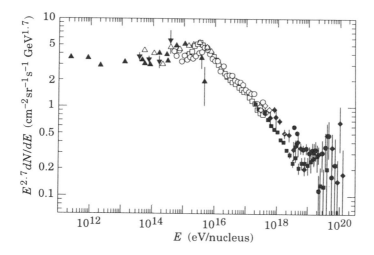

Figure 8.4. The spectrum of cosmic radiation from 10^{11}–10^{20} eV. The 'knee' at about 10^{15} eV is clearly visible. The behaviour above 10^{20} eV is controversial (from [PDG96]).

the interstellar medium according to the sequence $\pi^{\pm} \to \mu^{\pm} \overset{(-)}{\nu_{\mu}} \to e^{\pm} \overset{(-)}{\nu_{e}}$. As the observed abundance of positrons is only about 10% that of electrons, only about 10% of electrons should be produced from the latter process. Equally important information can be obtained from the observed anti-proton spectrum. A significant flux of anti-protons could indicate large amounts of anti-matter in the universe. Alternatively they could be produced from the interactions of protons with interstellar matter. The experimental data for the lower energy anti-proton flux are, however, still very unreliable (see e.g. [Ste87, Str89]). Attempts are underway to launch a spectrometer (AMS) on a space shuttle mission in 1998 which will be able to measure the matter/anti-matter ratio in cosmic rays at a level of precision of around 10^{-9}. This would represent an improvement over current sensitivity of about 4–5 orders of magnitude [Ahl94b, deR96].

8.1.2 Direct measurements of the primary radiation

In the region below about 100 TeV it is still possible to measure the primary cosmic radiation directly. As examples we will discuss two experiments.

8.1.2.1 The JACEE experiment

The JACEE experiment is a balloon experiment designed to examine charged particles in an energy region of about 1–100 TeV/nucleon [Bur83]. The principal method of detection is illustrated in figure 8.5. The underlying idea for the measurement of the charge is given by the energy loss dE/dx, which is given

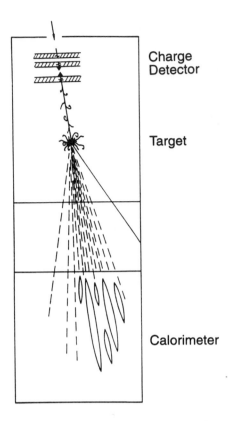

Figure 8.5. Schematic illustration of an event in the JACEE detector. A penetrating particle first crosses the tracking chambers, in which its charge can be determined. The interaction takes place in the interaction region, and the developing shower is measured with a calorimeter (from [Sok89]).

by the Bethe–Bloch formula as

$$\frac{dE}{dx} \sim \frac{Z^2}{\beta^2} \tag{8.2}$$

where $\beta = v/c$. Tracks may be detected via etching of passive plastic emulsions, as well as CR39-Lexan layers (see chapter 10). This method produces an accuracy for charge determination of $0.2e$ for protons and helium nuclei, which degrades to $2e$ for iron nuclei. The actual active detection region consists of a sandwich of emulsion plates and acrylic films (both about 50–75 μm thick). With the help of the emulsion tracks it is possible to reconstruct the vertex of the interaction, while the acrylic films are used to identify the nuclear fragments. The associated calorimeter has a depth of 7 radiation lengths and consists of alternate layers of lead foils, emulsion plates and X-ray sensitive film. Due to the density of the tracks on the X-ray film the energy of individual γ-quanta can be determined. The energy resolution is about 22%, as determined from examination of the invariant mass of the observed π^0 peak. It is possible to reconstruct the total energy of the event from the observed energy. An experimental simplification for measuring the energy occurs since a measurement of all the photon energies is sufficient to extrapolate back to the primary energy [Sok89]. Furthermore it can be shown that, for a primary spectrum obeying a power law which does not change in the observed energy region the same power behaviour is seen in the total gamma energy; i.e. the power behaviour of the gamma radiation directly determines that of the primary particles [Sok89]. A number of balloon flights using this technique and lasting several days were carried out in the upper atmosphere (at about 3–5 g cm^{-2}) [Bur87]. All data are consistent with the abundances discussed previously. As an interesting side effect these experiments permit the study of nucleus–nucleus interactions in energy regions not accessible to current accelerators.

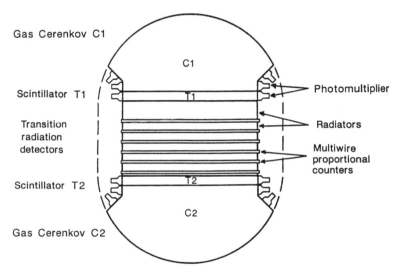

Figure 8.6. The construction of the Chicago egg experiment on board a space shuttle mission. It was used to directly examine the composition of the primary cosmic radiation (from [Sok89]).

8.1.2.2 The Chicago 'egg'

Observation in space results in complete freedom from atmospheric effects. Therefore the 'egg' of the University of Chicago (see figure 8.6) [Swo82] was recently included in a space shuttle flight. Cerenkov and transition radiation were used for the energy determination. Transition radiation occurs when a relativistic particle passes from one dielectric to an other. This radiation is typically in the

X-ray region, and the yield depends on the relativistic γ-factor. The detector consists of two gas Cerenkov detectors with a threshold of about 40 GeV/nucleon and two scintillators for the determination of the specific ionization, and therefore the charge, as well as acting as a coincidence trigger. The central section contains transition radiation detectors, where multi-wire proportional chambers are used for the detection of the produced X-rays. This experiment is sensitive to an energy region from 40 GeV/nucleon to a few TeV/nucleon. A spectrum taken in this experiment is shown in figure 8.7.

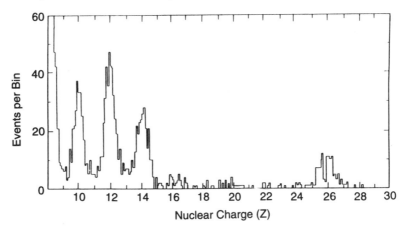

Figure 8.7. The charge resolution of a spectrum measured with the 'Chicago egg' (from [Sok89]).

8.1.3 Secondary products and showers

Above several TeV the fluxes become too small to use direct methods of detection. The indirect evidence rests on the interaction of primary radiation with our atmosphere. It is from the developing secondary products that the information is subsequently obtained (see figure 8.8). This technique of investigation ('extended air showers' (EAS)) not only applies to charged particle primaries, but also to the search for high energy gamma radiation from space. The latter will, however, be discussed in a separate section (see section 8.3.6).

When a primary proton meets the atmosphere a proton–nucleus interaction takes place, analogous to those at particle accelerators. Indeed, it was reactions of cosmic rays within newly developed detectors that led in the late 1940s and beyond to the discovery of a substantial number of new particles. At high energies ($E_p > 10$ GeV) the interaction takes place with the individual nucleons of a nucleus. Multiple scattering within a nucleus produces mainly pions, but also strange particles (K, etc), or anti-nucleons. As well as primary particles these high energy reaction products produce further hadronic interactions, until

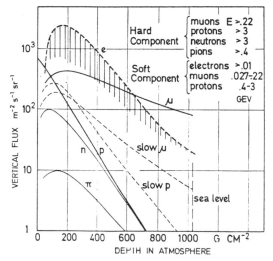

Figure 8.8. Composition of the secondary radiation as a function of height (from [Hil72], see also [Lon92, 94]).

the energy per particle has dropped below about 1 GeV, the energy necessary for multiple pion production. Such a reaction chain is called a *hadronic shower*.

Apart from the possibility of further interactions, there is also the possibility that the particles decay. Consider, for example, pions. Neutral pions decay with a lifetime of 1.78×10^{-16} s via

$$\pi^0 \to 2\gamma. \tag{8.3}$$

Charged pions decay with a lifetime of 2.55×10^{-8} s into

$$\pi^+ \to \mu^+ \nu_\mu \tag{8.4}$$

$$\pi^- \to \mu^- \bar\nu_\mu. \tag{8.5}$$

Analogous decays also take place for the K mesons produced. Their contribution to muon production is energy dependent. This means that about 8% of muons with energies of about 100 GeV originate from K decay, at 1000 GeV the fraction is about 19%, and the value for higher energies approaches 27% asymptotically [Gai90]. The main products of such showers to be detected, therefore, are highly energetic γ quanta, muons and neutrinos. For observations at sea level at vertical incidence the atmosphere presents a thickness of roughly 20 radiation lengths. High energy photons therefore produce e^+e^- pairs, which in turn produce photons via bremsstrahlung. This results in the development of an *electromagnetic shower*.

A complete shower therefore consists of three components: the electromagnetic, muonic and hadronic components (see figure 8.9). Directly

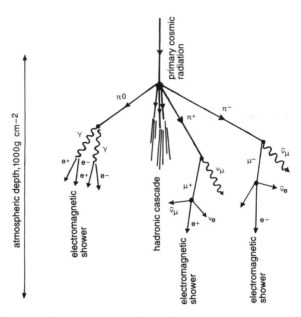

Figure 8.9. Development of an air shower. The various components of the shower are illustrated (from [Lon92, 94]).

after the interaction of the primary particles a multiplicative extension takes place in the shower, which usually reaches a maximum at a height of about 10 km. This height depends weakly (logarithmically) on the energy, i.e. the higher the energy, the deeper in the atmosphere lies the point of maximum extent. Afterwards a damping takes place, as more and more particles fall under the threshold for new particle production.

Three main methods are used to detect these showers:

- air Cerenkov technique
- particle observation in large area detectors
- detection of muons and neutrinos in underground laboratories.

8.1.3.1   Cerenkov technique

Cerenkov light is emitted as the speed of the particles reaches a value higher than

$$v = \frac{c}{n(H)} \tag{8.6}$$

where n is the atmospheric refractive index dependent on the height H. At standard atmospheric pressure on the Earth's surface $n = 1.00029$. The threshold energy depends on the height according to [Sok89]

$$E_{min} = \frac{0.511}{\sqrt{2\epsilon}} \text{ MeV} \qquad (8.7)$$

where $\epsilon = n - 1$, and $\epsilon \sim \exp(-H/H_S)$. At a height characteristic of the atmosphere of $H_S = 7.5$ km (where the density has decreased to $1/e$), the threshold for electrons is 35 MeV, while at sea level it is only 21 MeV. At sea level muons have a threshold of 4.3 GeV. Therefore practically all particles participate in the production of Cerenkov light. The angle of maximal emission can be estimated as [Sok89]

$$\theta_{max} \approx 81° \sqrt{\epsilon}. \qquad (8.8)$$

The horizontal expansion of the Cerenkov light as observed on the Earth's surface is dependent on the multiple scattering of the electrons, as well as on the Cerenkov angle itself. An intense Cerenkov beam is expected within 6 degrees around the shower axis, which can, however, extend to 25 degrees [Sok89]. Neglecting the multiple scattering of electrons, a primary shower produced at between 7 and 20 km height produces a cone of about 150 m radius at sea level. A primary particle of 1 TeV energy produces about 10^6 Cerenkov photons, which are detected with photomultipliers. One disadvantage of this method is that observations can only be carried out on moonless nights. We will return to this technique when we discuss high energy γ radiation (see section 8.3.6).

8.1.3.2 Large area detectors

Above primary energies of about 50 TeV, a significant fraction of secondary particles reaches the Earth's surface, so that their direct observation becomes possible. A typical shower from a 100 TeV primary produces at a height of about 1500 m about 30 000 electrons and positrons, approximately a factor of 5 more low-energy protons and many more than 1000 muons. In order to get an easily measurable signal for a primary particle of about 50 TeV, the experiment has to be carried out at altitude (a mountain), while above about 1 PeV ($\equiv 10^6$ GeV) sea level is sufficient. Table 8.1 gives an overview of existing large area air shower experiments. In the examination of these air showers a large number of detectors are used (generally more than 100), which are distributed over a very large area (typically over 1 km²). Figures 8.10 and 8.23 show two typical experiments, CASA-MIA and HEGRA. These air shower arrays usually contain scintillation counters [Sok89]. Information about the direction and energy of the primary particle can be obtained from the time development and the energy deposited in the various counters. The separate measurement of muons in the showers is also an advantage. It is in principle possible to determine the composition of the primary radiation from the ratio of muons and electrons in the shower. Assuming that the number of muons N_μ is correlated

Figure 8.10. The CASA-MIA air shower array in Utah, consisting of 1089 units. The detectors are distributed over an area of 500 m by 500 m. Inside the array the 'fly's eye' detector of the University of Utah is situated. Together they can detect cosmic radiation from 10^{13} eV up to the highest energies (with kind permission of the CASA-MIA collaboration).

with the number of electrons N_e via

$$N_\mu \sim N_e^\alpha \tag{8.9}$$

and further assuming the simple superposition model, namely that a nucleus of mass A and energy E_0 produces an identical shower compared with A individual showers from protons of energy E_0/A summed together (for example the shower of an iron nucleus with mass 56 and energy E is identical to 56 showers of protons, each with an energy $E/56$), an A dependence of

$$N_\mu \sim A^{1-\alpha} N_e^\alpha \tag{8.10}$$

results. The primary energy E_0 of the incoming particle can be roughly estimated as [Gai90]

$$(1 - \delta)E_0 \simeq \alpha \int_0^\infty dX N(X) \tag{8.11}$$

where δE_0 is the energy lost in invisible neutrinos, α the energy loss per unit length in the atmosphere and $N(X)$ is the number of charged particles in the shower at depth X relative to the shower axis.

Table 8.1. Summary of some current extended air shower experiments (from [Cro93]).

Experiment	Place	Depth (g cm^{-2})	Area (10^4 m^2)	Resolution (degrees)	E_{min} (TeV)	μ-area (m^2)
Akeno	36N, 138E	920	1	3	1000	225
NORIKURA	36N, 137E	738	≤ 1	2	200	—
JANZOS	41S, 170E	930	≥ 0.23	2	1000	—
BUCKLAND	35S, 138E	1030	1	2.5	1000	—
KGF	13N, 78E	915	1.66	1.5	500	210
Ooty	11N, 77E	785	0.5	3	100	—
Baksan	43N, 43E	840	0.5	1.5	300	—
Tien Shan	42N, 75E	690	0.5	3	100	35
EAS TOP	42N, 14E	800	10	1	100	—
Plateau Rosa	46N, 8E	675	1	5.5	100	—
GREX	54N, 1W	1030	≥ 1	1	500	40
HEGRA	29N, 18W	800	4	1	50	—
BASJE	16S, 68W	530	≥ 0.5	3	20	60
CYGNUS-II	36N, 106W	800	≈ 6.6	1	300	70
Mt Hopkins	32N, 111W	780	≈ 0.5	1	100	—
CASA-MIA	40N, 112W	870	25	1	70	2550
SPASE	90S	760	≈ 1	1	100	—
TIBET AS$_\gamma$	30N, 90E	600	2.0	0.8	10	—

8.1.3.3 Fluorescence radiation

A further method of detecting ionizing particles comes from fluorescence radiation from excited nitrogen molecules in the air. The fluorescence spectrum lies mainly in the UV region between 300–400 nm, at which the atmosphere is relatively transparent. A prototype of such a detector is the 'Fly's Eye' in Utah (USA) (see figure 8.11) [Bal85]. It consists of 880 phototubes in a total of 67 mirrors, each with a diameter of 1.5 m. Each phototube observes a specific angle of the sky, and the fluorescent light impinges on all tubes whose solid angle has been traversed by an EAS (see figure 8.12). Meanwhile a second 'eye' is situated at a distance of 3.4 km, so that a stereo observation is possible. In 1991 this detector observed the most energetic event to date, with an energy of $(3.2 \pm 0.9) \times 10^{20}$ eV (corresponding to 51 J!) [Bir93, Bir95]. The composition of the cosmic radiation above 2×10^{17} eV has also been examined with this detector [Bir93], while with JACEE the composition below about 3×10^{15} eV has been analysed. New detectors, such as KASCADE [Dol90, Kam97, Kla97], cover the intermediate region. The *Auger project* [Aug95, Bor96] is a planned new air shower array which will investigate cosmic rays at the highest energy. The detector consists of two separate elements (one in the northern hemisphere and the other in the southern hemisphere) each consisting of a detector of atmospheric

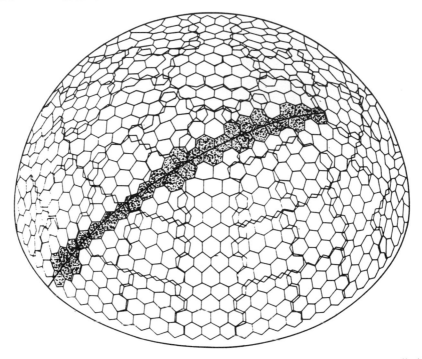

Figure 8.11. The 'Fly's Eye' detector for the detection of the fluorescence radiation from excited nitrogen molecules (schematic). The dotted surfaces represent the light of an extensive light shower, and the full line shows the trajectory of the shower across the sky (from [Sok89]).

fluorescence radiation as well as of a field of 1600 particle detectors distributed over 3000 km^2, so each of the elements will be sensitive both to the atmospheric fluorescence as well as to charged particles. The detector is expected to be able to observe more than 60 events per year with a primary energy above 10^{20} eV as well as 6000 events per year over 10^{19} eV in this so far poorly explored ultra high energy region. For a more detailed discussion of air shower experiments see [Sok89, Gai90, ICR95].

We now turn to the detection of muons and neutrinos in underground laboratories.

8.1.4   Atmospheric muons from cosmic radiation

Production mechanisms for muons have already been discussed in the previous section. For energies above about 10 GeV, practically all muons reach the Earth's surface prior to their decay. Below that muon decay into

$$\mu^- \rightarrow e^- \bar{\nu}_e \nu_\mu \tag{8.12}$$

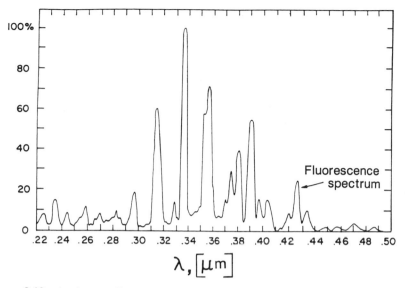

Figure 8.12. A nitrogen fluorescence spectrum in the near UV region, taken with the Fly's Eye detector. This is produced by the interaction of a shower with the Earth's atmosphere (from [Sok89]).

Table 8.2. Summary of some underground laboratories with shielding depths and threshold (i.e. minimal) energies for muons (from [Gai90]).

Place	Depth (km water equivalent)	Threshold energy (TeV)
KGF	≤ 7	10
Homestake	4.4	2.4
Mt Blanc	≈ 5	≈ 3
Frejus	≈ 4.5	≈ 2.5
Gran Sasso	≈ 3.5	≈ 2
IMB	1.57	0.44
Kamiokande	2.7	≈ 1
Soudan	1.8	0.53

and

$$\mu^+ \rightarrow e^+ \nu_e \bar{\nu}_\mu \qquad (8.13)$$

is important. The muon has a lifetime of $\tau = 2.2 \times 10^{-6}$ s. The expected muon spectrum results from the convolution of the decay kinematics of pions and kaons with the spectrum of the particles which produce them. The relation of the mean energy of the primary particles $\langle E_0 \rangle$ which is necessary to produce muons with energies greater than an energy E, depends on the observed energy. For example

$\langle E_0 \rangle / E \approx 37$ for $E_\mu > 14$ GeV, which means that for muon energies greater than 14 GeV a primary energy of about 500 GeV/nucleon is needed on average [Gai90]. For $E_\mu > 1$ TeV the ratio is about 10, and for $E_\mu > 6$ TeV about 8. These are typical muon energies which are seen in underground experiments. What can be detected from a muon event in underground experiments depends largely on the depth at which these experiments are carried out. Muons continually and discretely lose energy in passing through matter. Discrete energy losses indicate spatially isolated large energy deposits. The continuous energy loss dE/dx due to ionization is given by the Bethe–Bloch formula and is, e.g. for relativistic muons with energies smaller than 10 GeV in rock, about 2 MeV/g cm^{-2}. For $E_\mu > 10$ GeV it can be well approximated by [Gai90]

$$\frac{dE}{dx} = -\left(1.9 + 0.08 \ln \left(\frac{E_\mu}{m_\mu} \right) \right). \tag{8.14}$$

In addition to these continuous energy losses there are discrete losses by bremsstrahlung, electromagnetic interactions with nuclei and $e^+ e^-$ production. For energies of greater than about 500 GeV, these processes become more important than the continuously produced losses. The total energy loss can then be represented as

$$\frac{dE}{dx} = -\alpha - \frac{E}{\kappa} \tag{8.15}$$

where

$$\kappa^{-1} = \kappa_{\text{brems}}^{-1} + \kappa_{\text{pair}}^{-1} + \kappa_{\text{hadron}}^{-1} \quad \text{and} \quad \alpha \approx 2 \text{ MeV/g cm}^{-2}. \tag{8.16}$$

For rock κ is estimated to be $\approx 2.5 \times 10^5$ g cm^{-2} [Gai90]. The minimum energy E_0^{min}, which a muon at the surface has to have in order to penetrate to a vertical depth X with an energy $E(X) = 0$, can hence be calculated as

$$E_0^{\text{min}} = \alpha \kappa (e^{X/\kappa} - 1). \tag{8.17}$$

Table 8.2 shows a comparison of some underground laboratories, as well as the minimum energy for muons to reach these laboratories. Of course the intensity of the observed muon flux in underground detectors varies from one experiment to another, depending on their screening depth. Consider the MACRO detector (see chapter 10) in the Gran Sasso underground laboratory. Its angular resolution is about 1 degree. This means that the direction of the primary particle can be reconstructed from single tracks. This happens principally in three steps. With the help of complex simulations of the energy loss in the rock, the measured muon energy can be used to calculate the muon energy on the surface. The spectrum obtained in this way can then be brought into agreement with the expected muon spectrum, in which the primary energy of the cosmic particle plays the deciding role. More than 5 million single muon events have so far been observed in this way, and hence a 'muon sky map' can be produced, on which, however, no point-sources are visible [DiC93]. Equally interesting is the examination of multiple muon events (muon bundles). For several muons from a atmospheric shower to actually reach the underground experiment, the primary interaction has to be an even more energetic event. Above a certain energy lighter particles are less effective in producing multiple muons than heavier nuclei of the same energy. This is a statistical statement, and should not be applied to individual events. By comparing with simulations it is therefore in principle possible to obtain information about the composition of the cosmic radiation in the range of 10^{13}–10^{16} eV. For example, in order to produce an event of more than 3 muons in the MACRO detector, the energy of the primary particle must have been of order of 1000 TeV. Multiplicities of up to 40 have been recorded (see figure 8.13) [Pal93]. The ability of the air shower detector installed *on* the Gran Sasso (EAS-TOP) to measure shower events and the corresponding muons underground simultaneously will be advantageous, allowing a direct comparison of air showers with underground muons. It is hoped that statements about the composition in the PeV region can also be made from this apparatus. At an 'intermediate' depth with a shielding of 500 mwe, the COSMOLEP experiment [Bal94] is being installed. It is essentially an underground air shower array with an effective area of $\sim 60\,km^2$, allowing coincident detection of muons with the four LEP detectors. For a discussion of the physics of muons from cosmic radiation see [Bar52, Gai90, Gai94, Gai96b].

8.1.5 Atmospheric neutrinos

The decay of pions and kaons produces neutrinos as well as muons. The decay of the muons also produces neutrinos. Under the assumption that a major fraction of muons decays in transit through the atmosphere (the assumption is justified for the energy region of 0.1 to about 2 GeV), the relationship between the various neutrino flavours is expected to be

$$\bar{\nu}_\mu \simeq \nu_\mu \simeq 2\nu_e \tag{8.18}$$

and

$$\frac{\nu_e}{\bar{\nu}_e} \simeq \frac{\mu^+}{\mu^-} < 1 \tag{8.19}$$

as well as

$$\frac{\bar{\nu}_\mu + \nu_\mu}{\bar{\nu}_e + \nu_e} = 2. \tag{8.20}$$

The origin of the asymmetry in equation (8.19) is the excess of protons compared to neutrons in the primary radiation. For energies above about 2 GeV however, due to the Lorentz contraction the mean free path of muons becomes equal to or greater than the depth of the Earth's atmosphere, and hence the ratio ν_e/ν_μ decreases. Detection of neutrinos can only be performed in underground laboratories, where they form the main background for the proton decay we

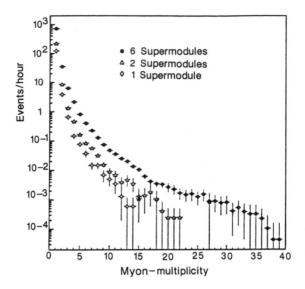

Figure 8.13. Muon multiplicities in underground laboratories, measured here with the MACRO detector in the Gran Sasso laboratory. Multiplicities up to 40 have been observed. Conclusions on the composition of the primary particles can be obtained from the multiplicity information (from [Pal93]).

discussed in chapter 2. The detection reactions happen mainly via charged weak currents, e.g.

$$\nu_e(\nu_\mu) + N \rightarrow e(\mu) + \ldots \tag{8.21}$$

where N represents a nucleus of the target. The cross section for the production of charged leptons in νN reactions in the region of 1–3000 GeV, averaged over neutrino and anti-neutrino, is given by [Gai90]

$$\sigma \simeq 0.5 \times 10^{-38} \text{cm}^2 \frac{E_\nu}{\text{GeV}}. \tag{8.22}$$

The neutrino flux at 1 GeV is about $\Phi_\nu \approx 1 \text{ cm}^{-2} \text{ s}^{-1}$, integrated over all directions. Hence the rate of events can be estimated as [Gai90]

$$R = 1 \text{ cm}^{-2} \text{s}^{-1} \times \frac{0.5 \times 10^{-38} \text{ cm}^2}{\text{nucleon}} \times \frac{6 \times 10^{32}}{\text{kton}} \times \frac{3.15 \times 10^7 \text{s}}{\text{y}}$$

$$\approx 100 \frac{\text{events}}{\text{y kton}}. \tag{8.23}$$

From these estimates it can be seen that enormous detectors are necessary in order to obtain a reasonable event rate. All existing detectors are essentially sensitive to neutrinos with a maximum energy of a few GeV. Due to the influence of the Earth's magnetic field on the low energy cosmic radiation individual

Monte Carlo simulations are necessary for each detector in order to estimate the observed neutrino flux and spectral form. These have typical uncertainties in the region of 20%. It is also important for the experimental sensitivity to have a good knowledge of the detector response to electrons and muons. Usually only those events which lie completely within the detector ('contained events') are used. In the Cerenkov detectors (for example Kamiokande and IMB) the difference between ν_e and ν_μ shows itself in the observed Cerenkov pattern. Only events with a single Cerenkov ring are used. Due to multiple Compton scattering the Cerenkov ring of electrons is more diffuse than a ring caused by a muon. After a careful examination of this 'sharpness' it is possible to distinguish between these events [Tot96]. In addition, the muon decay can be detected. Results from five experiments (Frejus, NUSEX, IMB, Soudan and Kamiokande) exist on observed atmospheric neutrino fluxes at about 1 GeV [Agl89, Ber90c, Bec92, Kaf94, Fuk94]. In order to be independent of the uncertainties in the absolute fluxes a ratio R of observed to expected events is defined as

$$R = \frac{(\mu/e)_{\text{Data}}}{(\mu/e)_{\text{Simulation}}}. \tag{8.24}$$

This should normally be 1. The analysis of the Kamiokande experiment with an exposure of 6.16 kton years, however [Kak93] yields

$$R = 0.60^{+0.07}_{-0.06} \pm 0.05. \tag{8.25}$$

In comparison the results from other experiments give

$$R = 0.54 \pm 0.05 \pm 0.12 \quad \text{(IMB)} \tag{8.26}$$

$$R = 0.99 \pm 0.13 \pm 0.08 \quad \text{(Frejus)} \tag{8.27}$$

$$R = 0.72 \pm 0.19^{+0.05}_{-0.07} \quad \text{(Soudan II)} \tag{8.28}$$

$$R = 0.99^{+0.35}_{-0.25} \quad \text{(NUSEX)} \tag{8.29}$$

(see [Hir92a, Goo95b, Kaf94, All96]). The exposures were 7.7 kton years (IMB), 1.53 kton years (Frejus), 1.52 kton years (Soudan) and 0.74 kton years (NUSEX). In IMB as well as at Kamiokande (confirmed at Superkamiokande [Tak97]) there seems to be a deficit of muon neutrinos. This discrepancy is particularly interesting in connection with the study of neutrino oscillations (see chapter 2). As most of the neutrinos are produced in the upper atmosphere, we are dealing with intermediate oscillation lengths (10–100 km) and are therefore testing regions between the accelerator and reactor data and solar neutrinos. If the deficit of muon neutrinos were to be interpreted as an effect of ν_μ–ν_x oscillations, the areas shown in figure 2.22 results. It can be described by

$$\Delta m^2 \approx 10^{-2}\,\text{eV}^2 \quad \text{and} \quad \sin^2 2\theta \approx 0.5 \tag{8.30}$$

Several experiments are planned (see chapter 2) which will explore this range allowed by Kamiokande and SuperKamiokande in the near future. The first amoung them will be the long-baseline neutrino experiment between KEK and Superkamiokande [Suz96] which will start taking data in 1999. Also the proposed GENIUS experiment (see section 2.4.2) could probe the problem of atmospheric neutrino oscillations (see figure 12.26).

For a detailed discussion of the measurements obtained so far see [Kos92, Gai94, Sta96, Gai96b]. Should the above parameters really be the solution to the deficit of atmospheric neutrinos, a reduction in muon events should also be apparent at higher energy. We will return to this subject in section 8.4.

8.2 Sources of cosmic radiation

In this section the astrophysical origin of cosmic radiation is discussed (for high energy photons and neutrinos see also sections 8.3.6 and 8.4). Due to the enormous energy range observed, several sources seem likely to exist. The observations seem, at least for energies below 10^{19} eV, to point to an origin within our galaxy. One strong indication comes from the observed power dependence of the electron spectrum. High energy electrons produce synchrotron radiation, which up to 10^{19} eV also obeys a power law. As the highest energy electrons experience significant Compton scattering from the cosmic background radiation, their lifetime only amounts to about 10^6 years, and therefore they have a range of about 300 kpc. At greater distances the power dependence would be destroyed by the scattering.

An interesting piece of information results from the consideration of ultra high energy events. Several experiments report the observation of events with an energy of more than 10^{20} eV [Efi88, Hay94, Bir95, Daw95, PDG96]. Due to photo-production with photons from the 3 K radiation by the inverse Compton effect (main effect for high energy electrons) and to pion production in interactions of protons with 3 K radiation photons, the range of such high energy particles is limited to about 50 Mpc in the case of protons (*Greisen–Zatsepin–Kuzmin* cut-off [Gre66, Zat66, Kuz97]. As such ultra high energy particles are almost uninfluenced by magnetic fields, their direction should reflect that of their source. Interestingly enough the events from [Efi88] and [Bir95] point in the direction of the radio galaxy 3C 134 and the event from [Hay94] in the direction of the radio galaxies NGC 315 and 3C 31, all at a distance of about 65 Mpc. The investigation of the highest energy particles will probably become more important in the future searches for sources, in particular by the Auger project (see section 8.1.3).

As the *charged* particles of cosmic radiation, with the exception of the highest energy events, have lost any directional information due to interactions with the interstellar medium and magnetic fields, neutrinos, neutrons (these have, however, only a life-time of about 887 s) and γ radiation are particularly useful in the search for sources (see sections 8.3.6 and 8.4).

The energy density in the cosmic radiation can be easily estimated to be about 1 eV cm^{-3}, which is comparable with that of the interstellar magnetic field. The power necessary to produce such an energy density in a volume the size of the Milky Way, in order to provide such an energy density in cosmic radiation, can also be easily estimated. If we assume a thickness of 300 pc and a radius of 15 kpc for the Milky Way, the result is a power of

$$P = \frac{V\rho}{\tau} \approx 5 \times 10^{40} \text{ erg s}^{-1} \tag{8.31}$$

where ρ is the energy density (about 1 eV cm^{-3}) and τ the time the particles remain inside the volume V of the galaxy (about 6 $\times 10^6$ years). Such energy releases can be brought about by supernova explosions (see chapter 13) [Gin64]. In these about 10^{51} erg are released in the form of kinetic and electromagnetic energy. Assuming a mean rate of one supernova per 30 years, this corresponds to an energy of 10^{42} erg s^{-1}. An energy transmission to cosmic radiation with an efficiency of 10% would already be sufficient. Stellar winds from the pre-supernova phase could also be a possible source, however, due to the lower power (about 10% of the supernova power) correspondingly higher efficiencies are necessary, which are physically difficult to realize. Young pulsars are also good candidates. Pulsars are rotating neutron stars which, according to the present interpretation, are the remnants of supernovae. In the beginning they have a rotational energy of about 10^{53} erg, and can therefore provide the necessary power. Binary star systems could also be a source. If one of the participating partners is a compact object, such as a neutron star or a black hole, it accretes mass, which is strongly accelerated, from its companion. The candidates discussed above are sources as well as accelerators and are of very small size. They are therefore called *point sources*. Another source that can be significant for the extra-galactic part of cosmic radiation are active galactic nuclei (AGN), (see also section 6.5). Many galaxies have a compact nucleus which carries a significant part of the total radiated energy of the galaxy across the entire spectral region. One peculiarity of these objects are short term variations in the luminosity within only a few hours to days. This can only be explained if the objects are no bigger than 10^{16} cm. The present explanation of the phenomenon of active galactic nuclei is the presence of a super massive black hole (of order $10^8 M_\odot$), which aquires the surrounding mass (accretion) (see e.g. [Dus92, Rob96]). These objects belong to the most powerful in the universe and are a strong source of X-ray and γ-radiation, as well as of ultra high energy neutrinos (see sections 8.3.6 and 8.4).

For an overview of sources of cosmic radiation see [Gin64, Hil84, Sok92, Gai95, Kir96, Bie96].

8.2.1 Acceleration of cosmic radiation

In this section we discuss the acceleration mechanism of particles up to 10^{20} eV and the generation of the observed power law. We assume that the place of production of cosmic rays and their acceleration in general are identical.

The principal mechanisms discussed are acceleration through shock waves and through moving magnetic plasmas, (Fermi acceleration of first and second order) [Gai90]. Consider a particle with initial energy E_0, which at every acceleration receives an increase in energy proportional to its own energy, i.e. $\Delta E = \epsilon E$. After n such accelerations it has an energy of

$$E_n = E_0(1 + \epsilon)^n. \tag{8.32}$$

The number of accelerations necessary to reach an energy E is given by

$$n = \ln\left(\frac{E}{E_0}\right) \bigg/ \ln(1 + \epsilon). \tag{8.33}$$

If in every acceleration there is an escape probability P_e, the probability to remain in the acceleration mechanism after n such accelerations is $(1 - P_e)^n$. The number of particles with an energy greater than E is therefore

$$N(> E) \sim \sum_{m=n}^{\infty} (1 - P_e)^m = \frac{(1 - P_e)^n}{P_e}. \tag{8.34}$$

Equations (8.33) and (8.34) give (compare with equation (8.1))

$$N(> E) \sim \frac{1}{P_e}\left(\frac{E}{E_0}\right)^{-\gamma} \tag{8.35}$$

with

$$\gamma = \ln\left(\frac{1}{1 - P_e}\right) \bigg/ \ln(1 + \epsilon). \tag{8.36}$$

In this way the required power dependence for the number of particles N is obtained.

By Fermi acceleration we understand acceleration of relativistic particles through statistically distributed magnetic clouds (second order) or through strong shock waves (first order). The order relates here to the dependence of the energy gain on $\beta = v/c$. We concentrate on first order Fermi acceleration [Lon92, 94]. The principle is based on a shock wave which propagates through the interstellar medium, in which some very energetic particles already exist (see figure 8.14). For strong shock waves, their velocity u is a great deal larger than the speed of sound in the gas. This is, for example, the case in supernova explosions. The density relation before and after the shock is given by:

$$\frac{\rho_2}{\rho_1} = \frac{\gamma + 1}{\gamma - 1} \tag{8.37}$$

where γ stands for the ratio of the specific heats and for a completely ionized gas is 5/3. Before the shock the particle distribution is isotropic, so that some

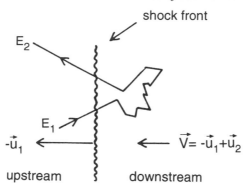

Figure 8.14. Schematic illustration of Fermi acceleration of the first order on a smooth shock front. The shock moves at a velocity of u_1. Particles with an energy E_1 arrive from the isotropic distribution in front of the shock into the region of the shocked material (right). There they gain kinetic energy with an isotropic distribution, so that some particles with an energy E_2 pass again in front of the shock front. In this cycle particles gain energy proportional to $u_1 - u_2$, where u_2 is the velocity of the backwards flowing gas. Particles can undergo this cycle several times and can be accelerated in this way to very high energies (from [Gai90]).

particles are able to traverse the shock. Here the distribution becomes also isotropic because of the interaction with the shocked gas. Particles gain kinetic energy from this gas behind the shock. Some particles are now left behind the shock and escape the acceleration mechanism, while others pass again back through the shock front. The scattering processes again produce an isotropic particle distribution. The total effect of this whole cycle is a net energy gain given by

$$\frac{\Delta E}{E} = \frac{4}{3} \frac{u_1 - u_2}{c} \tag{8.38}$$

where u_1 is the velocity of the shock wave and u_2 the velocity of the gas flowing back across the shock front, where $|u_2| < |u_1|$. Subsequently those particles which are situated before the shock are again caught by it, resulting in a further acceleration cycle. Such shock waves can appear in different places. A good example is supernova explosions. The escaping shock front (see chapter 13) has enough energy to accelerate both material discharged from the supernova as well as interstellar material. However, it can be shown that the maximum energy reached under realistic conditions is about 100 TeV. Since far higher energies have been observed, other sources and acceleration mechanisms are necessary. Systems in which strong accelerations can appear are, for example, young pulsars, binary star systems with a neutron star or even galactic winds. As yet, however, acceleration mechanisms which can reach more than 100 TeV are still uncertain. For a more detailed discussion of acceleration mechanisms see [Bla87, Gai90, Kuz97].

8.2.2 Propagation of the cosmic radiation

Having so far discussed the origin and acceleration mechanisms of cosmic rays, we now consider how particles propagate. The easiest case is that of the neutrinos, as due to their very weak interaction they travel in a straight line and are therefore particularly suitable for a search for sources (see section 8.4). For the classic cosmic radiation, i.e. charged particles, the ratio of spallation products, such as Be, B, to the primary nuclei C, N, implies that in the GeV region the cosmic radiation must on average have crossed about 5–10 g cm^{-2} of matter. Integrating the mass along a line through the galaxy, however, results in only 10^{-3} g cm^{-2}. This implies a long circulation time and, due to the much longer path, a diffusion effect within a closed volume. For higher energies the mass traversed decreases, which points to a lower circulation time within the volume, and shows furthermore that the accelerating mechanism comes before the actual propagation. The propagation of the cosmic radiation is determined mainly by magnetic fields. The propagation and acceleration is determined by several factors, and the development of the particle density $N(E, x, t)$ at position x with energy E can be adequately described by a transport equation [Gai90]:

$$\frac{\partial N}{\partial t} = \nabla \cdot (D\nabla N) - \frac{\partial}{\partial E}(B(E) \cdot N(E)) - \nabla \cdot uN + Q(E, t) - p \cdot N$$
$$+ \frac{v\rho}{m} \sum_{k \geq 1} \int \frac{d\sigma(E, E')}{dE} N(E')dE'. \tag{8.39}$$

The first term describes diffusion with a diffusion coefficient D, the second term describes the mean energy change $B = dE/dt$. The term can describe energy gain (for example through acceleration) as well as energy loss (e.g. through ionization). The third term describes convection with a velocity u, and Q represents the source strength. The fifth term describes the loss of a particular nuclide due to collision and decay, while the last term represents the so-called cascade term, and describes the change of abundance from both higher energy cascades and nuclear fragmentation processes.

   The simplest model for the description of propagation of cosmic radiation is the 'leaky box model' [Sha70]. This describes the free expansion of particles in a closed volume, with a time independent, energy dependent escape probability τ. Hence the diffusion term in equation (8.39) can be replaced by $-N/\tau$. Assuming a delta-like source function and the absence of convection and energy changing processes, equation (8.39) can be solved simply to give

$$N(E, t) = N_0(E) \exp(-t/\tau). \tag{8.40}$$

Unstable nuclei, such as ^{10}Be and ^{26}Al, are suitable to investigate the circulation time. Their lifetimes, of $\tau \approx 3.9 \times 10^6$ and $\tau \approx 1.0 \times 10^6$ years respectively, are exactly of the right order of magnitude. The experimentally observed data suggests that a larger volume than the galactic disk is required to act as a closed volume, for example the galactic halo. In contrast to 'leaky box models', which describe the ratio of primary to secondary nuclides with an energy dependent escape probability from the galaxy, 'nested leaky box' models use an energy dependent escape probability from the actual source volume [Cow73]. An example is supernovae in dense molecular clouds. Another, perhaps more realistic description, is provided by diffusion models [Gin80]. In these models, the diffusion operator in equation (8.39) is assumed to be variable. While the 'leaky box' model in general leads to a homogeneous, isotropic distribution, gradients and anisotropies can also be produced with diffusion models. These models are however not discussed further here. A summary of the propagation of cosmic rays can be found in [Ces80, Gai90].

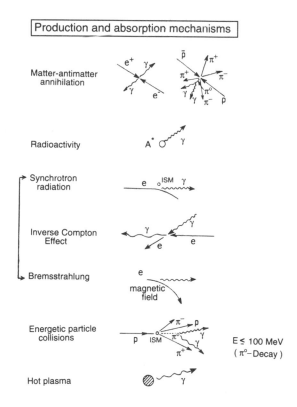

Figure 8.15. Examples of processes that can contribute to the production of gamma radiation in the universe (for [Mur93]).

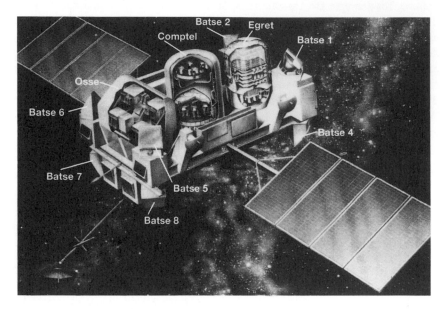

Figure 8.16. Schematic illustration of the Compton observatory (GRO). It has four detectors: EGRET (in an energy region of 30 MeV–20 GeV), COMPTEL (1–30 MeV) and OSSE (100 keV–10 MeV) and the BATSE experiment, consisting of eight detector units, (sensitive in an energy region of 20–600 keV) for the detection of gamma-ray bursts (from [Sch94]).

8.3   X-ray and γ-ray astronomy

Many of the objects discussed above, such as quasars, neutron stars, supernovae and black holes, allow the emission of high energy γ-radiation [Hip90] (see figure 8.15).   While earlier space missions such as the Einstein [Tuc85] and Uhuru satellites [Gia71], produced interesting information, this branch of astrophysics has recently made great advances through the ROSAT mission (Roentgen satellite) [Trü90, Trü93, Böh94, Böh95, Bec95, Has95] (X-ray region, figure 7.12 and the Compton-gamma-ray observatory (GRO) [Sch91a, Sch94, Sch95, Fic95] (γ-region (see figure 8.16)).   For example, the ROSAT all sky survey contains about 130 000 X-ray sources, about 150 times the number previously known [Trü97].   The X-ray emission from the *comet Hyakutake* could be mentioned as a recent and unexpected discovery.   Further satellites exploring the X-ray sky are the Russian GRANAT (launched 1989), the Japanese ASCA (launched 1993), the Italian–Dutch Beppo-SAX (launched 1994) and the American RXTE (launched 1995).   Further missions are the planned European XMM satellite and the American AXAF telescope.   AXAF together with the already launched HST and GRO, and the VLA (see figure 6.8(a)) and another future infrared telescope SIRTF form the basic skeleton of 'Great Observatories' of NASA for the next twenty years.

Due to the great diversity of phenomena and objects, we restrict ourselves here to a brief discussion of only a few topics. The production of γ radiation in supernova explosions is discussed for the example of SN 1987a in chapter 13. Very energetic γ radiation with energies of more than 100 MeV is discussed in a separate chapter (see section 8.3.6). For a more detailed discussion of X-ray and γ-ray astronomy see e.g. [Mur93, Mat94a, Cha95a, Ram95].

8.3.1 ^{26}Al in the Milky Way

^{26}Al is a radioactive isotope and decays via β^+ decay and electron capture

$$^{26}\text{Al} \rightarrow {}^{26}\text{Mg} + \gamma \ (1809 \text{ keV}) \tag{8.41}$$

with a half-life of $T_{1/2} = 1.04 \times 10^6$ years. ^{26}Al is mainly produced via the reaction $^{25}\text{Mg}(p, \gamma)^{26}\text{Al}$. We therefore assume that the isotope is mainly produced by novae and supernovae. This isotope is interesting for several reasons. Because of the similarity of the half-life with the expected circulation time of cosmic rays inside the Milky Way, it can act as a probe of the latter. Furthermore, the sources of the ^{26}Al can be examined. Its half-life is exactly in the region in which it has not gone too far from its point of origin, but on the other hand, is already far enough away that a significant interaction with the interstellar medium has taken place. The observation is thus independent of the details of the dynamics of the supernova. It is advantageous for observation that the isotope produces a narrow emission line; it was in fact the first γ line observed from nucleosynthesis in stars. For details of ^{26}Al observations in the Milky Way see [Ram95, Pra96].

The first observation with the HEAO-3 satellite showed a γ flux at the expected energetic position of the ^{26}Al line (the energy resolution of the detector did not allow the line to be clearly resolved) in the direction of the galactic centre of [Mah84]

$$\Phi(1809 \text{ keV}) = (4.8 \pm 1.0) \times 10^{-4} \gamma \text{ cm}^{-2} \text{ s}^{-1} \text{ rad}^{-1}. \tag{8.42}$$

From this value model calculations predict the entire amount of ^{26}Al in our Milky Way to be between 1.7–$3M_\odot$, somewhat more than had previously been thought [Ram77]. Computer simulations of supernova Type II explosions (see chapter 13) do however seem to be compatible with this value [Woo90b]. Due to the poor angular resolution, sources were previously not resolved. New observations with the COMPTEL detector on GRO show however a 'flaky' distribution of ^{26}Al in the galactic disk [Die93, Die95]. A comparison with possible sources within the disk shows a good correlation with Wolf–Rayet stars [Sch94]. These are massive stars such as those from which supernovae Type II's originate. The poor observed correlation disfavours supernova Type I's

Figure 8.17. The start of a balloon flight, here the GRIS experiment (collaboration between ESA, NASA, MPI for nuclear physics and Kurchatov Institute) with enriched ^{70}Ge detectors for the high resolution examination of γ lines from the centre of the galaxy. One of the authors is visible in the foreground (H V Klapdor-Kleingrothaus) (from [Kla94a], see also [Bar94]).

as a source. It is hoped that in further observations clues to the synthesis of elements in supernova can be obtained. Projects with germanium detectors (good energy resolution), especially those enriched in ^{70}Ge (reduced background), seem particularly suitable [Geh90, Kla91a], as has already been shown in balloon experiments [Bar94] (see figure 8.17). Such detectors should therefore be used possibly in future satellite projects at the beginning of the next millennium (see e.g. [Kla97g]).

Also of interest is the observation of the ^{44}Ti line at 1.156 MeV in the supernova remnant Cas A by COMPTEL [Iud94, Die95, Woo95b]. This line allows the search for supernova remnants less than 100 years old.

8.3.2 The 511 keV line in the Milky Way

The 511 keV line from e^+e^- annihilation is an important line in astrophysics. For example, because of accretion and acceleration of the surrounding mass, processes expected at the edge of a black hole give rise to massive amounts of e^+e^- annihilation, which should be reflected in an observable line at 511 keV (or red-shifted to lower energy) [Haw74] (see figure 8.18). Super-heavy black holes (10^6–$10^8 M_\odot$) can, moreover, be a good explanation for the processes in galactic nuclei. The central area of our Milky Way has been located near the radio source Sgr A* [Gen87], with a massive black hole of $\geq 2.45(\pm0.4) \times 10^6 M_\odot$ inside a radius of ≤ 0.015 pc around Sgr A*, and with a mass density of probably more than $10^{12} M_\odot \, pc^{-3}$ [Eck96, Eck97] (see figure 8.19). A line at 511 keV, as well as a red-shifted line, have been observed not far from this region. The observation of a narrow 511 keV line [Smi93c, Pur93] points to annihilation in the interstellar medium a significant distance away from any compact object. The source, connected with a broad line at 400 keV, is called 1E1740.7-2942 (the 'Great Annihilator') and is situated at an angular distance of 0.9° from the galactic centre [Sun91, Bou91]. It has a size of less than 0.3 pc and is therefore a good candidate for a *stellar* black hole. A similar observation of a line at about 480 keV also exists for Nova Muscae, another candidate for a black hole [Gol92]. More recent observations with GRO (OSSE) result in a flux in the 511 keV line from the direction of the galactic centre of [Pur93]

$$\Phi(511 \text{ keV}) = (2.5 \pm 0.3) \times 10^{-4} \gamma \, cm^{-2} \, s^{-1} \, rad^{-1} \qquad (8.43)$$

where the 511 keV line can be interpreted as consisting of two components, a Galactic bulge component which is concentrated in the direction of the Galactic center and a Galactic disk component producing significant emission at longitudes up to $\pm20°$. They are however also consistent with a constant point source located within a few degrees of the Galactic center and superposed on a diffuse Galactic distribution. These models are also consistent with the GRIS observations of the Galactic center in April 1992 (figure 8.17) [Lev93, Bar94], which were nearly simultaneous with one of the OSSE (GRO) Galactic center observations.

    Most of the positrons from the galactic plane probably originate from the radioactive decays of ^{56}Co, ^{44}Sc and ^{26}Al, which are produced in various galactic nucleosynthesis processes [Ram95].

8.3.3 Geminga

One of the greatest puzzles in astrophysics until recently was an object named Geminga (Gemini gamma ray) [Big96]. In missions in the early 1970s it could already be identified as one of the most intense gamma sources [Fic75], but it was, however, never seen in any other part of the spectrum. Only after a long and intensive search, lasting several years, could a weak object of the 25th magnitude be discerned in the visible, which was identified with the gamma source. Its origin was not however understood. ROSAT has now proved that the object is a neutron star [Hal92]. It has a period of 0.237 s. The observation could be verified with the EGRET experiment on GRO [Dek92], and with knowledge of the period the source could also be identified in the older COS-B data [Big92]. By measuring the proper motion of the optical counterpart, the distance to Geminga could be

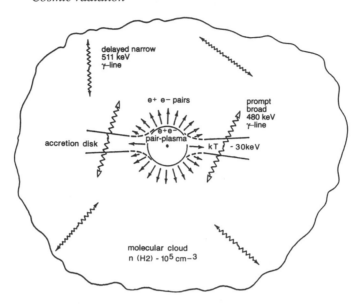

Figure 8.18. A schematic model of the 511 keV positron annihilation line production near a black hole. Both a short term intensity variation of the 511 keV line and continuous annihilation radiation can be understood in terms of different places of origin (see [Lin89, Ram92]).

established as a maximum of 380 pc [Big93]. It is therefore one of the closest known neutron stars. There is also the possibility that the associated supernova explosion could be responsible for the so-called 'local bubble' [Geh93]. This is an area of lowered interstellar gas concentration, at the edge of which our solar system is situated. The energy of a close supernova could have blown the gas away.

8.3.4   The Crab and Vela pulsars

One of the best researched examples for the late phases in the evolution of massive stars is the supernova in the constellation of Cancer. This event, recorded in ancient Chinese writings, took place in 1054 AD, and today we can observe at that place a large expanding gas cloud (Crab nebula) with a central pulsar (PSR 0531+21). It was the very first pulsar observed [Hew68]. The pulsar has a period of 0.0332 s and radiates from the radio to the gamma ray region. It is also seen by several TeV γ-experiments (some of them are listed in table 8.3) and is used as a calibration source in VHE γ astronomy. The Vela pulsar (PSR 0833−45) is a similar system at a distance of about 500 pc. This pulsar also radiates from the radio to the gamma ray region with a period of 0.089 s. Many of the phenomena of pulsars and the creation mechanisms for

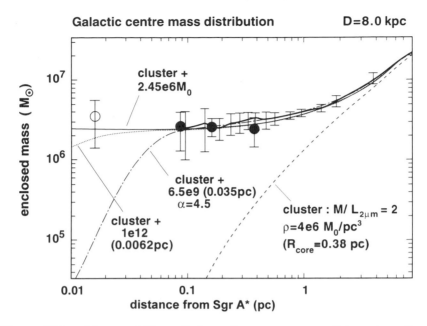

Figure 8.19. Mass modelling of the stellar proper and radial motions (for a Sun–galactic-centre distance of 8.0 kpc). The thick curve (with 1σ error bars) gives the enclosed mass as a function of distance from Sgr A* derived from the Jeans equation mass modelling of the stellar radial velocities, assuming anisotropy $\delta = 0$ ([Gen96] and references therein). Filled circles and 1σ error bars denote the masses estimated from independent Jeans modelling of the proper motions for 39 stars (78 motions) between 0.9 and 8.8 arcsec again for $\delta = 0$. The open circle (and combined statistical and systematic error bar) denotes a preliminary mass estimate from the motions in the Sgr A* (IR) cluster. The thick dashed curve represents the mass model of the stellar cluster observed in the near-infrared ($M/L(2\ \mu m) = 2$, $R_{core} = 0.38$ pc, $\rho(R = 0) = 4 \times 10^6 M_\odot$ pc^{-3}). The thin continuous curve is the sum of the visible stellar cluster plus a central $2.45 \times 10^6 M_\odot$ point mass. The thin dash–dotted curve is the sum of the visible cluster plus a dark cluster of core radius 0.035 pc and central density $6.5 \times 10^9 M_\odot$ pc^{-3}. For $R \gg 0.035$ pc the dark cluster density is proportional to $R^{-4.5}$. The dotted curve is the sum of the visible cluster and a dark cluster of a core radius 0.0062 pc and central density $10^{12} M_\odot$ pc^{-3} (from [Eck97]).

this radiation are presently still unclear, so that they are an interesting area of research (see e.g. [Tay86, Lyn90, Man93]).

8.3.5   Gamma-ray bursters

'Gamma-ray bursters' [Har91] are phenomena which are not completely understood at present. They are bursts in the gamma region with durations

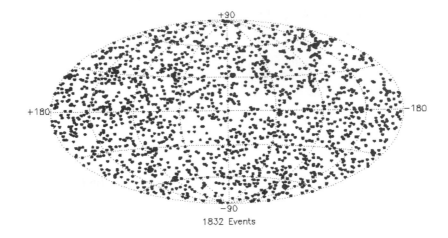

1832 Events

Figure 8.20. The origin of the gamma-ray bursters is at present one of the greatest mysteries of astrophysics. The more than 1832 bursts recorded by the Compton gamma-ray observatory GRO have an isotropic distribution. Currently either a cosmological source or a galactic source are discussed (with kind permission from the BATSE team and NASA, see also [Har95]).

of 10 ms up to 100 s and only a few generic characteristics. The energies are in the region 10 keV–100 MeV, and the intensity is roughly described by [Pis94]

$$I(> E) \approx 7 \times (E^{-1.25}/\text{MeV}) \quad \text{photons cm}^{-2}\,\text{s}^{-1}. \tag{8.44}$$

The general interpretation of these objects in the past was that they were neutron stars, on whose surface brief thermonuclear explosions of accreted material were taking place (see e.g. [Mur93]). These objects should then be mainly observed in the galactic disk. GRO has seen more than 1800 of these bursters, at a rate of about one burst per day [Mao92, Sch95, Har95], and surprisingly, however, the distribution across the sky is completely homogeneous (see figure 8.20). This makes the just mentioned view of their origin untenable. On the other hand neutron stars produced in asymmetric supernovae explosions could be accelerated in the galactic halo [Jan95a, Woo95] and produce an isotropic distribution [Pod95]. A galactic halo consisting of fast neutron stars which accrete planetoids and acting as a source of the isotropic component of gamma-ray bursters is discussed in [Col95]. Another scenario could be merging neutron stars [Pir94]. There are currently two explanations under discussion: objects in the galactic halo or extra-galactic, cosmological objects [Nar92, Woo93, Woo95]. A first follow-up detection of the burst gRB970228 with the BeppoSax satellite (X-ray region) [Par97] and the Hubble Space Telescope (optical region) [Sah97] seem to confirm the cosmological origin. A common source for gamma-ray bursters and the highest energy

Table 8.3. Observations of the Crab nebula (from [Cro93] and [Kon96]).

Energy (TeV)	Flux (10^{-12} cm^{-2} s^{-1})	Significance (Sigma)	Method	Group
0.2	170	5.8	Cerenkov	Gamma
0.4	70	45.5	Cerenkov	Whipple
0.6	27	5.7	Cerenkov	ASGAT
1	8	14	Air shower	HEGRA
3	4.4	8	Cerenkov	Themistocle
10	< 1.2		Air shower	Tibet
30	< 0.18		Air shower	Tibet
40	< 0.44		Air shower	Cygnus
75	< 0.126		Air shower	HEGRA
160	< 0.12		Air shower	CASA-MIA
190	< 0.021		Air shower	CASA-MIA

cosmic radiation is discussed in [Wax95]. For overviews of this subject see [Mur93, Pac93, Woo93, Pir94, Woo95, Wax95, Fis95a, Kla96].

8.3.6 Ultra high energy γ-radiation

Ultra high energy γ-radiation allows—similar to high-energy neutrinos—the search for sources of cosmic rays . It originates from the decay of neutral pions produced when high energy protons interact with matter. Another possibility is production through the interaction of electrons with matter (bremsstrahlung) or magnetic fields (synchrotron radiation), as well as through the inverse Compton effect. The γ-ray sky has been examined up to a few GeV with satellites. For example, with the EGRET detector on GRO the sky is being searched up to energies as high as 20 GeV. Information on ultra high energy γ-radiation in the region 0.1–10 TeV, on the other hand, is gained with the help of the air Cerenkov technique, and at even higher energies with the help of the air shower arrays [Wee88]. Due to the very small fluxes the positive identification of sources in the TeV region is currently of little statistical significance and partially contradictory. A clear source for gamma radiation in the TeV area is, however, the Crab nebula [Vac91, Kon96] (table 8.3). Recently a TeV gamma signal from the Seyfert galaxy Markarian 421 has been observed at the Whipple observatory (see figure 8.21) [Pun92] which is also seen by HEGRA [Pet96a]. The mean observed flux is six standard deviations above the background and amounts to

$$\langle \Phi \rangle = 1.5 \times 10^{-11} \quad \text{photons cm}^{-2}\,\text{s}^{-1} \quad \text{for} \quad E > 0.5 \text{ TeV}. \tag{8.45}$$

Figure 8.21. The 10 m optical reflector of the Whipple observatory for the detection of high energy γ radiation. The telescope has been operating since 1968 on Mount Hopkins in Arizona, and is the largest of its kind for the detection of air Cerenkov radiation. A second telescope with a diameter of 11 m now exists. The small detail shows the arrangement of the light sensitive photo-multipliers in the primary focus (from [Sky95b]).

This corresponds to about 30% of the signal from the Crab pulsar. Two further sources of TeV photons are the pulsars PSR 1706-44 [Kif94] and the galaxy Markarian 501 [Lor96, Aha97]. This is especially interesting, because the optical depth in the region around 10^{15} eV is rather small because of the process $\gamma\gamma \rightarrow e^+e^-$ with the 3K radiation and a so-far unknown infrared background, which can be limited in this way.

At least hints of an emission exist for the X-ray binary Cygnus X-3 (see [Bon88, Mur91]) (see figure 8.22). A possible signal from Cygnus X-3 in the PeV-area [Sam83] created great excitement in the field of high energy γ-ray astronomy, even if subsequent experiments could not confirm the signal.

Experimentally, the main difficulty is to distinguish a photon-initiated shower from those produced by particles. Some of the main differences in the behaviour of detected Cerenkov light are illustrated in table 8.4. Data taken with imaging Cerenkov telescopes, such as the Whipple observatory [Wee88, Cro93], have proved to be very successful. It consists of a 10 m optical reflector, with a camera consisting of 109 pixels in the form of photo-

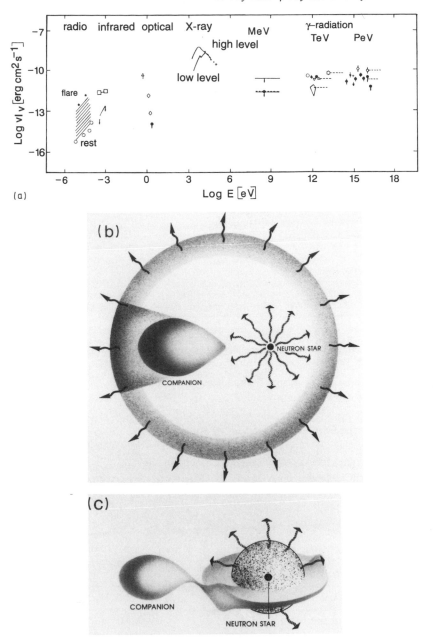

Figure 8.22. (*a*) Energy emission of the X-ray binary star Cygnus X-3. (*b*) and (*c*) Illustrations of the possible construction of the binary star system (from [Bon88]).

Figure 8.23. The HEGRA detector on La Palma. (*a*) The various types of detector in the HEGRA experiment are grouped around the central measurement region containing the readout electronics and computers. The small white and grey boxes house the scintillation counters. Beside some of them stand small cubes containing open photo-multipliers for measuring the Cerenkov light (AIROBICC). The large 'towers' are used for the identification of muons in air showers. Near the centre of the installation, which is 180×180 m^2, stands a prototype Cerenkov telescope with a mirror size of 5 m^2. Beside to the right is the central telescope of the system of 5 Ccrenkov telescopes with a mirror size of 8.5 m^2. (*b*) The central telescope with the high resolution camera. 271 pixels are summed together to form a visual field of nearly 5°. The receiver is constructed of 30 mirrors with a diameter of 60 cm each (with kind permission of the HEGRA collaboration).

Table 8.4. Characteristics of the Cerenkov light from γ and hadron induced showers (from [Cro93]).

Characteristic	Parameter	γ	Hadron
lateral spread	depth	broad	narrow
	uniformity	regular	irregular
angular distribution	image size	narrow	broad
duration	pulse width	short	long
colour	ratio UV/visible	small	big

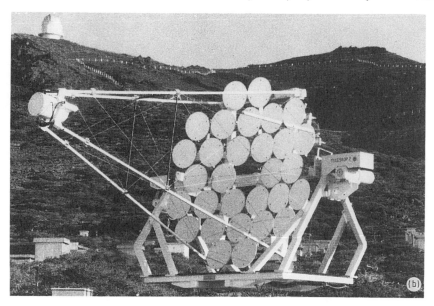

Figure 8.23. Continued.

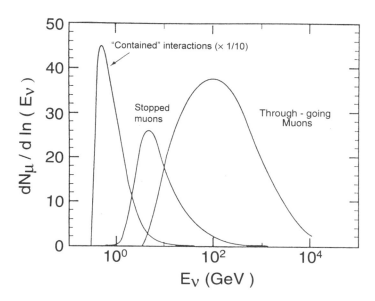

Figure 8.24. The number of muons as a function of the incident neutrino energy. Different detection strategies are necessary depending on the energy. While the low energy neutrinos produce 'contained' events, the higher energy neutrinos produce muons that cross the detector (from [Gai94a]).

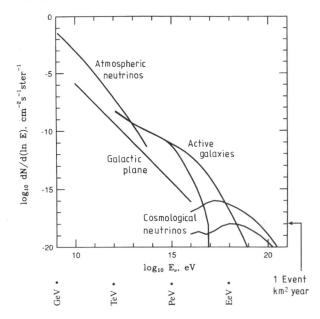

Figure 8.25. Composition of the high energy cosmic neutrino spectrum as a function of energy. Also shown is the experimental limit which, in a detector of one square kilometre, would result in one event per year (from [Hal95]).

multipliers situated at its focal plane. The separation of the tubes subtends $0.25°$, and the telescope has a threshold of 0.4 TeV. Further telescopes of this kind are currently being built [Cro93]. One, for example, is the HEGRA detector on La Palma [For95, Pan95, Lor96] (see figure 8.23). In its final form it will consist of four elements:

- A conventional air shower array consisting of 256 elements

- 17 large 'towers' for detecting the tracks of shower muons. The direction of the muon can be measured with a resolution of up to $1°$.

- A system of 5 imaging Cerenkov telescopes (IACT) each with a camera with 271 pixels and 8.5 m^2 mirror surface

- A Cerenkov detector (AIROBICC) consisting of 49 elements with an opening angle of about $60°$, with an angular resolution of about $0.15°$.

The lower thresholds of the detectors are at about 0.5 TeV (IACT), 15–30 TeV (AIROBICC) and 50–100 TeV (scintillators). Due to the combination of these various detector components HEGRA will after its completion be a particularly valuable instrument for γ-ray astronomy.

For a detailed exposition of high energy γ-ray astronomy see [Wee88, Mur93, Cro93, ICR95, Ann95].

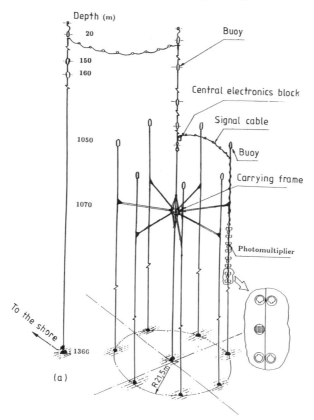

Depth (m)

Buoy

Central electronics block

Signal cable

Buoy

Carrying frame

Photomultiplier

To the shore

(a)

Figure 8.26. (*a*) The planned construction of the Baikal experiment NT-200. Eight rods, each 70 m long, are arranged in the form of a heptagon with a central rod. Several pairs of photo-multipliers are attached to each rod. The arrangement is held by an umbrella-like construction. The experiment is situated about 600 m off shore, at a depth of 1.1 km. (*b*) Installation of one of the rods of the NT-36 experiment in Lake Baikal. As the lake freezes over in winter, this season is ideal for installation (with kind permission of Ch Spiering).

8.4 High energy neutrinos

We now return once again to neutrinos. As already mentioned, a possible interpretation of the deficit in the number of atmospheric muon neutrinos at about 1 GeV would be neutrino oscillations. This would also influence the number of muon neutrinos with higher energies. Due to the very steeply falling neutrino spectrum at higher energy the detection of neutrino interactions in the detector becomes ever more difficult (see figure 8.24). There is, however, the possibility to enlarge the effective area of the detector, at least for ν_μ. 'Contained events', where, as already mentioned, the interaction and detection of all end

Figure 8.26. Continued.

products takes place within the detector, exist for both flavours (ν_e and ν_μ). For muon neutrinos there exists an additional possibility of detection, namely when they impinge on the detector from below after penetration of the Earth and interact in the rock beneath the experiment. The resulting muons ('upward going muons') cross the detector from bottom to top, contrary to atmospheric muons. The effective detection volume is now the product of the detector cross section and the range of muons in the rock. TeV muons, for example, have a range of about one kilometre in rock. The expected flux for such events lies in the region of 10^{-13} cm^{-2} s^{-1}sr^{-1} [Gai90] after penetration of the Earth. Neutrinos produced at the far side of the Earth can be used to check oscillation lengths of about 13 000 km, i.e. much larger than the atmospheric neutrinos discussed in section 8.1.5. Neutrino oscillations therefore should show a zenit angle dependence

Several experiments have thus far reported the observation of such muons [Bec92, Mor91a, Bol91, Mic94, Fuk94, Ahl95, Mon97]. Of these only the Kamiokande experiment [Fuk94, Tak97] shows clear evidence for oscillations (see figures 2.17 and 2.21). For a detailed discussion see [Gai94, Sta96].

High energy neutrinos can be produced not only in the atmosphere, but also in astrophysical processes (see figure 8.25) [Gai95, Gai96a]. As in particle accelerators, accelerated protons striking nuclei produce pions and kaons. Their decay produces a source of neutrinos. A further possibility is pion production from photo-production, when a highly accelerated proton reacts with a low energy photon (see section 8.2). Possible sources are identical with those

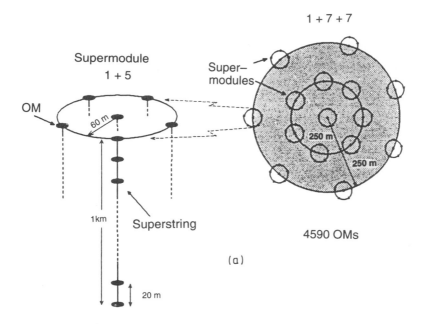

Figure 8.27. (*a*) An early sketch of the construction of the AMANDA experiment as a first stage for a planned 1 km³ detector (from [Hal95]). (*b*) Installation of photo-multipliers for the AMANDA experiment. Holes in which photo-multipliers are sunk are thawed into the ice of the Antarctic using hot water drilling. Afterwards the water above will re-freeze (with kind permission of F Halzen).

of cosmic ray particles. What makes the search for sources of high energy neutrinos so attractive is the fact that, on their way to Earth, they experience no significant interactions and therefore point directly to their source. Because of pion production, high energy γ-radiation and neutrinos should be produced together; however, the search for neutrinos has the advantage compared to photons, that there is no absorption from astrophysical objects (such as very dense media). This means that sources can be observed by neutrinos that remain undetected in the γ-region. For a discussion of astrophysical neutrino sources see figure 8.25 and [Gai95, Hal95, Gai96a, Gai96b]. A further source for high energy neutrinos would be the decay of massive exotic particles. We will return to this in chapter 9.

The detection of the highest energy neutrinos and muons uses natural water resources (lakes, oceans, ice) with large sensitive areas as Cerenkov counters (see i.e. [Bal92, Spi93]). These have a much greater amount of target material and therefore much larger event rates than conventional detectors; however, they also have very different background problems. It is difficult to give a theoretical

Figure 8.27. Continued.

estimate for the rate in such an experiment, since this requires knowledge of the nucleon structure function in a so-far unexplored region. On the other hand, the study of the highest energy neutrinos may offer the possibility to measure the structure functions in this unexplored region [Gan95].

Four experiments which were or are currently under construction are discussed here:

(i)   NT-200 experiment [Bel94, Spi96]: this experiment is being installed in Lake Baikal (Russia) (see figure 8.26). It is situated at a depth of about 1.1 km. One of the advantages of this experiment is that Lake Baikal is a sweet water lake, which contains practically no ^{40}K, which could produce a large background. The photo-multipliers used in the light detection have a diameter of 37 cm, and are fastened on rods over a length of about 70 m. The rods are arranged in the form of a heptagon, and are attached to an additional rod at the centre. The whole arrangement is supported by an umbrella-like construction, which keeps the rods at a distance of 21.5 m from the centre. The photo-multipliers are arranged in pairs, with one facing upwards and one downwards. The distance between two photo-tubes with the same orientation is about 7.5 m, and between two with opposite orientation about 5 m. In a prototype, NT-36, 6 pairs of these tubes were fixed on 3 rods, and first data have been recorded. Since April 1996, 96 photomultipliers on 4 strings are operating. Two very good candidates for high energy neutrinos have meanwhile been observed [Bel97].

(ii) DUMAND experiment [Sam94, Bos96]: this detector was to be installed off the western coast of Hawaii, at a depth of 4.8 km. It should consist of a total of nine rods arranged in the form of an octagon, with an additional unit in the centre. The distance of the rods in the octagon is 40 m. The optical detectors are also photo-multipliers and 24 are installed per rod, at a separation of 10 m and viewing downwards. The detector has a height of 230 m, a diameter of 105 m and an effective detection area for muons of about 20 000 m^2. First measurements with an installed string have been carried out. The project has been stopped in the meantime.

(iii) The NESTOR experiment [Res94, Nut96]: this experiment is planned for the Mediterranean, off the coast of Greece, at a depth of 3.8 km. It is planned that it will be constructed of seven rods in the form of a hexagon with a radius of 100 to 150 m and a central rod. In contrast to DUMAND the optical detectors are spaced in groups on the rod rather than linearly along it. Each group consists of a hexagon of radius 16 m, at each apex of which is installed a pair of phototubes (one pointing upwards, one downwards). Twelve such hexagons are attached to each rod with a separation of 20–30 m. The total effective area is about 10^5 m^2.

(iv) The ANTARES experiment [Bla97] is a large scale experiment planned near the French Mediterranean coast.

(v) The AMANDA (Antarctic muon and neutrino detector array) experiment [Low91, Hal95, Hal97, Hal97a]: instead of water the AMANDA detector uses the ice of the Antarctic as the Cerenkov material (see figure 8.27). The phototubes are arranged at a depth of between 1 and 3 km. The ice is extremely low in background counts and, because of cooling, strongly reduces the noise of the photomultipliers. The first test results were very promising [Bar93a]. 302 photo-multipliers in 10 strings at a depth of 1500–2000 m are soon to be in operation, which should grow to 800 between 1998 and 1999. Special electronics suited to a search for supernovae and gamma ray bursters has been installed. As is the case in the Baikal experiment, the first good candidate for a very high energy neutrino interaction has been observed.

The AMANDA detector with its effective area of about 0.1 km^2 which will be sensitive to neutrinos in the region between 100 GeV and 1 PeV, will be completed in the near future. There are concrete ideas as to how the area of this detector is to be increased to around 1 km^2 [Bar92b, Hal95, Hal97a].

Chapter 9

Dark matter

We now turn to what is probably the most interesting problem of modern astrophysics, the problem of dark matter. This problem shows how particularly important the co-operation between particle physics and astrophysics is. Roughly speaking, the problem is the realization that there seems to be a great deal more gravitationally interacting matter in the universe than is luminous. This chapter will deal with the questions of what leads to this assumption, the origin of the dark matter, whether it is baryonic, non-baryonic or perhaps exotic, and whether, if it truly exists, it can be detected experimentally.

9.1 Evidence for dark matter

9.1.1 Dark matter in galaxies

9.1.1.1 Rotational curves of spiral galaxies

Spiral galaxies are structures containing billions of stars, arranged in the form of a rotating disc with a central 'bulge' (see chapter 6). If we assume that the stars have a circular orbit around the galactic centre, the rotation velocities of single stars can be calculated from the equality of the gravitational and centrifugal forces, according to

$$F_G = \frac{GmM_r}{r^2} = \frac{mv^2}{r} = F_Z. \tag{9.1}$$

From this it follows that

$$v(r) = \sqrt{\frac{GM_r}{r}} \tag{9.2}$$

where M_r is the mass within the orbit of radius r. Here advantage was taken of the fact that for cylindrically and spherically symmetric distributions the forces due to the mass lying outside the orbit compensate exactly. Assuming the bulge is spherically symmetric with constant density ρ, then

$$M_r = \rho \cdot V_r = \rho \frac{4}{3}\pi r^3 \tag{9.3}$$

267

Hence for the innermost part of a galaxy a *rotation curve* (velocity as a function of the radial distance from the centre) of

$$v(r) \sim r \qquad (9.4)$$

follows. If one is situated outside the galaxy, M_r corresponds to the total mass of the galaxy. In this case therefore

$$v(r) \sim r^{-1/2} \qquad (9.5)$$

since $M_r = M_{\text{gal}}$. If the rotation curves of spiral galaxies are measured using the Doppler shift, the result in all galaxies observed so far is

$$v(r) = \text{constant}$$

for large r (also at distances corresponding to several times the extension of the optically visible disc), which means

$$M_r \sim r. \qquad (9.6)$$

This indicates the existence of an enormous mass extending far beyond the visible region, which, however, is invisible optically (see figure 9.1) [Ost74]. Thus, obviously nobody has ever seen an entire galaxy and therefore estimations of its total mass are highly uncertain. Even the mass of our Milky Way is unknown [Fic91, Kul92]. These observations lead to the hypothesis of a halo of dark matter. The distribution of globular star clusters suggests a spherical distribution [Bin87, Tho89]. Recently, during the examination of the spiral galaxy NGC 5907, strong indications for a dark halo have been found [Sac94]. There are more reasons for the hypothesis of a dark halo:

- Theoretical model calculations show that pure disc galaxies have a strong tendency to form bars, i.e. within the central nucleus a bar-like structure forms [Bin87]. Even though barred spiral galaxies do exist, they are relatively rare. A spherical halo with a significant amount of matter within the disc radius increases the stability of the pure disc structure thus producing a ratio of spiral galaxies compared to barred spirals consistent with observations [Ost73].
- The formation of galaxies probably originated from the gravitational collapse of a spherical proto-galaxy. This is still apparent today by the spherical halo of globular clusters, which are among the oldest objects in the universe.
- The observations of 'polar ring galaxies'. These have a ring of stars perpendicular to the disk surface. Measurement of the rotational velocities within the disc and within the ring results in flat curves in both cases. This implies that the velocity for large radii remains constant [Sch83]. This can be best understood by assuming a spherical mass distribution [Bin87].

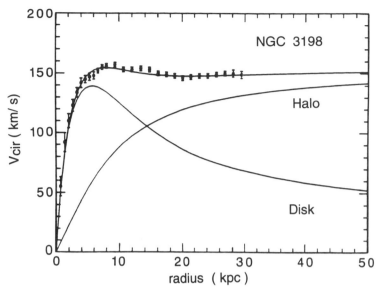

Figure 9.1. Rotation curve of the galaxy NGC 3198. The flat behaviour up to and far beyond the luminous edge can only be explained by a very massive dark halo. The points correspond to the observations, the solid line to a model of contributions from halo and disc (from [Alb85]).

- The Magellanic stream. This is a hydrogen bridge between the Magellanic clouds and our Milky Way. If it is explained by a common gravitational potential, it follows that a massive dark halo exists in our Milky Way [Fic91].

- The rotation velocity of the Magellanic clouds around the Milky Way. Most recent measurements for this imply the existence of an extensive, very massive halo [Lin95].

9.1.1.2 Elliptical galaxies

Considerations from stellar dynamics in elliptical galaxies imply that they also contain a significant fraction of dark matter. The velocity distribution in elliptical galaxies is, however, not so much determined by rotation, but rather by an anisotropic velocity field. Information about the inner regions of elliptical galaxies is obtained via measurements of the velocity dispersion and the luminosity profile of their surface. Assuming a spherically symmetric galaxy, hydrostatic balance and the ideal gas equation lead to a mass distribution of [Bin87]

$$M(<r) = \frac{k_B T r}{G \mu m_p} \left[-\frac{d \ln \rho}{d \ln r} - \frac{d \ln T}{d \ln r} \right] \tag{9.7}$$

where μ is the mean molecular weight, and m_p the proton mass. By measuring the density profile $\rho(r)$ and the temperature profile $T(r)$ it is in principle possible to determine the mass distribution. The density profile is obtained from the luminosity profile, since for an optically thin, fully ionized gas $L \sim \rho^2$. The galaxy best examined in this connection is M87. An analysis of its data shows an almost linear increase of mass up to above 300 kpc, which results in $M(r < 300 \text{ kpc}) \simeq 3 \times 10^{13} M_\odot$ [Ste92]. This would mean that more than 99% of M87 consists of dark matter. The question is whether M87 is a typical galaxy, since it is situated in the centre of the Virgo cluster, and therefore is in strong gravitational interaction with its neighbours. For further evidence of dark matter in elliptical galaxies see e.g. [Sag93]. A further possibility for the determination of mass appeared when it was realized that nearly all luminous elliptical galaxies contain about $10^{10} M_\odot$ of gas in the form of gas halos, with a size of at least 50 kpc [For85]. Due to the X-ray emission of this hot gas its temperature can be deduced to be about 10^7 to 10^8 K, which implies a velocity of the gas particles which lies far above the escape velocity derived from the visible mass. If this gas is really gravitationally bound, a great deal more mass is needed. The evidence for dark matter is supported by new observations of the X-ray halos with the ROSAT satellite [Trü93].

9.1.1.3   *Dark matter in dwarf spheroidals*

A particularly large amount of dark matter seems to be contained within dwarf spheroidals (see chapter 6). If these were really to be systems in dynamic equilibrium, they would require a central density of dark matter 10 times higher than expected from more luminous systems [DaC92, Sil93].

9.1.2   **Dark matter in clusters of galaxies**

We now turn to clusters of galaxies. The virial theorem

$$2\langle E_{\text{kin}} \rangle + \langle E_{\text{pot}} \rangle = 0 \tag{9.8}$$

is mainly used here for mass determination. In these estimates, however, it is necessary to remember that the virial theorem can only be used under certain conditions; i.e. a closed system in a state of mechanical equilibrium, and it applies to the time average of the system. Whether the observed clusters do indeed fulfil these conditions or, in other words, whether they are relaxed, is not easy to determine, and therefore the application of the virial theorem is questionable. Assuming nevertheless that it is applicable, the kinetic energy for N galaxies in a cluster is given by

$$\langle E_{\text{kin}} \rangle = \frac{1}{2} N \langle m v^2 \rangle \tag{9.9}$$

and with $\frac{1}{2}N(N-1)$ independent galaxy pairs the potential energy is

$$\langle E_{\text{pot}} \rangle = -\frac{1}{2}GN(N-1)\frac{\langle m^2 \rangle}{\langle r \rangle}. \tag{9.10}$$

With $(N-1) \approx N$ and $N\langle m \rangle = M$ an estimate of the dynamic mass is given by

$$M \approx \frac{2\langle r \rangle \langle v^2 \rangle}{G}. \tag{9.11}$$

By measuring the quantities r and v it is now possible to calculate M, so that for example for the Coma cluster values of [Ken83],

$$\frac{M}{L} \approx 300h\frac{M_\odot}{L_\odot} \tag{9.12}$$

result, to within an uncertainty of a factor two. Another method, probably better suited for testing the gravitational potential of clusters of galaxies, is to examine the hot X-ray gases. An examination with the help of ROSAT shows [Trü93, Böh94], that typically 10–40% of the total mass is in the form of this gas. However, it means also that since the percentage of visible galaxies is only 1–7%, two thirds of the entire cluster mass consists of dark matter. Another strong argument for dark matter in very small galactic clusters was given by ROSAT. Significant X-ray emission of the relatively small galaxy cluster NGC 2300, originating from the hot ($T = 10^7$ K) intergalactic gas was observed. This was found to be concentrated towards the cluster centre. This makes the assumption that this gas is gravitationally bound to the cluster seem sensible. From this a mass of the group can be derived, in which the baryonic part in the form of gas and galaxies is only 4% (maximum 15%), and the rest is dark matter [Mul93]. On the other hand the ROSAT observation of a further small galaxy group (HGC62) suggests that at least 13% of the total mass consists of baryonic matter [Pon93]. The observation of 13 such small clusters concluded that the hot X-ray gases amount to about 10–30% of the total mass of the clusters [Pil95, Sch95]. Further clarification can only come with more observations. For this the gravitational lens effect (see section 9.2.2.1) and the Sunyaev–Zeldovich effect (see chapter 7) may become more important in future (see also [Böh95]).

9.1.3 Dark matter and large scale structure

In chapter 6 it was seen that the modelling of large scale structures in the universe results in limits on Ω. The POTENT-method (see chapter 6), which calculates the underlying mass distribution from the observed velocity field, results in values for Ω of between 0.3 and 2.5, i.e. far above the value that corresponds to the *baryonic* density produced in the primordial nucleosynthesis (see chapter 4).

Further information about the composition of dark matter comes from the fluctuation spectrum we discussed in chapter 6, from which the large scale structure developed. The measurements of the anisotropies of the microwave background radiation through COBE and the IRAS galaxy survey allows this spectrum to be determined experimentally. The result is that for a consistent description both hot and cold dark matter are needed (see also chapter 9.2.3.3).

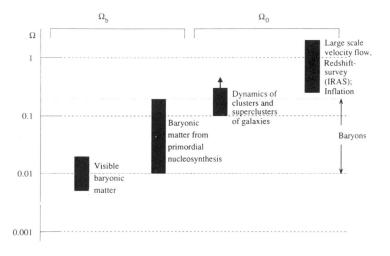

Figure 9.2. The value of Ω as a function of the distance scale of the observed region of the universe. The measurements seem to indicate that the greater the scale on which we observe, the larger the value of Ω (from [Kla95]).

9.1.4 Dark matter and cosmology

One cosmological aspect of the problem of dark matter is the destiny of the universe, characterized by the value of Ω. The theoretically attractive model of inflation which was discussed in chapter 3 predicts in general that $\Omega = 1$. The determination of q_0 described in chapter 5 does indeed seem to be close to this value (see figure 9.2). Considering the results of the primordial nucleosynthesis (see chapter 4), the maximum contribution of baryonic matter is $\Omega_B \leq 0.11$, which means that under the assumption of inflation the missing mass *has to be* of *non-baryonic* nature—as already deduced from the large scale structure. (This is only true for the case $\Lambda = 0$. The energy density of the vacuum connected with $\Lambda > 0$ on large scales could to a large extent produce the same effect as dark matter (see section 9.2.1)!) The observational results do indeed seem to be consistent with an evenly distributed, non-baryonic component, as all the *dynamic* tests discussed would not be sensitive to it. Its small contribution would simply be swamped by the strong density contribution due to normal matter in galaxies and clusters.

In summary it is clear that the observed visible matter is insufficient to close the universe. An explanation of the rotation curves of galaxies and the behaviour of galaxy clusters also does not seem possible. If inflation is taken seriously, dark matter is inevitable if one assumes that $\Lambda = 0$. Baryonic forms of dark matter seem to be able to explain the rotation curves, but fail in the large scale problems. Inflation and primordial nucleosynthesis require more than 80% of matter in the universe to be in an unknown, dark, non-baryonic form!

9.2 Candidates for dark matter

Having discussed the evidence for the existence of dark matter, and found its existence increasingly verified, we discuss in this chapter the nature of this dark matter. Before dealing with the candidates from elementary particle physics we discuss alternative explanations.

9.2.1 Alternative candidates: cosmological constant, MOND theory, time-dependent gravitational constant

- *The cosmological constant*
  As seen in chapter 5, the vacuum has a non-zero energy density, which manifests itself in the cosmological constant Λ. A non-vanishing Λ results in a re-definition (decrease) of the critical density, and the density parameter:

$$\rho_{c,0} = \frac{3H_0^2 - \Lambda c^2}{8\pi G} \tag{9.13}$$

and

$$\Omega_0 = \frac{8\pi G \rho_0}{3H_0^2 - \Lambda c^2}. \tag{9.14}$$

From this follows an estimate for Λ, based on the condition that the critical density must not be negative, of

$$\Lambda \leq \frac{3H_0^2}{c^2} \approx 3.5 \times 10^{-56} \text{ cm}^{-2}. \tag{9.15}$$

If we assume that $\Lambda \neq 0$ is not only needed to create an inflationary universe, but that it has also played a role in the further evolution of the universe (see i.e. [Pee84, Blo84, Kla86a]), then it can be shown that the $\Omega = 1$ needed in inflationary models can be produced, if the cosmological constant had a value of

$$\Lambda_{\Omega=1} = \frac{3H_0^2}{c^2}(1 - \Omega_{\Lambda=0}) \tag{9.16}$$

where $\Omega_{\Lambda=0}$ would be the apparent Ω determined from ρ_0. If we assume $H_0 = (75 \pm 25)$ km s^{-1} Mpc^{-1} and $\Omega_0 = 0.2 \pm 0.1$, then

$$\Lambda_{\Omega=1} = (1.6 \pm 1.1) \times 10^{-56} \text{ cm}^{-2} \tag{9.17}$$

which is compatible with all experimental limits (see chapter 5). A finite value of the cosmological constant therefore has an effect identical to that of a homogeneous distribution of dark matter. Such a solution would therefore obviate to some extent the necessity of an excursion into the zoo of exotic elementary particles.

- *Deviations from Newtonian dynamics*
  These MOND (modified Newtonian dynamics) theories change the law of gravity below a critical acceleration of $a_0 = 10^{-8}$ cm s^{-2}, so that (see e.g. [Mil83, Bec84, San90])

$$a_G = \frac{GM}{r} + \frac{\sqrt{GMa_0}}{r}.$$ (9.18)

  This could explain the flat rotation curves. For a discussion of deviations from the currently accepted law of gravity, see e.g. [Kla95].

- *A time-dependent gravitation constant*
  A time-dependent gravitation constant (i.e. $G \sim t^{-1}$ [Dir37]) can be of great importance in the calculation of the primordial element abundances, and therefore for the prediction of Ω_B [Sta92b]. However, precision measurements show no sign whatsoever of a time dependence, moreover such behaviour would finally lead to a non-conservation of energy, an extremely unpalatable possibility. For more discussions see [Fla75, Fla76, Nar93], and in context with dark matter [Tri97].

All of these models offer solutions to only some aspects of dark matter and cannot solve all the problems simultaneously. We therefore turn to other candidates.

9.2.2 Baryonic dark matter

These kind of candidates are astrophysical objects that are not too exotic, such as planets, brown dwarfs, white dwarfs or black holes [Car94]. These are bodies that have either never managed to become stars ($M < 0.08 M_\odot$, such as planets or brown dwarfs), or are the remnants of a star, such as the white dwarfs or black holes. Also, on larger scales low surface brightness galaxies or cold hydrogen clouds, which escape observation, could contribute to dark matter. However, we have to consider here that in the solution of the problem of dark matter we have the already mentioned limit from primordial nucleosynthesis on the fraction of baryonic matter. It is remarkable on the other hand that the rotation curves of galaxies imply values for Ω that correspond well to this upper limit. This means that baryonic matter seems adequate to at least explain the rotation curves of galaxies. For this reason a search for so-called MACHOs ('massive compact halo objects') has recently been started, utilizing the gravitational lens effect predicted by the general theory of relativity.

Gravitational Lens in Abell 2218   HST · WFPC2
PF95-14 · ST ScI OPO · April 5, 1995 · W. Couch (UNSW), NASA

Figure 9.3. HST image of the rich cluster Abell 2218, as a spectacular example of gravitational lensing. The arc-like pattern spread across the picture is an illusion and is caused by the gravitational field of the cluster. The arcs are the distorted images of a very distant galaxy population extending 5–10 times farther than the lensing cluster (with kind permission of R Ellis, V Couch, STSCI and NASA, see also [Pel92, Kne95]).

9.2.2.1 The gravitational lens effect

This effect is the production of multiple images of an object due to massive bodies between the source and the observer. There are three main manifestations of this effect:

- Multiple images of quasars, produced by galaxies or galaxy clusters
- Light bows or arcs, i.e images of galaxies with high red-shifts, which are produced by a galactic cluster lying along the path of the light
- Radio rings, i.e. images of point-like radio sources stretched out by galaxies along the line of sight.

The first object of this kind was the double quasar Q0957+561 discovered in 1979 [Wal79]. Table 9.1 shows some of the earliest known objects from the above three groups. Figure 9.3 shows a recent example. These phenomena are illustrated in figure 9.4, for which an application of geometrical optics is sufficient [Bla92]. This approximation implies the assumptions of a weak gravitational field, a small angle and a thin lens.

In a homogeneous Friedmann universe, consider the angular diameter distance $D(z_i, z_j)$ of an object which combines the distance ξ_j at a red-shift z_j with the angular size θ_i at observation. An observation at a red-shift z_i with

Table 9.1. Some early examples of observed gravitational lens effects of quasars, radio galaxies (arcs) and point sources (radio rings). z_s, z_d are the red-shift of the source and lens (from [Bla92]).

Sources	z_s	z_d	θ_{max}
Multiple quasars			
Q0957+561AB	1.41	0.36	6.1″
Q0142−100AB	2.72	0.49	2.2″
Q2016+112ABC	3.27	1.01	3.8″
Q0414+053ABCD	2.63	?	3″
Q1115+080A_1A_2BC	1.72		2.3″
H1413+117ABCD	2.55	?	1.1″
Q2237+031ABCD	1.69	0.039	1.8″
Arcs			
Abell 370	0.72	0.37	
Abell 963	0.77	0.21	
Abell 1352	?	0.28	
Abell 1525	?	0.26	
Abell 1689	?	0.18	
Abell 2163	?	0.17	
Abell 2218	?	0.17	
Abell 2390	0.92	0.23	
Cl 0024+17	?	0.39	
Cl 0302+17	?	0.42	
Cl 0500−24	0.91	0.32	
Cl 1409+52	?	0.46	
Cl 2244−02	2.23	0.33	
Radio rings			
MG1131+0456	?	?	2.2″
0218+357	?	?	0.3″
MG1549+3047	?	0.11	1.8″
MG1634+1346	1.75	0.25	2.1″
1830−211	?	?	1.0″

$z_i < z_j$ results in [Bla92]

$$D(z_i, z_j) = \frac{\xi_j}{\theta_i} = \frac{2c}{H_0} \frac{(1 - \Omega_0 - F_i F_j)(F_i - F_j)}{\Omega_0^2(1 + z_i)(1 + z_j)^2} \tag{9.19}$$

where $F_{i,j} = (1 + \Omega_0 z_{i,j})^{1/2}$. From figure 9.4 we can easily obtain the lens equation that connects source and image position:

$$\beta = \theta - \alpha(\theta). \tag{9.20}$$

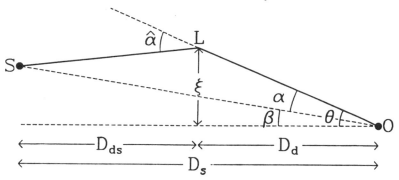

Figure 9.4. The geometrical basis of the gravitational lens effect. A beam of light from the source S with a red-shift z_s falls onto a lens L at a red-shift z_d with an impact parameter ξ relative to the centre of a suitable lens. Assuming a thin lens, its influence can be described by an angle of deviation $\hat{\alpha}(\xi)$. The beam reaches the observer O, who sees the image at a position θ in the sky. The true position of the source without the lens is, however, β. Also illustrated are the angular diameter distances D_d, D_s and D_{ds}, that separate the source, lens and observer (from [Bla92]).

The reduced angle of deflection $\alpha(\theta)$ and the real angle of deviation $\alpha(\xi = D_d\theta)$ are related by

$$\alpha(\theta) = D_{ds}\alpha(\xi)/D_s. \tag{9.21}$$

Equally this effect can be described as the curvature of the light path due to the gravitational potential as predicted by the general theory of relativity, thereby connecting the angle of deflection $\alpha(\xi)$ with the gravitational potential. For a point-like lens the angle of deflection is given by

$$\alpha = \frac{2R_S}{\xi} \rightarrow \alpha = \frac{4GM}{\xi c^2} \tag{9.22}$$

where M is the lens mass, and $R_S \approx 3M/M_\odot$ km is the Schwarzschild radius. For a galaxy with $10^{12}M_\odot$, corresponding to a Schwarzschild radius of 0.1 pc, a deflection of about $\alpha \approx 2 \times 10^{-4} = 4''$ takes place over a distance of 10 kpc. An object directly in the line of sight to the source provides a ring-like image (Einstein ring) with an angular diameter of [Bla92]

$$\theta_E = \left(\frac{4GM}{Dc^2}\right)^{1/2} = 3\left(\frac{M}{M_\odot}\right)^{1/2}\left(\frac{D}{1\text{ Gpc}}\right)^{-1/2} \mu\text{arcsec} \tag{9.23}$$

where

$$D = \frac{D_s D_d}{D_{ds}}.$$

More realistic models for lenses lead to a considerably more complex description, both in the number of images, as well as the corresponding magnification (see e.g. [Bla92, Wu95]).

It is also hoped that this effect may provide proof for the cosmic strings we discussed in chapter 6. It is also of quantitative importance. The angular separation of the quasar images depends on the total mass of the lens. It is therefore possible to determine the total mass of a galaxy or a cluster which makes up the lens. Due to differences in the transit time along the various light paths of the images any variations of brightness of the source, for example, show up at different times in the images, since all images have the same source. By using a realistic model for the lens it is therefore possible to measure the Hubble constant [Ref64]. For the above mentioned double quasar Q0957+561 a propagation time difference of (1.48 ± 0.03) year has indeed been measured, but the parameters are still too model dependent to determine the Hubble constant [Bla92, Pre92]. This method will, however, become more important in future. A great deal more information about cosmologically relevant quantities can be obtained from gravitational lenses, but all these projects are still in the initial stages [Car92].

Besides the lens effect of a structure consisting of many masses, there is also the effect of *micro-lensing* [Pac86, Pac96, Rou97]. Here modification and increase in the brightness of the image of a star are caused by massive compact objects crossing the line of sight. It is precisely this micro-lensing which should be used in the search for MACHOs [Pac86]. Such effects should be detectable from objects with masses of between 10^{-5} and $10^2 M_\odot$ in our own halo. A decisive quantity in such experiments is the *Einstein radius*, given by

$$R_e = \frac{2\sqrt{L}}{c}(Gmx(L-x))^{\frac{1}{2}} \tag{9.24}$$

where m is the lens mass, L the distance between source and observer and x the distance between lens and observer. The Einstein radius is therefore proportional to the square root of the lens mass. Since complete alignment is unlikely, the formation of multiple images, which cannot be resolved in a MACHO search, is more probable than the formation of a ring. Thus the effect manifests itself as a magnification of the brightness A given by

$$A(t) = \frac{u^2 + 2}{u\sqrt{u^2 + 4}} \tag{9.25}$$

where $u = b/R_e$ is the impact parameter (a measure of the distance from the lens to the line of sight) in units of the Einstein radius. If the MACHO moves with a transverse velocity $v_\perp$, the duration of the phenomenon is given by

$$t = \frac{R_e}{v_\perp} \approx 100\sqrt{\frac{M_{\text{MACHO}}}{M_\odot}} \quad \text{days.} \tag{9.26}$$

The signatures are relatively clear and can be used particularly to discriminate against variable stars:

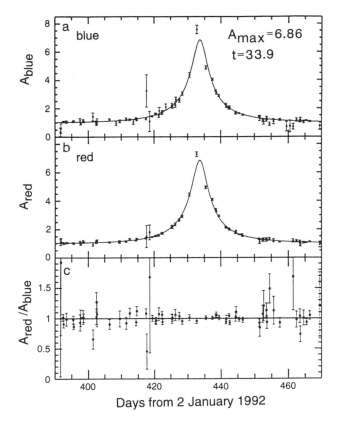

Figure 9.5. The brightness behaviour of one of the MACHO candidates found so far. The symmetry in all wavelengths in the curve can clearly be seen, also shown by the flat ratio. In the case of a variable star this would not be the case (from [Alc93]).

- A large light amplification can be achieved (more than $0^m.75$, an effect easily measurable in astronomy)
- The phenomenon has a clear, symmetrical light curve, characterized by only three parameters (maximal brightness, duration and time of the maximum)
- The change is achromatic, i.e. it takes place in all spectral regions in the same way (variable stars usually change colour)
- Such an event should, from statistical considerations, only take place once per star.

There are currently several groups involved in carrying out this programme [Uda92, Ben93, Mag93]. Through long observation of single stars in our neighbouring Large Magellanic Cloud (LMC), two groups (the MACHO and EROS collaborations) are attempting to observe this effect from the traversal of a MACHO in our halo. This requires an enormous amount of data, since many

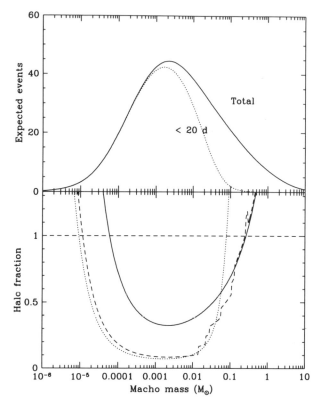

Figure 9.6. The limitations for MACHOs to account for dark matter (from [Sut96]). The upper panel shows the expected number of events for an all-MACHO halo with unique MACHO mass m. The lower panel shows derived limits on the halo MACHO fraction. Regions above the curves are excluded at 95% CL. The solid line is from 8 observed events, the dotted line from no events in $\hat{t} < 20$ days.

million stars must be observed every night in several spectral regions over many years. The OGLE collaboration [Pac95], on the other hand, concentrates more on the galactic centre, which is also being examined by the MACHO collaboration. In this way 18 events have been observed here [Uda94], which can, however, be explained by the lens effect of normal stars. In addition there are more than 120 events from the MACHO collaboration [Alc93, Alc95, Alc96]. An additional 10 events were observed by the DUO experiment [Ala95]. These studies show that these lens effects are not as rare as previously thought. Positive evidence from observations of the LMC by the MACHO (8 candidates) and EROS collaborations (2 candidates) recently created excitement [Alc93, Aub93, Alc95, Alc96]. A most likely mass for the body acting as the lens of about $(0.5^{+0.3}_{-0.2})M_\odot$ was determined from the duration of the effect (see figure 9.5). As well as providing evidence that there are indeed objects in our halo that produce the micro-lensing effect, studies on the observable frequency in the direction of the galactic centre and of the LMC allow the conclusion that MACHOs can only provide up to about 50% of the dark halo [Alc95, Gat95].

The absence of very short events leads to the limit that objects with masses between $10^{-4} M_\odot$ and $0.03 M_\odot$ contribute less than 20% to the halo [Alc96, Mil96] (see figure 9.6).

Motivated by these successes there are projects on the way to observe M31 (Andromeda) with the help of the micro-lensing effect (AGAPE and VATT–Columbia) [Ans95d, Ans96, Cro96]. To extract the maximum amount of information from microlensing events, a global alert network GMAN has been installed [Alc96b]. A first good candidate for a nontrivial lens might be the MACHO 95–30 event [Alc97]. Amoung others, effects of binary systems as lenses or of planetary systems can be investigated. Part of the GMAN project is the MOA collaboration, also observing the galactic bulge and both Magellanic clouds.

9.2.3 Non-baryonic dark matter

The possible candidates for this are limited not so much by physical boundary conditions as by the human imagination and the resulting theories of physics. Some of the most discussed candidates are shown in table 9.2. Consider first the abundance of relics (such as massive neutrinos) from the early period of the universe, which was then in thermodynamic equilibrium. For temperatures T very much higher than the particle mass m the abundance is similar to that of photons, while at low temperatures ($m > T$) the abundance is exponentially suppressed (Boltzmann factor). How long a particle remains in equilibrium depends on the ratio between the reaction rate and the Hubble expansion (see chapter 3). Pair production and annihilation determine the abundance of long-lived or stable particles. The particle density n is then determined by the Boltzmann equation [Kol90]

$$\frac{dn}{dt} + 3Hn = -\langle \sigma v \rangle_{\text{ann}} (n^2 - n_{\text{eq}}^2) \tag{9.27}$$

where H is the Hubble constant, $\langle \sigma v \rangle_{\text{ann}}$ the thermally averaged product of the annihilation cross section and velocity and n_{eq} is the equilibrium abundance. The annihilation cross section of a particle results from the consideration of all of its decay channels. It is useful to parametrize the temperature dependence of the reaction cross section as follows: $\langle \sigma v \rangle_{\text{ann}} \sim v^p$, where in this partial wave analysis $p = 0$ corresponds to an s-wave annihilation, $p = 2$ to a p-wave annihilation, etc. As furthermore $\langle v \rangle \sim T^{1/2}$, it follows that $\langle \sigma v \rangle_{\text{ann}} \sim T^n$, with $n = 0$ s-wave, $n = 1$ p-wave, etc. This parametrization is useful in the calculation of abundances for Dirac and Majorana particles. While the annihilation of Dirac particles only occurs via s-waves, i.e. independent of velocity, Majorana particles also have a contribution from p-wave annihilation,

Table 9.2. Summary of non-baryonic candidates for dark matter[a] (from [Pri88]).

Particle candidate	Approximate mass	Predicted by	Astrophysical effect
$G(R)$	—	non-Newtonian gravitation	apparent DM on large scales
Λ (cosmological constant)	—	general relativity	$\Omega = 1$ without DM
axion, majoron, Goldstone boson	10^{-5} eV	QCD; PQ-symmetry breaking	cold DM
normal neutrinos	10–100 eV	GUTs	hot DM
light Higgsino, photino, gravitino, axino, sneutrino[b]	10–100 eV	SUSY/SUGRA	hot DM
para-photon	20–400 eV	modified QED	hot, warm DM
right-handed neutrinos	500 eV	superweak interaction	warm DM
gravitino etc[b]	500 eV	SUSY/SUGRA	warm DM
photino, gravitino, axino, mirror particles, Simpson neutrino[b]	keV	SUSY/SUGRA	warm/cold DM
photino, sneutrino, Higgsino, gluino, heavy neutrino[b]	MeV	SUSY/SUGRA	cold DM
shadow matter	MeV	SUSY/SUGRA	hot/cold DM (baryon-like)
preon	20–200 TeV	Composite models	cold DM
monopole	10^{16} GeV	GUTs	cold DM
pyrgon, maximon, Perry pole, Newtorite, Schwarzschild	10^{19} GeV	higher dimensional theories	cold DM
supersymmetric strings	10^{19} GeV	SUSY/SUGRA	cold DM
quark-nuggets, nuclearites	10^{15} g	QCD, GUTs	cold DM
primordial black holes	10^{15-30} g	general relativity	cold DM
cosmic strings, domain walls	$10^{8-10} M_\odot$	GUTs	support for galaxy formation, but without large contribution to Ω

[a] DM dark matter; PC Peccei and Quinn; SUGRA, supergravitation; for others see text.

[b] Only one from the many different supersymmetric particles predicted by supersymmetric theories and/or supergravity can be the lightest and therefore stable, thereby contributing to Ω. However, at the present time the theories are unable to give any guidance as to the expected masses of these particles.

leading to different abundances. We now consider three models as examples.

9.2.3.1 Hot dark matter, light neutrinos

Light neutrinos remain relativistic and freeze out at about 1 MeV, so that their density is given by

$$\rho_v = \sum_i m_{vi} n_{vi} = \Omega_v \rho_c. \tag{9.28}$$

From this it follows that for $\Omega \approx 1$ a mass limit for light neutrinos (masses smaller than about 1 MeV) is given by [Cow72]

$$\sum_i m_{vi} \left(\frac{g_v}{2}\right) = 94 \, \text{eV} \Omega_v h^2. \tag{9.29}$$

Given the experimentally determined mass limits mentioned in chapter 2, and the knowledge that there are only three light neutrinos, v_e is already eliminated as a dominant contribution for closing the universe (but can still play an important contribution in mixed models of hot and cold dark matter—see sections 2.4 and 9.2.3.3) although v_μ and v_τ are still possible candidates [Car89]. If, on the other hand, we take this limit seriously, it provides the currently most stringent limit for the mass of the μ and τ neutrinos.

9.2.3.2 Cold dark matter, heavy particles, WIMPs

The freezing-out of non-relativistic particles with masses of GeV and higher has the interesting characteristic that the abundance is inversely proportional to the annihilation cross section. This follows directly from the Boltzmann equation, and implies that the weaker particles interact, the more abundant they are today. Such weakly interacting massive particles are generally known as WIMPs ('weakly interacting massive particles'). If we assume a WIMP with mass m_{WIMP} smaller than the Z^0 mass, the cross section is roughly equal to $\langle \sigma v \rangle_{\text{ann}} \approx G_F^2 m_{\text{WIMP}}^2$, [Kol90] i.e.

$$\Omega_{\text{WIMP}} h^2 \approx 3 \left(\frac{m_{\text{WIMP}}}{\text{GeV}}\right)^{-2}. \tag{9.30}$$

Above the Z^0 mass the annihilation cross section decreases as m_{WIMP}^{-2}, due to the momentum dependence of the Z^0 propagator, and hence a correspondingly higher abundance results. An upper limit is eventually given by the unitarity condition (i.e. the probability for a reaction cannot be greater than 1), which requires $\langle \sigma v \rangle_{\text{ann}} < 8\pi/m^2$ for point-like particles. This condition, together with the limitation that Ωh^2 must not be much greater than 1, results in an upper limit for the mass of any stable, point-like particle candidate for dark matter of 340 TeV [Gri90]. Figure 9.7 shows as an example the contribution of massive neutrinos to the mass density in the universe. Neutrinos of between 100 eV and about 2 GeV as well as beyond the TeV region, should, if they exist, be unstable according to these cosmological arguments [Lee77a]. In order to be cosmologically interesting, that is to produce a value of $\Omega \approx 1$, stable neutrinos must either be lighter than 100 eV or heavier than about 2 GeV (Dirac neutrinos) of 5 GeV (Majorana neutrinos).

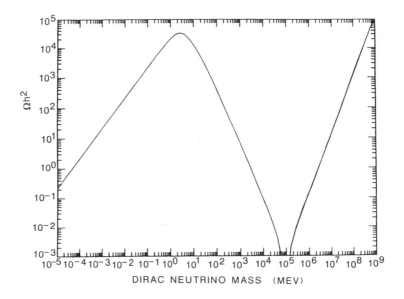

Figure 9.7. The contribution of stable neutrinos of mass m to the matter density of the universe. Only neutrino masses smaller than 100 eV and heavier than several GeV to TeV are cosmologically acceptable. Otherwise neutrinos have to be unstable (from [Tur92a]).

Another class of possible candidates for dark matter are supersymmetric particles. A calculation of their relic abundance (e.g. of neutralinos, heavy sneutrinos, ...) depends on various assumptions, due to the many free parameters in SUSY models, so that we refer to the corresponding literature [Ell84, Gri88, Ell93b, Bot94a, Bot94b, Fal94, Bed94a, Bed94b, Jun95, Bed97a, Bed97b]. The shadow matter, already discussed in chapter 2, which is motivated by superstring theories, could also in principle be a candidate. Other possible, even more exotic candidates are not considered further here.

The above statement notwithstanding, we mention again the possibility of topological defects. Magnetic monopoles (see chapter 10), cosmic strings, domain walls and textures (see chapter 6) all provide a contribution to the energy density.

9.2.3.3 Mixed models

Due to the observations of the microwave background radiation from the COBE satellite (see chapter 7) and, for example, the IRAS survey of galaxies (see chapter 6), the spectrum of the density fluctuations (see figure 6.13) has been restricted. This led to the introduction of mixed models, for example a combination of about 70% cold and 30% hot dark matter [Dav92a, Tay92, Pri97, Cal97], which describe the observed spectrum relatively well. These models provide very precise statements for the masses for the neutrinos favoured for the hot part, which are in the region of a few (2–21) eV. In this connection GUT models, which predict a an almost *degenerate* mass hierarchy of neutrinos of various flavours (see e.g. [Pet94, Moh94]), are of particular interest (see section 2.4.2), and can be verified via double beta-decay experiments.

9.3 Detection of dark matter

Before discussing the specific experiments searching for dark matter, we briefly examine the limits from accelerator experiments. Here it is particularly the properties of the Z^0 boson, discussed in more detail in chapter 4.3, which provide the limits. They limit the number of light neutrinos to three. The mass limits discussed in chapter 2 only allow neutrinos as hot dark matter, or require at least a mass of 45 GeV. Equally there are also limits on the lightest supersymmetric particle (LSP), which has also been discussed in chapter 2. All additional particles with masses less than 45 GeV, and coupling with the full electroweak coupling strength to the Z^0, are excluded.

In experiments searching for dark matter we distinguish *direct* and *indirect* detection [Pri88, Smi90, Ber95]. Direct experiments try to find evidence for dark matter through interactions in laboratory experiments, while indirect experiments try to detect the reaction products of interactions of dark matter outside the laboratory. Figure 9.8 shows the main ways discussed to find evidence for dark matter.

9.3.1 Reaction rates for WIMP–nucleus scattering

In order to build experiments with some chance of success it is necessary to have a rough idea of the expected reaction rates. The reaction rate can be calculated via

$$R = N \int \Phi(E)\sigma(E)\mathrm{d}E \tag{9.31}$$

where N is the number of target atoms, Φ the flux of dark matter particles and σ the cross section. The local density of dark matter is obtained from the comparison of theoretically expected and experimentally observed mass–luminosity relations. The halo density near the Sun is calculated from kinematic considerations of stellar movements to be about $\rho_D = n_D \cdot m_D \approx 0.18 M_\odot$

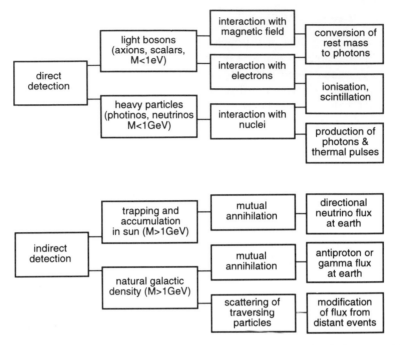

Figure 9.8. Main strategies of experimental detection methods for dark matter (after [Smi90]).

$pc^{-3} \approx 0.3$ GeV cm^{-3} with an uncertainty of 50% [Bin87]. The number of target atoms depends on the available detector material, and the cross sections can be calculated or treated as free parameters. We now concentrate on the search for WIMPs.

The search has so far been carried out mainly using direct experiments on elastic WIMP–nucleus scattering. The recoiling nucleus has an ionizing effect which can be detected. For example, in order to create an electron–hole pair in germanium, an energy of about 2.9 eV is needed. The recoil energy of the nucleus is given by

$$E_R = \frac{m_T \cdot m_D}{(m_T + m_D)^2} m_D v^2 (1 - \cos\theta) \tag{9.32}$$

where m_D indicates the WIMP mass and m_T the mass of the target nucleus, given by $m_T \approx 0.94$ A GeV. The maximum recoil energy is given by

$$E_R = \frac{2m_T \cdot m_D}{(m_T + m_D)^2} m_D v^2. \tag{9.33}$$

If the detector has a threshold energy E_T, a minimum velocity of

$$v^2 = \frac{(m_T + m_D)^2}{2m_T \cdot m_D} m_D^{-1} E_T \tag{9.34}$$

is required to give a detectable signal. This can be transformed into a minimum mass, given by

$$m = \frac{m_T}{(2m_T v_{\max}^2 / E_T)^{1/2} - 1}.$$ (9.35)

$v_{\max}$ is the local escape velocity in our galaxy, and is about 600 km s^{-1}. For incident particles in the GeV region the energy transfer is around a few keV. In this region the ionizing effect of the recoiling nucleus is, however, not the same as for, say, electrons and photons, but rather a large part of the recoil energy transforms itself into lattice vibrations (phonons). The theoretical description by Lindhard [Lin63] is in good agreement with the results obtained from neutron scattering [Ger90, Mes95]. Down to about 10 keV recoil energy the ionization can be described well by the function

$$f_0(E_R) = \frac{g(E_R)}{1 + g(E_R)}$$ (9.36)

where $g(E_R)$ is given by [Smi90]

$$g(E_R) \approx 0.66(Z^{5/18} / A^{1/2})(E_R \text{ keV})^{1/6}.$$ (9.37)

Below 10 keV threshold effects come into play and require a modification to

$$f_1(E_R) = f_0(E_R)(1 - \exp(-E_R / E_1))$$ (9.38)

where $E_1 \approx 0.3$ keV is the appropriate value for germanium [Smi90]. It can be seen that only about 20–30% of the recoil energy goes into ionization. The recoil spectrum is given by

$$\frac{dR}{dE_R} = \int v f(v) \frac{d\sigma}{dE_R} dv$$ (9.39)

which results in a functional form

$$\frac{dR}{dE_R} = \frac{R_0}{E_0 r} \exp\left(-\frac{E_R}{E_0 r}\right)$$ (9.40)

where r is the reduced mass, $E_0 = (1/2)m_D \beta^2$, and R_0 corresponds to the total event rate [Smi90]. Here the differential cross section is assumed to be isotropic in the centre of mass system. A Maxwell-Boltzmann velocity distribution results if an isothermal halo is assumed:

$$f(v)dv = \left(\frac{\sqrt{3/2\pi}}{v_{\text{rms}}}\right)^3 \exp\left(-\frac{|v - v_E|^2}{v_{\text{rms}}^2}\right) dv$$ (9.41)

with a root mean square velocity of about $v_{\text{rms}} \approx 250$ km s^{-1}.

R_0 depends on the cross section and therefore on the type of interaction.

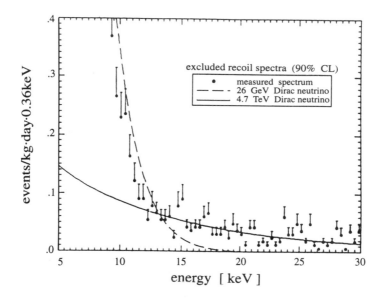

Figure 9.9. Low energy spectrum from a Ge detector of the Heidelberg–Moscow collaboration for the detection of WIMPs. Also shown are the theoretical recoil spectra of WIMPs with masses of 26 GeV and 4.7 TeV (from [Bec94b]).

9.3.2 Direct experiments

WIMPs which interact spin-independently, such as heavy Dirac neutrinos or sneutrinos, undergo coherent scattering ($\sigma \sim N^2$, where N represents the number of neutrons (e.g. for Dirac neutrinos) or nucleons (e.g. for neutralinos) in the target nucleus), while Majorana neutrinos or the LSP react to some extent also at least for light target nuclei, spin-dependently [Bed94a, Bed94b, Bed97a, Bed97b]. In both cases we are dealing with extremely low count rates.

9.3.2.1 Ionization in semiconductor detectors

9.3.2.1.1 Spin-independent interactions
In coherent scattering the rate of events is approximately given by [Smi90]

$$R = \frac{1}{2} \frac{4m_D m_T}{(m_D + m_T)^2} \left(\frac{N}{2}\right)^2 \text{ kg}^{-1} \text{ day}^{-1}. \tag{9.42}$$

Experimental results currently exist mainly from semiconductor detectors, especially Ge detectors, which are used in the search for the neutrinoless $\beta\beta$-decay (see chapter 2) [Cal90a, Reu91, Bec94b, Kla94]. In these detectors a WIMP scatters from a germanium nucleus. Figure 9.9 shows a spectrum. Together

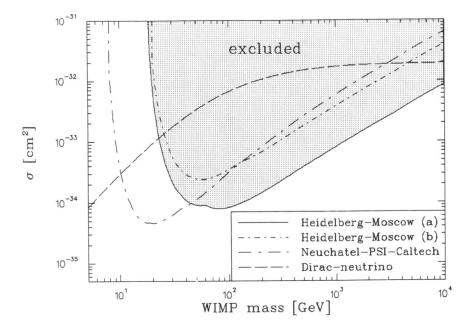

Figure 9.10. Conversion of the Ge results from figure 9.9 into an exclusion plot in terms of cross section and mass. The area above the curve is excluded. This excludes Dirac neutrinos as the dominant component for dark matter. The expectation for the latter for $\Omega = 1$ is shown by the dashed line (from [Bec94b]).

with the UK experiment using NaI detectors [Smi96, Ber97], it yields the most stringent limits at present (figure 9.16). Exclusion limits can be obtained by taking account of the theoretically expected recoil spectrum in the low energy region of the detector. Figure 9.10 shows the excluded area of the free parameters mass and cross section. For heavy Dirac neutrinos with a normal electroweak coupling the mass region of 26 GeV $< m_{\text{WIMP}} < 4.7$ TeV can be excluded. Therefore heavy Dirac neutrinos as a dominant part of dark matter are ruled out. The results are complementary to those of LEP (figure 9.14). As the cross section for sneutrino–nucleus interaction is four times as big as that for Dirac neutrino–nucleus interaction, the Heidelberg–Moscow experiment (see chapter 2) also rules out heavy sneutrinos of the minimal supersymmetric standard model as the dominant candidates for dark matter [Fal94] (figure 9.11).

As the energy transfer in the collision is most effective when both colliding objects have about equal mass, an alternative is to use silicon instead of germanium, in order to improve the sensitivity to lower masses. A spectrum taken with a silicon detector is shown in figure 9.12 [Cal92, Cal94]. It can be seen that the noise threshold lies at about 2 keV. This experiment leads to the almost complete exclusion of the cosmion hypothesis (see chapter 12).

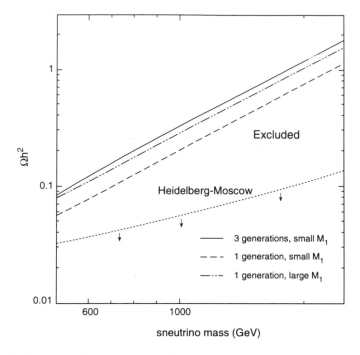

Figure 9.11. The residual density $\Omega_{\tilde{\nu}}h^2$ as a function of $m_{\tilde{\nu}}$. The upper three lines are the expectations from the minimal SUSY standard model (MSSM) of [Fal94]. The Heidelberg–Moscow experiment [Bec94b] excludes this scenario of heavy sneutrinos as dark matter (after [Fal94]).

The experimental signatures for an event caused by dark matter in a detector are the following (see e.g. [Smi90])

- A seasonal dependence of the signal. Due to the relative movement of the Earth in comparison to the halo different relative velocities result in summer and winter. This would show up in a periodic variation of the signal (see e.g. [Sar94]).
- The recoil spectra for different nuclei should be different.
- The rates of events at different nuclei should be different, due to the dependence of the cross section on, among other factors, the nuclear structure of the detector material.

A discussion of the minimal detector configurations necessary to investigate these signatures is given in [Kla97b, Kla97d]. Although a large part of the possible phase space for heavy Dirac neutrinos and sneutrinos is already excluded, the experimental situation with respect to probing the prediction of SUSY models for the LSP as dark matter is still only in the first stages of investigation. Only experiments just starting to operate, like the Heidelberg project HDMS

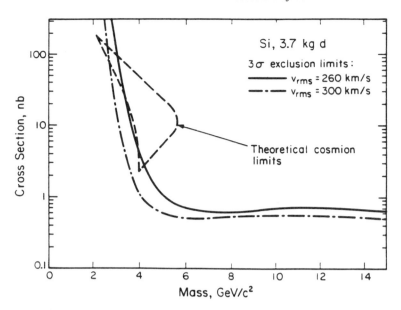

Figure 9.12. Exclusion curve in terms of cross section and mass for dark matter particles (different velocities) due to results from silicon detectors. The region above the curve is excluded including the theoretically allowed area of cosmions, which is lying inside the dashed line (see chapter 12), (from [Cal90b, Cal91, Cal94]).

[Bau97, Kla97c, Kla97d], the Berkeley project CDMS, or the Munich project CRESST (see section 9.3.2.2.2 and [Gai97]), will enter into *part* of the parameter space of the SUSY model predictions (figure 9.16). A large range of SUSY model predictions will be covered by the future Heidelberg project GENIUS (Germanium Nitrogen Underground Set-up) which can reach a sensitivity increased by further two orders of magnitude [Kla97b].

9.3.2.1.2 Spin-dependent interactions There are several candidates for dark matter that interact also spin-dependently. These are, among others, Majorana particles and the lightest supersymmetric particle (LSP)—at least for light target nuclei $A \leq 50$ [Bed94a, Bed94b]. In the case of R-parity conservation (see chapter 2) the LSP is stable and therefore a good candidate for dark matter. The most likely candidate for the LSP is the neutralino (see chapter 2). Due to the free parameters in SUSY models the predicted abundances depend on the particular choice of the neutralino. The same applies to the predicted event rates for the different detectors [Goo85, Sre88, Res93, Ros93, Bot94a, Bot94b, Bed94a, Jun95, Bed97a, Bed97b]. A detailed examination of the general possibility of restricting the free parameter space of SUSY models through the search for dark matter is given in [Bed94a, Bed94b, Jun95, Bed97a, Bed97b].

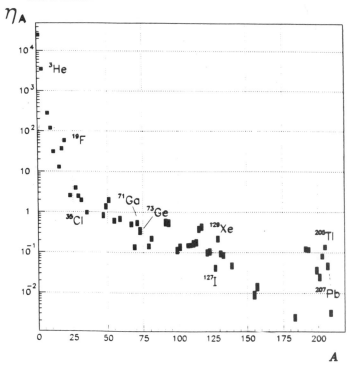

Figure 9.13. The nuclear factor η_A (a measure of the strength of the interaction with dark matter) as a function of the atomic weight A for nuclei with spin. The width of the bars represents the variation for interaction with neutralinos with masses M_χ of between 20 GeV and 500 GeV (from [Bed94b]).

Figure 9.13 shows a comparison of the nuclei which, from the point of view of simple nuclear models, are expected to give the best reaction rates for the interactions of neutralinos via spin-dependent interaction. In general obviously, for nuclei with mass number $A > 50$, the contribution of the spin-independent interaction dominates the reaction rate [Bed94a]. As well as this factor there are, of course, also limitations on the practical availability of a particular material in the amount necessary and the purity required for such experiments. The possibility of building a CaF_2 scintillation detector which makes use of the spin of the theoretically preferred nucleus ^{19}F, looks quite hopeful.

Of the semiconductor detectors discussed above, only ^{73}Ge ($I = 9/2$) and ^{29}Si ($I = 5/2$) have a nuclear spin. However, they unfortunately also have a relatively low natural abundance (^{73}Ge 7.8%, ^{29}Si 5%). For this reason it would be interesting to build isotopically enriched detectors (see e.g. [Kla91a, Kla91b]). An appropriate project using ^{73}Ge (HDMS) which will allow a background reduction to the region planned for the cryo projects currently under construction (see below and figure 9.16(*b*)) will begin operation soon [Kla96a, Bau97].

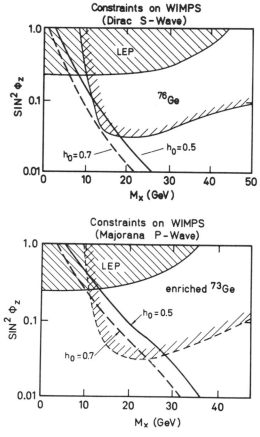

Figure 9.14. Comparison of experiments searching for dark matter with Ge detectors and high energy experiments at LEP. The fraction $\sin^2\phi_Z$ of a normal coupling to the Z^0 against the mass M_X of WIMPs is shown: (*a*) for spin-independently interacting particles, (*b*) for spin-dependently interacting particles. Both kinds of experiments are complementary, but the Ge detector experiments for dark matter are sensitive to considerably larger masses. The exclusion region shown for ^{76}Ge corresponds to the experiment of [Cal91], the one for ^{73}Ge corresponds to a planned experiment (from [Ell90, Sch90b, Kla91b]).

Until now information on spin-dependent interacting WIMPs was obtained from experiments with natural Ge [Cal90a, Cal94] and NaI detectors (^{23}Na and ^{127}I have nuclear spins of $I = 3/2$ and $I = 5/2$) [Ger94, Que95, Smi96], as well as ^{131}Xe [Bac94, Bel96c].

In contrast to the case of spin-independent interactions, it will however be very difficult to enter experimentally the predicted SUSY parameter space for spin-dependent interactions (figure 9.16(*b*)).

In conclusion, figure 9.14 demonstrates the potential of semiconductor

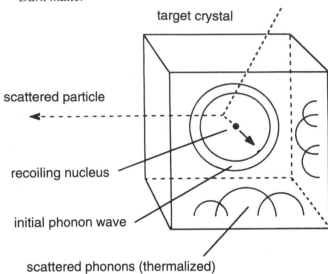

Figure 9.15. Schematic construction of a bolometric detector. The deposited energy leads to a temperature increase that can easily be measured with precise thermometers (from [Smi90]).

experiments to search for dark matter, in comparison to present high energy experiments. Their sensitivity is far higher than that of accelerator experiments for high WIMP masses. Only LHC will be able to cover the full cosmologically relevant region of the neutralinos as dark matter [Den97] (figure 9.17).

A further possibility for measuring the recoil rests on the use of superheated liquids in droplet form, functionally similar to bubble chambers [Zac94, Ham96]. This would have the advantage of being insensitive to α-, β- and γ-radiation.

9.3.2.2   Cryogenic detectors

As well as the more conventional detectors described above a new type of detector has been developed recently, which makes use of the techniques of low temperature physics. All reactions used for the detection of low energy neutrinos and dark matter require the detection of very low energy depositions E_r of a few keV in the detector. As already mentioned, the largest part of the energy deposition transforms into lattice vibrations (phonons) rather than ionization. It would therefore be of great advantage to measure the phonons, or better still both simultaneously.

9.3.2.2.1 Bolometers   If the phonons are thermalized, bolometric detection is appropriate. Such small energy depositions can best be observed at very low temperatures (< 1 K), since

$$\Delta T \simeq \frac{E_r}{V C_V(T)} \qquad (9.43)$$

where V is the volume of the detector, and C_V is the specific heat capacity of the material. The number of phonons is given by C_V/k_B, where the mean energy per degree of freedom is kT. This results in a mean square energy of

$$\langle E^2 \rangle = N(kT)^2 = C_V(T)kT^2. \qquad (9.44)$$

In order to obtain the smallest possible heat capacity isolators are used, on the surface of which thermistors are installed. One puts a small threshold voltage to these thermistors. If energy is now deposited in the isolator the temperature rises, which results in the reduction of the resistance of the thermistor. This results in a measurable voltage signal. Highly doped semiconductors (whose resistance is strongly temperature dependent), or superconducting transition bolometers are generally used as thermistors. At low temperatures the heat capacity is practically determined by the contribution of the phonons, which depends on the cube of the temperature, i.e.

$$C_V = \frac{\Delta Q}{\Delta T} \approx B \left(\frac{T}{\Theta_D} \right)^3. \qquad (9.45)$$

Θ_D corresponds to the characteristic Debye temperature of the material. The constant B is about $1940\,(\rho/A)$ J cm^{-3} K^{-1}. With a typical ratio of mass density ρ to atomic number A of about 0.08 the heat capacity becomes [Smi90]

$$C_V \approx 160 \left(\frac{T}{\Theta_D} \right)^3 \text{ J cm}^{-3} \text{ K}^{-1} \approx 1 \times 10^{18} \left(\frac{T}{\Theta_D} \right)^3 \text{ keV cm}^{-3} \text{ K}^{-1}. \qquad (9.46)$$

Hence for a Si crystal of 1 cm^3 it follows that

$$\frac{\Delta T}{E_r} \approx 3 \times 10^{-10} T^{-3}. \qquad (9.47)$$

If we now assume a working temperature of 55 mK and an energy deposit through an interaction with a particle of dark matter of about 6 keV, a temperature change of $\Delta T \approx 10^{-5}$ K results. In the milliKelvin region this is still a reasonably detectable temperature step. We have, however now entered the temperature range in which impurities and surface effects can produce a noticeable deviation from the temperature behaviour of heat capacities described so far. The principle of such an seen by is shown in figure 9.15. One special advantage of this kind of detector is the excellent energy resolution, given by

$$\Delta E = \kappa \sqrt{k_B T^2 C_V} \qquad (9.48)$$

where κ is a numerical factor of order one. If the Boltzmann constant is expressed as $k_B = 8.6 \times 10^{-11}$ MeV K^{-1}, it can be seen at once that resolutions in the eV range are possible. In this way a Munich group managed to reach an energy resolution of 220 eV at 6 keV energy with a 31 g sample of sapphire [Nuc94].

A variation of the method to measure the energy deposited in the form of thermalized phonons is carried out by measuring the phonons *prior* to the thermalization (*ballistic phonons*). As already discussed for normal semiconductor detectors, most of the energy at low recoil energies appears as phonons, and only a little as ionization. The recoil nucleus is typically stopped on a scale of 10^{-6}–10^{-7} cm, and therefore emits a spherical phonon wave. The energy of these phonons lies in the order of $10^{-3} - 10^{-2}$ eV, which corresponds to temperatures greater than 10 K. Below 1 K the mean free path of these ballistic phonons is so large that it is comparable to the size of the detector. Using a suitable crystal with sensors to measure the phonon energy, as well as for measuring the relative arrival times on the surface, the phonons could be measured directly. Moreover, an increase of the energy density in the phonons and a concentration onto the sensors seems possible via phonon focusing. Due to the intensity ratios seen by different sensors, even the location of the interaction might be pin-pointed. It is also possible to work at about 1 K and not, as previously discussed, at 100 mK. For a detailed discussion we refer to [Smi90, Boo92a, Möß93].

9.3.2.2.2 *Quasi-particles in superconductors*

A further method of measuring phonons is via the detection of quasi-particles in superconductors. The mean phonon energy due to a nuclear recoil is much higher than the energy of $2\Delta \times 10^{-3}$ eV which is needed to produce quasi-particles by the break-up of Cooper pairs. Contrary to semiconductors, the Debye temperature in a superconductor is very much higher than the binding energy 2Δ. In thermal equilibrium the breaking-up of Cooper pairs due to phonons with $E > 2\Delta$ is identical to the formation of Cooper pairs with the emission of phonons. A density of the quasi-particles of

$$n(T) \sim T^{1/2} e^{-\Delta/k_B T} \tag{9.49}$$

results. The detection of these quasi-particles is best achieved via superconducting tunnel diodes. This is a Josephson junction with a small magnetic field. It consists of two superconductors separated by a thin insulating film. With the help of a magnetic field the Josephson current, due to the tunnelling of Cooper pairs, is suppressed and only the contribution due to the tunnelling of the quasi-particles remains. It is, however, not possible to produce very large samples in this way, as the tunnel current is inversely proportional to the sample size. It is, however, possible to concentrate the phonons. For details see [Boo92b].

9.3.2.2.3 Superheated superconducting grains   Another way to detect energy deposition relies on the phase transition between the superconducting and normal conducting phases.   According to the Meissner–Ochsenfeld effect superconductors expel external magnetic fields from their interior. At a critical field intensity and corresponding temperature, however, the superconductivity breaks down, and the material becomes normal conducting. If a superconductor of type I is kept below its critical curve in a metastable phase, it takes only a small energy deposition to make it normal conducting. The corresponding change in magnetic field can then be detected. In order to realize this technique superheated superconducting grains (SSG), which are embedded in a regular way into a dielectric, are used [Pre93, Pal97]. The grains have a diameter of about 5–10 μm. One loop for about 1000 grains is sufficient to measure the phase transition in a grain. However, this method has its difficulties. In recent years the problem of getting as many equal sized grains as possible with relatively low fluctuation widths has become closer to being solved using lithographic techniques. Measuring the energy also proves to be extremely difficult. Energy deposition is a local effect, which affects only one grain.   The theoretical understanding of the expansion of such an energy deposition, and the dependence on various experimental parameters, is currently being studied. One solution is to create a kind of avalanche effect, in which the number of grains involved is correlated with the primary energy deposition. A realization of this technique is planned in the ORPHEUS experiment [Pal97].

9.3.2.2.4 Liquid ^{4}He   A very different kind of detector is based on superfluid ^{4}He, e.g. HERON [Ban92]. Due to the specific characteristics of this material there is practically no background. An energy deposition within ^{4}He leads to the excitation of *rotons*, which eventually all reach the surface. There their energy is transformed into vaporization of ^{4}He. As ^{4}He has also practically no vapour pressure, we can gain information about the deposited energy by measuring the vaporized ^{4}He, with the help of a detector installed just above the surface. Several experimental hurdles still have to be overcome, but a prototype of a few kilograms exists, with which several test experiments have already been carried out.

9.3.2.3 Experimental situation and perspectives of direct detection

The first cryogenic running experiment in particle physics is used in the search for double beta-decay (see chapter 2) of ^{130}Te, and uses semiconductor thermistors [Ale94]. Presently four 334 g TeO_2 crystals, operating at 10 K, are used [Ale94, Fio96]. This experiment uses only the bolometric principle. Because of its high background at low energies it is, however, not suitable for dark matter searches.

However, simultaneous measurement of the phonon contribution and ionization allows a higher background reduction to be reached, as the signals then have a clearer signature. A collaboration at Berkeley succeeded in measuring these two contributions separately in a 60 g germanium crystal [Shu92, Cal94]. A corresponding large-scale experiment (CDMS) is in progress [Gai97].

In the effort to get the best possible energy resolution, a group in Munich has succeeded in obtaining good results with calorimeters with superconducting transition detectors [Sei90, Coo93]. An experiment using this technique (CRESST) is currently under construction in the Gran Sasso underground laboratory [Sei97].

Future development of cryo-detectors is looking rather promising. However, all these detectors suffer from the problem of producing and operating the several kilograms required. The present state of the art can be found in [LTD96].

Conventional techniques have reached the most sensitive limits for dark matter—with the Heidelberg–Moscow ^{76}Ge [Bec94b], the UK NaI [Smi96] and the Italian NaI (DAMA) [Ber97] experiments. The Heidelberg project HDMS (Heidelberg Dark Matter Search) [Bau97] still using conventional ionization detector techniques will reach the projected sensitivity of the Berkeley CDMS experiment and will be in the WIMP mass range complementary to the Munich CRESST experiment. These last three experiments will be in fact the first to partly enter the SUSY parameter space for neutralinos as dark matter (for spin-independent interaction). The situation is illustrated in figure 9.16(a). The planned Heidelberg project GENIUS (see chapter 2 and [Kla97b]) will cover— with its sensitivity larger by two orders of magnitude—a very considerable range of SUSY neutralino predictions (see figure 9.16a and [Arn97]). Such a detector will be particularly sensitive in regions of large $\tan \beta$ in the minimal SUGRA parameter space, where many conventional signals for supersymmetry in collider experiments are difficult to detect. Thus, if the parameter $\tan \beta$ is large, then there is a significant probability that the first direct evidence for supersymmetry could come from direct dark matter detection experiments, rather than from collider searches for sparticles [Bae97]. Such a detector will of course also be sensitive to other dark matter candidates. Finally the large hadron collider (LHC) at CERN will cover the whole cosmologically relevant range for neutralinos [Den97] (figure 9.17). But also if SUSY were detected by collider experiments, it would still be fascinating to verify the existence and properties of neutralino dark matter.

9.3.3 Indirect experiments

Indirect experiments do not detect the interaction of dark matter in the laboratory, but the products of reactions of particles of dark matter taking place extra-terrestrially or inside the Earth. For dark matter this is mainly particle– antiparticle annihilation. Two main kinds of annihilation will be discussed here:

- Annihilation inside the Sun or Earth.
- Annihilation within the galactic halo.

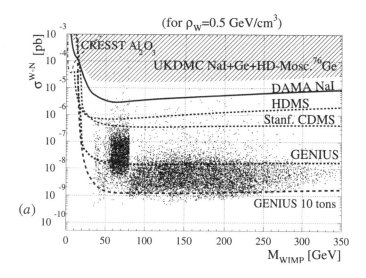

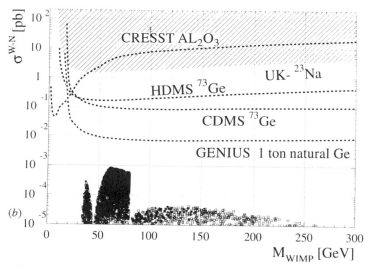

Figure 9.16. (*a*) Speculative view for obtainable cross section limits on WIMP–nucleon (spin-independent interaction) elastic scattering for future experiments, CRESST, CDMS, HDMS and GENIUS (1 and 10 tons of enriched ^{76}Ge). For comparison the present lowest-limit results, the Heidelberg–Moscow result [Bec94b], the results for NaI (UK and DAMA) [Smi96, Ber97], are also shown. Expected cross sections for neutralinos (dotted lines), for a SUSY–GUT scenario with and without universal scalar mass unification are also shown [Bed97a, Bed97b] (from [Kla97b]). (*b*) Same as (*a*) but for spin-independent interactions, indicated as WIMP–nucleon scattering cross sections. For GENIUS here 1 ton of natural ^{76}Ge is assumed. The difference between the SUSY–GUT (filled circles) and the relaxed SUSY–GUT scenario without scalar mass unification (open squares) (see [Bed97a, Bed97b]) is here nearly invisible (from [Kla97b]).

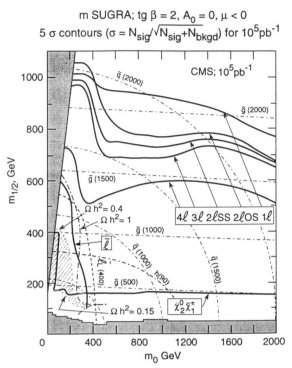

Expected reach in various channels

m SUGRA; tg β = 2, A_0 = 0, μ < 0

5 σ contours (σ = $N_{sig}/\sqrt{N_{sig}+N_{bkgd}}$) for 10^5pb^{-1}

Figure 9.17. Minimal SUGRA parameter space explorable through different channels ($\tilde{q}$, $\tilde{g}$, $\tilde{l}$, χ_i^0, $\chi_j^\pm$) at LHC in *pp* collisions at E_{cm} = 14 TeV and for maximum luminosity (areas below the various curves will be excluded by experiment). The neutralino mixed dark matter expectations—the hatched area corresponds to dark matter density contours with Ωh^2 from 0.15–0.4 and the upper limit of Ωh^2 = 1 is also shown—will be fully covered (from [Den97]). $m_{1/2}$ and m_0 denote the common gaugino and scalar masses respectively, at the SUSY-GUT scale, assumed in this example.

9.3.3.1 Annihilation inside the Sun or Earth

It is possible that dark matter may accumulate in stars and annihilates there with anti-dark matter. One signal of such an indirect detection would be high energy solar neutrinos. These would be produced through the capture and annihilation of dark matter particles within the Sun [Pre85, Gou92]. This concept of the cosmion will be discussed in more detail in chapter 12. They would give rise to a high energy neutrino flux, where the neutrino energy lies far above that of the actual solar neutrinos [Sil85, Ell87b], namely in the range GeV–TeV. An estimate of the expected signal due to photino (neutralino) annihilation results in about 2 events per kiloton detector material and year. These high energy neutrinos would show up in the large water detectors via both charged and neutral weak currents. The charged weak interactions are about three times as frequent as the neutral ones. Since no particular neutrino signatures were found in the IMB [LoS87], Kamiokande [Sat91] and Frejus detectors, photino (neutralino) masses of between 4 GeV and 12 GeV have been ruled out and also sneutrinos of all known flavours between 4 and 90 GeV [Sat91]. The capture of particles of dark matter in the Earth, and their annihilation, have also been discussed. Neutrinos from neutralino–anti-neutralino annihilation in both the Sun and the Earth are being searched for by looking for upward-going muons (see chapter 8). Such investigations are being carried out with the Kamiokande detector [Mor93], the Baksan scintillator telescope [Bol96] and with the MACRO detector [Mon96] (see chapter 10). The measured flux limits are beginning to constrain certain SUSY models [Bot95].

9.3.3.2 *Annihilation within the halo*

If the annihilation of dark-matter-δ–anti-dark-matter-$\bar{\delta}$ takes place within the halo, it should be noticed [Sil84, Ahl88] as a contribution to the γ-background radiation, or in a significant flow of anti-protons and positrons in the cosmic rays, according to

$$\delta\bar{\delta} \to q\bar{q} \to \pi^0 \to 2\gamma \tag{9.50}$$

or

$$\delta\bar{\delta} \to q\bar{q} \to \bar{p} + \ldots \tag{9.51}$$

The experimental data, however, do not yet permit any conclusions to be drawn.

The problem of dark matter turns out to be one of the most difficult in modern astrophysics. It continually provokes new astronomical observations in order to undermine its existence. It challenges the theory which has to define the possible candidates. In the end it does, however, also lead to experimental innovations, the uses of which are not yet obvious. In the next two chapters two further candidates for dark matter are considered, magnetic monopoles and axions; their possible existence also touches on other areas of modern physics.

Chapter 10

Magnetic monopoles

10.1 The Dirac monopole

In the Maxwell equations of classic electrodynamics there is an asymmetry between the electric and magnetic charges. While isolated electric charges can be observed in nature, a free magnetic charge has not been observed. Assuming the existence of such a magnetic charge ρ_m and a related current density j_m, the consequences for electrodynamics would be the modification of the Maxwell equations in a vacuum to become [Jac82]:

$$\operatorname{div} E = 4\pi \rho_e \tag{10.1}$$

$$\operatorname{div} B = 4\pi \rho_m \tag{10.2}$$

$$\operatorname{rot} B = \frac{1}{c} \frac{\partial E}{\partial t} + \frac{4\pi}{c} j_e \tag{10.3}$$

$$-\operatorname{rot} E = \frac{1}{c} \frac{\partial B}{\partial t} + \frac{4\pi}{c} j_m. \tag{10.4}$$

In addition both kinds of charge obey the continuity equation

$$\frac{\partial \rho}{\partial t} + \operatorname{div} j = 0 \tag{10.5}$$

These equations are extremely symmetrical with respect to E and B. It can easily be seen that transformations of E, B, j and ρ leave the Maxwell equations unchanged. Application of the orthogonal matrix

$$\begin{pmatrix} \cos\alpha & \sin\alpha \\ \sin\alpha & \cos\alpha \end{pmatrix} \tag{10.6}$$

to each of the four quantities leaves the Maxwell equations unchanged. The definition of what is magnetic and what electric charge is now largely a question of convention, because, if all particles had the same ratio between magnetic and electric charge, an α can be defined, such that

$$\rho_m = \rho'_e \left(-\sin\alpha + \frac{\rho'_m}{\rho'_e}\cos\alpha \right) = 0 \tag{10.7}$$

and therefore

$$j_m = j'_e \left(-\sin\alpha + \frac{j'_m}{j'_e}\cos\alpha \right) = j'_e \left(-\sin\alpha + \frac{\rho'_m}{\rho'_e}\cos\alpha \right) = 0. \tag{10.8}$$

In this case the Maxwell equations have their usual form with $q_e = e$ and $q_m = 0$. However, should particles exist with *different* ratios of magnetic and electric charge, the equations would have to be written in the above stated general form.

It was Dirac in 1931 who showed that the existence of magnetic charge implied inevitably the quantization of electric charge [Dir31].

If we assume a magnetic monopole of the strength g at the origin, its magnetic field at the point r is given by

$$\boldsymbol{B} = \frac{g}{r^3}\boldsymbol{r} = -g\nabla\left(\frac{1}{r}\right). \tag{10.9}$$

The resulting consequences can be discussed in two ways.

(i) Consider a particle with charge e and velocity v passing a magnetic monopole at a distance b (figure 10.1). At a point r it will experience a Lorentz force of

$$\boldsymbol{F} = \frac{ev}{c}\boldsymbol{B} = \frac{eg}{c}\frac{vb}{(b^2 + v^2 t^2)^{3/2}}. \tag{10.10}$$

From this there results a momentum transfer of

$$\Delta p = \int F \mathrm{d}t = \frac{2eg}{cb}. \tag{10.11}$$

A change in momentum is, however, also connected to a change of angular momentum, such that

$$\Delta L = b\Delta p = \frac{2eg}{c}. \tag{10.12}$$

The quantization of the electric charge follows directly from the quantization condition for the orbital angular momentum $L = n\hbar$, such that

$$e = \frac{n}{2}\frac{\hbar c}{g} \qquad n = 0, \pm 1, \pm 2, \dots. \tag{10.13}$$

Hence the size of the magnetic charge is given by

$$g = \frac{n}{2}\frac{e}{\alpha_{\mathrm{em}}} = \frac{137}{2}e. \tag{10.14}$$

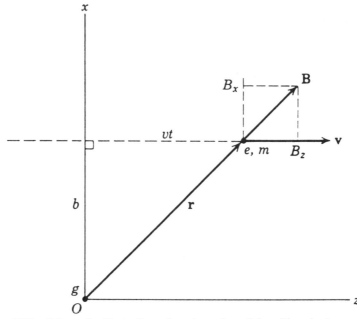

Figure 10.1. Schematic illustration of a charged particle with velocity v, passing a magnetic monopole with strength g at the origin (after [Jac75]).

It can be seen that although there is a formal symmetry in the Maxwell equations, there is no numeric symmetry.

(ii) A somewhat more theoretical derivation is as follows (see e.g. [Pre84, Kla95]): The field of a monopole is radially symmetric, so that the flux through a surface entirely enclosing the charge is given by:

$$\Phi = 4\pi r^2 B = 4\pi g. \tag{10.15}$$

Consider an electron in the magnetic field of this monopole. The wavefunction of the free particle is given by

$$\Psi = |\Psi| \exp\left(\frac{i}{\hbar}\,(\boldsymbol{p}\cdot\boldsymbol{r} - Et)\right). \tag{10.16}$$

The presence of the electromagnetic field changes the phase of the wavefunction such that $\alpha \to \alpha - \frac{e}{\hbar c}\boldsymbol{A}\cdot\boldsymbol{r}$. Here A is the vector potential. If we fix r and θ, and consider a loop in Φ, i.e. Φ between 0 and 2π, the total change in phase is given by

$$\Delta\alpha = \frac{e}{\hbar c} \oint \boldsymbol{A}\cdot d\boldsymbol{l} = \frac{e}{\hbar c}\Phi(r,\theta). \tag{10.17}$$

$\Phi(r, \theta)$ is the flux through the loop, as schematically shown in figure 10.2. For $\theta \to 0$ the flux $\Phi(r, 0) = 0$. For $\theta \to \pi$ the flux $\Phi(r, \pi) = 4\pi g$. Although

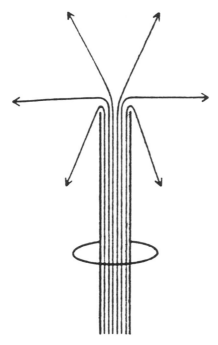

Figure 10.2. Illustration of a Dirac string. A semi-infinite (along the z-axis) singularity (from [Pre84]).

the loop has again shrunk to zero, the flux is still non zero. This means that for $\theta = \pi$ the vector potential A becomes singular. But since we can choose r arbitrarily, this means that A is singular along the whole negative z axis, the so called *Dirac string*. On it the wavefunction vanishes, and the phase is completely undefined. Due to the single valuedness of the wavefunction and the non-detectability of such an effect in electron interference experiments it follows that $\Delta\alpha$ must equal $2\pi n$, where n is an integer. From that follows Dirac's quantization condition (see (10.13)):

$$\frac{ge}{\hbar c} = \frac{n}{2} \qquad (10.18)$$

It can be seen that the existence of magnetic monopoles immediately implies the quantization of the electric charge. Using the fine structure constant $\alpha = e^2/\hbar c$, the size of the magnetic charge can be written as $g = e/(2\alpha) = (137/2)e$. There is no information, however, as to the mass of these Dirac monopoles. However, it can be roughly estimated in a classical way. If we arbitrarily assume that the radius of a monopole corresponds to the classical electron radius, it can be deduced that

$$\frac{g^2}{m_M} \approx \frac{e^2}{m_e} \qquad (10.19)$$

from which equation (10.14) immediately implies that

$$m_M = \left(\frac{137}{2}\right)^2 m_e = 4700 m_e \approx 2.4 \, \text{GeV} \tag{10.20}$$

However, it is obvious that this here is merely a rough order of magnitude estimate.

10.2 The 't Hooft Polyakov monopole

In 1974 the idea of magnetic monopoles experienced a revival, in the framework of unified theories, through the work of 't Hooft and Polyakov [t'Ho74, Pol74]. They were able to show that every higher gauge group, whose breaking would result in a U(1) subgroup, will inevitably predict the existence of magnetic monopoles. As an example we consider an SO(3) group which is spontaneously broken to a U(1) group with the help of a Higgs triplet Φ^a. The Lagrange density is given by [Kol90]

$$\mathcal{L} = \frac{1}{2} D_\mu \Phi^a D^\mu \Phi^a - \frac{1}{4} F^a_{\mu\nu} F^{a\mu\nu} - \frac{1}{8} \lambda (\Phi^a \Phi^a - \sigma^2)^2. \tag{10.21}$$

Due to the symmetry breaking two of the three gauge bosons receive mass of

$$M_V^2 = e^2 \sigma^2 \quad \text{and} \quad M_S^2 = \lambda \sigma^2. \tag{10.22}$$

The size of the vacuum expectation value is again given from the minimum of the potential $\langle \Phi^a \rangle = \sigma$. The direction in the SO(3) space is, on the other hand, undetermined. The solution of lowest energy is given by $\Phi^a = \text{constant}$, as here not only the potential, but also the kinetic energy is minimized. The spatial dependence can also be gauged away in the case that $\Phi^a \neq \text{constant}$, with the help of a suitable field A^a_μ of finite energy. However, there are also Higgs field configurations that can not be transformed through a gauge transformation of a finite energy into a configuration corresponding to a $\Phi^a = \text{constant}$. Such an example is the so called *hedgehog* solution, in which the direction of Φ^a in the group space is proportional to the unit vector in the normal space. The solution is spherically symmetrical and has for $r \to \infty$ the form

$$\Phi^a(r, t) \to \sigma \hat{r} \tag{10.23}$$

$$A^a_\mu(r, t) \to \epsilon_{\mu ab} \frac{\hat{r}_b}{er}. \tag{10.24}$$

Continuity requires the Higgs field to vanish for $r \to 0$. However, it is not possible to deform the hedgehog solution in such a way that the condition $\langle \Phi^a \rangle = \sigma$ applies everywhere. The topological stability of the solution is derived from this. The size of the hedgehog is determined by the region in which

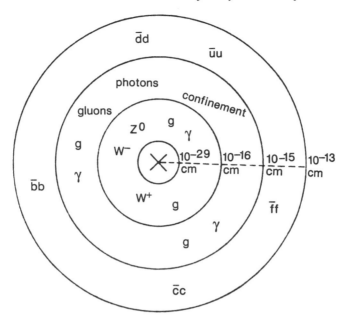

Figure 10.3. The onion structure of a GUT monopole. The monopole changes from the innermost region, in which the unbroken GUT symmetry prevails, via a region of the electroweak symmetry, into a fermion–anti-fermion-condensate. Through the GUT symmetry at its centre it should catalyse nucleon decay (from [Boe88]).

$\langle \Phi^a \rangle \neq \sigma$, i.e. of order σ^{-1}. At large distance the magnetic field connected to the solution is given by:

$$B_i^a = \frac{1}{2} \epsilon_{ijk} F_{jk}^a = \frac{\hat{r}_i \hat{r}^a}{er^2} \tag{10.25}$$

which corresponds exactly to the magnetic field of a magnetic monopole with a double Dirac charge, $g = 137e$! Solutions which have the characteristics of magnetic monopoles are implied from the topology of the vacuum expectation value of the Higgs field alone. The monopole mass is determined by $\sqrt{\lambda}/e = M_S/M_V$ (where λ is defined in equation (10.21) and M_S and M_V in equation (10.22)). It is not possible to derive a closed analytical formula for the mass. In the limit $M_S/M_V \to 0$, however,

$$m_M = \frac{4\pi\sigma}{e} = \frac{M_V}{\alpha_{em}}. \tag{10.26}$$

The mass is a monotonically increasing function of this ratio (roughly proportional to its square), and, for the other extreme case $M_S/M_V \to \infty$, reaches about 1.8 times the above value [Kol90].

As the standard model (see chapter 1) has a $SU(3) \otimes SU(2) \otimes U(1)$ symmetry, every unification should give rise to monopoles. In contrast to Dirac's ideas, these theories for the first time allow statements to be made about monopole masses. Assuming $M_V = M_X$, the following relation applies:

$$m_M \simeq \frac{M_X}{\alpha_{\text{em}}} \qquad (10.27)$$

where α is the fine structure constant, and M_X the typical mass of a vector boson connected to the scale of the symmetry breaking. For the SU(5) theory a unification is expected at about 10^{15} GeV, which then corresponds to a monopole mass of about 10^{17} GeV. This corresponds to the weight of a bacterium! Monopoles are therefore extremely heavy structures and cannot be produced in today's accelerators. The only place in which such heavy particles could be produced was the early universe. The structure of such a GUT monopole is shown in figure 10.3. Inside the monopole the GUT symmetry is still intact, which leads to the hypothesis that monopoles can catalyse nucleon decay.

10.3   Astrophysics of monopoles

The production of topological defects such as monopoles is caused by phase transitions in the early universe. As we have already discussed in the formation of cosmic strings, the particle event horizon is the maximum possible correlation length. On larger scales the value of the Higgs field is uncorrelated. The correlation length ξ is thereby dependent on details of the phase transition and on the temperature. We assume here a second order phase transition, such that $\xi \sim m_H^{-1}(T) \sim T^{-1}$ [Kol90]. Since the Higgs field, on scales larger than the Hubble volume (see chapter 3), points to different directions of group space, it is assumed that one monopole formed in each such domain. As the particle event horizon (see chapter 3) is $d_H \sim H^{-1} \sim m_{Pl}/T^2$, it follows that $n_M \sim d_H^{-3} \sim T^6/m_{Pl}^3$. This implies an initial value for the dimensionless relation n_M/T^3 of

$$\frac{n_M}{T^3} > \left(\frac{T}{Cm_{Pl}}\right)^3 \qquad (10.28)$$

where C is a constant. Since in unified theories such as SU(5) T typically has values of $T \approx 10^{15}$ GeV and $Cm_{Pl} \approx 10^{19}$ GeV, a relation of $n_M/T^3 \simeq 10^{-12}$ results. After the phase transition there are no more creation processes, and only monopole–anti-monopole annihilation will have any influence on the density. However, it appears [Kol90], that for

$$\frac{n_M}{T^3} < 10^{-9}\left(\frac{m_M}{10^{16} \text{ GeV}}\right) \qquad (10.29)$$

monopole annihilation plays practically no role. This means that the above ratio hardly changes. The contribution of magnetic monopoles to the critical density is therefore given by

$$\Omega_M h^2 \simeq 10^{24}\left(\frac{n_M}{T^3}\right)\left(\frac{m_M}{10^{16} \text{ GeV}}\right) \qquad (10.30)$$

and hence from equation (10.28)

$$\Omega_M h^2 \simeq 10^{11} \left(\frac{T}{10^{14} \, \text{GeV}} \right)^3 \left(\frac{m_M}{10^{16} \, \text{GeV}} \right). \tag{10.31}$$

This leads to a serious problem. With the typical GUT values for T and n_M, monopole densities result that are of order $10^{11} \rho_c$, an unacceptably high value (see chapter 3). This is what is known as the monopole problem. We now have to either introduce a phase transition at much lower temperatures, or the monopole density has to be drastically lowered. The limitation that Ω_M must not be much greater than 1 implies a critical temperature for the phase transition of order 10^{11} GeV ($m_M \sim T/\alpha$, where α is the coupling constant). The most elegant solution to this problem is the inflationary universe we discussed in chapter 3. Due to the strong expansion during inflation, a dilution of the monopoles took place, so that their contribution to the density is considerably decreased. However, in the post-inflationary phase a new heating up of the universe takes place, which could allow new monopoles to be created. Detailed estimates of the monopole density with inflation are so far not possible, but could lead to a change of the above mentioned density. If a strong GUT first order phase transition is employed to solve these problems, which would forbid inflationary scenarios, the problem would become even more drastic [Kol90]. This is because the Higgs fields inside the developing bubbles now become correlated. Since no correlation of fields in different bubbles is expected after their fusion and heating up a monopole density of

$$\frac{n_M}{T^3} \simeq \left[\left(\frac{T}{m_{Pl}} \right) \ln \left(\frac{m_{Pl}^4}{T^4} \right) \right]^3 \tag{10.32}$$

is produced. In this case the discrepancy between this and the observed density is even worse!

We now consider what information can be obtained on the properties of these monopoles. Once produced, their mutual annihilation with anti-monopoles plays only a minor role. They are in kinetic equilibrium with charged particles, such as electrons, through reactions such as $M + e^- \rightarrow M + e^-$. During $e^+ e^-$ annihilation they have an internal velocity dispersion of [Kol90]

$$\langle v_M^2 \rangle^{\frac{1}{2}} \simeq \sqrt{\frac{T}{m_M}} \simeq 30 \, \text{cm s}^{-1} \left(\frac{10^{16} \, \text{GeV}}{m_M} \right)^{\frac{1}{2}}. \tag{10.33}$$

As they subsequently expand relatively freely ($\sim R(t)^{-1}$), they have a relatively small velocity dispersion of

$$\langle v_M^2 \rangle^{\frac{1}{2}} \simeq 10^{-8} \, \text{cm s}^{-1} \left(\frac{10^{16} \, \text{GeV}}{m_M} \right)^{\frac{1}{2}}. \tag{10.34}$$

Having such a small root mean square velocity, such monopoles are unstable against a gravitational collapse on all astrophysically relevant scales, i.e. they collapse together with normal massive particles. When the universe became matter dominated, the monopoles took part in the formation of structure. However, they did not populate galactic discs, and hence should be rarely found there, but rather in galactic halos or clusters. However, monopoles are accelerated to higher velocities than implied in equation (10.34) by magnetic fields. Our Milky Way has a magnetic field of about 3 μG with a coherence length l of 300 pc. A passing monopole would be accelerated by this field to a speed of

$$v_M \simeq 3 \times 10^{-3} c \left(\frac{10^{16} \text{ GeV}}{m_M} \right)^{\frac{1}{2}}. \tag{10.35}$$

It is therefore expected that monopoles will have velocities in the region of a few per cent of the speed of light due to accelerations within the particular structure (galaxy, galaxy cluster) within which they find themselves. It is important to know the velocity in order to estimate the monopole flux and therefore have a basis for laboratory experiments. A first estimate of the monopole flux $\Phi_M = n_M v_M / 4\pi$ already follows from the fact that the monopole density must not be greater than the critical density. If we assume a velocity of $v \approx 10^{-3} c$, then

$$\Phi_M \leq 10^{-14} \left(\frac{10^{16} \text{ GeV}}{m_M} \right) \text{ cm}^{-2} \text{ s}^{-1} \text{ sr}^{-1}. \tag{10.36}$$

The local monopole flux can, however, be higher due to the galactic halo. The local density is about 10^{-23} g cm^{-3}, of which the halo can contribute only half at most, as stars, gas and dust make up the other half. A conservative upper limit of 10^{-24} g cm^{-3} implies a flux of

$$\Phi_M \leq 10^{-10} \left(\frac{10^{16} \text{ GeV}}{m_M} \right) \text{ cm}^{-2} \text{ s}^{-1} \text{ sr}^{-1}. \tag{10.37}$$

A very different flux limit follows from the magnetic field of our Milky Way. Assume a monopole at rest, i.e. $v_0 = 0$, and a magnetic charge $g = (137/2)e$. If this monopole travels a length l in this field, it will be accelerated to a kinetic energy E_M of

$$E_M \approx g B l \simeq 10^{11} \text{ GeV} \left(\frac{B}{3 \, \mu\text{G}} \right) \left(\frac{l}{300 \text{ pc}} \right) \tag{10.38}$$

resulting in a velocity v_{ma} of the magnetic monopole of

$$v_{ma} = \left(\frac{2 g B l}{m_M} \right)^{\frac{1}{2}}. \tag{10.39}$$

This acceleration results in an energy loss of the field within a volume V, given by

$$\Delta E_M = -\Delta \left[V \cdot \frac{B^2}{2} \right]. \tag{10.40}$$

If we now drop the assumption of monopoles at rest, there are in principle two possibilities. If the initial velocity is small compared to v_{ma}, the monopoles experience a strong change of their kinetic energy, and the above equations (10.38)–(10.40) are a good description. If, however, $v_0 \gg v_{ma}$, the monopoles experience only a slight effect from the magnetic field, which in addition depends on the direction of movement with respect to the orientation of the magnetic field. For an isotropic flux of both types of monopoles the net energy gain to first order even disappears, since some monopoles gain energy, while others lose it. Only at second order is there an energy gain, as the entire monopole distribution increases its kinetic energy, leading to [Kol90]

$$\Delta E_M \simeq \frac{1}{4}(g B l)\left(\frac{v_{ma}}{v_0}\right)^2 \quad \text{per monopole.} \tag{10.41}$$

Assuming typical velocities within our galaxy of $v_0 \approx 10^{-3}c$, it follows that monopoles with masses smaller than 10^{17} GeV experience very large perturbations, and are quickly spun out of our Milky Way. Even monopoles with masses up to about 10^{20} GeV are, due to the second order effect, still effectively lost during the age of the galaxy. This permanent energy loss of the galactic magnetic field has somehow to be compensated. As we imagine the galactic magnetic field as being created due to a dynamo effect, it follows that the typical regeneration time τ of the field is that of one rotation of the Milky Way, i.e. about 10^8 years. Otherwise the monopoles would lead to the existing field being neutralized after a time

$$\tau = \frac{B}{8\pi g \Phi_M}. \tag{10.42}$$

From the existence of the current field the limits

$$\Phi_M \leq 10^{-15} \text{ cm}^{-2} \text{ s}^{-1} \text{ sr}^{-1} \left(\frac{B}{3\,\mu G}\right)\left(\frac{300\text{ pc}}{l}\right)^{\frac{1}{2}} \tag{10.43}$$

$$\times \left(\frac{3 \times 10^7 \text{ years}}{\tau}\right)\left(\frac{r}{30\text{ kpc}}\right)^{\frac{1}{2}} \quad \text{for } m_M \lesssim 10^{17} \text{ GeV}$$

$$\Phi_M \leq 10^{-16} \text{ cm}^{-2} \text{ s}^{-1} \text{ sr}^{-1} \left(\frac{300\text{ pc}}{l}\right) \tag{10.44}$$

$$\times \left(\frac{3 \times 10^7 \text{ years}}{\tau}\right)\left(\frac{m}{10^{16} \text{ GeV}}\right) \quad \text{for } m_M \gtrsim 10^{17} \text{ GeV}$$

follow, where r represents the extent of the galactic magnetic field. The flux limit given by equation (10.43) is called the *Parker bound* [Par70, Tur82]. These limiting values correspond to a very small flux of the order of one monopole per year and a surface area of 2500m². An extended Parker limit of

$$\Phi_M \leq 10^{-16} \left(\frac{m}{10^{17} \text{ GeV}}\right) \quad \text{cm}^{-2} \text{ s}^{-1} \text{ sr}^{-1} \tag{10.45}$$

is given by [Ada93]. For very heavy monopoles ($m > 10^{20}$ GeV) the limits become more stringent, as their contribution to the galactic mass would otherwise be much too large. It would be a different situation if monopoles themselves act as sources of the galactic magnetic field. However, we will not discuss this case in any more detail (see e.g. [Kol90]).

A limit several orders lower is provided by neutron stars. Assuming that monopoles catalyse nucleon decay (see chapter 10.4.3), this process should take place at an enormous rate in neutron stars, and also contribute noticeably to their X-ray radiation [Kol84]. As an example, we mention here the relatively well known radio pulsar PSR 1929+10, which is situated at a distance of about 60 pc. Its luminosity, determined from the Einstein satellite, is about 3×10^{30} erg s^{-1}, which points to a surface temperature of only 30 eV [Kol84]. In the course of its life of about 3×10^6 years it should have accumulated about 10^{33} monopoles. The observed luminosity can, however, be converted into a monopole number

$$n_M \leq 10^{12} \left(\frac{\sigma v}{10^{-28} \text{ cm}^2} \right)^{-1} \tag{10.46}$$

which then corresponds to a flux of [Fre83]

$$\Phi_m \leq 10^{-21} \left(\frac{\sigma v}{10^{-28} \text{ cm}^2} \right)^{-1} \text{ cm}^{-2} \text{ s}^{-1} \text{ sr}^{-1} \tag{10.47}$$

where σv is the product of the cross section for catalysis of nucleon decay and velocity. The flux is further reduced by seven orders of magnitude if the possibility of monopole capture during previous combustion phases of the star is taken into consideration. This would in practice be a hopeless level to search for in terrestrial laboratories. However, since the understanding of neutron stars and their X-ray spectra is very model dependent, and also the assumption of nucleon catalysis is used, experimental searches should rather aim for levels of sensitivity corresponding to the Parker bound.

10.4 Experimental search for monopoles

As we have seen from mass estimates, it would not be possible to produce GUT monopoles in accelerators. However, as there are no mass predictions for Dirac monopoles, it is already traditional to search for such monopoles at every new accelerator. They could, for example, be produced as monopole–anti-monopole pairs, $M\bar{M}$, in reactions such as

$$\begin{aligned} e^+ + e^- &\rightarrow M + \bar{M} \\ p + p &\rightarrow p + p + M + \bar{M} \\ p + \bar{p} &\rightarrow M + \bar{M}. \end{aligned} \tag{10.48}$$

Table 10.1 shows the limits established in this way for Dirac monopoles.

Table 10.1. Monopole searches at accelerators. The limits for monopole production cross sections as well as for monopole masses are given for different energy regions. The square root of s denotes the center of mass energy.

σ_M (cm^2)	m_M (GeV)	Beam	$\sqrt{s}$ (GeV)	Events	Reference
$< 2 \times 10^{-35}$	< 1	p	6	0	[Bra59]
$< 1 \times 10^{-35}$	< 3	p	28	0	[Fid61]
$< 2 \times 10^{-40}$	< 3	p	30	0	[Pur63]
$< 5 \times 10^{-42}$	< 13	p	400	0	[Car74]
$< 4 \times 10^{-38}$	< 10	e^+e^-	34	0	[Mus83]
$< 3 \times 10^{-32}$	< 800	$p\bar{p}$	1800	0	[Pri87]
$< 1 \times 10^{-38}$	< 17	e^+e^-	35	0	[Bra88]
$< 1 \times 10^{-37}$	< 29	e^+e^-	50–61	0	[Kin89]
$< 2 \times 10^{-34}$	< 850	$p\bar{p}$	1800	0	[Ber90d]
$< 7 \times 10^{-35}$	< 44.9	e^+e^-	89–93	0	[Kin92]
$< 3 \times 10^{-37}$	< 45	e^+e^-	88–94	0	[Pin93]

We now concentrate on GUT monopoles. There are various detection strategies for the detection of cosmic magnetic monopoles in terrestrial laboratories, taking into account both the small monopole flux and their particular characteristics. The experiments with the clearest monopole signature are induction experiments [Sto84, Gro86a].

10.4.1 Induction experiments

If we assume that magnetic monopoles exist, the induction law becomes [Jac82]

$$\oint \boldsymbol{E} \mathrm{d}\boldsymbol{r} = -\frac{\mathrm{d}\Phi}{\mathrm{d}t} - \frac{\mathrm{d}Q_m}{\mathrm{d}t}. \tag{10.49}$$

The first term is the induced magnetic flux, and the second the monopole current. In the case that the integration region lies completely inside a superconducting coil, the electric field disappears along the path and it follows that

$$\Delta\Phi = -\Delta Q_m. \tag{10.50}$$

Therefore, if a monopole passes through an isolated superconducting loop or coil it produces a change of flux of $\Delta\Phi = hc/e$. In superconductors this corresponds to two elementary flux units and is therefore easily detectable. Figure 10.4 shows schematically the result of a monopole passing through a superconducting coil. A change of flux due to a passing monopole with a magnetic charge of $4\pi g = 4.14 \times 10^{-7}$ G cm^2 and an assumed coil inductance

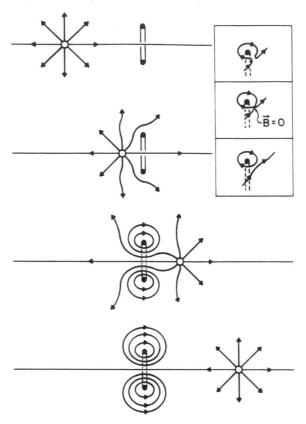

Figure 10.4. The behaviour of the field lines in a superconducting coil during the crossing of a monopole. It can clearly be seen how a current is induced (from [Gro86a]).

of 10 μH results in a current of 0.4 nA, equivalent to an energy deposition of $\approx 10^{-24}$ J. In principle this energy is not very difficult to measure. Very sensitive superconducting elements (SQUIDs, see e.g. [Smi90]), are used, whose sensitivity is around 10^{-27} J or lower. However, a background problem is caused by the fact that a change of the Earth's magnetic field of 10^{-11} produces the same effect as a monopole. These experiments are therefore very carefully screened against the Earth's magnetic field. The typical construction of such an experiment is shown in figure 10.5. Due to the required screening and the worsening signal to background ratios with detector size, these detectors cannot be operated at any desired size, and typical orders of magnitude are 1 m^2. It is now common to use several coils, one put inside the others, and then look for a coincidence signal (figure 10.6). A change in the magnetic field would affect all the coils in the same way, while a monopole event would only ever hit two. Figure 10.7 shows the positive signal of a group in Stanford [Cab82], although this has not

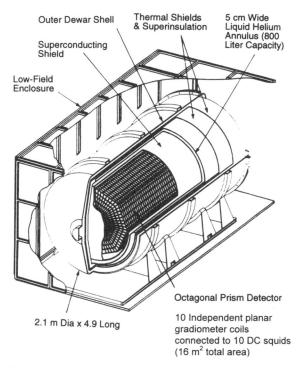

Outer Dewar Shell

Thermal Shields & Superinsulation

5 cm Wide Liquid Helium Annulus (800 Liter Capacity)

Superconducting Shield

Low-Field Enclosure

Octagonal Prism Detector

2.1 m Dia x 4.9 Long

10 Independent planar gradiometer coils connected to 10 DC squids (16 m² total area)

Figure 10.5. Illustration of the IBM BNL detector as an example of an experiment for the search for magnetic monopoles. Careful screening against the Earth's magnetic field is necessary (from [Gro86a]).

been verified by subsequent, more sensitive experiments. The present status of monopole searches using induction detectors results in a flux limit of [Ber90d]

$$\Phi_M < 3.8 \times 10^{-13} \text{ cm}^{-2} \text{ s}^{-1} \text{ sr}^{-1}. \tag{10.51}$$

Other experiments result in similar limits [Hub90, Gar91]. Table 10.2 gives a summary of the induction results.

10.4.2 Ionization experiments

In contrast to superconducting induction experiments, ionization experiments can be constructed with sizes of order 100 m^2. The method of detection here is the energy loss of the monopoles via their interaction with bound electrons. The atomic energy levels are strongly excited by the passing monopole and its magnetic field, resulting in an energy loss. Normally the energy loss dE/dx due to ionization is described using the Bethe–Bloch formula (see e.g. [Per87]). However, this is no longer applicable for monopole velocities of $\beta \approx 10^{-3}$. This is, rather, an area in which it has to be shown that a detectable signal is

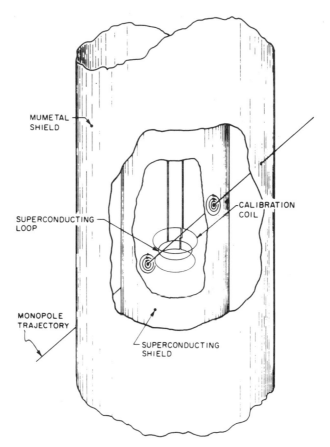

Figure 10.6. The superconducting induction detector of Cabrera for the detection of magnetic monopoles (from [Gro86a]).

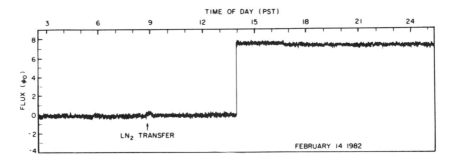

Figure 10.7. The famous event of Cabrera in 1982 shows a very clean signal of a magnetic monopole. It has however been contradicted by the flux limits determined subsequently (from [Cab82]).

Table 10.2. Flux limits for magnetic monopoles from induction detectors.

F_M (cm^{-2}sr^{-1}s^{-1})	Reference
$< 6.7 \times 10^{-12}$	[Inc84]
$< 5.5 \times 10^{-12}$	[Ber85]
$< 6.0 \times 10^{-12}$	[Cap85]
$< 5.0 \times 10^{-12}$	[Cro86]
$< 3.8 \times 10^{-13}$	[Ber90d]
$< 7.2 \times 10^{-13}$	[Hub90]
$< 4.4 \times 10^{-12}$	[Gar91]

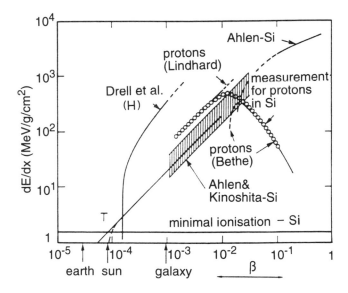

Figure 10.8. The behaviour of the energy loss dE/dx for very slow monopoles (typical velocities are marked with the astronomical object). The dots indicate measurements of the energy loss of slow protons in Si. Various theoretical extrapolations for small velocities β are shown. In this area we are mainly dependent on theoretical calculations and therefore on a precise knowledge of atomic and molecular physics (from [Bar84], see also [Gro86a]).

Figure 10.9. Front view of the MACRO detector in the Gran Sasso underground laboratory in Italy (with kind permission of the Gran Sasso laboratory).

produced at all, and for this special knowledge of the atomic and molecular physics involved is necessary (see figure 10.8). For such slow monopoles the velocity is very close to the velocity of the atomic electrons. Assuming a velocity of $v_e \approx \alpha c$ for the latter, it follows that for $g = e/2\alpha$, $g\alpha = e/2$. This means that the interaction of a slow monopole corresponds to a normal particle with charge $e/2$, and hence with an ionization of about a quarter of that of a proton with the same velocity. Scintillators are mainly used as the active element in detectors used to search for such evidence.

10.4.2.1 The MACRO detector

As an example of such an experiment we consider the MACRO detector (monopole astrophysics and cosmic ray observatory, figure 10.9) [Ahl93]. This experiment was constructed in the Gran Sasso underground laboratory (Italy) and consists of six so-called super-modules or twelve modules. The cross section of such a module is shown in figure 10.10. The MACRO detector consists in principle of three different detection elements. Scintillation counters, based on mineral oil, are attached to the top and bottom. In between there are ten streamer chambers, each separated by half a metre of absorber. The streamer tubes are filled with

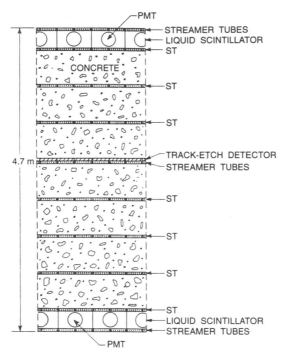

Figure 10.10. Cross section through a module of the MACRO detector. This detector is 72 m long, 12 m wide and about 10 m high. It is being constructed in a modular way, and liquid scintillators are installed both on the top and bottom. In between are several streamer tubes and in the centre a track-etch detector (from [Gro86a]).

a mixture of helium and n-pentane. Two effects are being made use of here. The strong magnetic field of a passing monopole leads to a perturbation of the energy levels of the helium (Zeeman effect). This leads to an increased number of transitions between the atomic energy levels and the helium stays in a metastable, excited state (He*). The excitation energy of the He* is about 20 eV. The energy loss of the monopoles due to the Zeeman effect is larger by about a factor of 10 than in the previously described ionization process [Dre83]. If we now add a gas with an ionization potential of less than 20 eV (in the case of the MACRO, n-pentane), a de-excitation of the helium takes place due to collisions with the n-pentane, which is thereby ionized (Penning effect). This forms the basis for the detection method. The ionization energy of n-pentane is about 10 eV. This is also the principle of gas detectors used in the search for monopoles. In the middle of each module is a layer of track-etch detectors (Lexan and CR39), which are arranged in the form of a sandwich, with a layer of aluminium at the centre. If a monopole crosses the detector, its path is clearly visible along the streamer chambers. Due to the relatively low velocity of the monopoles the time difference between the two scintillators can be used as a

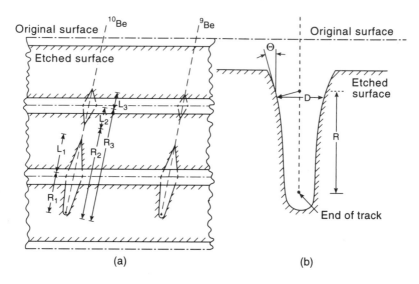

Figure 10.11. The principle of the track-etch method. Information about the ionization rate at different penetration depths is obtained either from the size of the etched cone or from the angle and diameter of the etched track (from [Pri71], see [Lon92, 94]).

signature for a crossing. The typical crossing time for a slow monopole is around 150 μs. If an interesting candidate is found, the track-etch plates are etched in order to examine the track created by the event (see figure 10.11). Very slow monopoles only damage the CR39 layers, as the threshold for Lexan is much higher. This detector has already lead to the following flux limits [Sto96]:

$$\Phi_M < 4.1 \times 10^{-15} \text{ cm}^{-2} \text{ sr}^{-1} \text{ s}^{-1} \quad (2 \times 10^{-4} < \beta < 3 \times 10^{-3})$$
$$\text{(scintillator)} \tag{10.52}$$

$$\Phi_M < 1.7 \times 10^{-15} \text{ cm}^{-2} \text{ sr}^{-1} \text{ s}^{-1} \quad (1.1 \times 10^{-4} < \beta < 5 \times 10^{-3})$$
$$\text{(streamer tubes)} \tag{10.53}$$

$$\Phi_M < 4 \times 10^{-15} \text{ cm}^{-2} \text{ sr}^{-1} \text{ s}^{-1} \quad (\beta = 1)$$
$$\text{(track-etch)} \tag{10.54}$$

$$\Phi_M < 6.2 \times 10^{-15} \text{ cm}^{-2} \text{ sr}^{-1} \text{ s}^{-1} \quad (\beta = 10^{-4})$$
$$\text{(track-etch).} \tag{10.55}$$

For fast monopoles the additional limit

$$\Phi_M < 3.8 \times 10^{-15} \text{cm}^{-2} \text{ sr}^{-1} \text{ s}^{-1} \quad (10^{-3} < \beta < 10^{-1}) \tag{10.56}$$

is obtained. These results correspond to a total measurement time of 3773.8 hours. In its final form the detector has an effective total area of 10 000 m^2 sr for isotropic fluxes, which allows sensitivities below the Parker bound (see

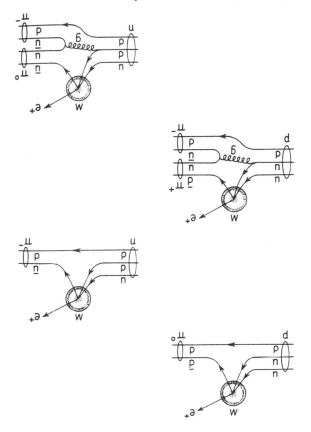

Figure 10.12. Schematic illustration of some processes of monopole-catalysed nucleon decay. A nucleon (left) decays with the help of a monopole M into different final states, which are predicted by the underlying GUTs (from [Err83]).

figure 10.15). A flux limit of $\Phi_M < 8.7 \times 10^{-15}\,\mathrm{cm}^{-2}\,\mathrm{sr}^{-1}\,\mathrm{s}^{-1}$ for $\beta > 2 \times 10^{-3}$ has been given by the Soudan collaboration [Thr92].

10.4.3 Catalysis of nucleon decay

A further detection method for the monopole is based on its already mentioned special structure. As the full GUT symmetry still exists inside the monopole, it should be possible for monopoles to catalyse nucleon decay (see figure 10.12). A purely geometrical estimate of the cross section results in [Kol90] $\sigma \simeq R^2 \simeq M^{-2} \simeq 10^{-56}\,\mathrm{cm}^2$ ($R \propto M^{-1}$ is the radius of the monopole core of mass M). This cross section is much too small to have any experimental consequences. However, this is modified since it has been shown that cross sections comparable with those of the strong interaction ($\sigma \simeq 10^{-26}\,\mathrm{cm}^2$) can

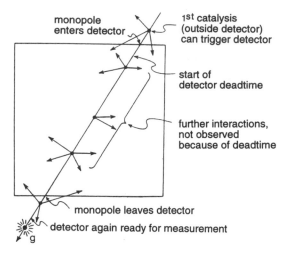

Figure 10.13. Signature of a monopole crossing a nucleon decay detector. Along its track it produces nucleon decay, but the dead time of the detector can be too large to record all these events (from [Err83]).

be reached due to fermion–anti-fermion condensates on the surface (*Callan–Rubakov* effect) [Cal82, Rub82]. For very small β the interaction cross section is given to first order by

$$\sigma = \frac{1}{\beta} \left(\frac{\sigma_0}{E_0^2} \right) \left(\frac{hc}{2\pi} \right)^2 \tag{10.57}$$

where E_0 corresponds roughly to the proton mass (1 GeV), and σ_0 lies in the region between 10^{-6} and 1, with a preferred value of 10^{-4}. A typical decay channel under the influence of a monopole M would be [Kol90]

$$p + M \rightarrow M + e^+ + \pi^0 \tag{10.58}$$

$$n + M \rightarrow M + e^+ + \pi^-. \tag{10.59}$$

If nucleon catalysis really takes place, it provides a *very effective alternative mechanism for energy production in stars*. Contrary to the *pp*-cycle (see section 11.3.2 and chapter 12), which proceeds at a weak interaction rate and only transforms about 0.7% of the rest mass into energy, 100% is transformed into energy in monopole catalysis. For example, only 10^{28} monopoles would be necessary inside the Sun to produce the present solar luminosity. From this we can again obtain information about monopoles indirectly, since such a process would produce high energy neutrinos which could be detected on Earth. On the other hand the results of the gallium solar neutrino experiments suggest that the fusion of hydrogen really does take place inside the Sun (see chapter 12).

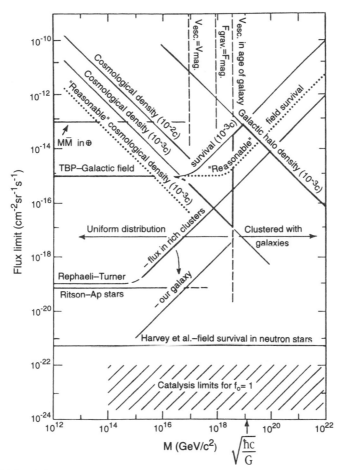

Figure 10.14. Different astrophysical limits and limitations for the monopole flux. The bold line is obtained from considerations on the lifetime of the galactic magnetic field. The dots indicate the cosmological considerations on monopole abundance. The limits become particularly restrictive in the case that the monopoles can catalyse nucleon decay (shaded area) (from [Gro86a]).

According to the discussion above (for example equations (10.58) and (10.59)), existing proton decay experiments should also be able to search for monopoles. In order to estimate the expected signal it is important to compare the mean free path λ between two catalysed nucleon decays with the dimension d of the detector. If $\lambda \gg d$, then it is most probable that only one decay occurs within the detector. If, on the other hand, $\lambda \ll d$, then several proton decays occur along the track, giving a very clear signal. One problem is that the trigger from the first decay creates a dead time in the detector of typically 600 ms, so that

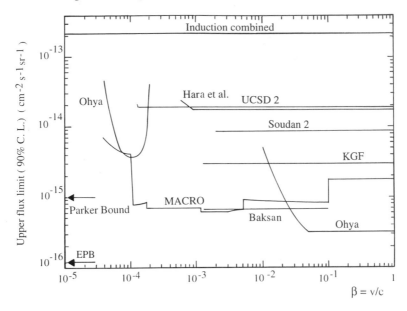

Figure 10.15. The 90% cl upper limits for an isotropic flux of bare $g = g_D$ magnetic monopoles, assuming a catalysis cross section smaller than 10 mb, obtained by MACRO and by other experiments: induction combined [Ber90d], UCSD2 [Buc90], Soudan2 [Thr92], Baksan [Ale82, Ale83], Orito [Ori91], KGF [Ada90], Hara *et al* [Har90]. Also shown in is the Parker bound and the extended Parker bound (EPB) (from [Amb96]).

subsequent decays can be either totally or partially undetected (see figure 10.13). Proton decay experiments have to date found no monopole candidates. For a search for monopoles with the BAIKAL detector (see chapter 8.4) we refer to [Bel96a, Spi96].

10.4.4   Other methods and conclusions

The track-etch method discussed above has also been used in other searches for monopoles using muskovite rock about 1 billion years old. If a monopole penetrates the Earth, bound monopole–nucleus systems could be created with nuclei of large magnetic moments (such as with ^{27}Al). This heavy object would damage the lattice structure of the rock it traversed. Such tracks would be a few Ångstroms wide and can be increased with appropriate chemical etching. Therefore muskovite from 5 km depth was treated in this way and microscopes used to search for crystal defects, but without success [Pri83, Pri87]. This led to limits for the monopole flux between 5×10^{-17} and 5×10^{-19} cm^{-2} sr^{-1} s^{-1} (for $\beta \approx 10^{-3}$). The advantage of this method is its very long exposure time. There are, however, also many assumptions in interpretation, such as that the muskovite has never been heated to such an extent that the tracks could be destroyed.

However, if the limits are taken seriously they would currently be the best (see figure 10.14). Trials are also being made with meteorites. The hope would be that monopoles were trapped in them, and they are examined for the previously described induction effects with the help of superconducting coils. Such an investigation with a total of 331 kg of material, of which 112 kg was meteorite material, also did not find any monopole, leading to a monopole/nucleon ratio of less than 1.2×10^{-29} (90% cl) [Jeo95].

Thus, all the searches to date have not led to a positive result (apart from the improbable Cabrera event in 1982, see above and figure 10.7). Flux limits obtained so far are shown in figure 10.15.

A limit below the Parker bound has been reached first by the Baksan group [Ale82, Ale83] in 1982 using the Baksan Underground Scintillating Telescope, and has been reached recently for $10^{-4} \leq \beta \leq 10^{-2}$ by MACRO [Amb96]. Limits near to the extended Parker bound will be reached with MACRO in future.

Chapter 11

Axions

11.1 Theoretical motivation

A long-standing unsolved problem in the theory of strong interaction is the strong CP problem. CP violation is observed in the weak interaction but has not been observed in the strong interaction. Due to the complex vacuum structure of non-Abelian gauge theories, such as the QCD, terms arise which are P and T as well as CP violating. QCD has an infinite number of vacuum states $|n\rangle$, characterized by a winding number n, similar to the discussion of the electroweak phase transition (see chapter 3). Any individual vacuum state is itself not gauge invariant, but rather a superposition of the various vacuum configurations.

$$|\Theta\rangle = \sum_n \exp(-in\Theta)|n\rangle \tag{11.1}$$

where Θ is an arbitrary parameter, and n a topological winding number, which characterizes the different vacuum gauge configurations that cannot be rotated into each other [Kim87, Pec89, Kol90, Pec96]. This vacuum is called the Θ-vacuum. The effects of this vacuum can be described via an additional term $\mathcal{L}$ in the Lagrange density of QCD (see chapter 1):

$$\mathcal{L}_{\text{QCD}} = \mathcal{L}_{\text{pert}} + \bar{\Theta}\frac{g^2}{32\pi^2}G^{a\mu\nu}\tilde{G}_{a\mu\nu}. \tag{11.2}$$

$\mathcal{L}_{\text{pert}}$ is here the Lagrange density of perturbative QCD, G is the gluon field strength tensor, given by

$$G^a_{\mu\nu} = \partial_\mu A_\nu - \partial_\nu A_\mu + gf^{abc}A^b_\nu A^c_\mu \tag{11.3}$$

and $\tilde{G}$ is the corresponding dual tensor. The factors f^{abc} are the structure constants of SU(3) and $\bar{\Theta}$ is defined as $\Theta + \arg \det M$, where M is the quark mass matrix. The effective $\bar{\Theta}$ term in equation (11.2) also contains the phase of the quark mass matrix in addition to the bare Θ term. For a more detailed discussion see e.g. [Kim87, Pec89]. It is the second term in equation (11.2) which violates the various symmetries. It changes sign under P and T transformations, and therefore also under CP. It can be seen that two independent terms are responsible for $\bar{\Theta}$. If one or more quarks were massless, the term would be without any importance, as it could be made to disappear by a chiral transformation of the quark fields, and therefore there would be no strong CP problem. But if all quarks are massive, there are two uncorrelated contributions to $\bar{\Theta}$, and no reason for them to cancel. This leads, among other things, to a contribution to the electric dipole moment of the neutron of [Bal79, Cre79]

$$d_n \simeq 5 \times 10^{-16} \bar{\Theta} \, e \, \text{cm} \tag{11.4}$$

while present experiments result in an upper limit of [Ram90, Pen93]

$$|d_n| < 1.2 \times 10^{-25} e \, \text{cm.} \tag{11.5}$$

This implies a $\bar{\Theta} \leq 10^{-10}$ or even exactly zero, and which has to be *a priori* made so small, since it is in principle an arbitrary parameter between 0 and 2π. This problem is known as the *strong CP problem*. The standard model discussed in chapter 1 predicts an electric dipole moment of the neutron of about $10^{-33} e$ cm. An experimental detection of an electric dipole moment of the neutron greater than this value would be a strong indication that Θ may be different from zero (since QCD is only used perturbatively in the standard model there is no necessity for Θ and it can be set to zero).

The question remains as to why Θ is so small, or even zero. One of the most promising suggestions comes from Peccei and Quinn [Pec77] (for an alternative solution by supersymmetry see [Moh95]). By introducing a further global, chiral $U_{PQ}(1)$ symmetry they succeeded in compensating the troublesome term. The decisive idea here was to turn Θ into a dynamic variable, whose minimum energy value lies at zero. However, this implies that the Peccei–Quinn symmetry has to be spontaneously broken. The Goldstone boson resulting from this breaking is called the axion [Wei78, Wil78]. In an exact symmetry this particle would be massless. However, because of the chiral anomaly it gets a mass of order Λ_{QCD}^2/f_{PQ}, where f_{PQ} represents the scale of the $U_{PQ}(1)$-symmetry breaking. The introduction of this additional field leads to a further term in the Lagrange density:

$$\mathcal{L} = \ldots + C_a \frac{a}{f_{PQ}} \frac{g^2}{32\pi^2} G_a^{\mu\nu} \tilde{G}_{\mu\nu}^a \tag{11.6}$$

where C_a is a model dependent constant. As both expressions from equations (11.2) and (11.6) contribute to the axion field, it can be minimized, and even set to exactly zero by

$$\langle a \rangle = -\frac{\bar{\Theta} f_{PQ}}{C_a}. \tag{11.7}$$

It is therefore possible to compensate the troublesome term in equation (11.2) by the introduction of an additional field, the axion field a. For an introductory exposition see [Sik95, Sik97].

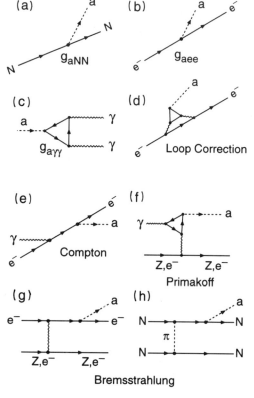

Figure 11.1. Feynman diagrams for the coupling of axions to normal matter, as well as the dominant axion emission processes in stars. (*a*) The axion–nucleon coupling. (*b*) The *direct* axion–electron coupling. This graph is only possible for DFSZ axions and provides a method of differentiation between this type and hadronic axions. The latter can only couple to an electron at higher order (case (*d*)). (*c*) The axion–photon coupling. (*e*) 'Compton' effect. (*f*) 'Primakoff' effect. (*g*) and (*h*) Axion bremsstrahlung (from [Tur90b]).

11.2 Characteristics of the axion

The Lagrange density of the free axion field is [Kol90]

$$\mathcal{L} = \frac{1}{2}\partial^\mu a \partial_\mu a + C_a \frac{a}{f_{PQ}} \frac{g^2}{32\pi^2} G_a^{\mu\nu} \tilde{G}_{\mu\nu}^a \qquad (11.8)$$

and its interactions with normal matter are described by

$$\mathcal{L}_{\text{int}} = \mathrm{i}\frac{g_{aNN}}{2m_N}\partial_\mu a(\bar{N}\gamma^\mu\gamma_5 N) + \mathrm{i}\frac{g_{aee}}{2m_e}\partial_\mu a(\bar{e}\gamma^\mu\gamma_5 e) + g_{a\gamma\gamma}a\boldsymbol{E}\cdot\boldsymbol{B}. \qquad (11.9)$$

g_{aNN}, g_{aee} and $g_{a\gamma\gamma}$ correspond to the coupling constants of the axion with nucleons N, electrons and photons, respectively. The corresponding Feynman diagrams are shown in figure 11.1. All diagrams have the coupling strength $g_{aii} \sim m_a \sim f_{PQ}^{-1}$ in common. There is only one free parameter in the theory, namely the axion mass, which is given by [Kol90]

$$m_a = \frac{\sqrt{z}}{1+z} \frac{f_\pi m_\pi}{f_{PQ} N_F} \approx 0.62 \, \text{eV} \left(\frac{10^7 \, \text{GeV}}{f_{PQ}/N_F} \right). \tag{11.10}$$

The quantities used are $z = m_u/m_d = 0.56$, $m_\pi = 135$ MeV and $f_\pi = 93$ MeV. The colour anomaly N_F of the Peccei–Quinn symmetry is given various notations according to convention, and is therefore not specified in more detail (for a discussion of this quantity see [Pec89, Raf90]). It was originally thought that the scale of the breaking lies at a similar point to the electroweak scale, corresponding to about 250 GeV, from which it follows that the axion mass is of order 200 keV. However, experiments at accelerators could quickly rule out axions with such a mass. The axion was then made 'invisible' by leaving the scale of the symmetry breaking and couplings open. Since these are now completely undetermined, axion masses of between 10^{-12} eV and 1 MeV result. The range to be experimentally examined is therefore 18 orders of magnitude!

The axion, like the pion, is a pseudo-scalar particle. We distinguish two kinds of axions. If a direct coupling to the electron is possible we talk about a *DFSZ-axion* (DFSZ stands for the physicists Dine, Fischler, Srednicki and Zhitnitsky) [Zhi80, Din81]. An axion which couples to the electron only via higher order corrections equivalent to a Peccei–Quinn charge $X_e = 0$, is known as a *hadronic axion* (or also as KSZV-axion [Kim79, Shi80]). As the coupling of the axion to electrons is particularly important to astrophysical considerations, those two options are discussed separately. The coupling of axions to photons, important for the experimental search, happens in a triangle graph similar to that for the π^0. Correspondingly the axion can also decay into two photons with a lifetime given by

$$\tau(a \to \gamma\gamma) \sim \tau_{\pi^0} \left(\frac{m_{\pi^0}}{m_a} \right)^5 \tag{11.11}$$

or

$$\tau(a \to \gamma\gamma) = 6.8 \times 10^{24} \frac{(m_a/\text{eV})^{-5}}{\left[\left(\frac{E_{PQ}}{N_F} - 1.95 \right) / 0.72 \right]^2} \, \text{s}. \tag{11.12}$$

In most GUTs the ratio E_{PQ}/N_F is 8/3, but it can also have other values, such as 2. This would then imply a strong suppression of the two-photon coupling of the axion [Kol90].

11.3 Axions and stellar evolution

11.3.1 Introduction

Because of their small interaction with normal matter, axions have a considerable influence on the evolution of stars (see also chapter 12) [Raf90]. Like neutrinos they easily carry off some of the energy created from the nuclear reactions from the interior of the star. The star has to adjust itself to this further means of energy loss and it therefore contracts, while simultaneously increasing its central temperature and luminosity, which again leads to a shortening of its lifetime due to the quicker use of fuel. Hence we can obtain information about axion masses from the observed nuclear burn off times in different stellar phases. As an example we consider the case of the Sun as a normal main sequence star.

11.3.2 Solar axions

The two main processes of DFSZ-axion production in main sequence stars are the 'Compton' effect

$$\gamma + e^- \to a + e^- \tag{11.13}$$

and axion bremsstrahlung

$$e^- + Z \to a + e^- + Z. \tag{11.14}$$

As hadronic axions have no direct coupling to electrons the Primakoff effect

$$\gamma + Z \quad (\text{or } e^-) \to a + Z \quad (\text{or } e^-) \tag{11.15}$$

is dominant here. We firstly consider a Sun with no axions. Neglecting neutrinos it follows from energy equilibrium that the energy Q_{nuc} released by nuclear reactions corresponds directly to the luminosity L_γ of the star, i.e.

$$L_\gamma^0 = Q_{nuc}^0. \tag{11.16}$$

The index 0 represents the Sun with no axions. The photon luminosity is related to the central temperature T_c by [Cha39, 67]

$$L_\gamma \sim (G\mu^7)M^5 T_c^{\frac{1}{2}} \tag{11.17}$$

where μ represents the average molecular weight. The energy creation rate by nuclear reactions $\dot{\epsilon}$ is given by

$$\dot{\epsilon} \sim \rho T^n \sim T^{n+3} \Rightarrow Q_{nuc} = \int \dot{\epsilon} \, \mathrm{d}M. \tag{11.18}$$

For hydrogen burning $n \simeq 4$. Consider now a Sun with an axion luminosity of L_a. The additional loss mechanism now has to be compensated by an increased energy production. Energy balance requires

$$L_\gamma + L_a = Q_{nuc} \tag{11.19}$$

where $L_\gamma = L_\gamma^0 + \delta L_\gamma$ and $Q_{nuc} = Q_{nuc}^0 + \delta Q_{nuc}$. With this equation (11.19) can be written as

$$\delta L_\gamma + L_a = \delta Q_{nuc}. \tag{11.20}$$

If the axion luminosity is written as a fraction of the original energy production rate ($L_a = \kappa Q^0_{nuc}$), the following behaviour of the star results:

$$\frac{\delta R}{R_0} = -\kappa/6.5 \tag{11.21}$$

$$\frac{\delta L_\gamma}{L^0_\gamma} = +\kappa/13 \tag{11.22}$$

$$\frac{\delta Q_{nuc}}{Q^0_{nuc}} = +14\kappa/13 \tag{11.23}$$

$$\frac{\delta T_c}{T^0_c} = +\kappa/6.5. \tag{11.24}$$

Here the relation $R \sim T_c^{-1}$, following from hydrodynamic equilibrium, has been used. The additional loss leads to a contraction, which results in a higher central temperature and increased nuclear reaction rate. At the same time the photon luminosity also increases.

A good measure for the solar axion emission is the solar neutrino flux. An increase of the central temperature in order to balance axion emission also results in an increased neutrino production. However, as we anyway observe less than the expected flux of neutrinos from the Sun, the problem is made rather worse by axions (see chapter 12). The accelerated combustion also results in a quicker evolution, so that we can draw conclusions on the axion mass from the age of the Sun. The helium content of our Sun, and assuming an age of 4.5 billion years, is only consistent with axion masses $m_a \leq 1$ eV, since otherwise the Sun with the current helium content would be too young. This only applies to DFSZ-axions. For hadronic axions the Primakoff effect is the most important, so from analogous considerations a mass limit of $m_a \leq 20$ eV$/[((E_{PQ}/N_F)-1.95)/0.72]$ follows [Kol90].

11.3.3 Axions and red giants

Even more stringent limits arise from the helium burning of red giants. After burning the major part of its hydrogen stock the core of the star begins to contract in order to increase its central temperature. The gravitational energy released by the contraction ($MR^3 = $ constant) is given by

$$\dot{E}_g \simeq \frac{d}{dt}\left(\frac{GM^2}{R}\right) \sim M^{\frac{4}{3}}\dot{M}. \tag{11.25}$$

Assuming that axion emission is the main cooling process of the star's core, the dominance of the Compton like interaction leads to $L_a \sim m_a^2 MT^6$. From the condition of energy balance it follows that [Kol90]

$$T_c \approx m_a^{-\frac{1}{3}} M^{\frac{1}{18}} \dot{M}^{\frac{1}{6}}. \tag{11.26}$$

The greater the axion mass, the lower the core temperature. This effect is contrary to the one discussed above, in which the core temperature increases in order to compensate the axion emission. This is now not possible, as no nuclear reactions take place in the ^{4}He core yet. We can easily obtain limits for possible axion masses for which helium burning can take place at all from equation (11.26). In the case of the DFSZ-axions no masses greater than 10^{-2} eV are allowed.

If red giants succeed in starting to burn helium, mass limits follow from the same considerations about the burning times as above. This burning phase has a typical time scale of $t_{\mathrm{He}} \approx 10^8$ years. For hadronic axions with masses

$$m_a \geq 2 \, \mathrm{eV} \left[\left(\frac{E_{\mathrm{PQ}}}{N_F} - 1.95 \right) \Big/ 0.72 \right]^{-1}$$

there can easily result a shortening of this time scale of an order of magnitude. If the burning time is reduced by an order of magnitude, we would expect a reduction in the number of observable red giants. The observation of large numbers of red giants in star clusters also excludes such axion masses from a comparison between the observed and expected abundances. These stellar arguments are, however, only applicable up to axion masses of about 200 keV, as such heavy axions can only be produced in a limited number inside the stars. However, higher mass regions are already ruled out from accelerator experiments.

11.3.4 Axions and SN 1987a

We can also gain very precise information about axion masses from SN 1987a (see chapter 13), by looking for a significant shortening of the neutrino emission due to axions, for which experimental bounds exist. In the post collapse phase nucleus–nucleus–axion bremsstrahlung is the dominant axion emission process. Here the axions carry away an energy of [Kol90]

$$L_a \simeq 10^{59} \left(\frac{m_a}{1 \, \mathrm{eV}} \right)^2 \, \mathrm{erg \ s^{-1}}. \tag{11.27}$$

For axion masses greater than about 10^{-3} eV this cooling down process becomes important, because it then provides a contribution equivalent to that of the neutrinos. This additional energy loss now results in more rapid cooling down of the proto-neutron star, provided the axions can escape without interaction. However, a quicker cooling also results (among other effects) in a shortening of the thermal neutrino emission. A particularly sensitive quantity is the time within which 90% of all neutrinos are measured [Tur90b]. The measured duration of this time period by Kamiokande and IMB can, in the case of free escape of axions, rule out masses of between 10^{-3} and 0.02 eV since otherwise the neutrino pulses produced would be shorter than observed. Above 0.02 eV a significant capture of axions takes place, and their emission occurs from an *axion-sphere* analogous to the photosphere. The luminosity is proportional to the fourth power of the temperature of the axion-sphere [Kol90]. The heavier the axion is, the further out in the star the axion sphere lies. This implies a decrease of luminosity with increasing mass, the opposite effect to that in the case of free escape. An acceptable axion emission rate can be obtained for masses larger than 2 eV. The observed duration of the neutrino pulse thus in total implies a forbidden mass region for axions of between 2 eV and 10^{-3} eV (see figure 11.2). A more detailed examination reveals that there is some model dependence of the lower limit so that only axion masses of down to 10^{-2} eV are excluded [Jan95]. For a more recent analysis see [Kei96].

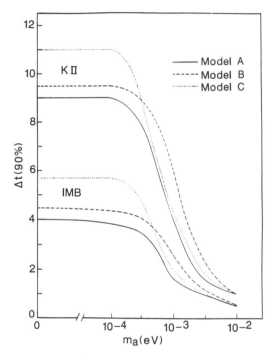

Figure 11.2. The length of the neutrino pulse of the supernova 1987a at Kamiokande and IMB, taking into account possible energy loss through emission of axions of mass m_a. Δt (90%) in seconds represents the time by which the number of neutrino events has reached 90% of its asymptotic value (from [Tur90b]).

11.4 Axions in cosmology

The appearance of axions in cosmology begins with the breaking of the Peccei–Quinn symmetry. This is a spontaneous symmetry breaking through a scalar, complex field, analogous to the Higgs mechanism discussed in chapter 1. Below $T \simeq f_{\text{PQ}}$ the symmetry is broken and a massless axion is produced. Through QCD instanton effects (see i.e. [Pec89]) the axion gains mass at $T \simeq \Lambda_{\text{QCD}}$. This is an analogous effect to the electroweak phase transition which was discussed in chapter 3. The temperature dependence of the mass can be approximated by [Kol90]

$$m_a(T) \simeq 0.1 m_a(T = 0)(\Lambda_{\text{QCD}}/T)^{3.7}. \tag{11.28}$$

The photo-production process $\gamma + q \rightarrow q + a$ (q = quark) and the pion–axion conversion $N + \pi \rightarrow N + a$ are the most important in the production of thermal axions. The process of pion–axion conversion is the dominant process for the period after the quark–hadron phase transition, since hadrons first exist here. Axions with masses greater than about 10^{-3} eV came almost in thermal equilibrium with the rest of the matter. For smaller masses the interaction is too weak to ever reach equilibrium. The number density of axions can be calculated to be [Kol90]

$$n_a \simeq 83 \text{ cm}^{-3} \left(\frac{10}{g_{\text{eff}}} \right) \simeq \frac{2}{11} n_\gamma \tag{11.29}$$

The contribution of these thermal axions to the mass density is

$$\Omega_{\text{therm}} h^2 = \left(\frac{10}{g_{\text{eff}}} \right) \left(\frac{m_a}{130 \text{ eV}} \right), \tag{11.30}$$

where g_{eff} is the number of relativistic degrees of freedom (see chapter 3). In order to create a Euclidean universe the axion mass has to be about 130 h^2 eV. According to the previous astrophysical considerations this is impossible. It can only be reached in the special case of a hadronic axion with a mass of about 30 eV and $h \approx 0.5$, and with additionally $E_{\text{PQ}}/N_F \simeq 2$. The only alternatives are non-thermal processes of axion creation in the early universe. One possibility is the initial shift of $\bar{\Theta}$ from its minimal value. The value of $\bar{\Theta}$ is at present close to the CP conserving values $\bar{\Theta} = 2\pi n$ (with $n = 0, 1, ..., N - 1$). Before the quark–hadron phase transition the axion is, however, massless and hence no value of $\bar{\Theta}$ is dynamically preferred. Hence the initial value of $\bar{\Theta}$ is arbitrary. However if $T \simeq \Lambda_{\text{QCD}}$, $\bar{\Theta}$ has to tend to zero. The equation of motion can be written in an analogous way to (3.89) as [Kol90]

$$\ddot{\bar{\Theta}} + 3H\dot{\bar{\Theta}} + m_a^2(T)\bar{\Theta} = 0 \tag{11.31}$$

The 'rolling down' to the minimum leads to an overshoot to above the minimum, and hence to oscillations around the minimum. From this a condensate of zero momentum axions develops, analogous to the effect of particle production in inflation. The difference here is that the axion oscillations do not decay. The initial freedom of $\bar{\Theta}$ leads to the problem of *axionic topological defects*, such as axion domain walls [Kol90]. In order to avoid this problem, we resort again to inflation, as our universe then only develops from *an arbitrary* initial value

$\bar{\Theta}_i$. The estimate of the axion contribution to the density due to this production mechanism has a large uncertainty, and is given by [Kol90]

$$\Omega h^2 = 0.85 \times 10^{\pm 0.4} \left(\frac{\Lambda_{QCD}}{200 \text{ MeV}} \right)^{-0.7} \left(\frac{m_a}{10^{-5} \text{ eV}} \right)^{-1.18}. \tag{11.32}$$

Another non-thermal contribution to axion production in the early universe is the decay of *axionic strings* (for a recent review see [Bat94, Bat97]). We have already discussed the existence of one-dimensional topological defects in chapter 6. Also in breaking a global U(1) symmetry, in our case the Peccei–Quinn symmetry, one-dimensional topological defects are produced. The discussion of global strings is a bit more complex. However, a network with corresponding loops is also crystallized out. The big difference compared to cosmic strings is that in the decay of the loops, it is not gravitational waves, but axions which are emitted. Since their spectrum is unknown, we easily find an uncertainty in the predicted number of a factor 100. A more detailed discussion shows that, according to the spectrum, their contribution to the number density ranges between being similar to the previously discussed production mechanism, and up to a factor 100 higher. It is given by [Kol90]

$$\Omega h^2 \simeq \left(\frac{m_a}{10^{-3} \text{ eV}} \right)^{-1.18}. \tag{11.33}$$

We therefore reach a signisicant density in the universe (see chapter 3) for axion masses of about 10^{-3} or 10^{-5} eV. The two non-thermal processes therefore, on the grounds of an acceptable contribution to the cosmological density, lead to a lower mass limit of about 10^{-3} or 10^{-5} eV.

To conclude, it is already possible, on the basis of astrophysical and cosmological considerations, to exclude nearly all of the allowed 18 orders of magnitude range of masses. We now consider what information can be provided by laboratory experiments.

11.5 Experimental search for axions

First ideas about the axion hypothesis assumed that the scale of the Peccei–Quinn symmetry breaking was similar to the electroweak scale. The first laboratory limits came from accelerator experiments. As some of the upper limits for branching ratios B of reactions such as the following [Asa81, Yam83, PDG96]

$$B(K^+ \to \pi^+ + a) < 1.7 \times 10^{-9}, \tag{11.34}$$

$$B(J/\Psi \to a + \gamma) \cdot B(\Upsilon \to a + \gamma) < 1.8 \times 10^{-10}, \tag{11.35}$$

$$B(\pi^+ \to a e^+ \nu_e) < 1 \times 10^{-9} \tag{11.36}$$

are lower by (several) orders of magnitude compared to predictions, this idea could be rapidly excluded. In fact, axions heavier than 10 keV are incompatible with these experimental limits [Kim87, Che88].

11.5.1 Cosmic axions

The main detection mechanism for cosmic axions uses the coupling of the axion to 2 photons. The two-body decay $a \rightarrow 2\gamma$ would result in mono-energetic photons of a wavelength of [Kol90]

$$\lambda_a = \frac{2hc}{m_a} = \frac{24800(\text{Å})}{m_a(\text{eV})} \tag{11.37}$$

and therefore for axions in the eV range would lie in the visible. Due to their late creation axions are non-relativistic particles ($p = 0$ condensate) (cold dark matter). They dissipate with the corresponding mass concentrations and take up the velocity determined by gravitational interaction. The expected line would therefore be Doppler broadened, i.e. the width would be determined by the observed system, where for galactic halos β is about 10^{-3}, and for clusters $\beta \approx 10^{-2}$. We expect from the above decay within a cluster an axion intensity of typically [Tur87]

$$I_a = 2 \times 10^{-20} \left(\frac{m_a}{\text{eV}}\right)^7 \left[\left(\frac{E_{\text{PQ}}}{N_F} - 1.95\right) \Big/ 0.72\right]^2 \text{ erg cm}^{-2} \text{ s}^{-1} \text{ arcsec}^{-2} \text{ Å}^{-1}. \tag{11.38}$$

A high resolution spectrum of the night sky is shown in figure 11.3. In order to eliminate the effects of the night sky itself, a group [Tur90b] has carried out observations of three galaxy clusters with different red-shifts. There were two possibilities of background reduction. One involved background reduction by subtraction between observation in the direction of the cluster and observation in any other direction. The other method involves further suppression of the background by exploiting the different red-shifts of the expected line. However, no positive signal of a possible axion decay has been detected, and therefore masses of between 3 and 8 eV have been ruled out [Tur90b].

11.5.2 Axions from the halo of our Milky Way

The detection of axions from the halo of our Milky Way uses a slightly different principle. If the halo has a density of $\rho \simeq 0.3$ GeV cm^{-3}, and would consist, e.g. only of 10^{-7} eV axions, the particle density would be $n_a \simeq 3 \times 10^{13}$ cm^{-3}, corresponding to a flux of

$$\Phi_a = n_a v_a \simeq 10^{21} \text{ cm}^{-2} \text{ s}^{-1}. \tag{11.39}$$

The lifetime of such axions would, according to equation (11.12), be $\tau_a \approx 10^{42}$ years, and therefore experimentally undetectable. However, Sikivie noticed [Sik83] that it should be possible to convert axions via a strong magnetic field into a mono-energetic photon with a frequency of

$$\nu_a = 2 \left(\frac{m_a}{10^{-5} \text{ eV}}\right) \quad \text{GHz} \tag{11.40}$$

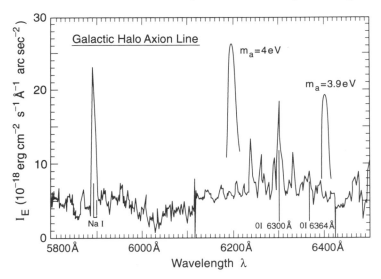

Figure 11.3. High resolution spectrum of the night sky at optical frequencies used in the search for lines from axion decay, and, as an example, from axion decay inside the galactic halo expected lines (for $m_a = 3.9$ eV and 4.0 eV) (from [Bro68, Tur90b]).

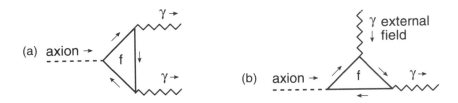

Figure 11.4. (*a*) The coupling of an axion to two photons via a charged fermion loop (triangle anomaly). (*b*) Sikivie suggested using a magnetic field to produce virtual photons in order to convert axions into mono-energetic photons (from [Smi90]).

(see figure 11.4). In this case a virtual photon is supplied by the magnetic field. A tunable microwave resonator is used to detect such photons. The coupling of axions to a particular eigenmode nl of the resonator is described by a term of the form

$$g_{a\gamma\gamma} a B \int_{\text{vol}} E_{nl} \mathrm{d}^3 x. \tag{11.41}$$

Hence it is clear that axions can only couple to TM modes of the cavity. The power fed into a mode is of order of magnitude 10^{-22} W. Three parameters are important: the volume V of the cavity, the magnetic field strength B and the cavity Q factor. Typical values for the three existing experiments are volumes of 10 litres, magnetic field strengths of 7 Tesla and Q factors of 10^5. An example of this type of experimental apparatus is shown in figure 11.5. A microwave resonator, tunable with the help of a movable rod, is situated in a strong magnetic field of several Tesla. In order to keep the thermal noise to a minimum the whole experiment is carried out at helium temperature (4.2 K). As the signal has a very small half-width ($\Delta E \approx 10^{-6} E$), a very precise tuning of the resonator, and therefore also a correspondingly long measuring time is necessary.

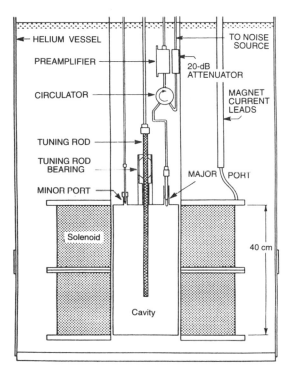

Figure 11.5. Construction of an experiment for the search for axions via conversion in a magnetic field. A microwave resonator for the detection of the produced photons is situated in a strong magnetic field at helium temperatures. The resonator can be tuned with the help of a movable rod, allowing the desired mass region to be probed (after [Wue89]).

Following two pilot experiments, [Wue89, Hag90], the cosmologically relevant parameter space will be tested in a new version of these experiments [Bib90, Hag95, Hag96] at the Lawrence Livermore National Laboratory (see figure 11.6). This experiment (figure 11.7) has a magnetic field of 8 T, a volume of the cavity of 200 l and is sensitive to the axion mass range between 1.3 and 13 μeV. This experiment began collecting data in autumn 1995.

A completely different kind of detection exploits Rydberg atoms, where the valence electrons have a very large main quantum number n (typically of the order 100). The principle idea is to convert the axions in a resonance cavity and to utilize the Rydberg atoms to detect the photons. An experiment CARRACK plans to use this method to explore axion masses around 10 μeV [Mat91, Oga96].

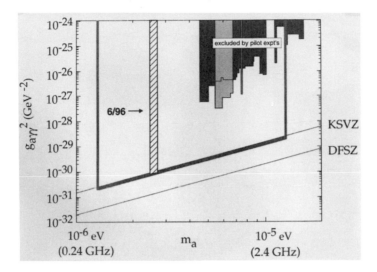

Figure 11.6. The limits obtained from [Wue89, Hag90] (dark areas) for the axion–photon coupling $g_{a\gamma\gamma}$ as a function of axion mass. The two lines describe the expected values for galactic halo axions of a particular kind (hadronic and DFSZ). The new version (figure 11.7) of the detector shown in the previous figure, with greater volume and higher magnetic field, is in operation since the end of 1995 and will be able (black frame) to probe the relevant area at least for hadronic axions. Until June 1996 the framed hatched area has been scanned (with kind permission of K van Bibber).

11.5.3 Direct laboratory production of axions

There are alternative strategies, but they are also based mainly on the above principle. For example, as in a 'beam dump' experiment, a strong laser beam can be sent into a magnetic field, where a fraction of the photons are converted into axions (see figure 11.8). All photons are absorbed in a shield, while the axions passed through are converted back to photons in another magnetic field and are then measured. Such an experiment has indeed been carried out [Cam93], and for axion masses smaller than 10^{-3} eV limits of $g_{a\gamma\gamma} < 3.6 \times 10^{-7}$ GeV^{-1} were obtained.

Figure 11.7. View of the new axion detector at LLNL. The cavity is on the floor with the cryostat system for cooling to liquid He temperatures above it. The entire cavity is placed in a magnetic field of 8 T (with kind permission of K van Bibber).

11.5.4 Solar axions

The Sun is a cosmic axion source. The spectral shape of a solar hadronic axion spectrum is well approximated by [Raf90]

$$\Phi_a(E) = 4.02 \times 10^{10} \frac{(E_a/\text{keV})^3}{e^{E_a/1.08} - 1} \quad \text{cm}^{-2}\,\text{s}^{-1}\,\text{keV}^{-1} \tag{11.42}$$

where the mean axion energy is about 4.2 keV. The constant C contains the dependence on the axion mass and on the relation E_{PQ}/N_F. The total axion luminosity is

$$L_a = 3.6 \times 10^{-3} L_\odot \left(\frac{m_a}{\text{eV}}\right)^2 \tag{11.43}$$

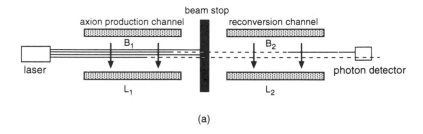

(a)

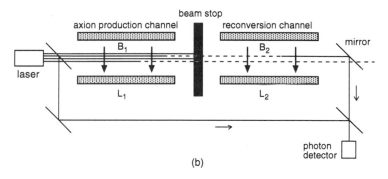

(b)

Figure 11.8. Two outlines for a 'Beam dump' experiment for axion search. Photons from a laser beam are converted into axions by a magnetic field and then converted back into photons in a second magnetic field. The laser beam is totally absorbed between the two magnetic field regions (from [Bib87, Smi90]).

from a solar luminosity of $L_\odot = 3.86 \times 10^{33}$ ergs^{-1}. A new idea for the detection of solar axions requires a tank filled with H_2 or He to be placed in a magnetic field, in order to detect the X-rays created by axion conversion [Bib90]. Typical dimensions of such an axion telescope would be a diameter of 4 m, a length 3 m and a magnetic field of 3 T (see figure 11.9). A multi-wire proportional chamber is planned as the detector. One advantage of this experiment is the directional information from the Sun's position. Such an experiment should be sensitive to axion masses in the region of 0.1 eV to 5 eV. In a prototype detector the proof of principle could be demonstrated, but due to the short observation time the limits deduced are no better than those from the previously discussed astrophysical arguments [Laz92]. Bragg scattering has been proposed as another method of detection [Buc90, Pas94]. An attempt to apply this idea with Ge detectors has been made recently [Avi97].

Experimental data for the DFSZ axion has been provided by germanium detectors used in the search for double beta-decay (see chapter 2). The detection principle here exploits the axio-electric effect in analogy to the photo-electric

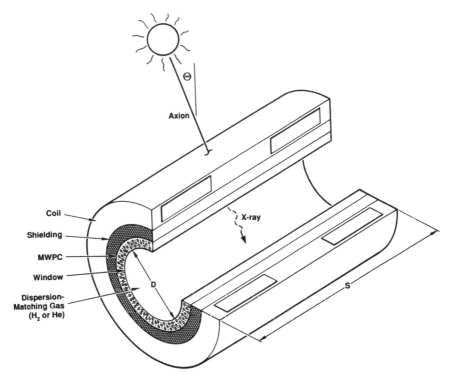

Figure 11.9. A proposed search for solar axions using an axio-helioscope. Solar axions are converted into X-rays in a magnetic field using a light gas (H or He), and can then be detected in a multi-wire proportional chamber (MWPC). One advantage of such an experiment is the use of directional information (from [Bib90]).

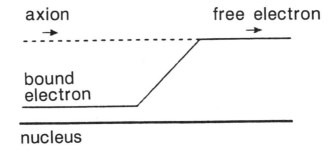

Figure 11.10. Illustration of the axio-electric effect. An axion knocks an electron out of a shell in an atom, similar to the photo-electric effect. This process can only be used for the detection of DFSZ axions.

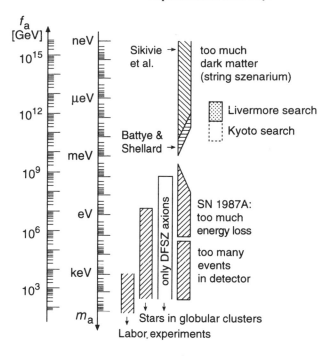

Figure 11.11. The open axion mass windows for axions remaining from experiment (see text and [Bat97]) and astrophysical considerations given in the text (i.e. mass regions which are not excluded): a region between $10^{-3}(10^{-2})$–10^{-5} eV and, possibly, also a small additional window for the KSVZ axion in the eV region still exist. All other ranges are excluded (from [Raf95a, Raf96]).

effect (see figure 11.10). The cross section is given by [Avi88]

$$\sigma_{ae} = \frac{g_{aee}^2}{4\pi\alpha_{em}}\left(\frac{E_a}{2m_e}\right)^2\sigma_{pe} \qquad (11.44)$$

From the observations it can be concluded that the axion mass must be [Avi87, Avi88]

$$m_a < 14.4\,\text{eV}. \qquad (11.45)$$

There is now a question of consistency, since for such massive axions the solar models would have to be heavily modified, which in turn would affect the predicted axion flux. This kind of experiment is only sensitive to DFSZ axions, as the couplings are too small for hadronic axions.

Presently there is no evidence for the existence of an axion or its connection to dark matter [Raf95a]. Taking all the limits discussed above seriously would limit the 18 orders of magnitudes to only one possible window for axions, the region between 10^{-5} eV and 10^{-3} eV (10^{-2} eV) for both kinds of axions.

For hadronic axions an additional window could possibly remain in the eV region (see figure 11.11). Because of its general importance the axion is a quite likely candidate for dark matter and accordingly there is substantial experimental activity attempting to detect it.

Having discussed two (still) hypothetical particles in the previous two chapters, we now turn to a particle which we know exists, and which is at present of particular interest in connection with solar physics.

Chapter 12

Solar neutrinos

Solar neutrinos are considered to be one of the most interesting problems in particle astrophysics today. It is assumed that significantly fewer solar neutrinos are observed than would be expected theoretically. It is however unclear to what extent this discrepancy points to 'new physics' like neutrino oscillations, rather than to an astrophysical problem such as a lack of knowledge of the structure of the Sun or of reactions in its interior, or a 'terrestrial' problem of limited knowledge of the capture cross sections in the neutrino detectors. In the following chapter the situation is discussed in some detail.

12.1 The standard solar model

12.1.1 Reaction rates

Before concerning ourselves with details of the Sun we first state some general comments on the reaction rates [Cla68, Rol88, Bah89, Raf96]. They play an important role in the understanding of energy production in stars. Consider a 2-particle reaction of the general form

$$T_1 + T_2 \rightarrow T_3 + T_4. \tag{12.1}$$

The reaction rate is given by

$$R = \frac{n_1 n_2}{1 + \delta_{12}} \langle \sigma v \rangle_{12} \tag{12.2}$$

where n_i is the particle density, σ the cross section, v the relative velocity and δ the Kronecker symbol to avoid double counting of identical particles. At typical thermal energies of several keV inside the stars and Coulomb barriers of several MeV, it can be seen that the dominant process for charged particles is that of the quantum mechanical tunnelling, which Gamow used to explain α-decay [Gam38]. Here it is usual to write the cross section in the form of

$$\sigma(E) = \frac{S(E)}{E} \exp(-2\pi \eta) \tag{12.3}$$

where the exponential term is Gamow's tunnelling factor, the factor $1/E$ expresses the dependence of the cross section on the de Broglie wavelength, and η is the so-called Sommerfeld parameter, given by $\eta = Z_1 Z_2 e^2 / \hbar v$. Nuclear physics now only enters into calculations through the so-called S-factor $S(E)$, which, as long as no resonances appear, has a relatively smooth behaviour. This assumption is critical, since we have to extrapolate from the values at several MeV, measured in the laboratory, down to the relevant energies in the keV region [Rol88, Dar96]. This energy region will be accessible for the first time in the LUNA experiments, (LUNA I and LUNA II), in the Gran Sasso laboratory [Gre94, Fio95, Arp96]. A first measurement of the reaction equation 12.8 is given [Arp97]. For the averaged product $\langle \sigma v \rangle$ we also need to make an assumption on the velocity distribution of the particles. In normal main sequence stars such as our Sun, the interior is not yet degenerate so that a Maxwell–Boltzmann distribution can be assumed. Due to the behaviour of the tunnelling probability and the Maxwell–Boltzmann distribution there is a most probable energy E_0 for a reaction, which is shown schematically in figure 12.1 [Bur57, Fow75]. This Gamow peak for the pp-reaction, which we will discuss later, lies at about 6 keV. If we define $\tau = 3E_0/kT$ and approximate the reaction rate dependence on the temperature by a power law $R \sim T^n$, then $n = (\tau - 2)/3$. For a detailed discussion of this derivation see e.g. [Rol88, Bah89].

12.1.2 Energy and neutrino production processes in the Sun

We now consider the Sun. According to our understanding, the Sun, like all stars, gets its energy via nuclear fusion [Gam38]. For a general discussion of the structure of stars see e.g. [Cox68, Cla68, Sti89]. Hydrogen fusion proceeds according to

$$4p \rightarrow {}^4\text{He} + 2e^+ + 2\nu_e + 26.73\,\text{MeV}. \tag{12.4}$$

There are two ways to bring this about: one is the pp-cycle [Bet38], the other the CNO-cycle [Wei37, Bet39]. Figure 12.2 shows the contribution of both processes to energy production as a function of temperature. The pp-cycle (see figure 12.3) is dominant in the Sun. Here the first reaction step is the fusion of hydrogen into deuterium

$$p + p \rightarrow {}^2\text{H} + e^+ + \nu_e \quad (E_\nu \leq 0.42\,\text{MeV}). \tag{12.5}$$

The primary pp fusion proceeds this way to 99.6%. In addition the following process occurs with a much reduced probability of 0.4%,

$$p + e^- + p \rightarrow {}^2\text{H} + \nu_e \quad (E_\nu = 1.44\,\text{MeV}). \tag{12.6}$$

The neutrinos produced in this reaction (pep neutrinos) are mono-energetic. The conversion of the created deuterium to helium is in both cases identical:

$$^2\text{H} + p \rightarrow {}^3\text{He} + \gamma + 5.49\,\text{MeV}. \tag{12.7}$$

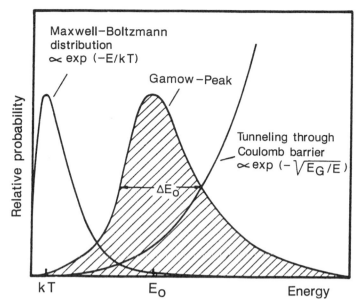

Figure 12.1. The most favourable energy region for nuclear reactions between charged particles at very low energies is determined by two effects which act in opposite directions. The first is the Maxwell–Boltzmann distribution with a maximum at kT, which implies an exponentially decreasing number of particles at high energies. The other effect is that the quantum mechanical tunnelling probability rises with growing energy. This results in the Gamow peak (not shown to true scale), at an energy E_0, which can be very much larger than kT (from [Rol88]).

Neutrinos are not produced in this reaction. From now on the reaction chain divides. With a probability of 85% the ^{3}He fuses directly into ^{4}He:

$$^3\text{He} + {}^3\text{He} \rightarrow {}^4\text{He} + 2p + 12.86\,\text{MeV}. \tag{12.8}$$

In this step, also known as the *pp* I-process, no neutrinos are produced. However, two neutrinos are created in total, as the reaction of equation (12.5) has to occur twice, in order to produce two ^{3}He nuclei which can undergo fusion. Furthermore, ^{4}He can also be created by with a probability of $2.4 \times 1\,0\text{–}5\%$

$$^3\text{He} + p \rightarrow {}^4\text{He} + \nu_e + e^+ + 18.77\,\text{MeV}. \tag{12.9}$$

The neutrinos produced here are very energetic (up to 18.77 MeV), but they have a very low flux. They are called *hep* neutrinos. The alternative reaction produces ^{7}Be:

$$^3\text{He} + {}^4\text{He} \rightarrow {}^7\text{Be} + \gamma + 1.59\,\text{MeV}. \tag{12.10}$$

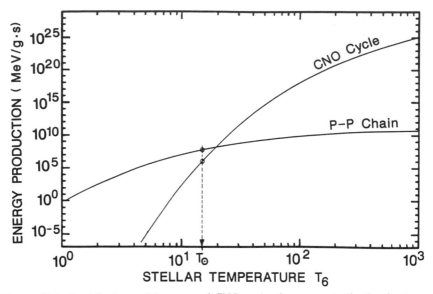

Figure 12.2. Contributions of the *pp-* and CNO-cycles for energy production in stars as a function of the central temperature. While the *pp*-cycle is dominant in the Sun, the CNO process becomes dominant above about 20 million degrees (from [Rol88]).

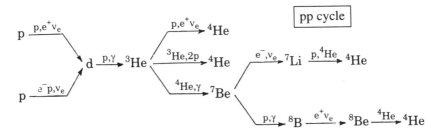

Figure 12.3. The route of proton fusion according to the *pp*-cycle. After the synthesis of ^{3}He the process branches into three different chains. The *pp*-cycle produces 98.4% of the solar energy.

Subsequent reactions again proceed via several sub-reactions. The *pp* II-process leads to the production of helium with a probability of 15% via:

$$^7\text{Be} + e^- \rightarrow {}^7\text{Li} + \nu_e \quad (E_\nu = 0.862\,\text{MeV} \quad \text{or} \quad E_\nu = 0.384\,\text{MeV}). \quad (12.11)$$

This reaction produces ^{7}Li in the ground state 90% of the time, and leads to the emission of mono-energetic neutrinos of 862 keV. The remaining 10% decay in an excited state by emission of neutrinos with an energy of 384 keV. Thus mono-energetic neutrinos are produced in this process. In the next reaction step

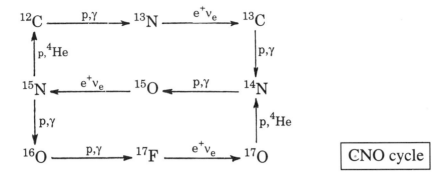

Figure 12.4. Representation of the CNO process. This also burns hydrogen to helium with C, N and O acting as catalysts, and is responsible for 1.6% of the solar energy.

helium is created via

$$^7\text{Li} + p \rightarrow 2\,^4\text{He} + 17.35 \text{ MeV.} \tag{12.12}$$

Assuming the ^{7}Be has already been produced, this pp II-branch has a probability of 99.98%. There is also the possibility of proceeding via ^{8}B (pp III-chain) rather than by ^{7}Li via:

$$^7\text{Be} + p \rightarrow \,^8\text{B} + \gamma + 0.14 \text{ MeV} \tag{12.13}$$

which undergoes β^+-decay via

$$^8\text{B} \rightarrow \,^8\text{Be}^* + e^+ + \nu_e \quad (E_\nu \leq 14.06 \text{ MeV).} \tag{12.14}$$

The neutrinos produced here are very energetic, but also very rare. Nevertheless they play an important role. ^{8}Be undergoes α-decay into helium:

$$^8\text{Be}^* \rightarrow 2\,^4\text{He} + 3 \text{ MeV.} \tag{12.15}$$

The CNO cycle only accounts for about 1.6% of the energy production in the Sun, which is why it is mentioned here only briefly. The main reaction steps are:

$$^{12}\text{C} + p \rightarrow \,^{13}\text{N} + \gamma \tag{12.16}$$

$$^{13}\text{N} \rightarrow \,^{13}\text{C} + e^+ + \nu_e \quad (E_\nu \leq 1.2 \text{ MeV)} \tag{12.17}$$

$$^{13}\text{C} + p \rightarrow \,^{14}\text{N} + \gamma \tag{12.18}$$

$$^{14}\text{N} + p \rightarrow \,^{15}\text{O} + \gamma \tag{12.19}$$

$$^{15}\text{O} \rightarrow \,^{15}\text{N} + e^+ + \nu_e \quad (E_\nu \leq 1.73 \text{ MeV)} \tag{12.20}$$

$$^{15}\text{N} + p \rightarrow \,^{12}\text{C} + \,^4\text{He.} \tag{12.21}$$

This process and its subsidiary cycle which is not discussed here because of its negligible importance are illustrated in figure 12.4.

We have now introduced the processes relevant for neutrino production. For predictions of the neutrino spectrum to be expected we need further information—in particular about the cross sections of the reactions involved [Pas94, Lan94].

12.1.3   The solar neutrino spectrum

The prediction of the solar neutrino spectrum requires detailed model calculation [Tur88, Bah89, Bah92, Tur93a, Tur93b, Ber93a, Bah95, Sha95, Bah96, Tur96, Dar96]. The simulations which model the operation of the Sun use the basic equations of stellar evolution (see [Cla68, Rol88, Bah89]).

(i)   Hydrodynamic equilibrium, i.e. the gas and radiation pressure balance the gravitational attraction:

$$\frac{dP(r)}{dr} = -\frac{GM(r)\rho(r)}{r^2} \tag{12.22}$$

with

$$M(r) = \int_0^r 4\pi r^2 \rho(r) dr.$$

(ii)   Energy transport by radiation or convection:

$$L(r) = -4\pi r^2 \left(\frac{ac}{3}\right) \frac{1}{\kappa\rho} \frac{dT^4}{dr}. \tag{12.23}$$

(iii)   Energy production by nuclear reactions:

$$\frac{dL(r)}{dr} = \rho(4\pi r^2)\left(\epsilon_{\text{nuc}} - T\frac{dS}{dt}\right) \tag{12.24}$$

where S is the stellar entropy.

Change in the chemical composition only takes place through nuclear reactions.

Input parameters include the age of the Sun and its luminosity, as well as the equation of state, nuclear parameters, chemical abundances and opacities κ. The opacity is a measure of the photon absorption capacity. It depends on the chemical composition and complex atomic processes. The influence of the chemical composition on the opacity shows up, for example, in different temperature and density profiles of the Sun. The ratio of the 'metals' Z (in astrophysics all elements heavier than helium Y are known as metals) to hydrogen X is seen to be particularly sensitive. The experimentally observable composition of the photo-sphere (see e.g. [And89]) is used as the initial composition of elements heavier than carbon. It is assumed here that the Sun has been a homogeneous star since joining the main sequence. In the solar core ($T > 10^7$ K) the metals do not play the central role for the opacity, which is more dependent on inverse bremsstrahlung and photon scattering on free electrons here. The diffusion of all the elements inside the Sun also plays an important role (see [Dar96] and table 12.2).

With all these inputs it is then possible to calculate a sequence of models of the Sun which finally predict values of $T(r)$, $\rho(r)$ and the chemical composition of its current state (see table 12.1). These models are called *standard solar models* (SSM) [Tur88, Bah89, Bah92, Tur93a, Tur93b, Bah95, Dar96]. From them the location and rate of the nuclear reactions which produce neutrinos can be predicted (see figure 12.5). Finally, these models give predictions for the expected neutrino spectrum and the observable fluxes on Earth (see figure 12.6 and table 12.2). It can clearly be seen that the largest part of the flux comes from the *pp* neutrinos. It can also be seen that there are considerable differences in the flux predictions, for example for the flux of ^{8}B neutrinos. In order to predict the signal to be expected in the various detectors, it is necessary to know the capture or reaction cross sections for neutrinos (see equation (12.26)). The former are governed by the distribution of the Gamow–Teller strength in the corresponding daughter nuclei (see for example [Gro89, Gro90]). The first realistic calculations were presented in [Gro84, Gro86c].

Table 12.1. Properties of the Sun according to the standard solar model (SSM) of [Bah88a].

	$t = 4.6 \times 10^9$ y (today)	$t = 0$
Luminosity $L_\odot$	$\equiv 1$	0.71
Radius $R_\odot$	696 000 km	605 500 km
Surface temperature T_S	5773 K	5 665 K
Core temperature T_c	15.6×10^6 K	—
Core density	148 g cm^{-3}	—
X(H)	34.1%	71%
Y(He)	63.9%	27.1%
Z	1.96%	1.96%

Although the fluxes on Earth have values of the order of 10^{10} cm^{-2} s^{-1}, their detection is extremely difficult because of the small cross sections. We now turn to the experiments, results and interpretations.

12.2   Solar neutrino experiments

In principle there are two kinds of solar neutrino experiment: radiochemical and real time experiments. The principle of the *radiochemical experiments* is the reaction

$$^A_N Z + \nu_e \rightarrow ^A_{N-1}(Z+1) + e^- \tag{12.25}$$

where the daughter nucleus is unstable and decays with a 'reasonable' half-life, since it is the radioactive decay of the daughter nucleus which is used in the

Table 12.2. Two examples of SSM predictions for the flux Φ_ν of solar neutrinos on the Earth (from [Bah95] and [Dar96], without (ND) and with taking diffusion into account).

Source	Φ_ν (10^{10} cm^{-2} s^{-1})			
	[Bah95] ND	[Dar96] ND	[Bah95]	[Dar96]
pp	6.01	6.08	5.91	6.10
pep	1.44×10^{-2}	1.43×10^{-2}	1.40×10^{-2}	1.43×10^{-2}
^{7}Be	4.53×10^{-1}	4.79×10^{-1}	5.15×10^{-1}	3.71×10^{-1}
^{8}B	4.85×10^{-4}	5.07×10^{-4}	6.62×10^{-4}	2.49×10^{-4}
^{13}N	4.07×10^{-2}	2.50×10^{-2}	6.18×10^{-2}	3.82×10^{-2}
^{15}O	3.45×10^{-2}	3.38×10^{-2}	5.45×10^{-2}	3.74×10^{-2}
^{17}F	4.02×10^{-4}	4.06×10^{-4}	6.48×10^{-4}	4.53×10^{-4}
$\sum(\Phi\sigma)_{cl}$ [sNU]	7 ± 1	7 ± 1	9.3 ± 1.4	4.1 ± 1.2
$\sum(\Phi\sigma)_{Ga}$	127 ± 6	128 ± 7	137 ± 8	115 ± 6

detection. The production rate of the daughter nucleus is given by

$$R = N \int \Phi(E)\sigma(E)dE \qquad (12.26)$$

where Φ is the solar neutrino flux, N the number of target atoms and σ the cross section for the reaction of equation (12.25). Given an incident neutrino flux of about 10^{10} cm^{-2} s^{-1} and a cross section of about 10^{-45} cm^2, about 10^{30} target atoms are required to produce one event per day. This corresponds to several ktons of the element in question. We therefore require very large detectors of the order of several tons and expect the transmutation of one atom per day. Detecting this is no easy matter. We define for convenience a new unit more suitable for such low event rates, the SNU (solar neutrino unit) where

$$1 \text{ SNU} = 10^{-36} \text{ captures per target atom per second.}$$

Any information about the time of the event, the direction and energy (with the exception of the lower limit, which is determined by the energy threshold of the detector) of the incident neutrino is lost in these experiments since only the average production rate of the unstable daughter nuclei over a certain time can be measured. The situation in *real time experiments* is different. The main detection method here is neutrino–electron scattering, in which Cerenkov light is created by the recoiling electrons, which can then be detected and is closely correlated with the direction of the incoming neutrino. In discussing the existing experimental data we will follow the historic sequence.

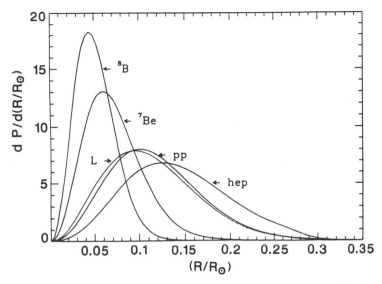

Figure 12.5. Production of neutrinos from different nuclear reactions as a function of the distance from the Sun's centre, according to the standard solar model. The luminosity produced in the visible (denoted by L) as a function of radius is shown as a comparison. It can be seen to be very strongly coupled to the primary *pp* fusion (from [Bah89]).

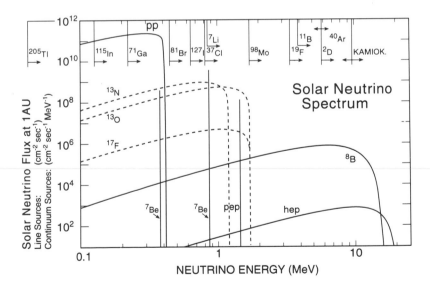

Figure 12.6. The solar neutrino spectrum at the Earth, as predicted by detailed solar model calculations. The dominant part comes from the *pp* neutrinos, while at high energy *hep* and ^{8}B neutrinos dominate. The threshold energies of different detector materials are also shown (see e.g. [Ham93]).

Figure 12.7. The chlorine detector of Davis for the detection of solar neutrinos in the about 1400 m deep Homestake Mine in Lead, South Dakota (USA) in about 1967. The 380 000 l tank full of perchloro-ethylene is shown. Dr Davis is standing above. (Photograph from Brookhaven National Laboratory).

12.2.1 The chlorine experiment

The first solar neutrino experiment, and the birth of neutrino astrophysics in general, is the chlorine experiment of Davis [Dav64, Row85b, Dav94a, Dav94b, Dav96], which has been running since 1968. The reaction used to the detect the neutrinos is

$$^{37}Cl + \nu_e \rightarrow {}^{37}Ar + e^- \qquad (12.27)$$

which has an energy threshold of 814 keV. The detection method utilizes the decay

$$^{37}\text{Ar} \rightarrow {}^{37}\text{Cl} + e^- + \bar{\nu}_e \tag{12.28}$$

which has a half-life of 35 days. Given the threshold of 0.81 MeV, this experiment is not able to measure the *pp* neutrino flux. The contributions of the various production reactions for neutrinos to the total flux are illustrated in table 12.3 according to one of the current solar models. The solar model calculations predict values of (9.3 ± 1.4) SNU [Bah95], (6.4 ± 1.4) SNU [Tur93b] and (4.2 ± 1.2) SNU [Dar96], where the major part comes from the ^{8}B neutrinos. A production rate of one argon atom per day corresponds to 5.35 SNU. The experiment (figures 12.7) operates in the Homestake gold mine in South Dakota (USA), where a tank with 615 tons perchloro-ethylene (C_2Cl_4) which serves as the target is situated at a depth corresponding to 4100 mwe (metre water equivalent). The natural abundance of ^{37}Cl is about 24 %, so that the number of target atoms is 2.2×10^{30}. The argon atoms which are produced are volatile in solution and are extracted about once per month. The produced argon is concentrated in several steps, and finally filled into special miniaturized proportional counters. These are then placed in very low activity lead shields and then the corresponding argon decay is observed. In order to further reduce the background, both the energy information of the decay, and the pulse shape are used. The results from a little more than 20 years measuring time are shown in figure 12.8. The average counting rate is [Dav96]

$$2.56 \pm 0.22 \quad \text{SNU.} \tag{12.29}$$

This is less than the value predicted by the standard solar models. This discrepancy is the source of the so-called *solar neutrino problem*. It should be remembered, however, that the theoretical value for the expected neutrino rate has changed considerably with time, as illustrated in figure 12.9. A conformation of the deficit is usually seen in the results of the Kamiokande experiment, which are discussed in the following section.

12.2.2 The Kamiokande and Superkamiokande detectors

The only currently operational real-time experiment for solar neutrinos is being carried out with the Kamiokande and Superkamiokande detectors [Hir91, Suz94, Suz95, Suz96]. These experiments are situated in a Japanese mine and have a screening depth of 2700 mwe. After Superkamiokande started operation on 1 April 1996, the other was shut down. The detectors are water Cerenkov counters (see figures 12.10 and 12.11). If a charged particle moves in a medium with refractive index n and a velocity $\beta = v/c$ which is greater than the velocity of light in the medium, it emits Cerenkov light. The opening angle θ of the cone produced is given by

$$\cos\theta = \frac{1}{\beta n} \tag{12.30}$$

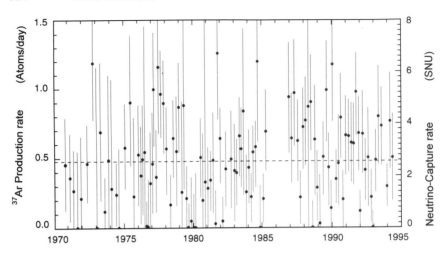

Figure 12.8. The neutrino flux measured from the Homestake ^{37}Cl detector since 1970. The average measured value (broken line) is significantly smaller than the predicted one. This discrepancy is the origin of the so-called solar neutrino problem (from [Dav96]).

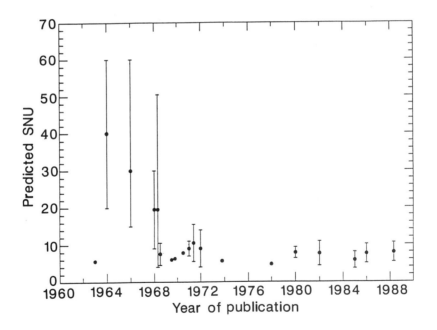

Figure 12.9. Time development of the event rate predicted for the chlorine detector during the last 30 years (error bars correspond to 1σ) (from [Bah89, Dav92b]).

relative to the direction of the particle. For $\beta = 1$ this angle is $42°$ for water with $n = 1.33$. The detection reaction in the detector is neutrino–electron scattering. The resulting recoil spectrum is given by [Bah89]

$$\frac{d\sigma}{dT} = \sigma_0 \left[g_l^2 + g_r^2 \left(1 - \frac{T}{E_\nu} \right)^2 - g_l g_r \left(\frac{T}{E_\nu^2} \right) \right] \qquad (12.31)$$

where T is the kinetic energy of the recoil electron, $\sigma_0 = 8.8 \times 10^{-45}$ cm^2 and

$$g_l = \left(\pm\frac{1}{2} + \sin^2\theta_W \right) \quad \text{and} \quad g_r = \sin^2\theta_W. \qquad (12.32)$$

The positive sign applies to ν_e scattering, while the negative sign applies to the other flavours. This means that the latter cross sections are smaller by about a factor 7, since no charged weak currents contribute. The total cross section for $\nu_e e$ scattering is

$$\sigma(\nu_e e) = 0.9 \times 10^{-43} \left(\frac{E}{10 \text{ MeV}} \right) \text{cm}^2 \qquad (12.33)$$

12.2.2.1 *Kamiokande II and III*

Kamiokande II and III (see figures 12.10 and 12.11) consisted of 3000 tons H_2O, of which, in order to reduce background, only a fiducial volume of 680 tons was actually used in the solar neutrino measurement. It is surrounded by 948 photo-multipliers, which cover about 20% of the surface. The whole tank is again surrounded by 1.5 m H_2O, which acts as an anti-coincidence.

Given the measured event rates, the methods necessary in order to reduce the background so that the solar neutrino fluxes can be observed at all must be briefly described (see figure 12.12). Originally the detector had a trigger rate of 1000 Hz. The main background was the decay of ^{222}Rn and ^{238}U. By re-circulating the water of the inner tank in a closed system, the rate could be reduced to 0.6 Hz using an ion-exchanger. From this rate, 0.37 Hz can be ascribed to muon events from cosmic rays. A muon veto is constructed by requiring that less than 30 photoelectrons are produced in the outer volume. As a further restriction a time difference of at least 100 μs is required between an event in the outer detector and one inside. With these two criteria it is possible to reduce the rate of muons. The energy information is also used. The Cerenkov light produced must fire at least 20 photo-multipliers within 100 ns in order to register as an event. This implies a threshold energy of about 7.5 MeV, which is relatively high. It is also required that less than 100 photoelectrons should be detected since this corresponds to an energy of 30 MeV, where solar neutrinos are not produced any more. Finally the direction of the electrons is used. The

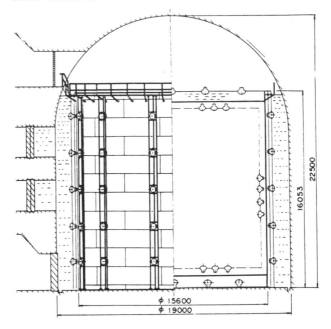

Figure 12.10. Construction of the high energy solar neutrino detector Kamiokande II and III. Water is used as the Cerenkov material. The light emitted in neutrino–electron scattering in the water is detected with photo-multipliers. The energy and direction of the neutrino can be reconstructed from the different detection times and the amount of light from each photomultiplier (from [Hir91]).

angular distribution of the electrons is given by

$$\cos\theta = \frac{1 + m_e/E_\nu}{(1 + \frac{2m_e}{T})^{1/2}} \tag{12.34}$$

where θ is the angle with respect to the direction of the Sun. As the neutrinos have a relatively high energy, the electrons point strongly in the forward direction (see figure 12.13(a)), i.e. the reconstructed track should be aligned with the direction of the Sun.

Given the threshold of 7.5 MeV, the detector can measure essentially only the ^{8}B neutrino flux. The experimental situation is shown in figure 12.13. From the measurements a time averaged flux of ^{8}B neutrinos of

$$\Phi(^8\text{B}) = (2.80 \pm 0.19 \pm 0.33) \times 10^6 \text{cm}^{-2}\,\text{s}^{-1} \tag{12.35}$$

has been obtained [Fuk96]. There is a discrepancy between the theories of [Bah92, Tur93a] and experiment [Suz95, Fuk96]:

$$\frac{\Phi(^8\text{B})_{\text{ex}}}{\Phi(^8\text{B})_{\text{th}}} = 0.50 \pm 0.04(\text{stat}) \pm 0.06(\text{sys}) \quad \text{[Bah92]} \tag{12.36}$$

Figure 12.11. A view into the Kamiokande II detector (with kind permission of Y Totsuka).

Table 12.3. Event rate for the gallium and chlorine detectors according to the standard solar model of [Bah92].

Source	Capture rate ^{71}Ga (SNU)	Capture rate ^{37}Cl (SNU)
pp	70.8	0
pep	3.1	0.2
^{7}Be	35.8	1.2
^{8}B	13.8	6.2
^{13}N	3.0	0.1
^{15}O	4.9	0.3
$\sum$	131.5 ± 19	8.0 ± 3.0

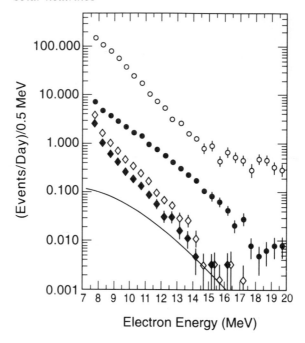

Figure 12.12. The data reduction of a typical low rate detector, Kamiokande II. (O) The raw measured spectrum. A correction for the fiducial volume (●) for cosmic rays (◊) and gamma radiation background leads to curve (♦). The solid line is a Monte Carlo prediction of the solar signal based on the SSM of [Bah88c]. The sensitivity to the level of the expected solar neutrinos is finally reached when the directional information is also used. The background could be considerably reduced in the construction of the Kamiokande III detector (from [Hir91]).

$$\frac{\Phi(^8B)_{ex}}{\Phi(^8B)_{th}} = 0.63 \pm 0.05(\text{stat}) \pm 0.15(\text{sys}) \quad [\text{Tur93a}]. \qquad (12.37)$$

A comparison of the experimental result with the theory of [Sha95a, Dar96] gives in contrast

$$\frac{\Phi(^8B)_{ex}}{\Phi(^8B)_{th}} = 1.13 \pm 0.16 \quad \text{and} \quad 1.02 \pm 0.15 \quad \text{respectively.} \qquad (12.38)$$

12.2.2.2 Superkamiokande

The largest water detector project is the Superkamiokande detector (see figure 12.14), which began taking data on 1st April 1996 [Suz94, Suz96]. The detector has a mass of 50 kton. The fiducial mass for solar neutrinos is 22 kton. The detector contains 11 000 photomultipliers which cover about 40% of the surface,

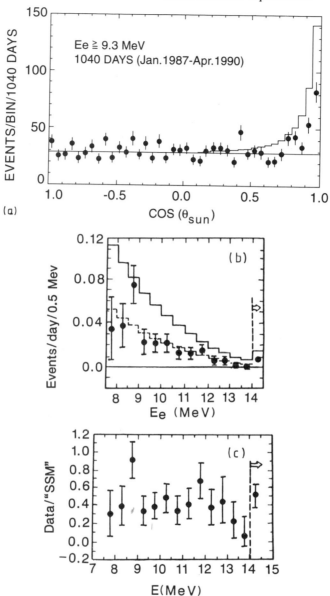

Figure 12.13. (*a*) Angular distribution of the events in the Kamiokande II detector, relative to the direction of the Sun (from [Hir90]). (*b*) Energy spectrum of the recoil electrons from the scattering of solar neutrinos in the Kamiokande experiment. The full line corresponds to the expectation of the SSM according to [Bah88c], (see however table 12.4); the dotted line shows the expectation for 46% of this model. (*c*) As (*b*) with the values normalized to the value expected from this SSM. The deficit compared to the expectation for this solar model of [Bah88c] can clearly be seen (from [Hir91]).

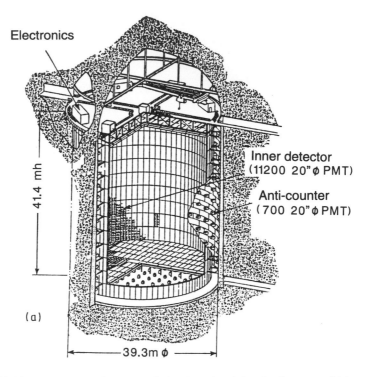

Electronics

Inner detector
(11200 20"ϕ PMT)

Anti-counter
(700 20"ϕ PMT)

41.4 mh

(a)

◄───── 39.3m ϕ ─────►

Figure 12.14. (*a*) Schematic view of the Superkamiokande detector which started operation in April 1996. It represents an increase of volume by a factor of 10 compared to the previous Kamiokande II and III experiments (from [Suz94]). (*b*) Cavern for the Superkamiokande detector. (*c*) During installation of the photomultipliers. The leader of the Superkamiokande project Professor Y Totsuka, with guests (Professor H V Klapdor-Kleingrothaus, right, and Dr V Gavrin, left). (*d*) Last cleaning of the photomultiplier tubes while filling the detector with water (with kind permission of Y Totsuka.)

about twice as much as the previous Kamiokande II and III detectors. The threshold is at about 5 MeV. A signal of between 20 and 30 events per day is expected.

Already after a running time of only a few months a much clearer directional dependence than is seen in figure 12.13(*a*) is observed (see figure 12.15). In a few months the new detector has already collected more data than the old versions did in several years.

12.2.3 The gallium experiments

Both the experiments described above are unable to measure directly the *pp* flux, which is the reaction directly coupled to the Sun's luminosity (figures 12.5 and

Figure 12.14. Continued.

12.6). A suitable material to detect these neutrinos is gallium [Kuz66]. There are currently two experiments operating which are sensitive to the *pp* neutrino flux, GALLEX (the gallium experiment, see figure 12.16) and SAGE (figure 12.17).

Both experiments use gallium as the target, and the detection relies on the reaction

$$^{71}\text{Ga} + \nu_e \rightarrow \,^{71}\text{Ge} + e^- \tag{12.39}$$

Figure 12.14. Continued.

with a threshold energy of 233 keV. The natural abundance of ^{71}Ga is 39.9%. The Soviet–American collaboration SAGE [Gav90, Abd95], uses 57 tons of gallium in metallic form as the detector and operates the experiment in the Baksan underground laboratory, while the predominantly European group GALLEX (the gallium experiment) [Kir90, Ans95b] uses 30 tons gallium in the form of 110 tons of GaCl$_3$ solution. This experiment is carried out in the Gran Sasso underground laboratory. The experiments differ by their method of germanium extraction. The germanium is concentrated in several stages, and subsequently transformed into germane (GeH$_4$), which has similar characteristics as methane (CH$_4$), which when mixed with argon is a standard gas mixture (P10) in proportional counters. The germane is therefore mixed with an inert gas (Xe), the mixture being optimized for detection efficiency, drift velocity and energy resolution. The detection of the ^{71}Ge decay (half-life 11.4 days, 100% via electron capture), is achieved using miniaturized proportional counters similar to the chlorine experiment. Both energy and pulse shape information are also used for the analysis.

In addition the first attempt was made to demonstrate the total functionality of a solar neutrino experiment using an artifical 2 MCi (!) ^{51}Cr source. This yields mono-energetic neutrinos, of which 81% have $E_\nu = 746$ keV.

First results from the SAGE collaboration (see figure 12.18) led to an estimate of the pp neutrino flux of [Aba91]

$$20^{+15}_{-20}(\text{stat}) \pm 35(\text{syst}) \text{ SNU}. \tag{12.40}$$

The first results from GALLEX gave a flux of [Ans92a]

$$83 \pm 19(\text{stat}) \pm 8(\text{syst}) \text{ SNU}. \tag{12.41}$$

Figure 12.14. Continued.

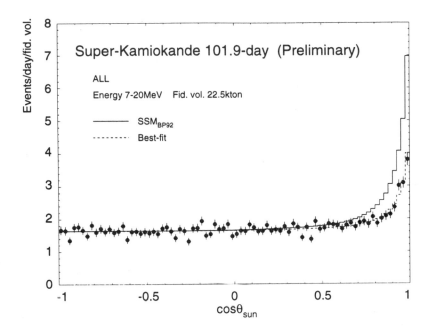

Figure 12.15. Angular distribution of the events in the Superkamiokande detector, relative to the direction of the Sun, after a measuring time of 102 days (from [Suz97]).

More recent results from GALLEX give a value of (see figures 12.18 and 12.19) [Ans94, Ans95b, Ans95c, Ham96a]

$$69.7 \pm 6.7(\text{stat})^{+3.9}_{-4.5}(\text{syst}) \text{ SNU.} \qquad (12.42)$$

SAGE has recently published new data taken since the summer of 1991 with 57 tons of gallium and obtains the result [Abd96]

$$69 \pm 10(\text{stat})^{+5}_{-7}(\text{syst}) \text{ SNU.} \qquad (12.43)$$

The quoted errors correspond to one standard deviation. The theoretical expectations correspond to 132^{+20}_{-17}, 123 ± 14 and 115 ± 6 SNU [Bah92, Tur93a, Dar96] (see tables 12.2, 12.3 and 12.4).

Both gallium experiments are in good agreement and show fewer events than expected in the standard solar models, although not by enough to solve the solar neutrino problem (see figure 12.18). Both experiments are considered the first evidence for *pp* neutrinos and as confirmation that the Sun's energy really does come from hydrogen fusion.

The results of the two calibrations of GALLEX with the artifical chromium neutrino source result in the following ratios between the observed number of

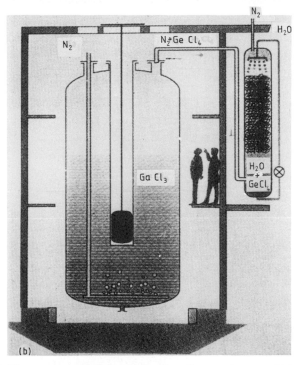

Figure 12.16. The GALLEX experiment in the Gran Sasso underground laboratory in Italy. (*a*) The front building contains the gallium tank and the extraction apparatus, while the rear building contains the low level counting apparatus. (*b*) Schematic cross section of the tank. The channel into which the Cr source is introduced can be clearly seen (from [Kir93]).

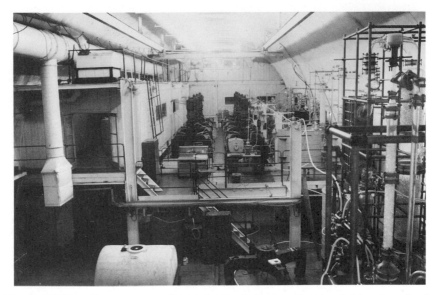

Figure 12.17. The SAGE experiment in the Baksan underground laboratory in the Caucasus. The 10 so-called reactors can be seen, 8 of which contain a total of 57 tons of metallic gallium (with kind permission of the SAGE collaboration).

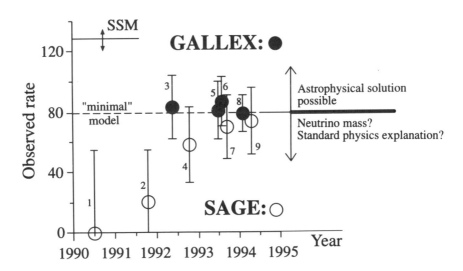

Figure 12.18. Time development of the GALLEX and SAGE results (after [Kir95]).

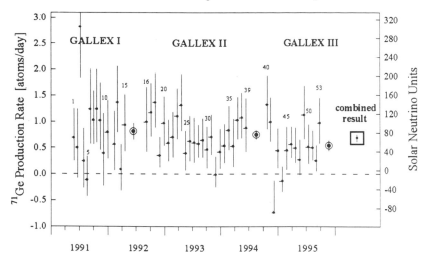

Figure 12.19. Results from the first 53 GALLEX runs (from [Ham96]).

^{71}Ge decays and the expectation from the source strength [Ans95a, Ham96a, Kir96a]

$$R = 1.04 \pm 0.12 \quad \text{and} \quad 0.83 \pm 0.08 \tag{12.44}$$

This result is of interest for several reasons. It represents the first experimental detection of terrestrial low energy neutrinos. Furthermore it proves for the first time that a radiochemical detection of neutrinos at the level of a few atoms is really possible. This confirms of the full functionality and sensitivity of the GALLEX experiment to solar neutrinos. A calibration of the SAGE experiment was performed in a similar way recently.

12.3 Attempts at theoretical explanation

The experiments so far indicate a deficit of solar neutrinos compared to the theoretically predicted flux (see table 12.4). If we accept that there is a real discrepancy between experiment and theory, there are two main solutions to the problem. One is that our model of the Sun's structure may not be correct or our knowledge of the neutrino capture cross sections may be insufficient, the other is the possibility that the neutrino has as yet unknown properties.

We first consider the astrophysical possibility (for a detailed review on the structure of the Sun's interior and possibilities for further research in this area, see [Tur93a]). Both of the long running experiments (chlorine and Kamiokande) are mainly sensitive to the ^{8}B neutrinos. This flux is however very strongly dependent on the core temperature of the Sun ($\sim T^{18}$). This means that even a slight change of the central temperature could explain the discrepancy to

Table 12.4. The current status of the solar neutrino problem (1 SNU $= 10^{-36}$ captures per atom of detector per second).

Experiment	Result	Theory
Homestake ^{37}Cl	2.56 ± 0.22 SNU	7.9 ± 2.6 SNU [Bah92]
		9.3 ± 1.4 SNU [Bah95]
		6.4 ± 1.4 SNU [Tur93b]
		7.43 ± 2.7 SNU [Ber93a]
		4.1 ± 1.2 SNU [Dar96]
Kamiokande	$(2.80\pm0.19\pm 0.30)$ $\times10^6$ cm^{-2} s^{-1}	5.66×10^6 cm^{-2} s^{-1} [Bah92]
		4.52×10^6 cm^{-2} s^{-1} [Tur93a]
		2.49×10^6 cm^{-2} s^{-1} [Dar96]
		$(2.77\pm0.55)\times10^6$ cm^{-2} s^{-1} [Sha95]
GALLEX ^{71}Ga	$69.7\pm6.7^{+3.9}_{-4.5}$ SNU	132^{+20}_{-17} SNU [Bah89, Bah92]
SAGE ^{71}Ga	$69 \pm 10^{+5}_{-7}$ SNU	123 ± 8 SNU [Tur93b]
		(115 ± 6) SNU [Dar96]

some solar models (see table 12.4) by reducing the flow. The *pp* flux, on the other hand, is much less temperature dependent ($\sim T^{-1.2}$, [Bah89]) and closely correlated with the Sun's luminosity. If therefore approximately the expected number of SNUs were to be observed in the gallium experiments for the *pp* flux, the solution of the problem could partly lie in the central temperature of the Sun. A change of the central temperature of around 5% seems in fact to be able to explain part of the the data (Kamiokande or chlorine data alone), but leads to contradictions from helio-seismology (see below and e.g. [Tur93a, Chr94, Chr96]). However, SNU rates below 70 for the Ga experiments could hardly be explained by changes in the solar model since such a reduction would conflict with the observed luminosity. We would in this situation have to resort to the neutrino for an explanation (assuming sufficient knowledge of the ν capture cross section) (see chapter 12.3.2).

12.3.1 Non-standard solar models, the ^{7}Be problem, cosmions and helio-seismology

12.3.1.1 Non-standard solar models and the ^{7}Be problem

Non-standard solar models attempt to find a possibility to somehow lower the central temperature in order to reduce the ^{8}B neutrino flux. One model is the so-called low Z model, i.e. with a low 'metal' content (astrophysicists call all elements heavier than helium metals) [Bah71]. The temperature gradient inside the Sun is directly proportional to the opacity for photons. Less metal implies

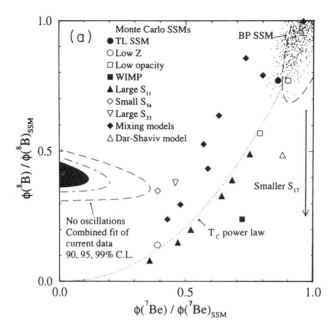

Figure 12.20. Flux constraints for solar neutrinos due to the combined Kamiokande, chlorine and gallium data, shown in (a) ^{7}Be–^{8}B; (b) pp–^{7}Be; (c) pp–^{8}B planes. The best fit parameters are $\phi(pp)/\phi(pp)_{SSM} = 1.095$, $\phi(Be)/\phi(Be)_{SSM} = 0$ and $\phi(B)/\phi(B)_{SSM} = 0.41$. However the fit is very poor (can be excluded with 93% confidence level). The dashed region describes the 90% confidence region of the Bahcall–Pinsonneault SSM (BP-SSM), the points are the Bahcall-Ulrich Monte Carlo SSMs and the solid circle the Turck–Chieze–Lopes SSM (TL-SSM). In addition there are further non-standard solar models indicated by different symbols, which are grouped along the dotted line (which describes the power dependence of the flux of the central temperature of the Sun). None of the models seems to describe even approximately the observations (black region) (from [Hat95], see also [Hat97]). For a discussion of the non-standard solar models shown see [Bah89]).

a smaller opacity. If the metal abundance in the solar interior is decreased, the core temperature can be lowered which therefore produces a lower neutrino flux. In order to explain the observed flux we require a Z/X ratio (X stands for hydrogen) of about 10% of the value at the surface. Other models assume a very quick rotation inside the Sun, which allows some of the gravitational force to be compensated by centrifugal forces, thereby reducing the required radiation pressure (i.e. lowering the core temperature). For an overview of this type of models, see [Bah89]. A comprehensive statistical survey of the different solar models and their effects (see figure 12.20), implies that they are unable to solve the solar neutrino problem [Hat94a, Hat94b, Hat95, Hat97].

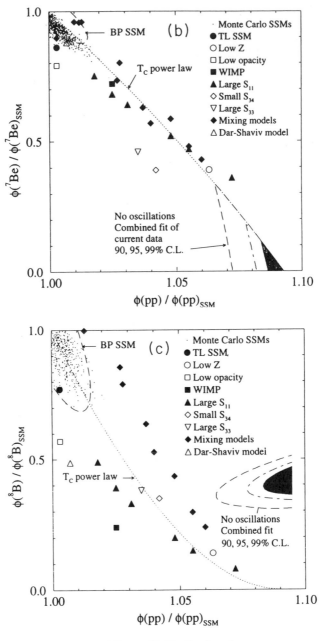

Figure 12.20. Continued.

Figure 12.20 emphasizes the problem of missing ^{7}B neutrinos. This results essentially from a comparison of the total rate seen in the gallium experiments of about 70 SNU and the observed rate of ^{8}Be neutrinos seen in Kamiokande of 7 SNU with the expectations of the solar models from *pp* neutrinos of about 70 SNU (see table 12.3). In this situation there is scarcely any room for ^{7}Be (or indeed any other) neutrinos. Such a ^{7}Be deficit could possibly be explained by standard physics, e.g. by a reduction of the capture cross section for *pp* neutrinos on ^{71}Ga and for ^{7}Be neutrinos in ^{37}Cl near threshold due to the screening effects of orbiting electrons on the nuclear charge, or through a reduction of the ^{7}Be neutrino signal in ^{71}Ga and ^{37}Cl from the influence of plasma effects on the branching ratios for electron capture by ^{7}Be in the Sun (see [Dar96]). The planned or currently under construction BOREXINO and HELLAZ experiments (see section 12.4) will be the first able to test the reduction of ^{7}Be neutrinos directly. A direct test of the neutrino oscillation hypothesis (MSW effect, see below) will be possible with the SNO experiment, (see section 12.4.2.1). Conclusions on the existence of neutrino oscillations or masses from the results of the present experiments seem premature.

12.3.1.2 Cosmions

One possibility to kill two birds with one stone is provided by the model of *cosmions* [Spe85, Pre85], a particular hypothetical kind of WIMP (see chapter 9). It is intended to solve both the solar neutrino problem and the dark matter problem simultaneously (see chapter 9). If dark matter consists of particles with as yet unknown characteristics, the Sun could have captured some of them during the course of its billions of years of existence. The actual capture process is via the scattering of such a cosmion on a nucleon inside the Sun, in which it loses so much energy that it can no longer reach escape velocity. In order to calculate this it is necessary to make assumptions about the number of incoming cosmions and their probability for such a successful capture collision. Furthermore we then have to sum over all possible orbits of cosmions within the Sun. The number of incoming particles depends on whether the cosmions as dark matter are arranged more in the galactic disc or in the halo. If they are situated in the disc they can have densities of $0.1 M_\odot$ pc^{-3} and a mean velocity of 30 km s^{-1}. Should they, on the other hand, be a large fraction of the halo, their density would only be of order $0.01 M_\odot$ pc^{-3} and their mean velocity about 300 km s^{-1}. The former possibility would result in a larger cross section; however, WIMPs are usually classified as being non-dissipative, which means they should not have taken part in the creation of the Milky Way disc, which would favour the latter assumption. The capture rate also depends significantly on the cross section of the WIMPs with the elements of the Sun's interior. The critical cross section above which theoretically everything is captured, is at about $\sigma \approx m_p R_\odot^2 M_\odot^{-1} \approx 4 \times 10^{-36}$ cm^2 [Pre85]. Here the mean free path of the WIMPs is roughly equal to their orbital radius in the Sun. It can be shown that the Sun in the course of billions of years

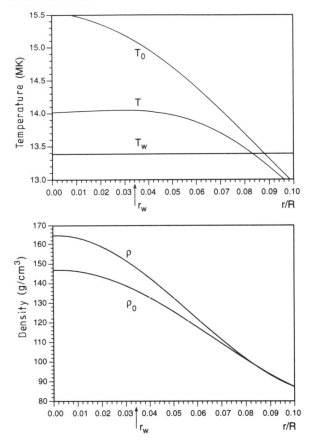

Figure 12.21. The effect of cosmions, assumed to have mass of 4 m_p and a cross section of $\sigma = 4.6 \times 10^{-37}$cm^2, on the temperature and density profile of the inner 10% of the solar radius. (*a*) The central temperature T in the presence of cosmions is reduced compared to the standard model T_0. The equilibrium temperature T_W of the cosmions is also shown, together with the radius r_W up to which their density is significant. (*b*) The central density ρ is in contrast increased in comparison to the standard model ρ_0 (from [Bou89]).

would have captured about enough WIMPs (cosmions) to produce a concentration of about 1 per 10^{12} nucleons [Pre85]. Surprisingly enough, this is exactly the concentration necessary to solve the solar neutrino problem.

The effects on stellar structure are the following: due to their low concentration, they have no direct influence on the hydrostatic equilibrium (see equation 12.22). They do however have an indirect influence on this equation, as they change the temperature, density and pressure profiles $T(r), \rho(r)$ and

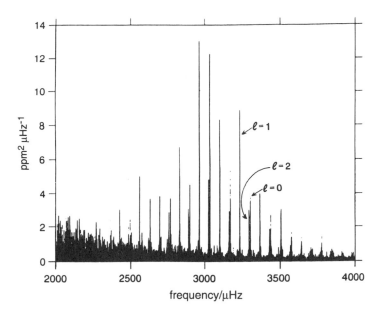

Figure 12.22. Plot of part of the frequency spectrum of the Sun, showing some of its eigenfrequencies. This spectrum was taken in 160 days of observation. The alternate series of double peaks and single peaks corresponds to the $l = 0, 2$ and $l = 1$ eigenfrequencies (from [Tou92]).

$p(r)$. The equation for the radiation equilibrium has to be modified to become:

$$L_N(r) = L_\gamma(r) + L_W(r), \tag{12.45}$$

where L_N and L_γ represent the luminosity created by the nuclear reactions and transported by the photons respectively. A new addition is L_W, the contribution to the energy transport to the exterior by the captured WIMPs. The mechanism for the energy transport through cosmions is that they interact once in the neutrino producing zones of the Sun, and due to the weak interaction, only interact again much further outside. As the temperatures are much lower here than in the core, an efficient energy transport from the interior to the outer parts is possible [Bou89]. In principle solar models should be recalculated using the modified equations. However, since this is very computation intensive, no numerical data are so far available. Analytic results which do show the effects of these phenomena exist (see figure 12.21), but they do not for example take into account the change of chemical composition in the course of time. However, it can clearly be seen that the central temperature reduces by incorporating the cosmions, while simultaneously the density in the core of the Sun increases. In order to solve the solar neutrino problem the cosmions must not be too heavy, since otherwise they would concentrate too much in the centre, and do not reach much into the area of the ^{8}B neutrino production. The radial extent r_W of the cosmions is approximately given by [Bah89]:

$$r_W = 0.13 R_\odot \left(\frac{m_p}{m_W} \right)^{1/2} \tag{12.46}$$

where m_p is the proton mass ($\simeq 1$ GeV) and m_W the cosmion mass. Outside this radius the cosmion density is negligible. This implies an upper mass limit for the cosmions of about 10 GeV. On the other hand they should not be too light, so that their interactions will not be limited to the production area of the ^{8}B neutrinos, but still allow them to carry energy effectively to the outside. This results in a lower mass limit of 2 GeV. With that we now have fixed, at least to a first approximation, the two essential parameters, the mass of the cosmions and their cross sections, i.e. in order that the solar neutrino problem and the dark matter problem can be solved simultaneously, the mass should be between 2 and 10 GeV, and the cross section around 10^{-36}cm^2.

However, three experimental results are in conflict with this model. One comes from underground experiments with silicon detectors (see chapter 9) [Cal90b, Cal94]. Cosmion collisions would give the silicon nuclei a recoil of several keV, which would result in a detectable signal in the detectors. The measurement practically excludes the cosmion hypothesis (see figure 9.12 and [Cal91, Cal92, Cal94]. Also the search for cosmions by looking for high-energy neutrinos produced by annihilation in the Sun did not yield a positive result (see chapters 8.4 and 9.3.3.1). The third kind of result in conflict with the cosmion model comes from *helio-seismology*.

12.3.1.3   Helio-seismology

The method of helio-seismology allows the possibility of examining the solar interior using solar oscillations (see figure 12.22). The Sun as a three-dimensional oscillator has normal vibrational modes characterized by three numbers n, l and m, whose lowest nodes reach very far into the Sun's interior. The acoustic modes (p-modes) are constantly reflected between the surface and the interior. The reflection point in the interior depends on the speed of sound. This in turn is determined by parameters such as the density, leading to information about the inner structure of the Sun. The frequency splitting between two modes:

$$\Delta\nu(l, n) = \nu_{ln} - \nu_{l+2,n+1} \tag{12.47}$$

has proved to be particularly useful for low (small l, n) modes. Deviations from the standard model show up in larger splittings which has to agree with the experimentally observable data. All observation results can, however, best be explained with a standard solar model [Chr94, Chr96, Har96]. The installation in 1996 of a world-wide network of observation centres (GONG) [Hil90, Chr94] together with making use of the recently started SOHO satellite [Hel96a], as well as other running experiments, will further refine the accuracy of these global oscillation measurements. The use of gravitational modes (*g*-modes) would be of particular interest, since they would be the best probe of the solar interior. Their observation is however extremely difficult [Tur93a, Gou96].

12.3.2 Neutrino oscillations in matter and the MSW effect

The most obvious possibility of solving the solar neutrino problem on the basis of neutrino properties would be neutrino oscillations, which requires massive neutrinos. In order to reach the required reduction from neutrino oscillations (chapter 2), the vacuum mixing angle would have to be almost maximal. This is not comfortable, since the comparable mixing angle in the quark sector, the Cabibbo angle, is only about 13°. However, Mikheyev, Smirnov and Wolfenstein (MSW) have shown that important modifications are produced inside matter [Wol78, Mik86a]. Matter influences the propagation of neutrinos by elastic, coherent forward scattering. The basic idea of this effect is the differing interactions of different neutrino flavours within matter. While interactions with the electrons of matter via neutral weak currents are possible for all kinds of neutrinos, only the ν_e can interact via charged weak currents (see figure 12.23). The charged current leads for the interaction with the electrons of the matter to a contribution to the interaction Hamiltonian of

$$H_{WW} = \frac{G_F}{\sqrt{2}} \left[\bar{e}\gamma^\mu (1 - \gamma_5)\nu_e \right] \left[\bar{\nu}_e\gamma_\mu (1 - \gamma_5)e \right]. \tag{12.48}$$

By a Fierz transformation this term can be brought to the form

$$H_{WW} = \frac{G_F}{\sqrt{2}} \left[\bar{\nu}_e\gamma^\mu (1 - \gamma_5)\nu_e \right] \left[\bar{e}\gamma^\mu (1 - \gamma_5)e \right]. \tag{12.49}$$

Calculating the four-current density of the electrons in the rest frame of the Sun, we obtain

$$\langle e|\bar{e}\gamma^i (1 - \gamma_5)e|e\rangle = 0 \tag{12.50}$$

$$\langle e|\bar{e}\gamma^0 (1 - \gamma_5)e|e\rangle = N_e. \tag{12.51}$$

The spatial components of the current must disappear (no permanent current density throughout the Sun), and the zeroth component can be interpreted as the electron density of the Sun. For left-handed neutrinos we can replace $(1 - \gamma_5)$ by a factor of 2, so that equation (12.49) can be written as

$$H_{WW} = \sqrt{2}G_F N_e \bar{\nu}_e\gamma_0\nu_e. \tag{12.52}$$

Thus the electrons contribute an additional potential for the electron neutrino $V = \sqrt{2}G_F N_e$. With this additional term, the free energy–momentum relation becomes

$$p^2 + m^2 = (E - V)^2 \simeq E^2 - 2EV \quad \text{(for } V \gg E). \tag{12.53}$$

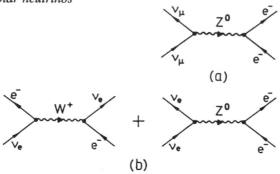

Figure 12.23. Origin of the Mikheyev–Smirnov–Wolfenstein effect. Whereas weak neutral current interactions are possible for all neutrino flavours, only ν_e also has the possibility of interacting via charged weak currents.

In more practical units this becomes [Bet86]

$$2EV = 2\sqrt{2}\left(\frac{G_F Y_e}{m_n}\right)\rho E = A \tag{12.54}$$

where ρ is the density of the Sun, Y_e is the number of electrons per nucleon and m_n is the nucleon mass. In analogy to the free energy–momentum relation an effective mass $m_{\text{eff}}^2 = m^2 + A$ can be introduced which depends on the density of the solar interior. In the case of two neutrinos ν_e and ν_μ in matter, the matrix of the squares of the masses of ν_e and ν_μ in matter has the following eigenvalues for the two neutrinos $m_{1m,2m}$ (for details see [Kla95])

$$m_{1m,2m}^2 = \frac{1}{2}(m_1^2 + m_2^2 + A) \pm [(\Delta m^2 \cos 2\theta - A)^2 + \Delta m^2 \sin^2 2\theta]^{1/2} \tag{12.55}$$

where $\Delta m^2 = m_2^2 - m_1^2$. The two states are closest together for

$$A = \Delta m^2 \cos 2\theta \tag{12.56}$$

which corresponds to an electron density of

$$N_e = \frac{\Delta m^2 \cos 2\theta}{2\sqrt{2}G_F E} \tag{12.57}$$

(see also [Bet86, Gre86a, Kla95]).

12.3.2.1 *Constant density of electrons*

The energy difference of the two neutrino eigenstates in matter is modified compared to that in the vacuum by the effect discussed in the previous section to become

$$(E_1 - E_2)_m = C \cdot (E_1 - E_2)_V \tag{12.58}$$

where C is given by

$$C = \left[1 - 2\left(\frac{L_V}{L_e}\right)\cos 2\theta_V + \left(\frac{L_V}{L_e}\right)^2\right]^{\frac{1}{2}} \tag{12.59}$$

and the neutrino–electron interaction length L_e is given by

$$L_e = \frac{\sqrt{2}\pi\hbar c}{G_F N_e} = 1.64 \times 10^5 \left(\frac{100 \text{ g cm}^{-3}}{\mu_e \rho}\right) \quad [\text{m}]. \tag{12.60}$$

The equations describing the chance of finding another flavour eigenstate after a time t correspond exactly to equation (2.71) with the additional replacements

$$L_m = \frac{L_V}{C} \tag{12.61}$$

$$\sin 2\theta_m = \frac{\sin 2\theta_V}{C}. \tag{12.62}$$

In order to illustrate this we consider the case of two flavours in three limiting cases. Using equations (2.71) and (12.59) the oscillation of ν_e into a flavour ν_x is given by:

$$|<\nu_x \mid \nu_e>|^2 = \begin{cases} \sin^2 2\theta_V \sin^2(\pi R/L_V) & \text{for} \quad L_V/L_e \ll 1 \\ (L_e/L_V)^2 \sin^2 2\theta_V \sin^2(\pi R/L_e) & \text{for} \quad L_V/L_e \gg 1 \\ \sin^2(\pi R \sin 2\theta_V/L_V) & \text{for} \quad L_V/L_e = \cos 2\theta_V. \end{cases} \tag{12.63}$$

The last case corresponds exactly to the resonance condition mentioned above. In the first case, corresponding to very small electron densities, the matter oscillations reduce themselves to vacuum oscillations. In the case of very high electron densities, the mixture is suppressed by a factor $(L_e/L_V)^2$. The third case, the resonance case, contains an energy dependent oscillatory function, whose energy average results typically in a value of 0.5. This corresponds to maximal mixing. These results imply that the solar neutrino problem cannot be solved with constant density (see also [Bah89]).

12.3.2.2 Variable electron density

A variable density causes a behaviour of the mass eigenstates which is shown in figure 12.24. Above the resonance density given by equation (12.57), the heavier mass eigenstate is mainly ν_e, while below that it is mainly ν_μ. A ν_e produced in the interior of the Sun therefore moves along the upper curve, and if it succeeds, at the resonance point, in avoiding a significant transition to the lower level, it remains on the upper level and leaves the Sun as ν_μ. However, these are not detectable in radiochemical experiments, and therefore a reduced ν_e flux can be explained in this way. The condition for this process is a relatively

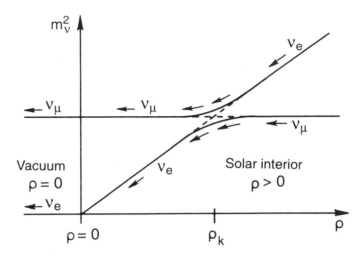

Figure 12.24. The MSW effect. The heavy mass eigenstate is almost identical to ν_e inside the Sun, in the vacuum however it is almost identical to ν_μ. If a significant jump at the resonance location can be avoided, the produced electron neutrino remains on the upper curve and therefore escapes detection in geochemical experiments (after [Bet86]).

slow density change at the resonance point, in order that the flavour eigenstate can follow the mass eigenstate (adiabatic condition). Assuming a density profile of the Sun of the form

$$\rho = \rho_0 \exp\left(-\frac{r}{R_s}\right) \tag{12.64}$$

which is a good approximation in the resonance region, the survival probability P of a ν_e is given by

$$P = \exp\left(-\frac{C}{E_\nu}\right) \tag{12.65}$$

where $C = \pi R_s \Delta m^2 \sin^2 \theta$ and $R_s = 6.6 \times 10^9$ cm. Since there are 2 unknowns, Δm^2 and $\sin^2 2\theta$, the solutions with reference to the solar neutrino experiments can only be given in contour plots. The contours for the different solar neutrino experiments which result from an analysis with the solar models of [Bah88c, Tur88] are shown in figure 12.25. Taking the results of GALLEX, Kamiokande and the chlorine experiment together, only very small parameter regions remain as possible MSW solutions [Ans92a, Lan92, Lan94]. The mixing parameters are either at around $\Delta m^2 = 6 \times 10^{-6}$ eV^2 and a mixing angle of $\sin^2 2\theta = 7 \times 10^{-3}$, or at around $\Delta m^2 = 8 \times 10^{-6}$ eV^2 and a mixing angle of $\sin^2 2\theta = 6 \times 10^{-1}$. By measuring the shape of the neutrino energy spectrum, SuperKamiokande should be able to distinguish between these two solutions. Also the upgraded GENIUS experiment (see section 2.4.2) could test at least the large angle solution (see figure 12.26). Vacuum solutions are also still

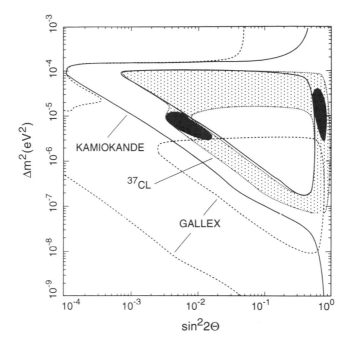

Figure 12.25. Only two very small parameter regions remain which can explain the solar neutrino problem using the MSW effect according to the combined results of the chlorine, Kamiokande and GALLEX experiments (black areas). 90% c.l. allowed regions by the individual experiments are also shown: Homestake chlorine (dotted region), Kamiokande (between solid lines) and GALLEX (between dashed lines) (from [Lan92, Hat97]).

possible with a $\Delta m^2 \approx 10^{-10}$ eV and $\sin^2 2\theta \approx 0.8$. Figure 12.26 shows the ranges probed in the $\Delta m^2 - \sin^2 2\theta$-plane by the solar neutrino experiments in comparison to those probed by other types of experiments as discussed in chapter 2.4.

The phenomenon of regeneration is also interesting. A slight reversal of the MSW effect takes place within the Earth. Electron neutrinos which are converted into muon neutrinos as they fly through the Sun, may be in part reconverted into electron neutrinos, as they pass through the Earth, since the Earth also has a non-zero electron density. As a consequence a day–night dependence of the effect is expected, which could be observed in real-time experiments. There is also the possibility of a seasonal variation since, as a result of the variation of the relative orientation of the axes of the Earth and the Sun, the average density of matter, which a neutrino must penetrate, depends on the season. Thus we expect maximum ν_e flux in spring and autumn. These fluctuations should also be detectable with a radiochemical detector. Neither effect have been observed to date. For details see [Mik86b, Bal94d].

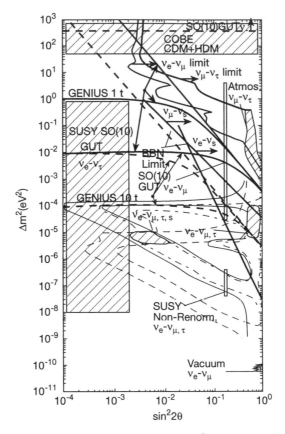

Figure 12.26. The ranges probed in the Δm^2–$\sin^2 2\theta$ plane by the solar neutrino experiments in comparison to constraints from other sources, together with predictions of some theoretical models (from [Hat94b]). Also shown is the potential of the Genius project. The three solid (and 3 dashed lines) correspond to different assumptions on the neutrino mass hierarchy—from up to down to $m_1/m_2 = 0$, 0.01 and 0.1, respectively (see [Kla97b]).

The first of the MSW solutions mentioned above could, assuming a mass hierarchy of neutrinos and a see-saw model, lead to $m_{\nu_\mu} \simeq 3 \times 10^{-3}$ eV and $m_{\nu_\tau} \simeq 10$ eV, which would make the ν_τ a candidate for hot dark matter. This is one of the main motivations for the oscillation experiments CHORUS and NOMAD (see section 2.4.5) [Win95]. On the other hand it was shown recently (see also section 2.4.2), that GUT models, which lead to almost degenerate neutrino masses for all flavours of about 1–2 eV, could solve both the problem of solar neutrinos and that of the atmospheric neutrino deficit, as well as the problem of a model of dark matter with a mixture of hot (light neutrinos) and cold dark matter [Lee94, Pet94, Moh94, Ion94, Moh96b, Pet96, Smi97]. Figure

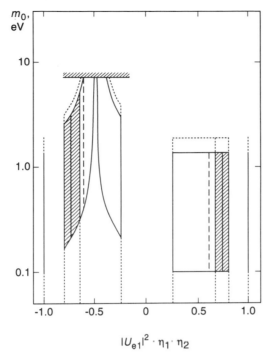

Figure 12.27. Values of the mass m_0 of the degenerate neutrinos and the mixing parameter $|U_{e1}|^2$ for which the MSW and vacuum oscillation solutions of the solar neutrino problem can be reconciled with observable Majorana mass $|m_{ee}| = (0.1 - 1.4)$ eV. Solid lines correspond to two-neutrino contributions in m_{ee} and to two-neutrino oscillations/conversions. The regions of the large mixing MSW solution are hatched; the small mixing solution is shown as a vertical line at $|U_{e1}|^2 \eta_1 \eta_2 \approx \pm 1$. For an appreciable contribution of the third neutrino state in m_{ee} the regions are larger: the dashed lines correspond to the case of three degenerate neutrinos, the dotted lines correspond to the case of a large m_3, so that $m_3 |U_{e3}|^2 = 0.5$ eV. The upper bound on the electron–anti-neutrino mass from the tritium experiments is also shown (from [Pet94, Pet96]).

12.27 shows such an explanation of the solar neutrino problem with effective neutrino masses in the range $m_\nu \approx 0.1 - 1.4$ eV. These models would imply that experiments searching for double beta-decay would play a key role, since they are able to test such a mass region for the electron neutrino (see section 2.4.2 and [Kla96a, Kla97b, Moh96, Smi97]).

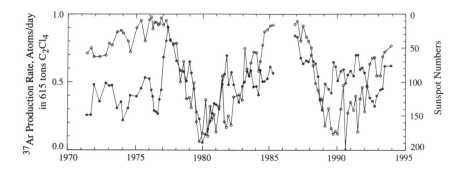

Figure 12.28. A plot of the five-point running averages of the data of figure 12.8 (small dots) compared to Sun-spot numbers (open circles). The Sun-spots are plotted on an inverted scale (from [Dav96]).

12.3.3   The magnetic moment of the neutrino

One possible explanation of the solar neutrino problem could be that as a result of a magnetic moment μ_ν of the neutrino, the magnetic field of the Sun leads to a spin precession which transforms left-handed neutrinos into *sterile* right-handed ones (sterile because the right-handed state does not take part in the weak interaction). The stronger the Sun-spot activity, the stronger the magnetic field and the more effective the transformation of neutrinos would be, which then do not show up in the experiments on Earth, i.e. the smaller the ^{37}Ar production. A significant transformation rate requires

$$\mu_\nu B x \approx 1. \tag{12.66}$$

Assuming the thickness of the convection zone x is about 2×10^{10} cm, a magnetic moment of about $10^{-10}\mu_B$ [Moh91] would be adequate to solve the solar neutrino problem via spin precession in a typical magnetic field of 10^3 G. However, Dirac neutrinos with normal weak interactions are expected from higher order corrections to have a magnetic moment only of order [Lee77b, Fuj80, Lin87]

$$\mu_\nu \simeq 3.1 \times 10^{-19}\mu_B \left(\frac{m_\nu}{1\,\text{eV}}\right). \tag{12.67}$$

There are also models which give μ_ν in the region 10^{-10} to $10^{-11}\mu_B$ (see e.g. [Vol86, Fuk87, Ste88a, Bab91a]).

This does not apply to Majorana neutrinos, whose magnetic moment is always identically zero due to CPT conservation (see e.g. [Kay89]). Laboratory experiments on the magnetic moments of neutrinos give limits in the region of $10^{-10}\mu_B$. Astrophysical observations such as the ν_e flux from SN1987a and others yield limits between 10^{-12} and $10^{-13}\mu_B$ (see e.g.

[Bar88, Gol88, Lat88, Lan92] and 13.4.3). These are only valid, however, for Dirac neutrinos.

As well as the left- to right-handed transformation, in the general case flavour transformations are also possible, for example transitions of the form $\nu_e \rightarrow \bar{\nu}_\mu$ or $\bar{\nu}_\tau$ for Majorana neutrinos, or $\nu_{e_L} \rightarrow \nu_{\mu_R}$ or ν_{τ_R} for Dirac neutrinos [Lim88]. This is by the effect of the so-called 'transition magnetic moments' which are non-diagonal, and are also possible for Majorana neutrinos. By analogy to the MSW effect, the probability of a spin reversal due to the magnetic moment can be increased in matter (see [Akh88, Lim88]). A complete description for large moments has therefore to consider both matter oscillations and effects of the magnetic moment.

The attempt to explain the solar neutrino problem via a magnetic moment does make two further experimentally testable predictions. A point often discussed is whether the production rate in the chlorine experiment shows an anti-correlation with Sun-spot activity over a period of 11 years. As Sun-spots are a phenomenon which is connected to the magnetic field of the convection zone, while solar neutrinos come from the inner regions, a connection seems to be comprehensible only if a magnetic moment for neutrinos is allowed (see above and also chapter 13). Statements concerning an anti-correlation between the number of detected solar neutrinos and Sun-spot activity are so far inconclusive [Hir90, Dav94a, Dav94b, Dav96]. Figure 12.28 shows the observations in the Cl experiment, which are the only ones to date which might indicate such an anti-correlation. In addition to this 11-year variation there should also be a half-yearly modulation [Oth95, Vol86], which is connected to the fact that the rotation axis of the Sun is tilted by about 7° to the vertical of the Earth's orbit, and that we therefore cross the equatorial plane twice a year. The magnetic field in the solar interior is nearly zero in the equatorial plane, which then leads to a considerably smaller precession.

In special scenarios [Bab91a, Ono91], the Gallium experiments should see only 25 SNU during Sun-spot maxima [Ans92a]. Since the published values correspond to a period of Sun-spot maximum, this is in contradiction to observation. On the other hand for suitable magnetic field strengths as a function of the solar radius, a rate varying with time of the ^{37}Cl detector, with at the same time time-independent rates of the Kamiokande and Gallium detectors could be understood [Akh97].

Another solution to the solar neutrino problem could be neutrino decay. One of the possible mechanisms [Fri88] would predict for example a value of under 45 SNU for the gallium experiment, which is ruled out by the experiment.

Which of the above attempts at explanation, or perhaps new explanations, will be correct can only be decided by further experimental results.

12.4 Future experiments

Great efforts are being made to improve measurements of the solar neutrino spectrum due to its importance to both astrophysics and elementary particle physics. This is done with the use of detectors with different threshold energies in radiochemical experiments and through direct measurements of the energy spectrum of solar neutrinos in real-time experiments. Some of the most important experiments planned are discussed below (see also e.g. [Bah89, Hel96, Mac96]).

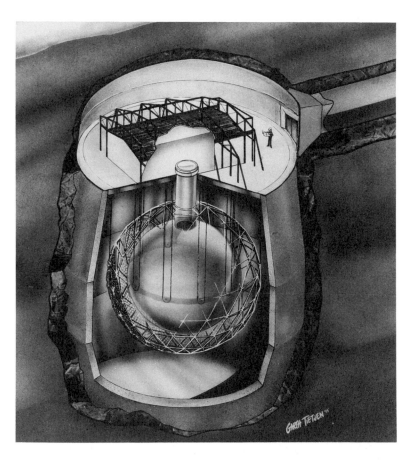

Figure 12.29. Construction of the Sudbury neutrino observatory (SNO) in a depth of 2070 m in the Craighton mine near Sudbury (Ontario). This Cerenkov detector uses heavy water rather than normal water. The heavy water tank is shielded by an additional 7300 tons of normal water. It will be possible in this experiment to test directly the oscillation hypothesis (from G Ewan, with kind permission of the SNO collaboration).

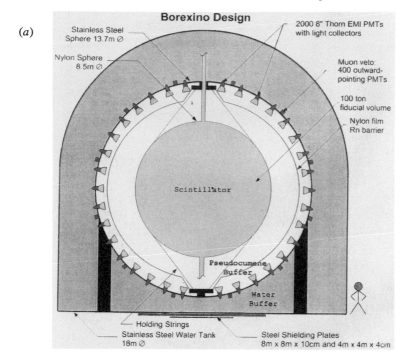

Figure 12.30. (*a*) Outline of the construction of the BOREXINO experiment in the Gran Sasso laboratory (with kind permission of G Bellini). Photo-multipliers to detect scintillation light are installed around a tank containing boron-loaded scintillator. The entire apparatus is screened by water. (*b*) Monte Carlo simulation for the background (dotted line) and standard solar model signal according to [Bah88c] (full line) for the BOREXINO detector with a fiducial mass of 100 tons. The plateau due to the ^{7}Be signal can clearly be seen. However, incredible material purities are essential for its detection. Also shown is the spectrum expected according to the SSM in the case that all the ν_e are converted into ν_μ by the MSW effect (from [Bor91]).

12.4.1 Radiochemical experiments

12.4.1.1 The ^{98}Mo experiment

One purely radiochemical experiment attempted solar neutrino detection using molybdenum [Wol85] and the reaction

$$\nu_e + {}^{98}\text{Mo} \rightarrow e^- + {}^{98}\text{Tc}. \tag{12.68}$$

Since the ground state transition with a threshold of 1.68 MeV is forbidden, only transitions into excited states are possible. This experiment is therefore only sensitive to ^{8}B and *hep* neutrinos. The expected event rate is $17.4^{+18.5}_{-11}$ SNU. Since ^{98}Tc decays with a half-life of 6×10^6 years, it should be possible

(b)

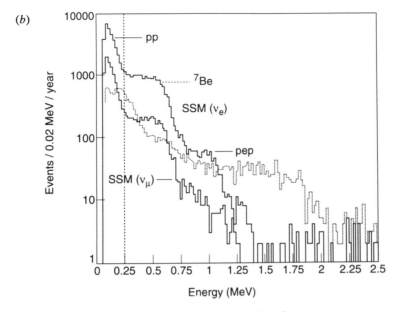

Figure 12.30. Continued.

to obtain information on the mean ^{8}B neutrino flux over the last few million years. For this reason a group at Los Alamos has extracted about 13 tons of molybdenite (MoS$_2$) from 2600 tons of an underground ore deposit. The isotope composition of the technetium contained in it was then determined by high resolution mass spectroscopy. Unfortunately the experimental background was too high to provide useful results [Bah89].

12.4.1.2   The ^{127}I experiment

Another radiochemical experiment using iodine [Hax88, Bel95a, Mac96], utilizes the following reaction:

$$\nu_e + \,^{127}\text{I} \rightarrow e^- + \,^{127}\text{Xe}. \tag{12.69}$$

The threshold energy for this reaction is 789 keV so that this experiment is mainly sensitive to ^{8}B and ^{7}Be neutrinos. This experiment is currently being constructed with 100 tons of iodine in the Homestake mine (USA) close to the Davis chlorine experiment. The first modules have recently started operation. This is appropriate since the extraction and processing is very similar to that of the chlorine experiment. Even the 36 day half-life of the xenon produced is very similar to that of the argon. The problem with this experiment is the theoretically poorly known neutrino capture cross section.

12.4.1.3 The Li experiment

A detector using 10 tons of metallic lithium with a threshold of 860 keV is proposed [Gal97] and a 300 kg prototype is currently being built as a first step. The expected rate is (60.8^{+7}_{-6}) SNU.

12.4.1.4 Other radiochemical experiments

The GALLEX experiment is continuing its operation under the new name of GNO (Gallium Neutrino Observatory) at the Gran Sasso laboratory, and a enlarged version is planned.

Many further target materials have been suggested, e.g. Br, Tl (which has the lowest energy threshold of 54 keV), but they are still far from being realized and will not be discussed further here (see e.g. [Bah89]). A particularly sensitive radiochemical detector for ^{7}Be neutrinos, ^{131}Xe, was recently proposed [Geo97] and could be useful complementary to BOREXINO (see below).

There is a plan to use several of the above mentioned isotopes simultaneously in the form of a LiI(Eu) detector [Cha94]. A prototype exists for a detector using Yb in a scintillator which provides a threshold energy low enough to be sensitive to pp neutrinos [Rag97]. The cryogenic detectors discussed in chapter 9 are also theoretically suitable for the solar neutrino search.

12.4.2 Real-time Cerenkov experiments

Two detectors of this type will operate in the near future: Superkamiokande (in operation since April 1996, see section 12.2.2.2) and the Sudbury neutrino observatory (SNO). Both should be able to measure directly the form of the solar neutrino spectrum above 5–6 MeV.

12.4.2.1 The Sudbury neutrino observatory (SNO)

In contrast to the other Cerenkov detectors, this experiment uses 1000 tons of heavy water D_2O in a transparent acrylic tank, which is surrounded by 9600 photo-multipliers [Ewa92, Ewa95, Mac96, Moo96] (figure 12.29). The main reaction is via charged weak currents

$$\nu_e + d \to e^- + p + p \tag{12.70}$$

with a threshold of 1.442 MeV. The number of events is about an order of magnitude higher than that of the equally possible neutrino–electron scattering (see section 12.2.2). Therefore above about 5–6 MeV (where the background is sufficiently small), a direct determination of the spectral form should be possible. The remarkable aspect, however, is the additional determination of the *total* neutrino flux, independent of any oscillations, due to the flavour-independent reaction due to neutral weak currents

$$\nu + d \to \nu + p + n \tag{12.71}$$

which has a threshold of 2.225 MeV. An admixture of NaCl or $MgCl_2$ is used to detect the created neutron, via the detection of the γ quanta produced in the ^{35}Cl (n, γ) process. The problem here is that any neutron with an energy larger than 2.2 MeV can produce this disintegration. By comparing the rate of these two processes (equations (12.70) and (12.71)) it will be directly possible to test the oscillation hypothesis. The SNO experiment offers for the first time the possibility of measuring the background directly—by replacing the heavy water by normal water, so that reactions (12.70) and (12.71) cease. This detector should start operation in 1998.

12.4.3 Real-time scintillator experiments

The advantage of using scintillators instead of the Cerenkov effect is that lower threshold energies are possible.

12.4.3.1 The C_6F_6 experiment

An experiment with 1000 tons of C_6F_6 is being proposed in the Baksan laboratory (see [Bah89]). The basic reaction is

$$\nu_e + {}^{19}F \rightarrow e^- + {}^{19}Ne \tag{12.72}$$

^{19}Ne decays with a half-life of 19 s. The signal is detected via the signature of the decay, i.e. the prompt electron is detected followed by the delayed positron which is produced in the ^{19}Ne decay. The detector would have a threshold of 3.24 MeV, so it would also be mainly sensitive to ^8B neutrinos.

12.4.3.2 The BOREX(INO) experiment

Comparison of the event rates predicted by the SSMs with those of the experiments leads to the conclusion (see section 12.3.1) that it is of the utmost importance to measure the mono-energetic (862 keV) ^7Be neutrinos. The most promising experiment for this is BOREX, and/or the smaller prototype BOREXINO [Rag86, Gia94, Mac96, Bel96a]. This experiment is designed to measure both the ^7Be and the ^8B neutrino fluxes. The BOREXINO experiment (figure 12.30(a)) will use 300 tons of a boron-loaded scintillator $(B(OCH_3)_3$, of which 100 tons can be used as a fiducial volume (i.e. in order to reduce background only this reduced volume is used for data taking). In its final form the experiment should contain about 2000 tons. Neutrino–electron scattering is used as the detection reaction. The mono-energetic ^7Be neutrinos produce a recoil spectrum which has a maximum energy of 665 keV ('Compton edge') (see figure 12.30(b)). The signal should hence be visible in an area around 250–800 keV, and, according to the SSM of [Bah88c], should amount to about 50 events per day. With such a counting rate it should be possible to search for the day–night effect, as well as for a $1/R^2$ dependence due to the eccentricity of the Earth's orbit. The latter would show itself in a 3.5% variation in the signal. Sufficient suppression of the radioactive background will be decisive for the success of this experiment. Extremely low radiation levels are necessary in the scintillators (less than 10^{-16} g (U, Th)/g) and in the materials used.

The ^{8}B flux is detected both via neutrino–electron scattering and via the reaction

$$\nu_e + {}^{11}B \to e^- + {}^{11}C. \tag{12.73}$$

The threshold should be at about 4 MeV. Due to the larger size of the target, it would also be possible using BOREX, to study neutral weak currents, which lead to excited states of ^{11}B. This would permit a possible comparison between charged and neutral weak currents similar to the situation at SNO. The BOREXINO experiment is currently being constructed in the Gran Sasso underground laboratory.

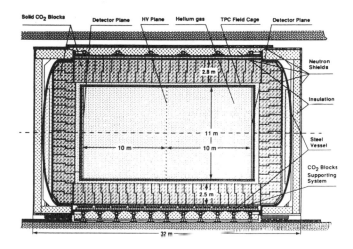

Figure 12.31. A longitudinal view of the heavily shielded, cylindrical HELLAZ time proportional counter solar-neutrino detector. Ionization tracks drift in an axial field generated by a field cage to modular multiwire detectors located at the endcaps which use the same low-temperature helium–methane gas mixture as the target volume (from [Yps96]).

12.4.4 The HELLAZ experiment

The basic idea of the HELLAZ (Helium at Liquid Azote), experiment is to measure the energy spectrum of the pp and ^{7}Be neutrinos using neutrino–electron scattering [Bon94, Yps94, Yps96]. It is planned to construct a time projection chamber (TPC) which contains 12 tons of helium at high pressure

Table 12.5 Parameters of the planned real time solar neutrino detectors in comparison with the Kamiokande III experiment.

experiment	Kamiokande	Super-K	SNO	BOREXINO	ICARUS
Shielding depth (mwe)	2700	2700	6200	4000	4000
Target	e^-	e^-	$e^-, {}^2H$	e^-	$e^-, {}^{40}Ar$
Material	H_2O	H_2O	D_2O	$B(OCH_3)_3$	liquid Ar
Active mass (tons)	680	22 000	1000	100	4700
Detection method	Cerenkov	Cerenkov	Cerenkov	Scintillation	Ionization (TPC)
$\Delta E/E$ (%)	30	14	12	4	5
$\Delta\Theta$ (Grad)[a]	≈ 28	≈ 25	≈ 25		≈ 10
$E_{\nu min}$	7.2	5.2	6.4	0.41	5.9

[a] for 10 MeV electrons, for ICARUS at 6 MeV.

(figure 12.31). The detection of the scattered electrons will provide information on the neutrinos. This would be the first real time *pp*-neutrino experiment.

12.4.5   The ICARUS experiment

This ionization experiment will be constructed in the Gran Sasso. The ICARUS, (Imaging Cosmic and Rare Underground Signals) collaboration plans to operate a 5000 tons liquid argon detector in the form of a time projection chamber (see chapter 2, figure 2.9 and [Rub96, Mac96]).

Besides neutrino–electron scattering (see section 12.2.2), which is mostly sensitive to electron neutrinos, although also to ν_μ and ν_τ, also the reaction

$$\nu_e +^{40}\text{Ar} \rightarrow e^- +^{40}\text{K}^\star \tag{12.74}$$

will be used.

The excited potassium decays via gamma emission with a total energy release of 4.3 MeV. The threshold energy for the above process is 5.9 MeV. The idea is to determine the sum of the ν_μ and ν_τ fluxes by comparison of elastic scattering with neutrino absorption according to equation (12.74).

One problem with this detector is that electrons must drift over a very long distance. On the other hand the detector has excellent energy, vertex and directional resolution. A functioning 3 ton test detector has been built at CERN [Ben94a]. It has a positional resolution of better than 100 μm and an energy resolution of 2 to 3% at a few MeV. In comparison Kamiokande II had an energy resolution at 10 MeV of about 20%. The first 600 ton module should be constructed in the year 1998 or 1999 in the Gran Sasso laboratory. In its final form the detector is mostly intended to search for proton decay and for long baseline neutrino experiments (see chapter 2). A particular possibility to study the shape of the ^{8}Be solar neutrino spectrum could be achieved by the introduction of CD_4 into the liquid argon. Varying the CD_4 concentration would allow discrimination between the charged weak current events and elastic scattering and background. The event rate with a few per cent of CD_4 would be around 1 event per day for a 600 ton module.

Table 12.5 gives a summary of the planned real-time experiments.

Having dealt in great detail with solar neutrinos, we now discuss another source of neutrinos, which has caused a lot of excitement and discussion over the last few years.

Chapter 13

Neutrinos from supernovae

A further source of neutrinos is supernovae. Supernova explosions are among the most spectacular events in astrophysics. They are phenomena from the late phase of stellar evolution (see figure 13.1). For more detailed information see [Tri82, Sha83, Tri83, Woo86a, Arn89, Woo90a, Whe90, Woo92, Woo94, Mül95b, Hay95, Bus95, Mül95c, Bur97]. The importance of supernovae in the synthesis of heavy elements will be dealt with in chapter 14.

13.1 Supernovae

A few historical astronomical events are known, at whose position in the sky glowing gas rings and sometimes also pulsars can be found (see table 13.1). The characteristic of these events is a very quick increase in the observed brightness to a maximum which then slowly decreases. The increase lasts typically a few hours, while the decrease takes weeks or months. Observations of the spectra during the maximum suggest a classification into two groups, depending on whether hydrogen lines can be detected or not. Supernovae with practically no hydrogen lines are classed as type I, those with H-lines as type II [Whe90]. A further subdivision of both types is given in figure 13.2. Type II supernovae are usually found in areas with young stars, and therefore mainly in spiral arms, while they are not observed in elliptical galaxies. This suggests the conclusion that they are the final stages of massive stars. Massive stars have a shorter life span. The behaviour of the light curve is strongly variable in type II supernovae, while in type I the behaviour is very constant (see figure 13.3). Two recent supernovae with light curves deviating from the described behaviour are supernova 1987a, discussed in more detail in section 13.4, and supernova 1993J (the numbering scheme for supernovae contains the year of their discovery and another letter which indicates the order of occurrence). Supernova 1993J was discovered in March 1993 in the spiral galaxy M81 in the constellation Ursa Major at a distance of 11 million light years (see figure 13.4) (see e.g. [Fil93]). Except for SN 1987a this is the brightest supernova seen since SN 1972e and the brightest supernova in the northern hemisphere for 40 years. Its light curve does not match the known forms.

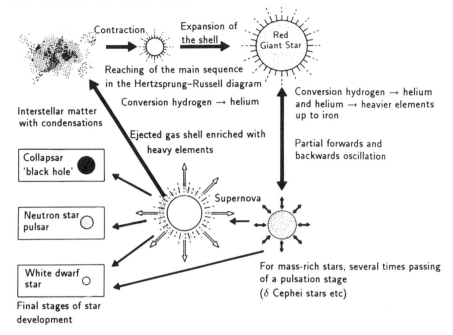

Figure 13.1. Schematic illustration of the evolution of stars and their final states (after [Her80] and [Gro89, Gro90]).

The similarities of the type I supernovae in their behaviour lead to the conclusion that they are caused by one and the same mechanism (see chapters 3.1.1.1 and 5.2.1.1). It is currently assumed that they are caused by a thermonuclear explosion of a white dwarf within a binary system

Table 13.1. Galactic supernovae visible with the naked eye over the last 1000 years (from [Ber91a]).

Name	Year	Distance (kpc)	Type
Lupus	1006	1.4	Ia
Crab	1054	2.0	II
3C 58	1181 ?	2.6	II ?
Tycho	1572	2.5	Ib ?
Kepler	1604	4.2	Ib/II ?
Cas A	1658 ± 3	2.8	Ib

MAXIMUM LIGHT SPECTRA

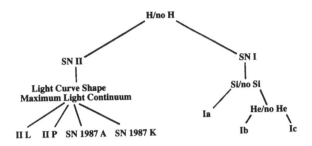

LATE SPECTRA ~ 6 MONTHS (SUPERNEBULAR)

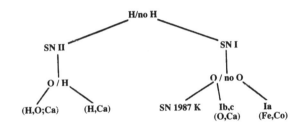

Figure 13.2. Classification of supernovae according to their spectra during maximal brightness and at a later stage. The appearance of particular spectral lines is used as a means of classification. The old classic type I corresponds today to type Ia (from [Whe90]).

[Whe73, Woo95a]. The compact star accretes matter from its main sequence companion until it is above the critical mass. In most computer simulations no supernova remnant is created in this process, and since also no significant number of neutrinos is produced, we refer to the literature on this subject and concentrate on type II. For details about supernova explosions see e.g. [Sha83, Woo86a, Woo92, Woo94, Mül95b, Mul95c, Bur95, Hay95, Woo95a].

13.1.1   The evolution of massive stars

The longest phase in the life of a star is hydrogen burning. After the hydrogen is burned off in the interior the energy production is no longer sufficient to withstand gravitation. The star does not compensate for this by a reduced luminosity, but by contraction. According to the virial theorem, only half of the energy released in this process produces an internal temperature rise, while

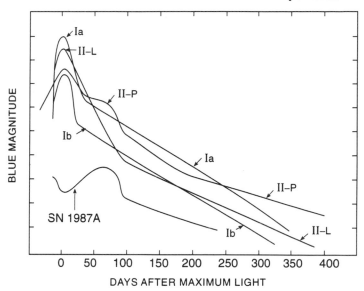

Figure 13.3. Brightness in the B-band for different supernova types. The deviation of supernova 1987a from the standard schemes can clearly be seen. Type II supernovae which have an almost linear decline after the maximum (II-L) are distinguished from those which remain almost constant over a longer time and display a form of plateau (II-P). SN 1987a appears from its characteristics to be a new form (from [Whe90]).

Table 13.2. Hydrodynamic burning phases during stellar evolution (from [Gro89, Gro90]).

Fuel	T (10^9 K)	Main product	Burning time for $25M_\odot$	Main cooling process
^{1}H	0.02	^{4}He, ^{14}N	7×10^6 a	Photons, neutrinos
^{4}He	0.2	^{12}C, ^{16}O, ^{22}Ne	5×10^5 a	Photons
^{12}C	0.8	^{20}Ne, ^{23}Na, ^{24}Mg	600 a	Neutrinos
^{20}Ne	1.5	^{16}O, ^{24}Mg, ^{28}Si	1 a	Neutrinos
^{16}O	2.0	^{28}Si, ^{32}S	180 d	Neutrinos
^{28}Si	3.5	^{54}Fe, ^{56}Ni, ^{52}Cr	1 d	Neutrinos

Figure 13.4. M81 in the constellation Ursa Major has an integrated apparent luminosity of $6^m.8$ and is therefore one of the brightest known galaxies in the Messier Catalogue. The galaxy became particularly interesting in March 1993 with the discovery of supernova 1993 J (see arrow), about 3 arc minutes south-west of the galactic nucleus. This is the brightest supernova to be seen in the northern hemisphere for 40 years (from [Fil93]).

the other half is released. Stellar evolution can be discussed with the help of the equation of state $p = p(\rho, t)$ [Cox68, Cla68, Arn77, Arn78, Wea80, Woo82, Woo86a, Woo86b, Hil88, Gro89, Gro90]. For example, the equation of state for a non-degenerate ideal gas is $p \sim \rho \cdot T$. A pressure increase is always connected to a corresponding temperature increase. If a sufficiently high temperature has been reached, the burning of helium ignites (helium flash). This causes the outer shells to inflate, and a red giant develops. The star moves to the top right in the Hertzsprung–Russel diagram (see chapter 2). After a considerably shorter time than taken by the hydrogen burning phase, the helium in the core has been fused, mainly into carbon, oxygen and nitrogen, and the same cycle, i.e. contraction with an associated temperature increase, now leads to a successive burning of these elements. The lower temperature burning phases move towards the surface in this process. From the burning of carbon onwards, neutrinos become the dominant energy loss mechanism, which leads to a further reduction of the burning time scales. The last possible reaction is the burning of silicon to nickel and iron. The silicon burning corresponds less to fusion than to the photo-disintegration of ^{28}Si with simultaneous building up of a kind of chemical equilibrium between the reactions of the strong ($n, p, \alpha, \ldots$ induced reactions) and the electromagnetic interactions. The equilibrium occurs from about $T \approx 3.5 \times 10^9$ K, and the distribution of elements created is, at that time,

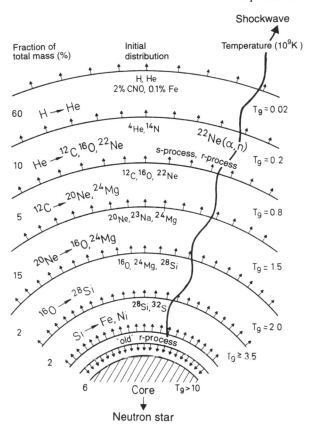

Figure 13.5. Schematic representation of the structure, composition and development of a heavy star (about $25M_\odot$). In the hydrostatic burning phases of the shells, elements of higher atomic number, up to a maximum of Fe and Ni, are built up from the initial composition (whose major components are labelled). The gravitational collapse of the core leads to the formation of a neutron star and the ejection of $\approx 95\%$ of the mass of the star (supernova explosion). The ejected outer layers are traversed by the detonation shock wave which initiates explosive burning (from [Gro89, Gro90]).

dominated mainly by ^{56}Ni, and later, at higher temperatures, by ^{54}Fe. This last burning process lasts only for the order of a day (see table 13.2). This ends the hydrostatic burning phases of the star. A further energy gain from fusion of elements is no longer possible, as the maximum binding energy per nucleon, of about 8 MeV/nucleon, has been reached in the region of iron. The number of burning phases which a star goes through depends on its mass. Brown dwarves, stars with less than $0.08M_\odot$, which did not succeed in reaching the temperature necessary for hydrogen fusion, have already been mentioned. On the other hand, stars with more than about $8M_\odot$ carry out burning up to iron. Such stars then

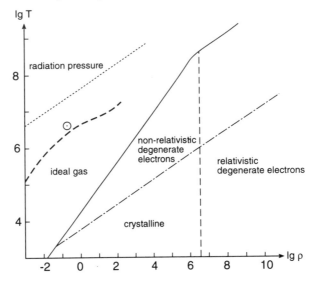

Figure 13.6. Schematic $\rho - T$ phase diagram for the characterization of matter inside stars. The areas shown are those in which the equation of state is dominated either by radiation pressure (above the dotted line) or by a degenerate electron gas (below the full line). The latter can be relativistic or non-relativistic (to the right or left of the vertical dotted line). The dot-dashed line characterizes a temperature below which the ions prefer a crystalline state. The heavy dotted line shows the standard solar model and the present Sun (from [Kip90]).

have a small, dense, iron core and an extended shell mainly of hydrogen. The burning regions form shells over each other and give the interior of stars a form of onion structure (see figure 13.5). Calculations [Arn77, Wea80] show that for stars with mass $\geq 12M_\odot$, the hydrostatic burning phases take place under *non*-degenerate conditions (i.e. non-degenerate electron gas). If the mass of the iron core exceeds a certain critical mass, the *Chandrasekhar limit*, it also becomes gravitationally unstable and collapses into a neutron star. This collapse is the pre-condition for a supernova explosion. The Chandrasekhar mass is given by ([Cha39, 67], also see [Hil88])

$$M_{Ch} = 5.72 Y_e^2 M_\odot \qquad (13.1)$$

where Y_e is the number of electrons per nucleon. The essential reason for the collapse is both the photo-disintegration of nuclei of the iron group and electron capture on free protons and nuclei, which becomes possible due to the growth of the Fermi energy of the now *degenerate* electron gas. The electron capture rates determine the initial dynamics of the collapse.

The origin of the degenerate electron gas can be understood from figure 13.6. This shows a phase diagram which characterizes the state of the material.

For very large densities the Pauli principle has to be taken into account. This implies that each cell in phase space of size h^3 can contain a maximum of 2 electrons. The entire phase space volume is then given by

$$V_{Ph} = \frac{4}{3}\pi R^3 \frac{4}{3}\pi p^3. \tag{13.2}$$

Higher densities at constant radius produce an increased degeneracy. The pressure is now no longer determined by the kinetic energy, but by the Fermi energy, or Fermi momentum p_F. This implies that $p_e \sim p_F^5$ (non-relativistic) and $p_e \sim p_F^4$ (relativistic). In addition the pressure no longer depends on the temperature, but only on the electron density n_e, according to [Sha83]

$$p = \frac{1}{m_e}\frac{1}{5}(3\pi^2)^{\frac{2}{3}}n_e^{\frac{5}{3}} \quad \text{non-relativistic} \tag{13.3}$$

$$p = \frac{1}{4}(3\pi^2)^{\frac{1}{3}}n_e^{\frac{4}{3}}. \quad \text{relativistic} \tag{13.4}$$

This has the fatal consequence of gravitational collapse. The previous cycle of temperature increase $\rightarrow$ pressure increase $\rightarrow$ ignition $\rightarrow$ expansion $\rightarrow$ temperature drop now no longer functions. Released energy leads only to a temperature increase and thereby to unstable processes. For a better understanding, consider the total energy of a star in thermodynamic equilibrium, which is given by [Lan75, Sha83]

$$E = \frac{3\gamma - 4}{5\gamma - 6}\frac{GM^2}{R} \tag{13.5}$$

$\gamma = \partial \ln p / \partial \ln \rho$ is the adiabatic index. For a bound star $E \leq 0$ and therefore $\gamma \geq 4/3$. The adiabatic index for a non-relativistic degenerate electron gas is, for example, 5/3, while it is 4/3 in the relativistic case. The stability of the iron core is, however, mainly guaranteed by the pressure of the degenerate electrons. We now consider in some more detail what happens when the nucleus of a star exceeds the Chandrasekhar mass.

13.1.2 The actual collapse phase

The structure of a star at the end of its hydrostatic burning phases can only be understood with the help of complex numerical computer simulations [Bet82, Woo86a, Woo92, Hay95, Mül95b, Bur95]. The typical parameter values for a $15 M_\odot$ star with a core mass of $M_{Ch} \approx 1.5 M_\odot$ are a central temperature of approximately 8×10^9 K, a central density of 3.7×10^9 g cm^{-3} and a Y_e of 0.42. The Fermi energy of the electrons is roughly 4–8 MeV. These are typical values at the start of the collapse of a star. The cause for this collapse is—as mentioned before—the photo-disintegration of nuclei of the iron group, such as via the reaction

$$^{56}\text{Fe} \rightarrow 13^4\text{He} + 4n - 124.4\,\text{MeV} \tag{13.6}$$

and electron capture by free protons and heavy nuclei

$$e^- + p \rightarrow n + \nu_e \qquad e^- + {}^Z A \rightarrow {}^{Z-1} A + \nu_e. \qquad (13.7)$$

The latter process becomes possible because of the high Fermi energy of the electrons. The number of electrons is strongly reduced by this process (13.7). Mainly neutron rich, unstable nuclei are produced. Since it was the degenerate electrons which balanced the gravitational force, the core collapses quickly. The lowering of the electron concentration can also be expressed by an adiabatic index under 4/3. Through corresponding contraction the core keeps the density of the electrons and therefore the pressure roughly constant. Thus approximately the inner half collapses homologeously ($v \sim r$), while the outer part collapses at supersonic speed. Homologeous means that the density distribution during the collapse remains self-similar. The behaviour of the infall velocities within the core is shown in figure 13.7 at a time of 2 ms before the total collapse. The matter outside the sonic point defined by $v_{coll} = v_{sound}$, collapses with a velocity characteristic of free fall. The outer layers of the star do not notice this collapse, due to the low speed of sound. In the core nuclei which are more and more neutron rich are produced, which is reflected in a further decrease of Y_e. The emitted neutrinos can initially leave the core zone unhindered. The neutrino diffusion time finally becomes larger than the collapse time (*neutrino trapping*). The dominant process for the neutrino opacity is coherent neutrino–nucleus scattering via neutral weak currents with a cross section of (see e.g. [Fre74, Hil88])

$$\sigma \simeq 10^{-44} \, \text{cm}^2 N^2 \left(\frac{E_\nu}{\text{MeV}} \right)^2 \qquad (13.8)$$

where N is the neutron number in the nucleus. Neutrinos are produced mainly in the density region between 10^{11} and 10^{12} g cm^{-3} in which typical nuclei have masses of between 80 and 100 and about 50 neutrons. This layer is also known as the deleptonization shell. The mean free path λ_ν of the neutrinos is given by [Hil88]

$$\lambda_\nu \simeq \frac{1}{n_A \sigma} \simeq 10^7 \, \text{cm} \left(\frac{\rho}{10^{12} \, \text{g cm}^{-3}} \right)^{-1} \frac{A}{N^2} \left(\frac{E_\nu}{10 \, \text{MeV}} \right)^{-2} \qquad (13.9)$$

where n_A is the density of an average heavy nucleus of mass A. The typical diffusion time of the neutrinos is then given by

$$\tau_d \simeq \frac{d^2}{\frac{1}{3} c \lambda_\nu} \qquad d \simeq 10^7 \, \text{cm}. \qquad (13.10)$$

Matter becomes opaque for neutrinos at about $\rho \simeq 5 \times 10^{11}$ g cm^{-3}, as the diffusion time is already 2 s and is therefore considerably larger than the collapse time of a few milliseconds. For this reason the neutrinos are trapped and move with the collapsing material (neutrino trapping). The transition between the 'neutrino optical' opaque and the transparent defines a *neutrino sphere* (a neutrino photosphere similar to the photosphere for photons). The electron capture comes to a halt at densities of $\rho \simeq 10^{12} \, \mathrm{g\,cm^{-3}}$. The increase of neutrino trapping by neutrons acts as an inverse process to that of equation (13.7) and consequently stabilizes electron loss. The core no longer cools through neutrino loss, but rather the neutrinos thermalize and an *equilibrium also with respect to the weak interaction* quickly forms, mainly via electron–neutrino scattering, so that no essential further decrease of the electrons takes place. Up to this point $Y_{\mathrm{lepton}} = Y_e + Y_\nu$ has dropped, according to [Bet86a], to about 0.41, ($Y_\nu \rightarrow 0.09$), i.e. the typical capture rate is about 0.07 electrons per nucleon (but see also figure 13.8).

Hence forward the collapse progresses adiabatically. This is equivalent to a constant entropy, as now neither significant energy transport nor an essential change of composition takes place. Figure 13.8 shows the mean mass numbers and nuclear charge number of the nuclei formed during neutronization, together with the mass fractions of neutrons, protons, as well as the number of electrons per nucleon. We thus have a gas of electrons, neutrons, neutrinos and nuclei whose pressure is determined by the relativistic degenerate electrons. In contrast to earlier assumptions a complete transition to a degenerate neutron gas does *not* take place before nuclear densities are attained. Thus the 'neutron star' begins

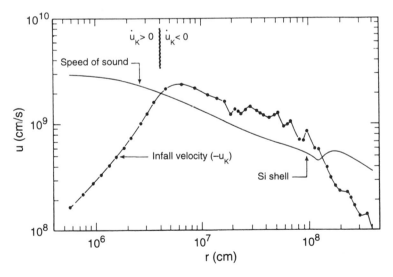

Figure 13.7. Infall velocity of the material in the core of a supernova about 2 ms before the complete collapse of the star. Within the homologous inner core ($r < 40$ km) the velocity is smaller than the local velocity of sound. In the region $r > 40$ km (outer core), the material collapses with supersonic speed (from [Arn77]).

as a hot lepton-rich quasi-static object, which develops into its final state via neutrino emission, i.e. it starts off as a *quasi-neutrino star* [Arn77].

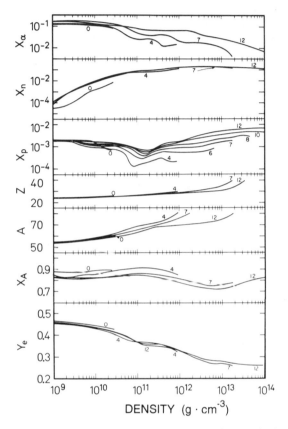

Figure 13.8. Change in the core composition during the gravitational collapse (the numbers correspond to various stages of the collapse). X_n, X_p, X_α, X_A denote the *mass fraction* (not the *number* densities!) of the neutrons, protons, α-particles and nuclei. Y_e denotes the electrons per nucleon (from [Bru85]).

The collapsing core finally reaches densities that normally appear in atomic nuclei ($\rho > 10^{14}\,\mathrm{g\,cm^{-3}}$). For higher densities, however, the nuclear force becomes strongly repulsive, matter becomes incompressible and bounces back (equivalent to $\gamma > 4/3$) (see e.g. [Lan75, Bet79, Kah86]). It then collides with further infalling matter and produces a pressure wave. This has an energy of about 10^{51} erg and passes through the iron core to the exterior. Figure 13.9 shows the main features of the iron core. As already mentioned above, the pressure wave cannot cross the sonic point, while the density and velocity are steadily changing. As material continues to flow in, bringing kinetic energy from outside, a discontinuity forms in the pressure at the sonic point, which develops into an outgoing shock wave (velocity greater than the speed of sound). This is illustrated in figure 13.10.

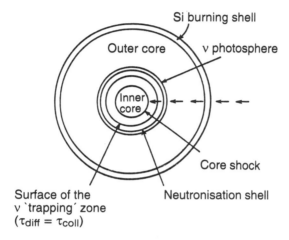

Figure 13.9. Structure of the iron core of a heavy star after completion of its hydrostatic burning (from [Arn87]).

Exactly how much energy this shock wave contains depends, among other factors, on the equation of state of the very strongly compressed nuclear matter [Kah86, Bet88]. The energy depends on whether the bounce back of the core is hard or soft. A soft bounce back provides the shock with less initial energy. Unfortunately the equation of state is not very well known, since extrapolation into areas of supernuclear density is required. The outgoing shock dissociates the incoming iron nuclei into protons and neutrons. This has several consequences. The shock wave loses energy by this mechanism, and if the mass of the iron core is sufficiently large the shock wave does not penetrate the core and a supernova explosion does not take place. In the other case the dissociation into nucleons leads to an enormous pressure increase, which leads to a reversal of the direction of motion of the incoming matter in the shock region. This transforms a collapse into an explosion. Furthermore, in the dissociated region the mean free path for neutrinos is again longer, so that they build up behind the shock wave. If the shock now enters regions in which the density is below 10^{11} g cm^{-3}, all neutrinos are released immediately. Once the shock wave penetrates the iron core, the outer layers represent practically no obstacle and are blown away, which results in an optical supernova in the sky. Such a mechanism is known as a *prompt explosion* [Coo84, Bar85].

Computer simulations have shown that for massive cores the shock does indeed not have the necessary energy to penetrate to the surface. However, it can be reactivated by the large number of subsequent neutrinos. The deposition of only 1% of their energy behind the shock is sufficient to let the shock break

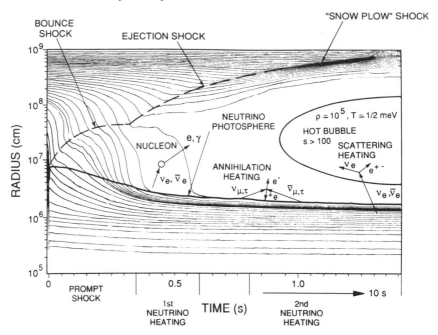

Figure 13.10. The development of a supernova explosion of type II, according to Wilson and Mayle. As the nuclear matter is over-compressed in the collapse, a rebound occurs and produces a shock wave. However, this is weakened by thermal decomposition of the incoming matter during the collapse. Thereafter a strong heating up of the incoming matter takes place because of the large number of available neutrinos and an explosive shock forms. Shortly thereafter, due to $\nu_\mu \bar{\nu}_\mu$ and $\nu_\tau \bar{\nu}_\tau$ –annihilation outside the neutrino sphere another heating takes place. As a result a hot bubble forms, compressing the expelled material and forcing it further outwards (from [Col90]).

through. Such a mechanism is called a *delayed explosion* [Bet85, Wil86]. For stars with $10{-}16 M_\odot$ the prompt mechanism was assumed to work usually, while for more massive stars the delayed explosion mechanism was required [Wil86].

In more recent work, a further, decisive boost is given to the shock by the formation of neutrino-heated hot bubbles [Bet85, Col90, Col90a] (figure 13.10), which furthermore produce a considerable mixing of the emitted material. This strong mixing has been observed in the Supernova 1987a (see section 13.4.1.2) and has been confirmed by two-dimensional computer simulations [Mül95b, Hay95, Mül95c, Bur95], which allow these effects to be described for the first time (see figure 13.11). The problem during the last 10 years has been that the shock stopped about 100–150 km from the centre of the star, and in general the inclusion of neutrino absorption and the implied energy deposition permitted only a moderate explosion. The newest supercomputers recently permitted the simulation of dying stars in *two* dimensions, following both the radial and lateral directions. Formerly only simulations in the radial direction had been possible. Typical explosions show asymmetrically ejected matter, resulting in a recoil of the core which gives the remaining core a speed of hundreds of kilometres per second (rocket effect) [Bur95, Bur95a, Jan95a].

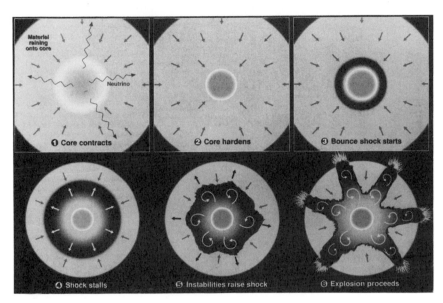

(a)

Figure 13.11. (a) Schematic picture of a supernova explosion. When the nuclear fuel is exhausted the core begins to collapse with the emission of neutrinos (1) until its density reaches that of nuclear matter (2). This is incompressible, which leads to the build up of an outward going shock wave (3), which stops about 100–150 km away from the star's centre (4). Convective movement from the neutrino heating (5) revives the shock wave and finally leads to the explosion (6). (b) Two-dimensional computer simulation which shows very clearly the convective movement caused by the Rayleigh–Jeans instability. The edge of each picture corresponds to a distance of 1000 km, and the time sequence shows the shock 101, 116, 131 and 161 ms from the start (from [Hay95]).

The energy released in a supernova results from the difference in the binding energy of the star and the neutron star produced. It is given by

$$\Delta E = \left(-\frac{GM^2}{R}\right)_{\text{star}} - \left(-\frac{GM^2}{R}\right)_{\text{neutron star}} \tag{13.11}$$

which, due to the dominance of the second term (difference in radii!), can be

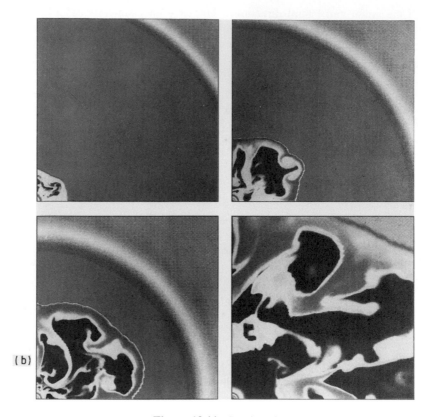

(b)

Figure 13.11. Continued.

written as

$$\Delta E = 5.2 \times 10^{53} \text{ erg} \left(\frac{10 \text{ km}}{R_{\text{neutron star}}} \right) \left(\frac{M_{\text{neutron star}}}{1.4 M_\odot} \right)^2 . \tag{13.12}$$

This is the basic picture of an exploding, massive star. For a detailed account see [Arn77, Sha83, Woo86a, Woo90a, Col90, Mül95b, Hay95, Bur95].

The neutron star can, in certain circumstances, transform into a *'strange star'* [Bay85, Alc86, Oli87], which consists of 'strange matter' (about the same number of u, d and s quarks, and a smaller number of electrons which guarantee charge stability). There are ideas [Wit84], that such strange matter could form a ground state of hadronic matter for objects with a baryon number of between 100 and 2×10^{57} ('strangelets'). The upper limit corresponds to $2M_\odot$ and is determined by the gravitational collapse. Such objects can also be considered as candidates for the cold dark matter which was discussed in chapter 9.

13.2 Neutrino emission in supernova explosions

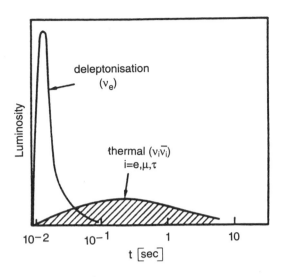

Figure 13.12. Time evolution of the expected neutrino spectrum of a supernova type II (schematic). After the first few ms long pulse of ν_e originating from the deleptonization, follows over a period of several seconds the emission of all neutrino and anti-neutrino flavours originating from the thermal cooling of the neutron star (from [Per88], see also [Bru87]).

We now consider the neutrinos which could be observed from supernova type II. First we consider the radius within the supernova from which the neutrinos are emitted, and the typical temperature in this region. The following derivation comes from [Sch90a]. The highest temperature occurs at the point from which the neutrinos can still just escape unhindered. Let us assume for simplicity that the mean free path λ_ν corresponds approximately to the core radius R

$$\lambda_\nu \simeq \frac{1}{n\langle\sigma\rangle} \simeq R \tag{13.13}$$

where $n = \rho/m_n$ is the particle density, and the effective cross section $\langle\sigma\rangle$ can be parametrized by [Sch87c]

$$\langle\sigma\rangle = \sigma_0 \frac{E_\nu^2}{2} \approx \frac{12\sigma_0}{2T_0^2}\left(\frac{T_\nu}{T_0}\right)^2. \tag{13.14}$$

In the second step averaging over a thermal Boltzmann spectrum is exploited (see also chapter 12). T_0 is a characteristic temperature of the order of MeV, and σ_0 is a typical weak cross section of about 10^{-42} cm^2. As the collapse is

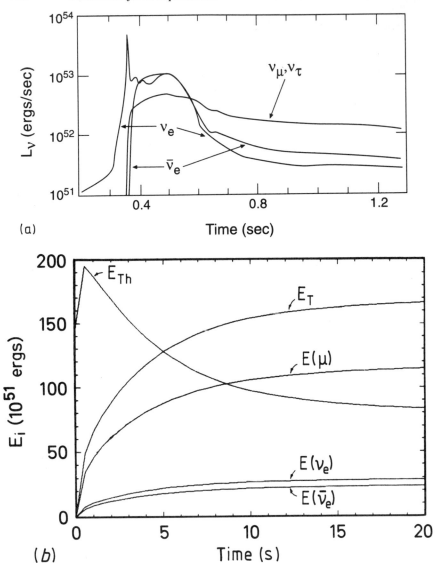

Figure 13.13. (*a*) Calculated luminosity of a $2M_\odot$ 'Fe' core of a $\approx 25M_\odot$ main sequence star as a function of the time from the start of the collapse for the various neutrino flavours (from [Bru87]). (*b*) Cooling of a hot proto-neutron star of $1.4M_\odot$ in the first 20 s after gravitational collapse. E_{Th} denotes the intergrated internal energy, E_T is the total energy released and E_{ν_e} and $E_{\bar{\nu}_e}$ are the total energies emitted as ν_e and $\bar{\nu}_e$, respectively. E_μ is the energy emitted as ν_μ, $\bar{\nu}_\mu$, ν_τ and $\bar{\nu}_\tau$. All energies are in units of 10^{51} ergs (from [Bur86]).

adiabatic, i.e.

$$\rho = \rho_0 \left(\frac{T}{T_0} \right)^3 \qquad (13.15)$$

and the radiated energy follows a Fermi–Dirac distribution to a good approximation, for which $\langle E_\nu \rangle = 3.15 T_\nu$, by substitution we obtain

$$\frac{T_\nu}{T_0} = \left(\frac{2 m_n}{12 \sigma_0 T_0 \rho_0 R} \right)^{1/5}. \qquad (13.16)$$

It can be seen that the temperature of the neutrinos is relatively insensitive to the initial parameters. With sensible assumptions ($\sigma_0 = 1.7 \times 10^{-44}$ cm^2, $\rho_0 = 10^{10}$ g cm^{-3}, $T_0 = 1$ MeV and $R = 5 \times 10^6$ cm)

$$T_\nu \approx 3.2 \,\text{MeV} \Rightarrow \langle E_\nu \rangle \approx 10 \,\text{MeV} \qquad (13.17)$$

results. As ν_μ and ν_τ interact only via neutral currents, even at the values of ρ_0 in the range assumed here, their neutrino spheres are further inside, and the mean energies are therefore higher (about 15 MeV). The spectrum of the $\bar{\nu}_e$ corresponds initially to that of the ν_e, but as more and more protons vanish, they also react more and more via neutral currents, which manifests itself in a temperature increase. The spectral form of all emitted flavours can be quite well approximated by a Fermi–Dirac distribution, even if detailed calculations conclude that an excess of higher energy states are present [May87]. Furthermore it must be noted that the cross section of equation (13.14) depends quadratically on the energy, which leads to a greater diffusion of the lower energy neutrinos from the core, and therefore distorts the spectrum. It can also be shown [May87] that practically all neutrino flavours carry away the same energy, so that due to the higher energy the flux of ν_μ and ν_τ is correspondingly lower.

In principle the detectable spectrum consists of two parts. Here the neutrinos that are produced before the core becomes opaque to neutrinos are neglected, i.e. those ν_e that can still escape at the very beginning of the actual collapse, as their number is low and should not be detectable. One main component consists of all the ν_e that follow from the deleptonization (these are the processes described in equation (13.7)). As soon as the shock breaks through the neutrino sphere, they are all released within one pulse of about 10 ms. Calculations have shown that these carry away about 5% of the total energy released in the collapse.

However, by far the largest number of neutrinos comes from the thermal cooling of the produced proto-neutron star, through processes of the kind

$$\gamma \rightarrow e^+ e^- \rightarrow \nu_i \bar{\nu}_i \qquad i = e, \mu, \tau \qquad (13.18)$$

where neutrinos and anti-neutrinos of all flavours are produced in a time range of several seconds (the production of ν_μ, ν_τ can only occur via neutral currents). Detailed simulations of this can be found, for example, in [Bur86, Bru87,

May87]. Figure 13.12 shows schematically the time development of the different neutrino luminosities and figure 13.13 shows a realistic calculation.

While the mean neutrino energy, the mean neutrino luminosity, and the total radiated energy depend only on the initial mass of the 'Fe' core, the most important property which is heavily dependent on the explosion mechanism is the *time structure* of the luminosity (see [May87]).

Neutrinos carry away the dominant part, about 99%, of the energy released in a supernova explosion.

13.3   Detection methods for supernova neutrinos

The detection in water Cherenkov detectors occurs essentially via three reactions:

$$\bar{\nu}_e + p \rightarrow n + e^+ \tag{13.19}$$

$$\nu_e + {}^{16}O \rightarrow {}^{16}F + e^- \tag{13.20}$$

$$\bar{\nu}_e + {}^{16}O \rightarrow {}^{16}N + e^+ \tag{13.21}$$

$$\nu_i(\bar{\nu}_i) + e^- \rightarrow \nu_i(\bar{\nu}_i) + e^- \quad (i = e, \mu, \tau). \tag{13.22}$$

The second of these reactions has a threshold of 13 MeV and the cross section in the interesting energy region is two orders of magnitude smaller than that of the first reaction (see e.g. [Sch90a]) and can therefore be ignored. The first reaction has a threshold of 1.8 MeV. Under realistic assumptions the ratio of anti-neutrino to neutrino reaction rates [Hax87, Per88] can be calculated as

$$\frac{N(\bar{\nu}p \rightarrow ne^+)}{N(\nu e \rightarrow \nu e)} = 9.26 \left(\frac{E}{10\,\text{MeV}} \right). \tag{13.23}$$

This means that one neutrino is detected for every 10 anti-neutrinos. Therefore it was mainly $\bar{\nu}_e$ which was detected in the supernova 1987a which is discussed in the next section.

13.4   Supernova 1987a

One of the most important astronomical events of this century must have been the supernova 1987a [Arn89, Hen92, Che92, Kos92, Kir97, Woo97] (see figure 13.14). This was the brightest supernova since Kepler's supernova in 1604 and provided astrophysicists with a deluge of new data and insights, as it was for the first time possible to observe it at all wavelengths. Furthermore it will be possible for the first time to examine a supernova from the explosion to the late phases of the interaction with the interstellar medium. However, this is not all, since for the first time neutrinos could be observed from this spectacular event. This first detection of neutrinos which do not originate from the Sun, marked for many the birth of neutrino astrophysics. Further details can be found e.g. in [Arn89, Bah89, Che92, Kos92].

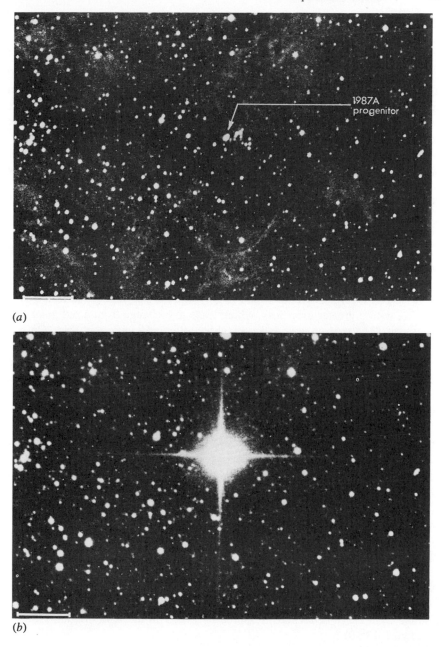

Figure 13.14. The supernova 1987a. (*a*) The large Magellanic cloud before the supernova on the December 9, 1977. The precursor star Sanduleak −69° 202 is shown. (*b*) The same field of view as (*a*) on 26 February 1987 at $1^h\ 25^{min}$ where the $4^m.4$ brightness Supernova 1987a can be seen. The length of the horizontal scale is 1 arcminute (from [Büh87] and with kind permission of ESO).

13.4.1 Characteristics of supernova 1987a

13.4.1.1 Properties of the progenitor star and the event

Supernova 1987a was discovered on 23rd February 1987 at a distance of 150 000
light years (corresponding to 50 kpc) in the large Magellanic cloud (LMC), a
companion galaxy of our own Milky Way [McN87]. It had several characteristics
which would necessitate a modification of the simple classification scheme for
supernovae. The evidence that the supernova was of type II, an exploding star,
was confirmed by the detection of hydrogen lines in the spectrum. However, the
identification of the progenitor star Sanduleak $-69°$ 202 was a surprise since
it was a blue B3I supergiant with a mass of about $20M_\odot$. Until then it was
assumed that only red giants could explode.

   A comparison of the bolometric brightness of the supernova in February
1992 of $L = 1 \times 10^{37}$ erg s^{-1} shows that this star really did explode. This is more
than one order of magnitude less than the value of $L \approx 4 \times 10^{38}$ erg s^{-1}, which
was measured *before* 1987, i.e. the original star has vanished. The explosion of
a massive blue supergiant could be explained by the smaller 'metal' abundance
in the large Magellanic cloud, which is only one third of that found in the Sun,
together with a greater mass loss, which leads to a change from a red giant to a
blue giant. The oxygen abundance plays a particularly important role. On one
hand, the oxygen is relevant for the opacity of a star, and on the other hand less
oxygen results in less efficient catalysis of the CNO process, which results in a
lower energy production rate in this cycle. Indeed it is possible to show with
computer simulations that blue stars can also explode in this way [Arn91, Lan91].
A schematic illustration of the collapse is shown in figure 13.15. The supernova
1987a also went through a red giant phase but developed back into a blue giant
about 20 000 years ago. The large mass ejection in this process was recently
discovered by the Hubble space telescope, as a ring around the supernova (see
figure 13.16). The course of evolution is shown in figure 13.17. The relatively
low luminosity in relation to other type II supernovae points to a compact star
(only about 50 solar diameters), as much of the initial energy was transformed
into kinetic energy of the outer shell, instead of into light.

   The total explosive energy amounted to $(1.4 \pm 0.6) \times 10^{51}$ erg [Che92].
The development of the bolometric light curve is shown in figure 13.18. It
is initially determined by the interaction of the escaping shock wave with the
surrounding matter, and after about four weeks after explosion is dominated by
the radioactive decay of the isotopes created in the explosion. Of interest here is
also the light-curve behaviour measured by ROSAT in the X-ray region [Has96]
(see figure 13.19).

13.4.1.2 γ radiation

γ radiation could also be observed for the first time. It seems that the doubly
magic nucleus ^{56}Ni is mostly produced in the explosion. This has the following

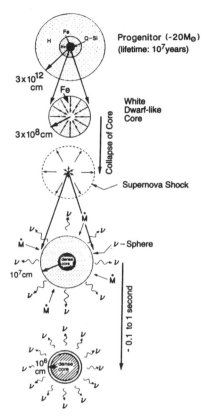

Figure 13.15. Schematic development of the processes in the core of supernova 1987a (from [Bur92]).

decay chain:

$$^{56}\text{Ni} \xrightarrow{\beta^+} {}^{56}\text{Co} \xrightarrow{\beta^+} {}^{56}\text{Fe}^* \rightarrow {}^{56}\text{Fe}. \tag{13.24}$$

^{56}Co decays with a half-life of 77.1 days, which is very compatible with the decrease of the light curve [Che92]. Two gamma lines at 847 and 1288 keV were detected by the solar maximum satellite (SMM) at the end of August 1987 [Mat88], which are characteristic lines of the ^{56}Co decay. From the intensity of the lines the amount of ^{56}Fe produced in the explosion can be estimated as about $0.075 M_\odot$. In addition, these lines appeared much earlier (about 200 days after the explosion) than expected (about 600 days after the explosion), which points to a strong mixing of the supernova after or during the explosion. Furthermore, the line profiles of the γ lines indicate that there is an asymmetrical distribution of the emission source, which implies a corresponding mass asymmetry either in the initial shell or in the core explosion. The line profile mirrors the velocity distribution of the supernova, as its Doppler broadening is produced by the

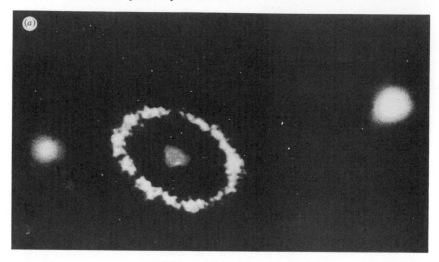

Figure 13.16. (*a*) False colour photograph of supernova 1987a taken by the Hubble space telescope in 1992. A ring of matter surrounds the debris from the supernovae (inner part). The ring is assumed to be a relic of the expelled outer layers of the progenitor red giant star. It has a diameter of about 1.4 light years (from [Cha92]). (*b*) A Hubble space telescope picture of three rings of glowing gas encircling the site of Supernova 1987a, taken in February 1994 in visible light. The small bright ring first identified in August 1990 (see (*a*) lies in the plane containing the supernova, the two larger rings lie in front and behind it. The rings are a surprise because the astronomers expected to see an hourglass shaped bubble of gas being blown into space by the supernova progenitor star. One possibility is that the two rings might be 'painted' on the invisible hour glass by a high-energy beam of radiation that is sweeping across the gas. The source of the radiation might be a previously unknown stellar remnant of the progenitor of SN1987a (with kind permission of C Burrows, ESA, STScI and NASA)

velocity dispersion of the radio-nuclides in the expanding shell. The observed fluxes of these lines [Tue90] can be made to agree with the model calculations assuming mixing (the model assumes: $16M_\odot$ star, $10M_\odot$ ejected as the shell of a blue supergiant, $6M_\odot$ remaining as He core) [Pin88] although significant differences remain. This and the latest observations using COMPTEL on GRO (see chapter 8) of ^{56}Co in a supernova of type Ia (SN 1991T) at a distance of 13 Mpc [Mor95], show the future potential of γ-astronomy in research into models of supernovae and for the analysis of supernova structure.

The measurement of the light curve can also provide other information. After about 1000 days it can already be seen that it deviates from the clear exponential decay of ^{56}Co [Sun92]. Other radioactive isotopes, such as primarily ^{57}Co (lifetime 392 days) or ^{44}Ti (lifetime about 80 years) now play a dominant role. The 122 keV line of ^{57}Co, which was created through neutron capture on ^{56}Co, has been observed by OSSE on GRO [Kur92]. In addition, the light curve

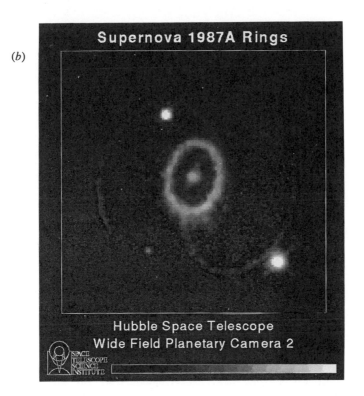

(b)

Figure 13.16. Continued.

seems to imply a deviation from expectations as according to nucleosynthesis calculations a ratio of $^{57}Co/^{56}Co$ of roughly 1.5–2.5 times that of the solar $^{57}Fe/^{56}Fe$ ratio is expected. However, the light curve can be adequately explained with a ratio of about 5, although the statistical significance is still small. It is hoped that an indication of the existence of a pulsar can be obtained from the behaviour of the light curve. The pulsar could remain as a remnant from the core of the supernova and could contribute considerably to the luminosity by heating of the surrounding material. However, it seems that radioactive sources will remain dominant for some years to come.

A direct search for a pulsar at the centre of SN 1987a with the Hubble space telescope has still been unsuccessful [Per95a]. From the duration of the neutrino pulse (see section 13.4.2) it is clear that it survived at least 12.5 s [Bur95]. This could indicate the rapid collapse of the created pulsar into a black hole. On the other hand, either the neutron star could have been formed without the strong magnetic field which is necessary to produce pulsar radiation, or the

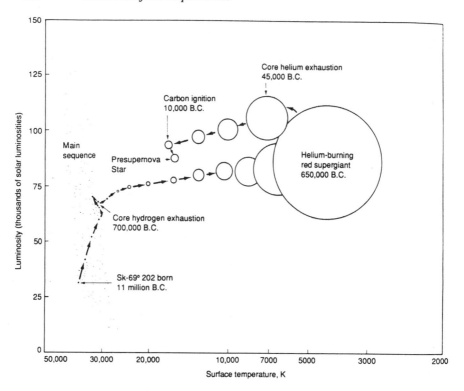

Figure 13.17. According to the theoretical model of S Woosley *et al* [Woo88a], Sanduleak−62°202 was probably born some 11 million years ago, with a mass about 18 times that of the Sun. Its initial size predetermined its future life, which is mapped in this diagram showing the luminosity against surface temperature at various stages, until the moment immediately before the supernova explosion. Once the star had burned all the hydrogen at its centre, its outer layers expanded and cooled until it became a red supergiant, on the right of the diagram. At that stage helium started burning in the core to form carbon, and by the time the supply of helium at the centre was exhausted, the envelope contracted and the star became smaller and hotter, turning into a blue supergiant, (with kind permission from T Weaver, S Woosley and J Maduell.)

narrow cone of radiation during the rotation could be pointing away from the Earth. The evidence for a pulsar in SN 1987a would be interesting in so far as it has never been possible to observe a pulsar and supernova directly from the same event.

13.4.1.3 Distance

The determination of the distance of the supernova has some interesting aspects. The Hubble telescope discovered a ring of diameter (1.66 ± 0.03)" around SN

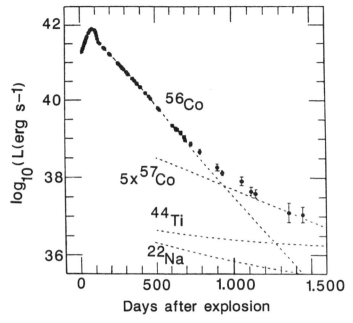

Figure 13.18. Behaviour of the light curve of supernova 1987a. After a long period which followed exactly the half-life of the decay of ^{56}Co, clear deviations can now be seen. Other isotopes, together with the possible pulsar, are now important in determining the behaviour of the light curve (from [Sun92], see also [Che92]).

1987a in the UV region in a forbidden line of doubly ionized oxygen [Pan91] (see figure 13.16(a)). Using the permanent observations of UV lines from the International Ultraviolet Explorer (IUE), in which these lines also appeared, the ring could be established as the origin of these UV lines. The diameter was determined to be $(1.27 \pm 0.07) \times 10^{18}$ cm, from which the distance to SN 1987a can be established as $d = (51.2 \pm 3.1)$ kpc [Pan91]. Correcting to the centre of mass of the LMC leads to a value of $d = (50.1 \pm 3.1)$ kpc [Pan91]. This value is not only in good agreement with those of other methods, but has a relatively small error, which makes the use of this method for distance measurements at similar events in the future very attractive.

13.4.1.4 Summary

In the above we have discussed only a small part of the observations and details of SN 1987a. Many more will follow, mainly concerned with the increasing interaction of the ejected layers with the interstellar medium. The detection of the neutron star (pulsar) created in the supernova, or even a black hole, are eagerly awaited. Even though supernova 1987a has provided a huge amount of new information and observations (e.g. light echoes, i.e. optical reflections of

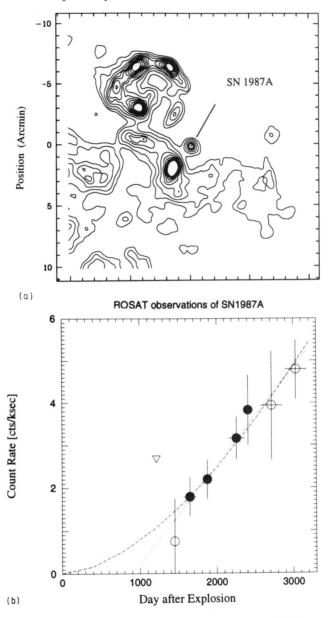

Figure 13.19. (*a*) X-ray image of the region of the SN 1987a with the PSPC detector on the ROSAT mission. (*b*) Time evolution of the X-ray emission of SN 1987a. The data points marked with filled cycles correspond to PSPC observations on ROSAT (energies 0.5 − 2 keV). The dashed line indicates a quadratic fit to the data. A linear increase starting on approximately day 900 can fit the data as well (dashed line) (from [Has96a]).

the explosion on surrounding matter, etc.) in all regions of the spectrum, the most exciting event was, however, the first detection of neutrinos from the star's collapse, which we now discuss.

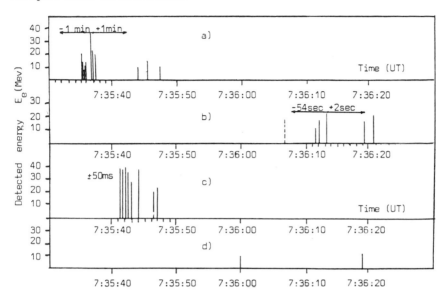

Figure 13.20. Time and energy spectrum of the four detectors which saw neutrinos from SN1987a as mentioned in the text. (*a*) The Kamiokande detector, (*b*) the Baksan detector, (*c*) the IMB detector, (*d*) the Mont Blanc detector (although this had no events at the time seen by the other experiments) (see text) (from [Ale87a, Ale88]).

13.4.2 Neutrinos from SN 1987a

A total of four detectors claim to have seen neutrinos from SN 1987a [Agl87, Ale87, Ale88, Bio87, Hir87, Hir88, Bra88a]. Two of these are water Cerenkov detectors (Kamiokande, Irvine–Michigan–Brookhaven (IMB) detector) and two are liquid scintillator detectors (Baksan and Mont Blanc). The Cerenkov detectors have a far larger amount of target material. All the detectors are sensitive to the reaction

$$\bar{\nu}_e + p \rightarrow n + e^+. \tag{13.25}$$

The observed events are listed in table 13.3. Within a certain timing uncertainty, three of the experiments agree on the arrival time of the neutrino pulse, while the Mont Blanc experiment detected them about 4.5 hours before the other detectors. Since its five events all lie very close to the trigger threshold of 5 MeV, and as the larger Cerenkov detectors saw nothing at that time, it is generally thought that these events are a statistical fluctuation and are not related to the supernova signal. The other three experiments also detected the neutrinos before the optical

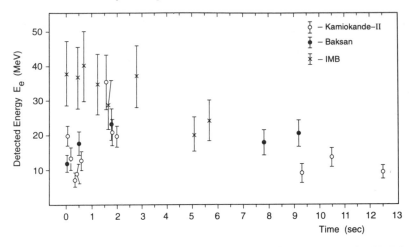

Figure 13.21. Energies of all neutrino events detected at 7.35 UT on the 23 February 1987 versus time: The time of the first event in each detector has been set to $t = 0.0$ (from [Ale88, Ale88a]).

signal arrived, which is after all to be expected. The relatively short time of a few hours between neutrino detection and optical discovery points to a compact star, as we have already hinted. The time structure and energy distribution of the neutrinos is shown in figure 13.20. If *assuming* the first neutrino events have been seen by each detector at the same time, i.e. setting the arrived times of the first event to $t = 0$ for each detector, then within 12 seconds, 24 events were observed (Kamiokande + IMB + Baksan). The number of analyses, however, far exceeds this number! Figure 13.21 shows the spectrum obtained under this assumption. It should be compared with calculations as shown in figure 13.13. We first consider the Kamiokande data. Here 11 events were detected within 12 s [Hir88]. Assuming that 8 anti-neutrino events arrived in about 2 seconds (the first two of them actually might have been neutrinos according to their angular distributions [Hir87]) allows the estimation of the integrated $\bar{\nu}_e$ flux to be about $0.5 \times 10^{10} \bar{\nu}_e \, \mathrm{cm}^{-2}\mathrm{s}^{-1}$. Assuming a mean energy of 15 MeV per neutrino an energy release in the form of $\bar{\nu}_e$ of about 8×10^{52} erg is obtained. This is in good agreement with the binding energy of a neutron star (about 5×10^{53} erg), given that the $\bar{\nu}_e$ form around one-sixth of the total neutrino flux from a supernova. This observation also for the first time experimentally verified that indeed more than 90% of the total energy released is carried away by neutrinos, and that the visible signal corresponds to only a minute fraction of the energy. The results from IMB and Baksan lead to similar conclusions, so that the total energy in the form of all neutrinos is estimated to be [Gol88, Ale88]

$$E_{\nu_{\mathrm{tot}}} \approx 3 \times 10^{53} \text{ erg.} \tag{13.26}$$

The observed neutrinos confirm to a large extent general features of supernovae models. The statistics are, however, not large enough to draw detailed conclusions on the explosion mechanism. The analysis is complicated further by the fact that the temporal scattering of the events could in addition to the time scale of the neutrino emission be affected by a dispersion resulting from a finite neutrino mass, and by neutrino oscillations. For further discussion of neutrinos from SN 1987a we refer to [Sch90b, Kos92, Raf96].

13.4.3 Neutrino properties from supernova 1987a

Some interesting conclusions for particle physics can be drawn from the fact that practically all neutrinos were detected within 12 s.

13.4.3.1 Lifetime

The first point relates to the lifetime of neutrinos. As the expected flux of anti-neutrinos has been measured on Earth, no significant number could have decayed in transit, which leads to a lifetime for the $\bar{\nu}_e$ of [Moh91]

$$\left(\frac{E_\nu}{m_\nu}\right)\tau_{\bar{\nu}_e} \geq 5 \times 10^{12} \text{ s.} \tag{13.27}$$

The interesting aspect of this result, beside the determination of the neutrino lifetime, is that assuming $\tau_{\nu_e} = \tau_{\bar{\nu}_e}$, neutrino decay is ruled out as a solution to the solar neutrino problem. One solution to this which has been discussed is the decay into a Majoron χ (see chapter 2) [Bah86], which however requires that

$$\left(\frac{E_\nu}{m_\nu}\right)\tau_{\nu_e} < 500 \text{ s} \tag{13.28}$$

if it is to solve the solar neutrino problem.

In particular the radiative decay channel for a heavy neutrino ν_H:

$$\nu_H \to \nu_L + \gamma \tag{13.29}$$

can be limited independently. No signal was detected in the gamma spectrum measured by the Solar Maximum Satellite (SMM) [Chu89]. The photons emanating from neutrino decay would arrive with a certain delay, as the parent heavy neutrinos do not travel at the speed of light. The delay is given by

$$\Delta t \simeq \frac{1}{2} D \frac{m_\nu^2}{E_\nu^2}. \tag{13.30}$$

For neutrinos with a mean energy of 12 MeV and a mass smaller than 20 eV the delay is about 10 s, which should be reflected in the arrival time of any photons

Table 13.3. Table of the neutrino events registered by the four neutrino detectors Kamiokande II [Hir87], IMB [Bio87], Mont Blanc [Agl87] and Baksan [Ale87a, Ale88]. T gives the time of the event, E gives the visible energy of the electron (positron). The absolute uncertainties in the given times are: for Kamiokande $\pm$ 1 minute, for IMB 50 ms and for Baksan -54 s, $+2$ s.

Detector	Event number	T (UT)	E (MeV)
Kamioka	1	7 : 35 : 35.000	20 ± 2.9
	2	7 : 35 : 35.107	13.5 ± 3.2
	3	7 : 35 : 35.303	7.5 ± 2.0
	4	7 : 35 : 35.324	9.2 ± 2.7
	5	7 : 35 : 35.507	12.8 ± 2.9
	(6)	7 : 35 : 35.686	6.3 ± 1.7
	7	7 : 35 : 36.541	35.4 ± 8.0
	8	7 : 35 : 36.728	21.0 ± 4.2
	9	7 : 35 : 36.915	19.8 ± 3.2
	10	7 : 35 : 44.219	8.6 ± 2.7
	11	7 : 35 : 45.433	13.0 ± 2.6
	12	7 : 35 : 47.439	8.9 ± 1.9
IMB	1	7 : 35 : 41.37	38 ± 9.5
	2	7 : 35 : 41.79	37 ± 9.3
	3	7 : 35 : 42.02	40 ± 10
	4	7 : 35 : 42.52	35 ± 8.8
	5	7 : 35 : 42.94	29 ± 7.3
	6	7 : 35 : 44.06	37 ± 9.3
	7	7 : 35 : 46.38	20 ± 5.0
	8	7 : 35 : 46.96	24 ± 6.0
Baksan	1	7 : 36 : 11.818	12 ± 2.4
	2	7 : 36 : 12.253	18 ± 3.6
	3	7 : 36 : 13.528	23.3 ± 4.7
	4	7 : 36 : 19.505	17 ± 3.4
	5	7 : 36 : 20.917	20.1 ± 4.0
Mt Blanc	1	2 : 52 : 36.79	7 ± 1.4
	2	2 : 52 : 40.65	8 ± 1.6
	3	2 : 52 : 41.01	11 ± 2.2
	4	2 : 52 : 42.70	7 ± 1.4
	5	2 : 52 : 43.80	9 ± 1.8

from the decay. A limit for this decay of neutrinos in the mass region 20–100 eV of [Kol89, Moh91]

$$\frac{\tau_\nu}{B_\gamma} \geq 3.4 \times 10^{16} N_{\mathrm{GRS}}^{-1/2} \text{ s} \qquad \text{for} \quad \tau_\nu \gg Dm_\nu/E_\nu \qquad (13.31)$$

$$B_\gamma \leq 1.4 \times 10^{-11} N_{\mathrm{GRS}}^{1/2} \left(\frac{m_\nu}{1 \text{ eV}}\right) \qquad \text{for} \quad \tau_\nu \ll Dm_\nu/E_\nu \qquad (13.32)$$

can be obtained from the SMM data [Moh91]. Here B_γ is the branching ratio of a heavy neutrino into the radiative decay channel, $N_{\mathrm{GRS}}^{-1/2}$ is the instrumental background of the gamma spectrometer, and D is the distance of the supernova. Thus there is no hint of a neutrino decay, and therefore of possible neutrino masses.

13.4.3.2 Mass

Direct information about the mass is obtained from the observed spread in propagation time (see the discussion in [Gro89, Gro90, Kla95]). Using $E = mc^2$ and $p = mv$, the propagation time of a neutrino with mass m and energy E from its point of production is given by

$$t_{\mathrm{obs}} - t_{\mathrm{em}} = t_0 \left(1 + \frac{m^2}{2E^2}\right) \qquad (13.33)$$

where t_{obs} is the time of observation, t_{em} the time of emission and $t_0 = 5.3 \times 10^{12}$ s is the propagation time of the light. Hence the time difference between two events is

$$\Delta t_{\mathrm{obs}} - \Delta t_{\mathrm{em}} \approx \frac{t_0 m^2}{2} \left(\frac{1}{E_1^2} - \frac{1}{E_2^2}\right). \qquad (13.34)$$

Depending on which events arc combined and assuming simultaneous emission (or emission within a short time of up to 4 s), mass limits of a maximum of 30 eV (or even somewhat smaller at the price of model dependence) are obtained (see figure 13.22) [Arn87, Kol87]. These limits were as good as the laboratory limits at that time, which had been obtained after many years of research!

13.4.3.3 Magnetic moment and electric charge

A quantity which is also of interest to the solution of the solar neutrino problem is the magnetic moment of the neutrino [Bar88]. If a magnetic moment existed, precession could convert left-handed neutrinos into right-handed ones due to the strong magnetic field of the neutron star and the long journey through the galactic magnetic field. Such right-handed neutrinos would then be sterile and would escape detection, thereby forming an additional loss mechanism. In this case also, the agreement of the observation with the expected number of neutrinos implies an upper limit of $\mu_\nu < 10^{-12} \mu_B$ [Lat88, Bar88]. This value is certainly

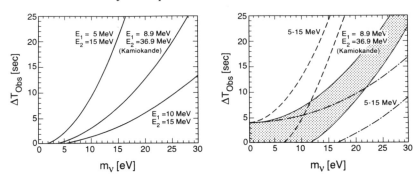

Figure 13.22. Neutrino masses deduced from SN 1987a exploiting the flight time and energy differences. (*a*) Neutrino masses from SN 1987a as a function of the time interval ΔT_{obs} between the observation of *simultaneously* emitted neutrinos of different energies E_1, E_2 on Earth. Within a time interval of $\Delta T_{obs} = 12.349$ s (1.915 s) 11 (8) neutrinos between 8.9 and 36.9 MeV energy were observed in the Kamiokande experiment. (*b*) Neutrino masses from SN 1987a as a function of ΔT_{obs} on the assumption that neutrinos were emitted within a time interval of $\Delta t = 4$ s. For every value of ΔT_{obs} we now have an area of possible m_ν between two related curves. For $\Delta T_{obs} = 1.915$ s, for example, m_ν must be less than 13.5 eV (from [Gro89, Gro90, Kla95]).

too small to solve the solar neutrino problem. However, many assumptions have been made in arriving at this conclusion, such as those of isotropy and equally strong emission of different flavours, as well as the magnetic field of the possibly created pulsar, so that this limit has to be treated with some caution. A further limit on the electric charge of the neutrino follows:

$$Q_\nu < 10^{-17} e \tag{13.35}$$

(see e.g. [Bah89, Moh91]). This is based on the fact that low energy neutrinos with charge would be influenced more strongly by the galactic magnetic field, and therefore had to travel a larger distance.

13.4.3.4 Equivalence principle

A test of the equivalence principle is also possible. Due to the observation of photons and the neutrinos within a few hours of each other, an upper limit can be placed on any new force which couples to neutrinos with a strength g and extends beyond galactic sizes, of [Moh91]

$$g < 5 \times 10^{-3} G \tag{13.36}$$

13.4.3.5 Emission of unknown particles

The emission of unknown particles is also restricted by these observations [Raf88, Raf96], for example, the effects on axions are discussed in chapter 11.

13.4.3.6   Conclusion

Our knowledge of supernova explosions has grown enormously in recent years because of supernova 1987a, not least from the confirmation for the first time that supernovae type II really are phenomena from the late phase in the evolution of massive stars, and that the energy released corresponds to the expectations. Also, the first detection of neutrinos has to be rated as a particularly remarkable event. As to how far the observed data is specific to supernova 1987a, and to what extent they have general validity, only further supernovae of this kind can show. Supernova 1987a initiated a great deal of experimental activity in the field of detectors for supernova neutrinos, so that we now turn to the prospects for future experiments. Besides describing some of the experiments themselves, we also have to consider the likelihood of supernovae occurring.

13.5   Supernova rates and future experiments

Over 860 supernovae have been observed to date [Ber94]. Although this number seems to be very large, many questions still remain unanswered. Less than 10% have been found in systematic searches. We define a mean supernova rate [Ber91a]

$$\omega(y^{-1}) = \frac{N_{SN}}{N_{Gal}\Delta t} \tag{13.37}$$

where N_{SN} is the number of supernovae that have been observed in a spot check of N galaxies N_{Gal} during an observation time Δt. The observation time is a function of the galaxy distance, as supernovae in far distant galaxies have to spend a different period of time above the defined brightness limit in order to qualify as supernovae and therefore be confirmed. In addition, the rate also depends on the type of galaxy, as well as on its brightness. Defined galaxies are kept under automatic observation by several observatories. This has resulted in an increase of the annual discovery rate from 20 to about 60. As we are primarily interested in the detection of neutrinos, we restrict ourselves to a discussion of the supernova rate in our own galaxy [Ber91a]. The simplest estimate rests on the historical fact that during the last millennium three supernovae (the Crab, 3C 58 and Tycho) have been observed between 100° and 260° of the galactic length. This results in a rate of 0.30 ± 0.17 per century in the direction of the anti-centre (the point opposite to the galactic centre in the sky). If the existing data from systematic search for supernovae is used, a rate of supernovae type II within 4 kpc of about 2.3 $(H_0/75)^2$ per thousand years is obtained [Ber94]. This agrees with the observation that in the last 2000 years 4 supernovae (SN 185, SN 1054, SN 1181 and SN 1670) have taken place within 4 kpc. Recent calculations of supernova rates using model galaxies can reproduce this [Tim95].

   Using a quantity known from radio astronomy, namely that there are 46 supernova remnants in our galaxy above a certain flux strength, and assuming that the probability for an explosion is the same throughout the entire galaxy, a rate of 3.4 ± 2.0 per century is obtained. Other estimates using the observed pulsar abundance also result in rates of the order of 1–3 supernovae per century [Ber91a]. Rather more theoretical estimates follow from stellar evolution models. Using a particular mass distribution for population I stars and their calculated lifetimes, as well as the global distribution within the Milky Way, it can be shown that the death rate for stars with $M > 8 M_\odot$ within 3 kpc from the Sun contributes about 0.3 per thousand years [Ber91a]. This seems at first glance to contradict the observed historical events, but in view of the small numbers and the general uncertainties we can conclude that between about 0.5 and 3 supernovae should occur in our galaxy per century.

A rather interesting future candidate is the red giant Betelgeuse (α Ori) in the constellation Orion (see figure 13.23).

Figure 13.23. A good candidate for a future supernova in our galaxy is Betelgeuse ($\rightarrow$) in the constellation Orion. A red giant with a mass of more than $20 M_\odot$, it fulfils all the criteria for a supernova. Because of its small distance of only 310 light years, its brightness as a supernova would be greater than that of the full moon (from [Hay95])!

Table 13.4 Existing and forthcoming detectors of galactic supernova neutrinos (from [Cli92].)

Detector	Process					
	$\bar{\nu}_e p \to e^+ n$ $\bar{\nu}_e d \to ppe$	$\bar{\nu}_e e \to \bar{\nu}_e e$	$\bar{\nu}_x e \to \bar{\nu}_x e$ $x = \mu, \tau$	$\nu_e N \to N^* \nu_x$ $\hookrightarrow n$	$\nu_\mu N \to N^* \nu_x$ $\hookrightarrow n$	ν_e prompt
ICARUS (3 kT)	—	~ 140	25	—	—	4[a]
SNO (1 kT)	~ 500	60	20	~ 200	~ 400	5[a]
(D$_2$O+H$_2$O)	(H$_2$O shield + D$_2$O)					
LVD/MACRO (3 kT)	~ 1000	—	—	—	—	$5 - 20^a$
Kamiokande II/IMB	(~ 480)	(~ 60)	(~ 20)	—	—	—
Superkamiokande (30 kT) H$_2$O	~ 4000	~ 600	200	—	—	$\sim 5^a$
SNBO (100 kT)	100's			~ 100's	10000	—
Remarks	measures t_ν $E_\nu \sim E_e$ no direction	t_ν E_ν from E_e measures Θ_ν	E_ν from E_e measures Θ_ν	only t_ν no E_ν no Θ_ν	only t_ν no E_ν no Θ_ν	$\Delta t_\nu \simeq 10$ ms

[a] Dependent on the energy spectra of the prompt ν_e and the detector threshold.

Table 13.5 Overview of the schedule of laser interferometers for gravtiational wave search. All institutes concerned have already collected experience from prototypes in the seventies and eighties. The two US projects should allow detection of gravitational waves with high probability. Originally GEO should also have had a 3 km long arm (from [Auf97]).

Project	LIGO		VIRGO		GEO600		TAMA300	
Country	USA	USA	France	Italy	Germany	UK	Japan	Japan
Institute	MIT	Caltech	CNRS	INFN	MPQ	Glasgow	ISAS	NAO
Start of planning	1982	1984	1986	1986	1985	1986	1987	1994
Armlength	4 km	4 km	3 km		600 m		300 m	
Location	Hanford (WA)	Livingstone (LA)	Pisa Italy		Hannover Germany		Mitaka Japan	

We now consider experiments for the detection of future supernova neutrinos [Bur92, Cli92]. Much essential information will be gained from the solar neutrino detectors discussed in chapter 12. The most important are the SNO detector, due to its use of heavy water, and SuperKamiokande, which with its enormous size is expected to detect about 5000 events from a supernova 10 kpc away. Another detector is the Large Volume Detector (LVD) in the Gran Sasso underground laboratory [Gal94, Gal96]. This detector in its complete version consists of 190 large steel tanks, each of which contains 8 smaller tanks containing scintillator material. When complete it will contain 1800 tons of scintillator material, arranged in 5 towers. The detection reactions are

$$\bar{\nu}_e + p \rightarrow n + e^+ \tag{13.38}$$

$$\nu_e + e^- \rightarrow \nu_e + e^- \tag{13.39}$$

$$\nu_e + {}^{12}\text{C} \rightarrow {}^{12}\text{C}^* + \nu_e \tag{13.40}$$

$$\nu_e + {}^{12}\text{C} \rightarrow {}^{12}\text{N} + e^- \tag{13.41}$$

$$\nu_e + {}^{12}\text{C} \rightarrow {}^{12}\text{B} + e^+. \tag{13.42}$$

The most frequent reaction by far will be the inverse β-decay. The first tower of this detector has been in operation since the summer of 1992 and the second since 1994. Also detectors like the AMANDA and the BAIKAL lake detector (see section 8.4) will contribute to supernovae detection.

Table 13.4 lists some of the planned detectors and those currently under construction or already in operation which will be used for the detection of supernova neutrinos. The number is impressive and will, at the time of the next supernova, provide much new information.

A completely different way to detect supernovae utilizes the vibrations of space-time due to the powerful explosion. This appears in the form of gravitational waves. Up to now only indirect evidence exists: the slow down of the period of the pulsar PSR1913+16 can be described exactly by assuming the emission of gravitational waves [Tay94]. For direct detection several detectors in the form of laser interferometers (e.g. VIRGO [Abr92], LIGO [Bra90]), GEO 600 and others [Auf97] which search for the small vibrations of space time of which supernovae are a good source, are currently being constructed (see table 13.5 and [Tho95, Auf97]). The ultimate goal would be a laser interferometer in space. Such a project (LISA) is proposed and may become a cornerstone project of ESA in the next century. In spite of their sensitivity, unfortunately even these detectors are still far from being able to see an expected stochastic graviton background of primordial origin (relic gravity waves from an early pre-big-bang phase typical of string cosmology models) [Gri75, Gas97] or gravitational waves excited during inflation [Tur97]. For a recent review see [Sch97].

Chapter 14

The creation of heavy elements

14.1 Introduction

The synthesis of elements in the universe occurs mainly in three different ways (see e.g. [Bur57, Cla68, Rol88, Gro89, Cop95]). First and foremost is primordial nucleosynthesis, which is responsible for the creation of the lightest elements (see chapter 4). The second important process which creates elements is the process of nuclear fusion in stars. From this elements up to iron in the periodic table can be formed. However, iron has the largest binding energy per nucleon of any element (about 8 MeV/nucleon), so that heavier elements can no longer be produced by nuclear fusion. Since these heavier elements could also not have been created in sufficient amounts via charged particle reactions, due to the increased Coulomb barriers, other mechanisms must have been at work. According to the theory of Burbidge, Burbidge, Fowler and Hoyle (B^2FH) [Bur57], these mechanisms are neutron capture reactions and β-decay (see also [Kla85, Gro89]). Several double maxima can be seen in the experimentally determined abundances of the elements in the solar system, one component lying at around the magic neutron numbers of 50, 82 and 126 respectively, the other about 10–15 mass numbers below (see figure 14.1). These features must be explained by any theory, and are an expression of two different reaction mechanisms. Understanding the creation of the heavy elements is the key to determine the age of the universe via cosmic chronometers [Kla83, Kla86b]. The age of the universe is one of the boundary conditions in cosmological models, and therefore, as shown in chapter 5, is important in obtaining the value of the cosmological constant [Kla86a].

432

14.1.1 Neutron capture

The reaction rate λ for neutron capture at a neutron density n_n is given by

$$\lambda = n_n \langle \sigma(v)v \rangle \tag{14.1}$$

where the average thermal cross section is given by

$$\langle \sigma v \rangle(T) = \left(\frac{8}{m\pi} \right)^{1/2} (kT)^{-3/2} \int_0^\infty E\sigma(E)e^{-E/kT} dE. \tag{14.2}$$

In the case of neutron–nucleus reactions, the unknown cross sections in the energy area around $T \approx 10^9$ K lead to model dependent predictions. At high temperatures and neutron densities these can however be avoided by assuming equilibrium between (n, γ) and (γ, n) reactions. In this way a Saha equation for the abundances of two isotopes can be derived as

$$\frac{n(A+1, Z)}{n(A, Z)} = \frac{n_n}{2} \left(\frac{\hbar^2}{2\pi mkT} \right)^{3/2} \frac{G_{A+1}}{G_A} e^{-S_n/kT} \tag{14.3}$$

where G_{A+1}, G_A are the statistical weights and S_n the neutron separation energy. The neutrons created in a star thermalize very quickly and then obey a Maxwell–Boltzmann distribution. If we assume a most probable energy $E_0 = kT$ and furthermore that the process takes place in the helium burning shell of a star $(T = (0.1$–$0.6) \times 10^9$ K), the energy region of interest is at about 30 keV.

14.1.2 β-decay

By plotting all known isotopes in a diagram of A versus Z, it can be seen that there is a most stable nucleus for all isobars which has Z_0 protons, where

$$Z_0 = \frac{A}{1.98 + 0.015A^{2/3}}. \tag{14.4}$$

This defines a line of stability within the nuclidic chart (see figure 14.2). The further away an isotope is from this line, the shorter the half-life against β-decay. The region of very neutron-rich nuclei which are essential for element synthesis is at present experimentally unknown to a great extent, so that we are dependent on theoretical extrapolations [Kla86b, Gro89, Gro90, Kla91d, Sta92b]. The temperature dependence of the half-lives is particularly important for the production of light nuclei [Ful82a, Ful82b, Ful82c, Oda94].

Before turning to element synthesis beyond iron, we briefly discuss possible scenarios which provide an adequate source of neutrons, and sketch the production of light nuclei in explosive scenarios, particularly supernovae.

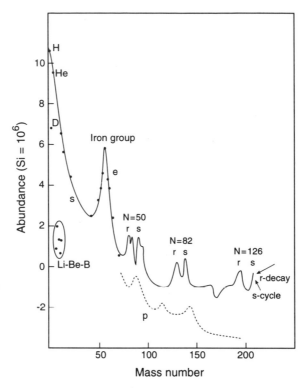

Figure 14.1. Simplified illustration of the observed solar (cosmic) element abundances as a function of the mass number A (normalized to Si $= 10^6$). The maxima close to the magic neutron numbers $N = 50, 82, 126$ can be attributed to the r- and s-processes. p denotes rare proton-rich nuclei, not produced in the main line of the r- and s-processes, but in the p-process (from [Bur57, Gro89, Gro90]).

14.2 Explosive scenarios and element synthesis up to iron

The quiet, hydrostatic burning phases of massive stars and the onion-like structure of stars which results have already been discussed (see chapter 13). This picture forms the starting point for explosive scenarios (e.g. a supernova explosion). The latter phenomena occur on time scales of typically less than 1 s. Above about 5×10^9 K the Coulomb barrier can be neglected and an equilibrium is quickly created in which the major product is ^{56}Ni. At temperatures higher than 10^{10} K photo-disintegration processes become important. In order to calculate these processes a precise knowledge of the cross sections is again required. Because of the higher energies, it is mostly possible to use the measured cross sections without extrapolation. The astrophysical aspects, on the other hand, are much more difficult. In order to carry out realistic calculations, a complete, three-dimensional hydrodynamic model would have, in principle,

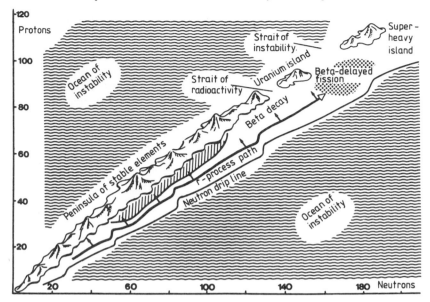

Figure 14.2. A schematic illustration of element synthesis in the r-process. The nuclides which are produced during 0.4 s in the supernova explosion are on the r-process path, far off from the peninsula of stability and near the neutron drip line, on which the neutron separation energy tends to 0, i.e. the nuclei are fully saturated with neutrons. In the β-decay back to the stability valley the uranium island is still reached; however, at higher masses the r-process path is broken off due to β delayed fission, which prevents the production of superheavy nuclei. The β decay characteristics of the nuclei between the r-process path and the β-stability line (about 6000 nuclides, of which the majority are not accessible in laboratory experiments) significantly effect the final distribution of stable nuclides. The shaded area indicates the residual heat released by the β-decay of the fission products when a nuclear reactor is switched off (see [Kla88] from [Gro89, Gro90]).

to be constructed, which so far has not proved possible. Considering the two types of supernova, there is a higher production of ^{56}Ni in type I since these are much lighter than type II. An estimate from the observed light curves results in about 0.2–1$M_\odot$ of ^{56}Ni from type I supernovae, in comparison to 0.07 $M_\odot$ for supernova 1987a. A comparison of the type I data with the observed values in the solar system shows a relative excess of elements of the iron group in the latter. Therefore, type I supernovae alone are inadequate to explain the observed abundances of heavy elements, which necessitates the consideration of type II supernovae. It this case it is the outgoing shock front, which provides high temperatures in the shells, for the silicon shell around 500 keV, in the O and Ne shells above 100 keV and in the H shell only about 10 keV. In this way explosive burning is reached only for the Si, Ne, O and He shells and the outer H shell scarcely takes part in explosive nucleosynthesis before it is blown away.

This also implies that within fractions of a second elements from Si to Fe are produced in the interior, while the abundances of light elements from O to Mg only moderately change [Wea80, Woo82, Woo86a, Woo86b, Thi90].

In the calculation of the abundances the so-called *mass cut* is very important. It marks the border between expelled matter and that which remains in the neutron star. Its position governs for example the size of the neutron excess in the innermost region of expelled matter. For details of the latter see e.g. [Hil78].

New model calculations on the galactic chemical evolution of the elements between hydrogen and zinc give a good description of the observed solar abundances [Tim95]. Sixty type II supernova models with varying masses and metallicity were used. Type Ia supernova contribute about 1/3 of the solar Fe abundance.

14.3 Element synthesis beyond iron

The above leads to the obvious question of how element synthesis beyond iron proceeds. The fundamental process is neutron capture in nuclei (A, Z)

$$(A, Z)(n, \gamma)(A + 1, Z). \tag{14.5}$$

The decisive factors are the number of available neutrons as well as the values of the β-half-lives of the created nuclei (see figures 14.2 and 14.3). For low neutron fluxes it is more likely that the created nucleus, if it is unstable, decays before another neutron is captured. To put it another way, the lifetime $\tau_{n\gamma}$ for neutron capture is much larger than the lifetime τ_β for β-decay, which is why the process is called the *s-process* (slow). In contrast, in very high neutron fluxes several neutron captures take place before the nuclei β-decay. This process is known as the *r-process* (rapid). In addition there is a third process, which is responsible for the creation of neutron depleted nuclei, e.g. via (p, n) or (γ, n) reactions (*p-process*) (see also [Mey94]).

14.3.1 The *s*-process

The β-half-lives of nuclei close to the valley of stability are typically of order days or years, which means that $\tau_{n\gamma}$ has to be correspondingly larger. An estimate of typical neutron densities can be obtained from a consideration of the typical reaction cross section at 30 keV of 0.1 barn (corresponding to 10^{-25} cm^2) and a neutron velocity $v = 3 \times 10^8$ cm s^{-1}. With $\tau_{n\gamma} N_n = 1/\langle \sigma v \rangle$ and for a lifetime for neutron capture of 10 years, this implies a neutron density of $N_n \approx 10^8$ cm^{-3} [Rol88]. As the β-decay dominates in this situation, the produced nuclei follow the course of the β-stability line and do not stray far from it. The *s*-process stops at ^{209}Bi, the last stable nucleus, due to the now very strong α-decay. Quantitative estimates of the expected abundance for an isotope N_A are given by [Cla68]

$$\frac{dN_A}{dt} = N_n(t)N_{A-1}(t)\langle\sigma v\rangle_{A-1} - N_n(t)N_A(t)\langle\sigma v\rangle_A - \lambda_\beta(t)N_A(t). \quad (14.6)$$

The first term describes the creation of the isotope through neutron capture. The remaining terms describe loss of the isotope either through further neutron capture or through β-decay. The neutron flux is proportional to the neutron density N_n. This is a complex network of differential equations which is not solvable analytically. It is also important to note that the variables are time dependent due to the temperature dependence of the individual processes. It is therefore common to assume a constant temperature in the s-process and to use the initial conditions $N_A(0) = N_{56}(0)$ for $A = 56$ and $N_A(0) = 0$ for $A > 56$, i.e. ^{56}Fe is used as the initial (or seed) nucleus. The solutions are then self regulating in the sense that they tend towards a local equilibrium

$$\sigma_A N_A = \sigma_{A-1} N_{A-1} = \text{constant.} \quad (14.7)$$

However, there is no equilibrium along the whole path between Fe and Bi. Although equilibrium is essentially maintained in the neighbourhood of the plateaux, it is not the case near full neutron shells. If the lifetimes for neutron capture and β-decay are roughly equal at a particular point, the s-process splits, i.e. further development can proceed via two alternative paths.

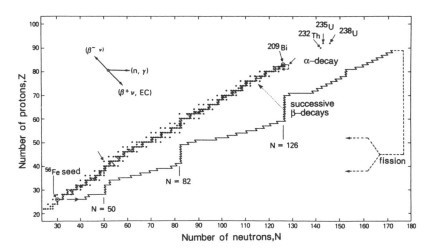

Figure 14.3. Illustration of the s- and r-process paths. Both processes are determined by (n, γ) reactions and β-decay. In the r-process the neutron rich nuclei decay back to the peninsula of stable elements by β-decay after the neutron density has fallen. Even the uranium island is still reached in this process. The s-process runs along the stability valley (from [Rol88]).

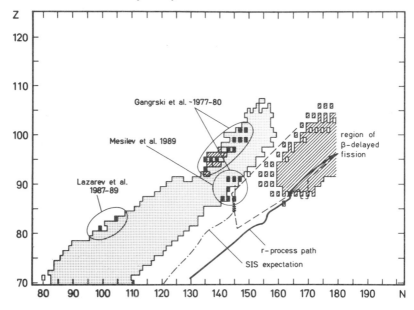

Figure 14.4. Part of the table of nuclides. The shaded area shows the regions where β-delayed fission is expected (on the neutron rich side the dashed area corresponds to the expectations of [Wen76], the dashed line shows the expectations of [Thi83]). The black squares show the parent nuclei of this process which have been experimentally examined to date. The dashed-dotted line shows the expectations for the experimental possibilities of SIS at GSI (Darmstadt). The *r*-process path is also shown (from [Sta92a]).

The observed element abundances indicate that there is an anti-correlation between the *s*-element abundances and the neutron capture cross section [Rol88]. This is to be expected since isotopes with large capture cross sections will transform quickly and therefore have small residual abundance. Examination of the cosmic *s*-process abundances indicates agreement with the theoretical models at about the 3% level [Käp91]. However, it must be remembered that some of the isotopes can also be produced via the *r*-process discussed below. The necessary neutron density can be determined quite precisely from the observed abundances as $N_n = (3.4 \pm 1.1) \times 10^8$ cm^{-3}. The production of isotopes is subdivided by the requirement for different levels of neutron flux into two main classes, from about zirconium to bismuth (main *s*-process) and from iron to yttrium (weak *s*-process) [Käp91]. The main process is characterized by the requirement for exposition to an exponential time distribution of neutrons. This leads us immediately to the question of the site of the *s*-process. Helium shell burning in intermediate mass ($3-6M_\odot$) red giants is generally assumed [Käp89, Käp91]. It is assumed that in such stars there is a CNO core surrounded by a convective shell in which hydrogen is burnt. The helium thus created sinks down to the core until at a sufficiently high pressure the thin helium shell ignites. This causes the star to expand, and thus extinguishes the hydrogen burning. Once the helium burning stops, the star contracts again and hydrogen burning restarts. This process repeats itself, with the helium burning shell lying further out each time and having a slight overlap with the previous shell. In this way regions can be subjected to repeated s-processes. The neutrons come from the reactions ^{22}Ne$(\alpha, n)^{25}$Mg [Ibe75] and ^{13}C$(\alpha, n)^{16}$O [Ibe82]. Furthermore, every cycle introduces new burning material into the helium shell by convective mixing, as well as transporting s-process elements to the surface. This picture is confirmed by the observation of technetium in a class of red giants, which can only be explained by the mixing from deeper layers. Another class of production sites are low mass stars on the asymptotic giant branch, which can also undergo thermal pulsed burning of helium [Käp89]. Here two neutron pulses occur of approximately 20 and 2.5 years duration. The first pulse comes from the ^{13}C$(\alpha, n)^{16}$O reaction, which after a break of about 15 years is followed by the shorter neutron pulse from the reaction ^{22}Ne$(\alpha, n)^{25}$Mg [Käp89].

The s-process produces isotopes with a very large range of half-lives, thereby enabling an examination of very different time scales. Just as particular branching points are suitable to determine the neutron flux, for example at the points $A = 147/148$ and $A = 185/186$, so technetium is suitable to investigate the transport to the surface. The isotope pair ^{187}Re/^{187}Os is of some interest in investigations of the age of our galaxy [Kir78, Yok83, Käp91] (but see also [Thi83, Gro89, Gro90]).

14.3.2 The p-process

From astrophysical considerations, the most plausible place for the creation of heavy *neutron depleted* isotopes ($Z \geq 34$), seems to be the interior of massive stars at a late stage of evolution. Such isotopes can be created there by (γ, n) photo-disintegration of neutron rich nuclei at temperatures of $(2\text{–}3.2) \times 10^9$ K. Two suitable places seem to be the O–Ne layer during a supernova type II explosion, and the hydrostatic oxygen burning in the pre-supernova phase [Pra89]. A s-process distribution created during helium burning is generally taken as the initial distribution of nuclei. Detailed calculations give agreement to within a factor of 3 with 60% of the p-process abundances observed in the solar system [Pra89].

14.3.3 The r-process

In order that several neutron captures are more likely than a β-decay, the lifetimes $\tau_{n,\gamma}$ must be of order 10^{-3} s. This requires very high neutron densities of order $N_n \approx 10^{20}$ cm^{-3}. Such densities can only be reached in explosive scenarios, such as supernova explosions, so that the r-process is probably occuring dominantly in such events [Hil78, Kla81, Gro89, Gro90, Cow92]. Rapid neutron capture can take the nucleus away from the β-stability line by

10–20 units until the β-half-lives become competitive (see figures 14.2 and 14.3). In the 'classic' r-process it is assumed that a (n, γ)–(γ, n) equilibrium takes place along a horizontal line in figure 14.3 until a dominating β-decay appears ('waiting point approximation'). The nucleus $(Z, A + i)$, where i indicates the number of captured neutrons, is now at a waiting point where further development can only proceed by β-decay

$$(Z, A + i) \rightarrow (Z + 1, A + i) + e^- + \bar{\nu}_e. \tag{14.8}$$

The daughter nucleus created occasionally captures j neutrons, up to a nucleus $(Z + 1, A + i + j)$, where the next waiting point is reached. The neutron rich nuclei created plotted as a function of Z produce the so-called 'r-process path' (see figures 14.2–14.4). The abundance of an isotope along this route is given by

$$\frac{dN_Z(t)}{dt} = \lambda_{Z-1}(t)N_{Z-1}(t) - \lambda_Z(t)N_Z(t). \tag{14.9}$$

The solutions of these equations also have a tendency to result in equilibrium. However, the abundances are now correlated with the β-half-lives at the waiting point with charge Z, in contrast to the s-process, where the abundance is correlated with the inverse of the capture cross section σ_A. The effect becomes particularly dramatic at magic neutron numbers. Here the neutron binding energies are particularly low, which means that the neutron capture cross sections are low and the β-half-lives are relatively long which creates a bottle-neck for the r-process. If now a decay such as

$$(Z, N) \rightarrow (Z + 1, N - 1) \tag{14.10}$$

takes place, the nucleus is still relatively stable as it still has an almost magic neutron number. This results in a slow increase of Z until the nucleus comes close enough to the line of stability that the neutron capture probabilities again exceed the β-decay rates and the breakthrough succeeds. Due to the long waiting time the isotopes close to magic neutron numbers are created with large relative abundance. However, for equal neutron numbers, the initial nuclei have considerably less protons compared to the s-process (the s-process moves 'diagonally', the r-process 'horizontally'), so that the maxima in the distribution of the stable final nuclei from the r-process are found 10–15 mass units below the magic numbers and hence below the s-process maxima.

The r-path and the corresponding waiting points are dependent on several parameters, for example temperature, neutron density, binding energies and β-half-lives. Higher neutron densities produce stops at more neutron rich nuclei, while higher temperatures have the opposite effect. Since the neutron binding energies enter exponentially in the condition for the (n, γ) and (γ, n) equilibrium (equation 14.3), they have considerable influence on the waiting points. However, in order to determine them a mass formula for nuclei far from stability, and therefore not accessible in the laboratory, is required. One is therefore very sensitive to theoretical extrapolations in this area.

If the neutron flux ceases, those nuclei which were formed up to then, decay via β-decay back in the direction of the stability line. The β-half-lives for nuclei far away from stability in most cases have not been observed in the laboratory and thus have to be calculated theoretically. Only recently great progress was made here by microscopic calculations [Kla84, Sta90b, Sta92a, Hir92b, Hom96]. The β-half-lives determine the mass region up to which the r-process extends as well as the abundances of the created isotopes and also in particular the production rates for the cosmo-chronometers which are discussed later [Kla89]. The masses at which the r-path stops also depend on nuclear physics parameters, since processes such as β-delayed fission and delayed neutron emission play an important role [Kla83, Kla85, Kla86b, Kla89, Kla91a, Kla91d, Hir92b, Sta92a] (see figures 14.2, 14.4 and 14.5). In order to correctly consider these effects information about the fission barriers is required, which in this region can only be obtained by theoretical extrapolation [Kla79, Sta92a]. The end of the path is expected at $A \simeq 270$. This conclusion is within a wide range independent of the heights of the fission barriers adopted in the calculations [Sta92a] (figure 14.6). For very heavy nuclei on the journey back to the line of stability the probability for spontaneous fission can become larger than for β-decay. The fission barriers also determine whether or not it will be possible to create super-heavy nuclei [Sta92a].These are expected in the neighbourhood of magic proton ($Z = 114$) and neutron ($N = 184$) numbers. The r-process in particular demonstrates how important an effect nuclear physics calculations can have on astrophysics. For an overview see [Gro89, Gro90].

The classic picture of the r-process starts with iron nuclei as seeds for the successive build-up of heavier elements. It was assumed that the outer layers of the neutron star created in supernova explosions would be very suitable for this. However, the coherent elastic neutrino–iron scattering was overlooked [Fre74, Fre93]. This leads to an increased neutrino density and therefore to a smaller neutron density due to the increased $\nu n \rightarrow ep$ rate. Also, the mass cut (see section 14.2) is very important in these calculations. In general such calulations yield a massive overproduction of r-elements [Hil78].

The model of explosive helium burning seems to work better [Kla81, Kla83, Mat90b, Kla91a], even though not all aspects of it are completely clear yet. The outward going shock wave produces in the helium-burning shell a very high neutron density for about 0.5 s, which allows the r-process to take place. The neutrons come mainly from the ^{22}Ne(α, n)^{25}Mg reaction, and, with an initial distribution of elements which corresponds to that in the Sun modified by the CNO-cycle and the s-process, the observed abundances can be reasonably well reproduced (see figure 14.7) [Kla83, Kla86a, Gro89, Gro90, Kla91d]. In order to compare with the *observed* r-process isotope abundances, the s-process contributions have to be subtracted. The remaining spectrum can then be attributed to the r-process. Nucleosynthesis in the r-process is therefore more

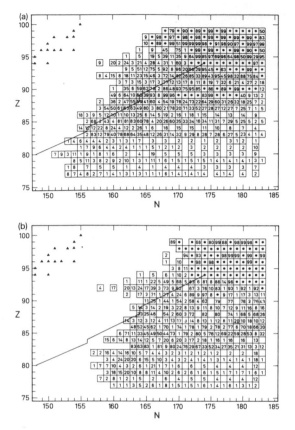

Figure 14.5. Calculated β-delayed fission probabilities with the mass formula of (*a*) Hilf *et al* [Hil76] and (*b*) Groote *et al* [Gro76]. Fission barriers are taken from [How80]. The daughter nuclei reached in β^--decay are labelled by the fission probability P_f in %; dots denote $P_f = 100\%$, triangles denote β-stable nuclei. The *r*-process path is also indicated (from [Sta92a]).

similar to a transition from an *s*-distribution to an *r*-distribution, than to the successive transformation of nuclei of the iron group. Relative to the total mass of the exploding star ($\sim 25 M_\odot$), explosive burning of He leads to an enrichment of *r*-material by a factor between 10 and 20, which is comparable with the enrichment factor which follows for light elements from the explosive burning of the inner shells (see section 14.2) [Wea80, Woo82].

Further, it can be shown that models with a primordial *r*-process in an inhomogeneous primordial nucleosynthesis (see chapter 4), cannot explain the observed abundances of heavy elements [App85, Kaj88].

More recent models have returned to a picture of the *r*-elements being built up again more in the *interior* of supernovae [Mey92, Woo94]. Production of

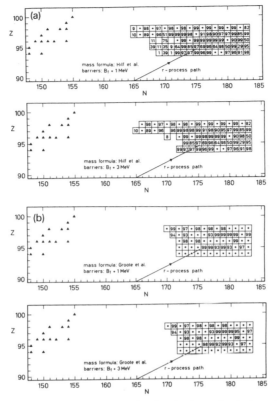

Figure 14.6. (*a*) β-delayed fission probabilities for the isotopic chains with $94 \leq Z \leq 98$ calculated with the mass formula of (*a*) [Hil76], (*b*) [Gro76] and with all fission barriers raised by 1 and 3 MeV, respectively, compared to figure 14.5 (from [Sta92a]).

the *r*-elements and transport to the surface then happens via neutrino-heated hot bubbles [Mey92, Woo94] (see section 13.1.2). Figure 14.8 shows the *r*-process abundances calculated in *this* way—still in a one-dimensional calculation—as a function of the mass number. Neutrino oscillations (MSW effect) could not only affect the supernova dynamics, but in *such* a process also nucleosynthesis [Ful93, Ful95]. on the other hand restrictions on neutrino mixing parameters could be derived from the requirement of the possibility of such an *r*-process [Qia93].

A problem with these and more recent two-dimensional calculations is that too much neutron rich material is emitted. In particular the synthesis of the elements around Fe is overestimated and does not agree with observation [Woo04, Bur95, Mül95c]. One solution could be a delay of the explosion above the 50 to 100 ms delay currently envisaged, which would allow a further reduction in the mass and density of the emitted material because it could fall back into the star [Bur95, Bur95a]. Model calculations of this type using the

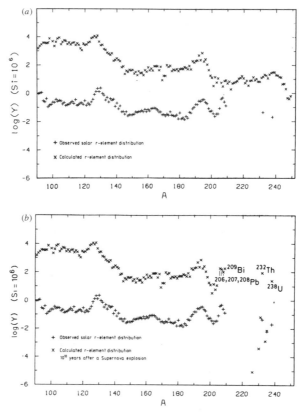

Figure 14.7. (*a*) Calculated abundances of the stable elements after β-decay of the temporary element distribution formed in the *r*-process by explosive helium burning as a function of mass number *A*. The observed solar abundances are also shown. (*b*) as in (*a*), but after further α-decay of the heavy β-stable nuclei over 10^{10} years (from [Thi83, Kla83, Kla91d]).

latest nuclear data, e.g. for β-decay, are however not yet available.

14.3.4   Cosmo-chronometers and the age of the universe

Some of the isotopes created in the *r*- or *s*-process are particularly suited to probing the history of the chemical evolution and age of our galaxy and the age of universe. We have seen in chapter 4 that only light nuclei were created during primordial nucleosynthesis. After the creation of our Milky Way heavier elements were able to form since the creation of the first generation of stars. Later the pre-solar nebula collapsed and uncoupled from the general chemical evolution of the Milky Way. All heavy elements within our planetary system thus had to have been created in the nucleosynthesis before the creation of the

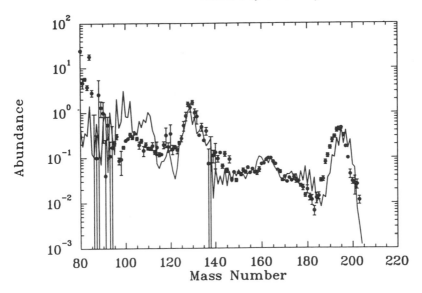

Figure 14.8. Calculated r-process abundances (solid line) as a function of mass number in an r-process in hot bubbles in a supernova explosion. The calculation is compared to the measured solar r-process abundances (points) (from [Käp89]). The calculation is normalized to the solar abundance of ^{129}Xe (from [Woo94]).

solar system. It is possible to obtain an estimate for the age of the universe from theoretical nucleosynthesis calculations, independent of other methods such as the age of globular clusters or the Hubble expansion. The method of using long lived radioactive isotopes for this purpose, analogous to the use of ^{14}C in archaeology, is called nucleo-cosmo-chronology [Sym81, Gro89, Gro90]. This method gains information about the duration of the nucleosynthesis period, and therefore about the age of the galaxy and the universe, through comparison of the calculated production ratios of sufficiently long lived nuclides in the r-process (the cosmo-chronometer) with the ratios found at the time of the condensation of the solar system (which are 'frozen' in meteorites).

Assuming that the duration T of the continuous synthesis of r-nuclei due to supernova explosions in the galaxy up to the point of isolation of the pre-solar cloud is known, the age of the universe can be written as

$$t_0 = (T + t_{SS} + \Delta + 10^9) \text{ years} \qquad (14.11)$$

where $t_{SS} = (4.55 \pm 0.07) \times 10^9$ years is the age of the solar system [Kir78], $\Delta \approx 10^8$ years is the time between the isolation of the pre-solar cloud (or the last passage of a spiral arm through the pre-solar cloud) and its condensation. The additional factor of 10^9 in the sum takes account roughly of the time between the time of the big bang and the beginning of the galactic r-process nucleosynthesis. In order to obtain information about T, it is necessary to observe nuclei with $T_{1/2} \gg \Delta$. The isotopes ^{232}Th ($T_{1/2} = 14.05 \times 10^9$ years), and ^{238}U ($T_{1/2} = 4.47 \times 10^9$ years) are therefore particularly useful. If, on the other hand, one is more interested in information about the last synthesis events and in Δ, the shorter lived isotopes such as ^{244}Pu ($T_{1/2} = 8.26 \times 10^7$ years) or ^{129}I ($T_{1/2} = 1.57 \times 10^7$ years) are adequate. ^{235}U ($T_{1/2} = 7.04 \times 10^8$ years) takes an intermediate position.

The quantity from which T can in principle be calculated is given by [Thi83, Kla83]

$$R_{ij} = \frac{Y_i^p / Y_j^p}{Y_i(T + \Delta)/Y_j(T + \Delta)}, \tag{14.12}$$

where Y_i^p and Y_j^p are the production rates of two chronometers in the r-process, and $Y_{i,j}(T + \Delta)$ are their abundances at the time of condensation of the pre-solar cloud. Generally it is necessary to make assumptions about the evolution of the galaxy in order to get precise answers for T. We first assume for simplicity the unrealistic case in which a single r-process at time t_n produces the chronometer. To determine the time of production we consider the ratio ^{232}Th/^{238}U. From the investigation of meteorites this is found to be 2.5 ± 0.2 [Sym81]. This is in good agreement with the expectation that since the condensation of the pre-solar cloud the solar system has been a closed system, since the *current* observed ratio [Fow78] of

$$\frac{^{232}\text{Th}}{^{238}\text{U}} = 4.0 \pm 0.2 \tag{14.13}$$

corresponds precisely to the assumption of a subsequent uninterrupted decay of the isotopes contained in the meteorites. For the production ratio of ^{232}Th/^{238}U in the r-process a value of 1.39 is used which is obtained by applying for the first time microscopic nuclear structure calculations in nucleosynthesis calculations [Kla83, Thi83, Gro89, Gro90]. Using this starting ratio, the time of free decay t_n is determined from the ratio of 2.5 observed in meteorites from the formula

$$\frac{\lambda_{232}}{\lambda_{238}} = \left(\frac{^{232}\text{Th}}{^{238}\text{U}}\right)_s \frac{\exp(t_n/\tau_{232})}{\exp(t_n/\tau_{238})} \tag{14.14}$$

which leads to a value of $t_n \approx 5.5 \times 10^9$ years before the time of the condensation of the solar system, for the moment of nucleosynthesis. $(^{232}\text{Th}/^{238}\text{U})_s$ in equation (14.14) denotes the isotope ratio at the time of condensation of the solar system. More interesting than t_n is $2t_n$, since this corresponds in a more-or-less model independent way to the time $T + \Delta$ [Sym81]. For sufficiently long lived chronometers continuous production can be considered as a *single* event which occurs in the middle of the production period (for the ^{232}Th/^{238}U pair this is not quite true so that the derived ages are too short). The final result of this simple estimate is an age of the galaxy of about 15.6×10^9 years.

We now consider a more realistic model of continuous production of r-nuclides (such as in many supernova explosions) as shown in figure

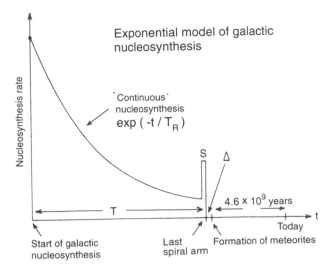

Figure 14.9. Model for nucleosynthesis in the galaxy (from [Gro89, Gro90]). For details see text.

14.9 (see [Fow60, Fow72, Fow78, Gro89, Gro90]). It is based on the following assumptions: continuous, exponentially decreasing nucleosynthesis characterized by a parameter T_R, describing the decrease of the matter which takes part in the synthesis through condensation into white dwarfs, neutron stars, black holes etc; the explicit assumption of a last 'spike' in nucleosynthesis due to the last 'passage' of a spiral arm (density wave) through the pre-solar cloud (or a supernova, which may have triggered the collapse of the pre-solar cloud) and which contributes a fraction S to the total galactic nucleosynthesis; finally, a subsequent time Δ for free decay up to the condensation of the solar system 4.55×10^9 years ago. The abundance Y_i of a nuclear species i at time $(T + \Delta)$ of the condensation of the solar system in this model is connected simply to its production rate Y_i^P in the r-process, its half-life τ_i and the parameters T, S, T_R, Δ (see [Thi83, Kla83]).

Therefore if four pairs of chronometers and their *production ratios* Y_i^P / Y_j^P and their relative abundances $Y_i(T + \Delta)/Y_j(T + \Delta)$ at the time $(T + \Delta)$ are known, the four parameters T, S, T_R, Δ can be determined from a system of four simple equations:

$$\frac{Y_i(T + \Delta)}{Y_j(T + \Delta)} = \frac{Y_i^P}{Y_j^P} \frac{f(T, S, T_R, \Delta, \tau_i)}{f(T, S, T_R, \Delta, \tau_j)} \tag{14.15}$$

Even if a 'continuous' model such as that outlined above for the period *before* some 10^8 years before the condensation of the solar system should be sensible, allowing for the possibility of *several* 'spikes' would certainly be desirable. On the other hand, however, this would require good experimental knowledge of the ratios of more chronometric pairs. The use of the chronometric pairs ^{232}Th/^{238}U, ^{235}U/^{238}U, ^{244}Pu/^{238}U, ^{129}I/^{127}I results in a value for the duration of nucleosynthesis of $(14.6^{+2}_{-5}) \times 10^9$ years [Thi83, Kla83]. This value is the mean value of the two results, 16.1 and 13.2×10^9 years, obtained using two different experimental ^{232}Th/^{238}U ratios.

This number is thus sensitive both to theoretical extrapolation of the β-decay rates and nuclear mass formulae for neutron rich nuclei and to the experimental determination of the abundances of the chronometers in meteorites. For a detailed discussion see [Thi83, Kla83, Kla85, Kla89, Gro89, Gro90, Kla91d]. Equation (14.11) leads to an age of the universe of 20.2^{+2}_{-5} billion years. The Hubble time (see chapter 3) calculated from a Hubble constant of 50 km s^{-1} Mpc^{-1}, leads to $t_0 = H_0^{-1} = 19.5 \times 10^9$ years. The chronometers therefore appear to prefer a lower value of the Hubble constant. Investigations of globular clusters result in an age between 14 and 20 billion years (see chapter 3) [Tam86, Buo89, San90, Ber95a, Ber96].

The importance of reliable research on the age of the universe for the determination of the cosmological constant has already been discussed in chapter 5.

References

[Aba91]   Abazov A I *et al* (SAGE Collaboration) 1991 *Phys. Rev. Lett.* **67** 3332
[Aba95]   Abachi S *et al* (D0 Collaboration) 1995 *Phys. Rev. Lett.* **74** 2632
[Aba95a]   Abachi S *et al* (D0 Collaboration) 1995 *Phys. Rev. Lett.* **75** 618
[Aba97]   Abachi S *et al* (D0 Collaboration) 1997 *Preprint* hep-ex/9703008
[Abb88]   Abbott L 1988 *Sci. Am.* **258** 82
[Abd95]   Abdurashitov J N *et al* (SAGE Collaboration) 1995 *Nucl. Phys. B (Proc. Suppl.)* **38** 60, *Proc. 16th Int. Conf. on Neutrino Physics and Astrophysics, NEUTRINO '94* eds A Dar, G Eilam and M Gronau (Amsterdam: North Holland)
[Abd96]   Abdurashitov J N *et al* 1996 *Nucl. Phys. B (Proc. Suppl.)* **48** 299, *Proc. 4th Int. Workshop on Theoretical and Phenomenological Aspects of Underground Physics, TAUP '95, Toledo, Spain* eds A Morales, J Morales and J A Villar (Amsterdam: North-Holland)
[Abe58]   Abell G O 1958 *Astrophys. J. Suppl.* **3** 211
[Abe84]   Abela R *et al* 1984 *Phys. Lett.* B **146** 431
[Abe90]   Abe F *et al* (CDF Collaboration) 1990 *Phys. Rev. Lett.* **65** 2243
[Abe93]   Abele H *et al* 1993 *Phys. Lett.* B **316** 26
[Abe94]   Abe F *et al* (CDF Collaboration) 1994 *Phys. Rev.* D **50** 2966
[Abe95a]   Abe F *et al* (CDF Collaboration) 1995 *Phys. Rev. Lett.* **74** 2626
[Abe95b]   Abe F *et al* (CDF Collaboration) 1995 *Phys. Rev. Lett.* **75** 11
[Abe96]   Abe F *et al* (CDF Collaboration) 1996 *Phys. Rev. Lett.* **76** 2006
[Abr92]   Abramovici A *et al* (LIGO Collaboration) 1992 *Science* **256** 325
[Ach77]   Achiman Y and Stech B 1977 *Phys. Lett.* B **77** 389
[Ach95]   Achkar B *et al* 1995 *Nucl. Phys.* B **434** 503
[Act97]   Actarelli G *et al* 1997 *Preprint* hep-ph/9703276
[Ada90]   Adarkar H *et al* 1990 (Kolar gold mine monopole experiment) *21st Int. Cosmic Ray Conf. (Adelaide, 1990)* p 95
[Ada93]   Adams F C *et al* 1993 *Phys. Rev. Lett.* **70** 2511
[Ade90]   Adeva B *et al* (L3 Collaboration) 1990 *Phys. Lett.* B **248** 227
[Adl95]   Adler R *et al* (CPLEAR Collaboration) 1995 *Phys. Lett.* B **363** 243
[Adl97]   Adloff C *et al* (H1 Collaboration) 1997 *Z. Phys.* C **74** 191
[Adl97a]   Adler S L 1997 *Preprint* hep-ph/9702378
[Afo85]   Afonin A I *et al* 1985 *JETP Lett.* **42** 285
[Agl87]   Aglietta M *et al* (Frejus Collaboration) 1987 *Europhys. Lett.* **3** 1315
[Agl89]   Aglietta M *et al* (Frejus Collaboration) 1989 *Europhys. Lett.* **8** 611

450   *References*

[Aha91]   Aharonian F *et al* (HEGRA Collaboration) 1991 *Proc. 22nd Int. Conf. on Cosmic Rays, (Dublin)* **4** 452
[Aha95]   Aharonov Y *et al* 1995 *Phys. Rev.* D **52** 3785
[Aha97]   Aharonian F A *et al* (HEGRA Collaboration) 1997 *Preprint* astro-ph/9702059
[Ahl88]   Ahlen S P *et al* (MACRO Collaboration) 1988 *Phys. Rev. Lett.* **61** 145
[Ahl93]   Ahlen S P *et al* (MACRO Collaboration) 1993 *Nucl. Instrum. Methods* **324** 337
[Ahl94a]   Ahlen S P *et al* (MACRO Collaboration) 1994 *Phys. Rev. Lett.* **72** 608
[Ahl94b]   Ahlen S P *et al* (MACRO Collaboration) 1994 *Nucl. Instrum. Methods* A **350** 351
[Ahl95]   Ahlen S P *et al* (MACRO Collaboration) 1995 *Phys. Lett.* B **357** 481
[Ahm94]   Ahmed T *et al* (H1 Collaboration) 1994 *Z. Phys.* C **64** 545
[Ahr85]   Ahrens L *et al* 1985 *Phys. Rev.* D **31** 2732
[Aid95]   Aid S *et al* (H1 Collaboration) 1995 *Phys. Lett.* B **353** 578
[Aid96]   Aid S *et al* (H1 Collaboration) 1996 *Nucl. Phys.* B **472** 3
[Aid96a]   Aid S *et al* (H1 Collaboration) 1996 *Phys. Lett.* B **380** 461
[Aid96b]   Aid S *et al* (H1 Collaboration) 1996 *Z. Phys.* C **71** 211
[Aid96c]   Aid S *et al* (H1 Collaboration) 1996 *Preprint* DESY96-163, see also Aid S *et al* (H1 Collaboration) 1997 *Nucl. Phys.* B **483** 44
[Ait89]   Aitchison I J R and Hey A J G 1989 *Gauge Theories in Particle Physics* (Bristol: Adam Hilger)
[Aka97]   Akama K and Katsuura K 1997 *Preprint* hep-ph/9704327
[Akh88]   Akhmedov E Kh and Khlopov M Yu 1988 *Mod. Phys. Lett.* A **3** 451
[Akh97]   Akhmedov E Kh 1997 *Proc. 4th Int. Solar Neutrinos Conf., Heidelberg*
[Akr91]   Akrawy M Z *et al* (OPAL Collaboration) 1991 *Z. Phys.* C **49** 49
[Ala95]   Alard C *et al* (DUO Collaboration) 1995 *ESO Messenger* **80** 31
[Alb82]   Albrecht A and Steinhardt P J 1982 *Phys. Rev. Lett.* **48** 1220
[Alb85]   van Albada T S *et al* 1985 *Astrophys. J.* **295** 305
[Alb87]   Albrecht H *et al* (ARGUS Collaboration) 1987 *Phys. Lett.* B **192** 245
[Alb89]   Albrecht A and Turok N 1989 *Phys. Rev.* D **40** 973
[Alb92]   Albrecht H *et al* (ARGUS Collaboration) 1992 *Phys. Lett.* B **292** 221
[Alb94]   Albrecht H *et al* (ARGUS Collaboration) 1994 *Phys. Lett.* B **324** 249
[Alc86]   Alcock C, Fashi E and Olinto A 1986 *Astrophys. J.* **310** 261
[Alc87]   Alcock C, Fuller G and Mathews G 1987 *Astrophys. J.* **320** 439
[Alc93]   Alcock C *et al* (MACHO Collaboration) 1993 *Nature* **365** 621
[Alc95]   Alcock C *et al* (MACHO Collaboration) 1995 *Phys. Rev. Lett.* **74** 2867
[Alc96]   Alcock C *et al* (MACHO Collaboration) 1996 *Preprint* astro-ph/9606165, submitted to *Astrophys. J.*
[Alc96b]   Alcock C *et al* (MACHO Collaboration) 1996 *Astrophys. J.* **463** 267
[Alc97]   Alcock C *et al* (MACHO and GMAN Collaborations) 1997 *Preprint* astro-ph/9702199
[Ale82]   Alekseev E N *et al* 1982 *Nuovo Cimento Lett.* **35** 413
[Ale83]   Alekseev E N *et al* 1983 *Proc. XVIII Int. Conf. on Cosmic Rays (Bangalore, India)* **5** 52
[Ale87]   Alekseev E N, Alexeyeva L N, Krivosheina I V and Volchenko V I 1987 *JETP Lett.* **45** 589

[Ale87a] Alekseev E N, Alexeyeva L N, Krivosheina I V and Volchenko V I 1987 *Proc. ESO Workshop SN1987A, (Garching near Munich, 6–8 July 1987)* 237

[Ale88] Alekseev E N Alexeyeva L N, Krivosheina I V and Volchenko V I 1988 *Phys. Lett.* B **205** 209

[Ale88a] Alekseev E N, Alexeyeva I N, Krivosheina I V and Volchenko V I 1988 *Neutrino Physics* eds H V Klapdor and B Povh (Heidelberg: Springer) 288

[Ale92] Alessandrello A *et al* 1992 *Phys. Lett.* B **285** 176

[Ale94] Alessandrello A *et al* 1994 *Nucl. Phys.* B *(Proc. Suppl.)* **35** 366, *3rd Int. Workshop on Theoretical and Phenomenological Aspects in Astroparticle and Underground Physics (TAUP'93), Assergi, Italy* eds C Arpesella, E Bellotti and A Bottino (Amsterdam: North-Holland)

[Ale95] ALEPH Collaboration 1995 *Preprint* CERN-PPE 95-03, see also 1995 *Phys. Lett.* B **349** 585

[Ali92] Alitti J *et al* (UA2 Collaboration) 1992 *Phys. Lett.* B **276** 354

[All90] Allen B and Shellard E P S 1990 *Phys. Rev. Lett.* **64** 119

[All96] Allison W W M *et al* (SOUDAN 2 Collaboration) 1996 *Preprint* hep-ex/9611007v2

[Alp48] Alpher R A and Herman R C 1948 *Nature* **162** 774

[Als93] Alston-Garnjost M *et al* 1993 *Phys. Rev. Lett.* **71** 831

[Alt94] Althaus T 1994 *Sterne und Weltraum* **3** 181

[Alt97] Altarelli G, Ellis J *et al* 1997 *Preprint* hep-ph/9703276

[Ama91] Amaldi U, de Boer W and Fürstenau H 1991 *Phys. Lett.* B **260** 447; 1991 *Proc. Joint Int. Lepton–Photon Symp. and Europhys. Conf. on High Energy Physics, LP-HEP'91, Geneva, Switzerland* eds S Hegarty, K Potter and E Quercigh (Singapore: World Scientific) **1** 690

[Amb96] Ambrosio M *et al* (MACRO Collaboration) 1996 *MACRO-PUB 96/2*

[And89] Anders E and Grevesse N 1989 *Act. Geochim. Cosmochim.* **53** 197

[Ang86] Angelini C *et al* 1986 *Phys. Lett.* B **179** 307

[Ann95] 1995 *Ann. NY Acad. Sci.* **759** 450 ff, *Proc. 17th TEXAS Symp. on Relativistic Astrophysics and Cosmology, Munich, Germany* eds H Böhringer, G E Morfill and J E Trümper

[Ano93] Anosov O L *et al* 1993 *Nucl. Phys.* B *(Proc. Suppl.)* **31** 111, *Proc. NEUTRINO '92, Granada, Spain* ed A Morales (Amsterdam: Elsevier)

[Ans92] Anselmann P *et al* (GALLEX Collaboration) 1992 *Phys. Lett.* B **285** 376

[Ans92a] Anselmann P *et al* (GALLEX Collaboration) 1992 *Phys. Lett.* B **285** 390

[Ans94] Anselmann P *et al* (GALLEX Collaboration) 1994 *Phys. Lett.* B **327** 377

[Ans95a] Anselmann P *et al* (GALLEX Collaboration) 1995 *Phys. Lett.* B **342** 440

[Ans95b] Anselmann P *et al* (GALLEX Collaboration) 1995 *Nucl. Phys.* B *(Proc. Suppl.)* **38** 68, *Proc. 16th Int. Conf. on Neutrino Physics and Astrophysics, NEUTRINO '94* eds A Dar, G Eilam and M Gronau (Amsterdam: North-Holland)

[Ans95c] Anselmann P *et al* (GALLEX Collaboration) 1995 *Phys. Lett.* B **357** 237

[Ans95d] Ansari R *et al* 1995 *Preprint* astro-ph/9502102, see also *Proc. 17th TEXAS Symp. on Relativistic Astrophysics and Cosmology* eds H Böhringer, G E Morfill and E Trümper *Ann. NY Acad. Sci.* **759** 608

[Ans96] Ansari R *et al* 1996 *Preprint* astro-ph/9607040

[Ant88]   Antoniadis A *et al* 1988 *Phys. Lett.* B **194** 231
[Ant89]   Antoniadis A *et al* 1989 *Phys. Lett.* B **208** 209
[App85]   Applegate J H and Hogan C J 1985 *Phys. Rev.* D **31** 3037
[App87]   Appelquist T, Chodos A and Freund P G O 1987 *Modern Kaluza–Klein Theories, Frontiers in Physics* (New York: Addison-Wessley)
[App88]   Applegate J H 1988 *Phys. Rep.* **163** 141
[Arm95]   Armbruster B *et al* (KARMEN Collaboration) 1995 *Nucl. Phys. B (Proc. Suppl)* **38** 235, *Proc. 16th Int. Conf. on Neutrino Physics and Astrophysics, NEUTRINO '94* eds A Dar, G Eilam and M Gronau (Amsterdam: North Holland)
[Arn77]   Arnett W D 1977 *Astrophys. J.* **218** 815; 1977 *Astrophys. J. Suppl.* **35** 145
[Arn78]   Arnett W D 1978 *The Physics and Astrophysics of Neutron Stars and Black Holes* eds R Giacconi and R Ruffins (Amsterdam: North-Holland) p 356
[Arn80]   Arnett W D 1980 *Ann. NY Acad. Sci.* **336** 366
[Arn83]   Arnison G *et al* 1983 *Phys. Lett.* B **122** 103
[Arn87]   Arnett W D and Rosner J 1987 *Phys. Rev. Lett.* **58** 1906
[Arn89]   Arnett W D *et al* 1989 *Ann. Rev. Astron. Astrophys.* **27** 629
[Arn91]   Arnett W D 1991 *Astrophys. J.* **383** 295
[Arp91]   Arp H C and Block D L 1991 *Sky and Telescope* **4** 373
[Art95]   Artemiev V *et al* 1995 *Phys. Lett.* B **345** 564
[Arp96]   Arpesella C *et al* (LUNA Collaboration) 1996 *Nucl. Phys. B (Proc. Suppl.)* **48** 375, *Proc. 4th Int. Workshop on Theoretical and Phenomenological Aspects of Underground Physics, Toledo, Spain 1995* eds A Morales, J Morales and J A Villar (Amsterdam: North-Holland)
[Arp97]   Arpesella C *et al* (LUNA Collaboration) 1997 *Preprint* nucl-ex/9707003
[Arn97]   Arnowitt R and Nath P 1997 *Preprint* hep-ph/9701301 and submitted to *Phys. Rev.* D
[Asa81]   Asano Y *et al* 1981 *Phys. Lett.* B **107** 159
[Ash89]   Ashman J *et al* (EMC Collaboration) 1989 *Nucl. Phys.* B **328** 1
[Asr81]   Asratyan A E *et al* 1981 *Phys. Lett.* B **105** 301
[Ass94]   Assamagan K *et al* 1994 *Phys. Lett.* B **335** 231
[Ass96]   Assamagan K *et al* 1996 *Phys. Rev.* D **53** 6065
[Ath95]   Athanassopoulos C *et al* (LSND Collaboration) 1995 *Phys. Rev. Lett.* **75** 2650
[Ath96]   Athanassopoulos C *et al* (LSND Collaboration) 1996 *Phys. Rev. Lett.* **77** 3082; 1996 *Preprint* nucl-ex 9605003
[Ath97]   Athanassopoulos C *et al* 1997 *Preprint* LA-UR-97-1998
[Aub93]   Aubourg E *et al* (EROS Collaboration) 1993 *Nature* **365** 623
[Auf97]   Aüfmuth P *et al* 1997 *Phys. Bl* **53** 205
[Aug95]   (AUGER Collaboration) 1995 *The Pierre Auger Project Design Report*
[Avi87]   Avignone F T *et al* 1987 *Phys. Rev.* D **35** 2752
[Avi88]   Avignone F T 1988 *Neutrinos* ed H V Klapdor (Berlin: Springer) p 147
[Avi97]   Avignone F T *et al* 1997 *Proc. Workshop on Non-accelerator New Physics (NANP), Dubna, Russia*
[Bab91a]   Babu K S, Mohapatra R N and Rothstein I Z 1991 *Phys. Rev.* D **44** 2265
[Bab95]   BaBar experiment 1995 *Tech. Design Report SLAC-R-95-437*

[Bac94]	Bacci C *et al* 1994 *Nucl. Phys. B (Proc. Suppl.)* **35** 165, 1993 *Proc. 3rd Int. Workshop on Theoretical and Phenomenological Aspects in Astroparticle and Underground Physics (TAUP'93), Assergi, Italy* eds C Arpesella, E Bellotti and A Bottino (Amsterdam: North-Holland)
[Bae95]	Baer H *et al* 1995 *Preprint* hep-ph/9503479
[Bae97]	Baer H and Brhlik M 1997 *Preprint* hep-ph/9706509
[Bah71]	Bahcall J N and Ulrich R K 1971 *Astrophys. J.* **170** 593
[Bah83]	Bahcall N A and Soneira R N 1983 *Astrophys. J.* **270** 20
[Bah86]	Bahcall J N *et al* 1986 *Phys. Lett.* B **181** 369
[Bah88a]	Bahcall N A 1988 *Ann. Rev. Astron. Astrophys.* **26** 631
[Bah88b]	Bahcall J N and Tremaine S 1988 *Astrophys. J. Lett.* **326** L1
[Bah88c]	Bahcall J N and Ulrich R K 1988 *Rev. Mod. Phys.* **60** 297
[Bah89]	Bahcall J N 1989 *Neutrino Astrophysics* (Cambridge: Cambridge University Press)
[Bah92]	Bahcall J N and Pinsonneault M H 1992 *Rev. Mod. Phys.* **64** 885
[Bah95]	Bahcall J N and Pinsonneault M H 1995 *Rev. Mod. Phys.* **67** 781
[Bah96]	Bahcall J N 1996 *Nucl. Phys. B (Proc. Suppl.)* **48** 309, *Proc. 4th Int. Workshop on Theoretical and Phenomenological Aspects of Underground Physics, Toledo Spain* eds A Morales, J Morales and J A Villar (Amsterdam: North Holland)
[Bah96a]	Bahcall N 1996 *Preprint* astro-ph/9611148, to appear in the *Proceedings of 'Formation of Structure in the Universe, Jerusalem Winter School (1995)*
[Bah97]	Bahcall J N and Ostriker J P (eds) 1997 *Unsolved Problems in Astrophysics* (Princeton, NJ: Princeton University Press)
[Bak84]	Baker N J *et al* 1984 *Phys. Rev.* D **28** 2705
[Bal79]	Baluni V 1979 *Phys. Rev.* D **19** 2227
[Bal85]	Baltrusaitis R M *et al* 1985 *Nucl. Instrum. Methods* A **240** 410
[Bal89]	Baldo-Ceolin M 1989 *Proc. Theoretical and Phenomenological Aspects of Underground Physics (TAUP '89)* eds A Bottino and P Monacelli (Gif-sur-Yvette: Editions Frontieres) p 25
[Bal92]	Baldo-Ceolin M (ed) 1992 *Neutrino Telescopes, Proc. 4th Int. Workshop, Venice, 1992* (Padua University)
[Bal93]	Balysh A *et al* (HEIDELBERG–MOSCOW Collaboration) 1993 *Phys. Lett.* B **298** 278
[Bal94]	Ball A *et al* 1994 *Proposal, CERN/LEPC 94-10*
[Bal94a]	Baldo-Ceolin M *et al* 1994 *Z. Phys.* C **63** 409
[Bal94b]	Baldo-Ceolin M 1994 *Nucl. Phys. B (Proc. Suppl.)* **35** 450, *Proc. 3rd Int. Workshop on Theoretical and Phenomenological Aspects in Astroparticle and Underground Physics (TAUP'93), Assergi, Italy* eds C Arpesella, E Bellotti and A Bottino (Amsterdam: North-Holland)
[Bal94c]	Balysh A *et al* (HEIDELBERG–MOSCOW Collaboration) 1994 *Phys. Lett.* B **322** 176
[Bal94d]	Baltz A J and Weneser J 1994 *Phys. Rev.* D **50** 5971
[Bal94e]	Balser D S *et al* 1994 *Astrophys. J.* **430** 667
[Bal95a]	Balysh A *et al* 1995 (HEIDELBERG–MOSCOW Collaboration) *Proc. 27th Int. Conf. High Energy Physics, (Glasgow)* eds P J Bussey and I G Knowles (Bristol: IOP Publishing) p 939

[Bal95b]   Balysh A *et al* (HEIDELBERG–MOSCOW Collaboration) 1995 *Phys. Lett.* B **356** 450

[Bam95]   Bamert P, Burgess C and Mohapatra R N 1995 *Nucl. Phys.* B **449** 25

[Ban87]   Bania T M, Rood R T and Wilson T L 1987 *Astrophys. J.* **323** 30

[Ban92]   Bandler S R *et al* 1992 *Phys. Rev. Lett.* **68** 2429

[Bar52]   Barret P H *et al* 1952 *Rev. Mod. Phys.* **24** 133

[Bar64]   Barnes V E *et al* 1964 *Phys. Rev. Lett.* **12** 204

[Bar80]   Barabanov I R 1980 *Sov. Phys. JETP* **32** 359

[Bar84]   Barish B C 1984 *MONOPOLE '83* ed J L Stone (New York: Plenum) 367

[Bar85]   Baron E, Cooperstein J and Kahana S 1985 *Phys. Rev. Lett.* **55** 126

[Bar88]   Barbieri R and Mohapatra R N 1988 *Phys. Rev. Lett.* **61** 27

[Bar89]   Barger V, Guidice G F and Han T 1989 *Phys. Rev.* D **40** 2987

[Bar92]   Barr G D 1992 *Proc. Joint Int. Lepton–Photon Symp. and Europhys. Conf on High Energy Physics 1991 (Geneva)* eds S Hegarty, K Potter and E Quercigh (Singapore: World Scientific) 179

[Bar92a]   Barloutaud R 1992 *Proc. 2nd Workshop on Theoretical and Phenomenological Aspects of Underground Physics TAUP'91, (Toledo, Spain)* (Amsterdam: North-Holland) *Nucl. Phys.* B **28A** 522

[Bar92b]   Barwick S *et al* 1992 *J. Phys. G: Nucl. Phys.* **18** 225

[Bar93a]   Barwick S *et al* (AMANDA Collaboration) 1993 *Proc. XXVI Int. Conf. on High Energy Physics, (Dallas)* ed J R Sanford AIP Conf. Proc. Series **272** 1250

[Bar93b]   Baron G D *et al* (NA31 Collaboration) 1993 *Phys. Lett.* B **317** 233

[Bar93c]   Bartelt J *et al* (CLEO Collaboration) 1993 *Phys. Rev. Lett.* **71** 1680

[Bar94]   Barthelmy S D *et al* 1994 *Astrophys. J.* **427** 519

[Bar95]   Barklow T, Dawson S, Haber H E *et al* 1995 in *Particle Physics* eds R D Peccei *et al*

[Bat94]   Battye R A and Shellard E P S 1994 *Preprint* astro-ph/9403018

[Bat97]   Battye R A and Shellard E P S 1997 *Preprint* astro-ph/9706014

[Bau86]   Baumann N *et al* 1986 *Proc. 12th Int. Conf. on Neutrino Physics and Astrophysics, (Sendai)* eds T Kitagaki and H Yuta (Singapore: World Scientific)

[Bau97]   Baudis L, Hellmig J, Klapdor-Kleingrothaus H V, Petry F and Ramachers Y 1997 *Nucl. Instrum. Methods* A **385** 265

[Bay85]   Baym G *et al* 1985 *Phys. Lett.* B **160** 181

[Bec83]   Becher P, Böhm M and Joos H 1983 *Eichtheorien* (Stuttgart: Teubner)

[Bec84]   Beckenstein J P and Milgram M 1984 *Astrophys. J.* **286** 7

[Bec92]   Becker-Szendy R *et al* (IMB Collaboration) 1992 *Phys. Rev. Lett.* **69** 1010

[Bec93a]   Beck M *et al* (HEIDELBERG–MOSCOW Collaboration) 1993 *Phys. Rev. Lett.* **70** 2853

[Bec93b]   Beck M *et al* (HEIDELBERG–MOSCOW Collaboration) 1993 *Proc. 16th Texas Symp. on Relativistic Astrophysics 3rd Particles, Strings and Cosmology, (Berkeley) Ann. NY Acad. Sci.* **688** 509

[Bec94a]   Beck M *et al* (HEIDELBERG–MOSCOW Collaboration) 1994 *Nucl. Phys.* B *(Proc. Suppl.)* **35** 150,   *Proc. Theoretical and Phenomenological Aspects in Astroparticle and Underground Physics (TAUP'93), Assergi, Italy* eds C Arpesella, E Bellotti and A Bottino (Amsterdam: North-Holland)

[Bec94b]   Beck M *et al* 1994 (HEIDELBERG–MOSCOW Collaboration) *Phys. Lett.*
             B **336** 141
[Bec95]   Becker M 1995 *Ann. NY Acad. Sci.* **759** 250
[Bed94a]   Bednyakov V, Klapdor-Kleingrothaus H V and Kovalenko S 1994 *Phys.
             Lett.* B **329** 5
[Bed94b]   Bednyakov V, Klapdor-Kleingrothaus H V and Kovalenko S 1994 *Phys.
             Rev.* D **50** 7128
[Bed97a]   Bednyakov V, Klapdor-Kleingrothaus H V, Kovalenko S and Ramachers
             Y 1997 *Z Phys.* A **357** 339
[Bed97b]   Bednyakov V, Klapdor-Kleingrothaus H V and Kovalenko S 1997 *Phys.
             Rev.* D **55** 503
[Beg84]   Begelman M C, Blanford R D and Rees M J 1984 *Rev. Mod. Phys.* **56** 255
[Bel91]   Bellotti E *et al* 1991 *Phys. Lett.* B **266** 193
[Bel92]   Belli P *et al* (DAMA Collaboration) 1992 *Phys. Lett.* B **295** 330
[Bel94]   Belolaptikov L A *et al* 1994 *Nucl. Phys. B (Proc. Suppl.)* **35** 290, *Proc.*
             3rd Int. Theoretical and Phenomenological Aspects in Astroparticle and
             Underground Physics (TAUP'93), Assergi, Italy eds C Arpesella, E
             Bellotti and A Bottino (Amsterdam: North-Holland)
[Bel95]   Belesev A I *et al* 1995 *Phys. Lett.* B **350** 263
[Bel95a]   Bellotti E 1995 *Nucl. Phys. B (Proc. Suppl.)* **38** 90, *Proc. 16th Int. Conf. on*
             Neutrino Physics and Astrophysics, NEUTRINO '94 eds A Dar, G Eilam
             and M Gronau (Amsterdam: North Holland)
[Bel95b]   Belle experiment 1995 *Tech. Design Report KEK-Rep 95-1*
[Bel96a]   Belolaptikov I A *et al* (BAIKAL Collaboration) 1996 *Preprint* astro-
             ph/9601160 and *Proc. 2nd Workshop on the Dark Side of the Universe:*
             Experimental Efforts and Theoretical Framework, Rome, Italy
[Bel96b]   Bellini G 1996 *Nucl. Phys. B (Proc. Suppl.)* **48** 363, *Proc. 3rd Int.*
             Workshop on Theoretical and Phenomenological Aspects of Underground
             Physics (TAUP'95), Toledo, Spain eds A Morales, J Morales and J A
             Villar (Amsterdam: North-Holland)
[Bel96c]   Belli P *et al* 1996 *Nucl. Phys.* B **48** 62
[Bel97]   Belolaptikov I A *et al* 1997 *Proc. DARK'96, Int. Workshop on Dark Matter*
             in Astro- and Particle Physics (Heidelberg, 16–20 September, 1996)
             eds H V Klapdor-Kleingrothaus and Y Ramachers (Singapore: World
             Scientific)
[Ben88]   Bennett D P and Bouchet F R 1988 *Phys. Rev. Lett.* **60** 257
[Ben93]   Bennett D P *et al* (MACHO Collaboration) 1993 *Ann. NY Acad. Sci.* **688**
             612, *Proc. TEXAS/PASCOS'92 Conf. on Relativistic Astrophysics and*
             Particle Cosmology eds C W Akerlof and M A Srednicki
[Ben94a]   Benetti P *et al* (ICARUS Collaboration) 1994 *Nucl. Phys. B (Proc. Suppl.)*
             35 276 *Proc. 3rd Int. Workshop on Theoretical and Phenomenological*
             Aspects in Astroparticle and Underground Physics (TAUP'93), Assergi,
             Italy eds C Arpesella, E Bellotti and A Bottino (Amsterdam: North-
             Holland)
[Ben94b]   Bennett C L *et al* 1994 *Astrophys. J.* **436** 423
[Ben95]   Bennett C L 1995 *Nucl. Phys. B (Proc. Suppl.)* **38** 415, Proc. 16th Int. Conf.
             on Neutrino Physics and Astrophysics (NEUTRINO'94) eds A Dar, G
             Eilam and M Gronau (Amsterdam: North-Holland)

[Ber72]   Bergkvist K E 1972 *Nucl. Phys.* B **39** 317
[Ber85]   Bermon S *et al* 1985 *Phys. Rev. Lett.* **55** 1850
[Ber89a]   van den Bergh S 1989 *Astron Astrophys. Rev.* **1** 111
[Ber89b]   Bernstein J, Brown L S and Feinberg G 1989 *Rev. Mod. Phys.* **61** 25
[Ber90a]   van den Berg D A 1990 *Astrophysical Ages and Dating Methods* eds E
            Vangioni-Flan *et al* (Paris: Edition Frontieres) 241
[Ber90b]   Berezinskii V S *et al* 1990 *Astrophysics of Cosmic Rays* (Amsterdam:
            North-Holland)
[Ber90c]   Berger C *et al* (FREJUS Collaboration) 1990 *Phys. Lett.* B **245** 305
[Ber90d]   Bermon S *et al* 1990 *Phys. Rev. Lett.* **64** 839
[Ber90e]   Berry M 1990 *Kosmologie und Gravitation* (Stuttgart: Teubner)
[Ber90f]   Bertani M *et al* 1990 *Europhys. Lett.* **12** 613
[Ber90g]   Bertschinger E *et al* 1990 *Astrophys. J* **364** 370
[Ber91]   Berger C *et al* (FREJUS Collaboration) 1991 *Z. Phys.* C **50** 385
[Ber91a]   van den Bergh S and Tammann G A 1991 *Ann. Rev. Astron. Astrophys.* **29**
            363
[Ber91b]   Bertin G *et al* 1991 *Proc. II DAEC-Meeting, The Distribution of Matter in
            the Universe, Meudon, France*
[Ber91c]   Bertola F *et al* 1991 *Astrophys. J.* **373** 369
[Ber91d]   Bershady M, Ressel M T and Turner M S 1991 *Phys. Rev. Lett.* **66** 1398
[Ber92]   Bertsch D L *et al* 1992 *Nature* **357** 306
[Ber93a]   Berthomieu G *et al* 1993 *Astron. Astrophys.* **268** 775
[Ber93b]   Bernatowicz T *et al* 1993 *Phys. Rev.* C **47** 806
[Ber93c]   Bertschinger E 1993 *Ann. NY Acad. Sci.* **688** 297, *Proc. TEXAS/PASCOS'92
            Conf. on Relativistic Astrophysics and Particle Cosmology* eds C W
            Akerlof and M A Srednicki
[Ber94]   van den Bergh S and McClure R D 1994 *Astrophys. J.* **425** 205
[Ber94a]   Berger C 1994 *Teilchenphysik* (Heidelberg: Springer)
[Ber95]   van den Bergh S 1995 *Preprint* astro-ph/9506027 and 1995 *J. R. Astron.
            Soc. Canada* **89** 6
[Ber95a]   Bernabei R 1995 *Riv. Nuovo Cimento* **18** 1
[Ber96]   van den Berg D A, Statson P B and Bolte M 1996 *Ann. Rev. Astron.
            Astrophys.* **34** 461
[Ber97]   Bernabei R *et al* 1997 *Phys. Lett.* B **389** 757
[Bet38]   Bethe H A and Critchfield C L 1938 *Phys. Rev.* **54** 248, 862
[Bet39]   Bethe H A 1939 *Phys. Rev.* **55** 434
[Bet79]   Bethe H A, Brown G E, Applegate J and Lattimer J M 1979 *Nucl. Phys.*
            A **324** 487
[Bet82]   Bethe H A 1982 *Essays in Nuclear Astrophysics* eds C A Barnes, D D
            Clayton and D N Schramm (Cambridge: Cambridge University Press)
[Bet85]   Bethe H A and Wilson J F 1985 *Astrophys. J.* **295** 14
[Bet86]   Bethe H A 1986 *Phys. Rev. Lett.* **56** 1305
[Bet86a]   Bethe H A 1986 *Proc. Int. School of Physics 'Enrico Fermi', Course XCI
            (1984)* eds A Molinari and R A Ricci (Amsterdam: North-Holland)
            p 181
[Bet88]   Bethe H A 1988 *Ann. Rev. Nucl. Part. Sci.* **38** 1
[Bet92]   Bethke S and Pilcher J E 1992 *Ann. Rev. Nucl. Part. Sci.* **42** 251
[Bib87]   van Bibber K *et al* 1987 *Phys. Rev. Lett.* **59** 759

[Bib90]   van Bibber K *et al* 1990 *Proc. Cosmic Axions, (Brookhaven)* eds C Jones and A Melissinos (Singapore: World Scientific) 98

[Bie96]   Biermann P L 1996 *MPI für Radioastronomie Preprint* 694 (to be published in *J. Phys. G: Nucl. Part. Phys.*)

[Big92]   Bignami G F and Caraveo P A 1992 *Nature* **357** 287

[Big93]   Bignami G F, Caraveo P A and Mereghetti S 1993 *Nature* **361** 704

[Big96]   Bignami G F and Caraveo P A 1996 *Ann. Rev. Astron. Astrophys.* **34** 331

[Bil87]   Bilenky S and Petcov S T 1987 *Rev. Mod. Phys.* **59** 671

[Bin87]   Binney J and Tremaine S 1987 *Galactic Dynamics* (Princeton, NJ: Princeton University Press)

[Bio87]   Bionta R M *et al* (IMB Collaboration) 1987 *Phys. Rev. Lett.* **58** 1494

[Bir84]   Birkinshaw M, Gull S F and Hardebeck H 1984 *Nature* **309** 34

[Bir91]   Birkinshaw M, Hughes J P and Arnaud K A 1991 *Astrophys. J.* **379** 466

[Bir93]   Bird D J *et al* (Fly's Eye Collaboration) 1993 *Phys. Rev. Lett.* **71** 3401

[Bir95]   Bird D J *et al* (Fly's Eye Collaboration) 1995 *Astrophys. J.* **441** 144

[Bla87]   Blanford R D and Eichler D 1987 *Phys. Rep.* **154** 1

[Bla90]   Blanford R D, Netzer H and Woltjer L 1990 *Active Galactic Nuclei* (Berlin: Springer)

[Bla92]   Blanford R D and Narayan R 1992 *Ann. Rev. Astron. Astrophys.* **30** 311

[Bla97]   Blanc F *et al* 1997 Proposal ANTARES experiment, 1997, *Preprint* astro-ph/9707136

[Blo84]   Blome H J and Priester W 1984 *Naturwissenschaften* **71** 456, 515, 528

[Blu84]   Blumenthal G R *et al* 1984 *Nature* **311** 517

[Boe85]   Boesgaard A M and Steigman G 1985 *Ann. Rev. Astron. Astrophys.* **23** 318

[Boe88]   Boerner G 1988 *The Early Universe* (Berlin: Springer)

[Boe92]   Boehm F and Vogel P 1992 *Physics of Massive Neutrinos* (Cambridge: Cambridge University Press)

[Boe92a]   Boehm F 1992 *Trends in Astroparticle Physics* eds D Cline and R Peccei (Singapore: World Scientific) 533

[Böh94]   Böhringer H 1994 *Phys. in unserer Zeit* **3** 114

[Böh95]   Böhringer H 1995 *Ann. NY Acad. Sci.* **759** 67, *Proc. 17th TEXAS Symp. on Relativistic Astrophysics and Cosmology, Munich, Germany* eds H Böhringer, G E Morfill and J E Trümper

[Bol87]   Boldt E 1987 *Phys. Rep.* **146** 215

[Bol91]   Boliev M M *et al* 1991 *Proc. 3rd Int. Workshop on Neutrino Telescopes, Venice, Italy* ed M Baldo-Ceolin (Padova University) 235

[Bol96]   Boliev M M *et al* 1996 *Nucl. Phys. B (Proc. Suppl.)* **48** 83, *Proc. 4th Int. Workshop on Theoretical and Phenomenological Aspects of Underground Physics (TAUP'95), Toledo, Spain* eds A Morales, J Morales and J A Villar (Amsterdam: North-Holland)

[Bon88]   Bonnet-Bidaud J M and Chardin G 1988 *Phys. Rep.* **170** 325

[Bon94]   Bonvicini G 1994 *Nucl. Phys. B (Proc. Suppl.)* **35** 441, *Proc. 3rd Int. Workshop on Theoretical and Phenomenological Aspects in Astroparticle and Underground Physics, Assergi, Italy* eds C Arpesella, E Bellotti and A Bottino (Amsterdam: North-Holland)

[Bon96]   Bonn J 1996 *Proc. Neutrino '96 (Helsinki, June 1996)*

[Boo92a]   Booth N E and Salmon G (eds) 1992 *Proc. 4th Int. Conf. on Low Temperature Dark Matter and Neutrino Detectors, Oxford, 1991* (Paris: Edition Frontieres)

[Boo92b]   Booth N E *et al* 1992 *Nucl. Instrum. Methods* A **315** 201

[Bor85]   Boris S *et al* 1985 *Phys. Lett.* B **159** 217

[Bor86]   Borge M J G (ISOLDE Collaboration) 1986 *Phys. Scripta* **34** 591

[Bor87]   Boris S *et al* 1987 *Phys. Rev. Lett.* **58** 2019

[Bor91]   BOREXINO Collaboration 1991 *Proposal for a Solar Neutrino Detector at Gran Sasso*

[Bor92]   Borodovsky L *et al* 1992 *Phys. Rev. Lett.* **68** 274

[Bor96]   Boratav M 1996 *Preprint* astro-ph/9605087 and *Proc. 7th Int. Workshop on Neutrino Telescopes, Venice, Italy* ed M B Geolin

[Bos96]   Bosetti A 1996 *Nucl. Phys. B (Proc. Suppl.)* **48** 466, *Proc. 4th Int. Workshop on Theoretical and Phenomenological Aspects of Underground Physics (TAUP'95), Toledo, Spain* eds A Morales, J Morales and J A Villar (Amsterdam: North-Holland)

[Bot94a]   Bottino A *et al* 1994 *Astropart. Phys.* **2** 67

[Bot94b]   Bottino A *et al* 1994 *Astropart. Phys.* **2** 77

[Bot95]   Bottino A *et al* 1995 *Astropart. Phys.* **3** 65

[Bou89]   Bouquet A, Kaplan J and Martin F 1989 *Astron. Astrophys.* **222** 103

[Bou91]   Bouchet L 1991 *Astrophys. J.* **383** L45

[Bra59]   Bradner H and Isbell W M 1959 *Phys. Rev.* **114** 603

[Bra84]   Bracci L *et al* 1984 *Phys. Lett.* B **143** 357: 1985 *Phys. Lett.* B **155** 468 (Erratum)

[Bra88]   Braunschweig W *et al* 1988 *Z. Phys.* C **38** 543

[Bra88a]   Bratton C B *et al* (IMB Collaboration) 1988 *Phys. Rev.* D **37** 3361

[Bra90]   Bradaschia C *et al* (LISA Collaboration) 1990 *Nucl. Instrum. Methods* A **289** 518

[Bre97]   Breitweg J *et al* (ZEUS Collaboration) 1997 *Z. Phys.* C **74** 207

[Bro68]   Broadfoot A L and Kendall K R 1968 *J. Geophys. Res. Space Phys.* **73** 426

[Bro90]   Broadhurst *et al* 1990 *Nature* **343** 726

[Bru78]   Bruzual G A and Spinrad H 1978 *Astrophys. J.* **220** 1

[Bru85]   Bruenn S W 1985 *Astrophys. J. Suppl.* **58** 771

[Bru87]   Bruenn S W 1987 *Phys. Rev. Lett.* **59** 938

[Buc90]   Buckland K N *et al* 1990 *Phys. Rev.* D **41** 2726

[Buc90a]   Buchmüller W and Hoogeveen F 1990 *Phys. Lett.* B **237** 278

[Büh87]   Bührke T 1987 *Sterne und Weltraum* **4** 199

[Buo89]   Buonanno R, Corsi C E and Fusi Pecci F 1989 *Astron. Astrophys.* **216** 80

[Bur57]   Burbidge E M *et al* 1957 *Rev. Mod. Phys.* **29** 547

[Bur83]   Burnett T H *et al* (JACEE Collaboration) 1983 *Phys. Rev. Lett.* **51** 1010

[Bur85]   Burrows A 1985 *Solar Neutrinos and Astronomy* eds M L Cherry, W A Fowler and K Lande, AIP Conf. Proc. No **126** 283

[Bur85a]   Burrows A, Hayes J and Fryxell B A 1995 *Astrophys. J.* **450** 830

[Bur86]   Burrows A and Lattimer J M 1986 *Astrophys. J.* **307** 178

[Bur87]   Burnett T H *et al* (JACEE Collaboration) 1987 *Phys. Rev.* D **35** 824

[Bur88]   Burkhardt H *et al* (NA31 Collaboration) 1988 *Nucl. Instrum. Methods* A **268** 116

[Bur90]   Burstein D 1990 *Rep. Prog. Phys.* **53** 241

[Bur92]   Burrows A, Klein D and Gandhi R 1992 *Phys. Rev.* D **45** 3361
[Bur92a]   Burrows A 1992 *Trends in Astroparticle Physics* eds D Cline and R Peccei
             (Singapore: World Scientific) 463
[Bur93]   Burgess C P and Cline J M 1993 *Phys. Lett.* B **298** 141
[Bur94]   Burgess C P and Cline J M 1994 *Phys. Rev.* D **49** 5925
[Bur95]   Burrows A and Hayes J 1995 *Ann. NY Acad. Sci.* **759** 375
[Bur95a]   Burrows A, Hayes J and Fryvell B A 1995 *Astrophys. J* **450** 830
[Bus95]   Buskulic D *et al* (ALEPH Collaboration) 1995 *Phys. Lett.* B **349** 585
[Bus95a]   Buskulic D *et al* (ALEPH Collaboration) 1995 *Phys. Lett.* B **349** 238
[Bur97]   Burrows A 1997 *Preprint* astro-ph/9703008
[But93]   Butterworth J and Dreiner H 1993 *Nucl. Phys.* B **397** 3
[Byr90]   Byrne J *et al* 1990 *Phys. Rev. Lett.* **65** 289
[Cab82]   Cabrera B 1982 *Phys. Rev. Lett.* **48** 1378
[Cal82a]   Callan C 1982 *Phys. Rev.* D **25** 2141
[Cal90a]   Caldwell D O *et al* 1990 *Mod. Phys. Lett.* A **5** 1543
[Cal90b]   Caldwell D O *et al* 1990 *Phys. Rev. Lett.* **65** 1305
[Cal91]   Caldwell D O 1991 *J. Phys. G: Nucl. Phys.* **17** S 325
[Cal92]   Caldwell D O 1992 *Nucl. Phys.* B *(Proc. Suppl.)* **28A** 148, *Proc. 2nd Int.*
             Workshop on Theoretical and Phenomenological Aspects of Underground
             Physics 1991 (TAUP'91), *Toledo, Spain* eds A Morales, J Morales and
             J A Villar (Amsterdam: North Holland)
[Cal94]   Caldwell D O 1994 *Prog. Part. Nucl. Phys.* **32** 109
[Cal95]   Caldwell D O and Mohapatra R N 1995 *Preprint* UMD-PP-95 and 1995
             Phys. Lett. B **354** 371
[Cal96]   Caldwell D O 1996 *Nucl. Phys.* B *(Proc. Suppl.)* **48** 158, *Proc. 4th Int.*
             Workshop on Theoretical and Phenomenological Aspects of Underground
             Physics (TAUP'95), Toledo, Spain eds A Morales, J Morales and J A
             Villar (Amsterdam: North-Holland)
[Cal97]   Caldwell D 1997 in *Proc. Int. Conf. on Particle Physics Beyond the Standard*
             Model, 'BEYOND THE DESERT '97', Castle Ringberg, Germany eds
             H V Klapdor-Kleingrothaus and H Päs (Bristol: IOP)
[Cam91]   Campbell B A *et al* 1991 *Phys. Lett.* B **256** 457
[Cam93]   Cameron R *et al* 1993 *Phys. Rev.* D **47** 3707
[Can85]   Candelas P *et al* 1985 *Nucl. Phys.* B **258** 46
[Cap85]   Caplin A D *et al* 1985 *Nature* **317** 234
[Car74]   Carrigan R A Jr, Nezrick F A and Strauss B P 1974 *Phys. Rev.* D **10** 3867
[Car89]   Carlson E D and Hall L J 1989 *Phys. Rev.* D **40** 3187
[Car90]   Cariosi R *et al* (NA31 Collaboration) 1990 *Phys. Lett.* B **237** 303
[Car92]   Carroll S M, Press W H and Turner E L 1992 *Ann. Rev. Astron. Astrophys.*
             30 499
[Car93]   Carone C D 1993 *Phys. Lett.* B **308** 85
[Car94]   Carr B 1994 *Ann. Rev. Astron. Astrophys.* **32** 531
[Car95]   Carlson C E, Roy P and Sher M 1995 *Phys. Lett.* B **357** 99
[Car96]   Caraveo P A *et al* 1996 *Astrophys. J.* **461** L91
[Car96a]   Caron B *et al* 1996 *Nucl. Phys.* B *(Proc. Suppl.)* **48** 107, *Proc. 4th Int.*
             Workshop on Theoretical and Phenomenological Aspects of Underground
             Physics (TAUP'95), Toledo, Spain eds A Morales, J Morales and J A
             Villar (Amsterdam: North-Holland)

[Cas48]   Casimir H G B 1948 *Proc. Kon. Akad. Wet.* **51** 793
[Cav84]   Cavaignac J F *et al* 1984 *Phys. Lett.* B **148** 387
[Ces80]   Cesarsky C J 1980 *Ann. Rev. Astron. Astrophys.* **18** 289
[Cha39, 67] Chandrasekhar S 1939 and 1967 *An Introduction to the Study of Stellar Structure* (New York: Dover)
[Cha92]   Chaisson E J 1992 *Sci. Am.* **6** 18
[Cha94]   Chang C C *et al* 1994 *Nucl. Phys. B (Proc. Suppl.)* **35** 464, *Proc. 3rd Int. Workshop on Theoretical and Phenomenological Aspects in Astroparticle and Underground Physics (TAUP'93), Assergi, Italy* eds C Arpesella, E Bellotti and A Bottino (Amsterdam: North-Holland)
[Cha95]   Chaboyer B 1996 *Proc. 2nd UCLA Conf. on Sources and Detection of Dark Matter in the Universe, Santa Monica, CA, 1995, Nucl. Phys. B (Proc. Suppl.)* **51** 10 ed D B Cline (Amsterdam: North-Holland)
[Cha95a]   Charles P A and Seward F D 1995 *Exploring the X-ray Universe* (Cambridge: Cambridge University Press)
[Che80]   Chen J H and Wasserburg G J 1980 *Lunar Planetary Sci.* **11** 131
[Che88]   Cheng H 1988 *Phys. Rep.* **158** 1
[Che92]   Chevalier R A 1992 *Nature* **355** 691
[Chi80]   Chikashige Y, Mohapatra R N and Peccei R D 1980 *Phys. Rev. Lett.* **45** 1926
[Cho97]   Choudhury D and Raychaudhuri S 1997 *Preprint* hep-ph/9702392
[Chr64]   Christenson J H *et al* 1964 *Phys. Rev. Lett.* **13** 138
[Chr72]   Christensen C J *et al* 1972 *Phys. Rep.* D **5** 1628
[Chr94]   Christensen-Dalsgaard J 1994 *Europhys. News* **25** 71
[Chr96]   Christensen-Dalsgaard J 1996 *Nucl. Phys. B (Proc. Suppl.)* **48** 325, *Proc. 4th Int. Workshop on Theoretical and Phenomenological Aspects of Underground Physics (TAUP'95), Toledo, Spain* eds A Morales, J Morales and J A Villar (Amsterdam: North-Holland)
[Chu89]   Chupp E L, Vestrand W T and Reppin C 1989 *Phys. Rev. Lett.* **62** 505
[Chu97]   Chung D J H, Farrar G R and Kolb E W 1997 *Preprint* astro-ph/9707036
[Cin93]   Cinabro D *et al* (CLEO Collaboration) 1993 *Phys. Rev. Lett.* **70** 3701
[Cla61]   Clayton D D *et al* 1961 *Ann. Phys.* **12** 331
[Cla68]   Clayton D D 1968 *Principles of Stellar Evolution* (New York: McGraw-Hill)
[Cle95]   Cleveland B T *et al* 1995 *Nucl. Phys. B (Proc. Suppl.)* **38** 47, *Proc. 16th Int. Conf. Neutrino Physics and Astrophysics, NEUTRINO '94* eds A Dar, G Eilam and M Gronau (Amsterdam: North-Holland)
[Cli92]   Cline D 1992 *Proc. Fourth Int. Workshop on Neutrino Telescopes (Venezia, 10–13 March 1992)* eds M Baldo-Ceolin (Padova University) p 399
[Coh93]   Cohen A G, Kaplan D B and Nelson A E 1993 *Ann. Rev. Nucl. Part. Sci.* **43** 27
[Col73]   Coleman S and Weinberg E 1973 *Phys. Rev.* D **7** 1888
[Col88]   Coleman S 1988 *Nucl. Phys.* B **310** 643
[Col89]   Collins P D B, Martin A D and Squires E J 1989 *Particle Physics and Cosmology* (London: Wiley)
[Col90]   Colgate S A 1990 *Supernovae, Jerusalem Winter School* **6** eds J C Wheeler, T Piran and S Weinberg (Singapore: World Scientific) 249
[Col90a]   Colgate S A 1990 *Supernovae* ed S E Woosley (Berlin: Springer) p 352

[Col95]   Colgate S A and Leonard P J T 1995 *Proc. 24th Int. Cosmic Ray Conf. (Rome, 28 August–8 September 1995)* **2** 152

[Com83]   Commins E D and Bucksham P H 1983 *Weak Interaction of Leptons and Quarks* (Cambridge: Cambridge University Press)

[Coo84]   Cooperstein J, Bethe H A and Brown G E 1984 *Nucl. Phys.* A **429** 527

[Coo93]   Cooper S *et al* 1993 *Proposal MPI-PhE Preprint* **93-29**

[Cop95]   Copi C J, Schramm D N and Turner M S 1995 *Science* **267** 192

[Cow72]   Cowsik R and McClelland J 1972 *Phys. Rev. Lett.* **29** 669

[Cow73]   Cowsik R and Wilson L 1973 *Proc. 13th Int. Conf. on Cosmic Rays, (Denver)* **1** 577

[Cow91]   Cowan J J, Thielemann F K and Truran J W 1991 *Phys. Rep.* **208** 267

[Cow92]   Cowan J J, Thielemann F K and Truran J W 1992 *Ann. Rev. Astron. Astrophys* **29** 447

[Cox68]   Cox J P and Guili R T 1968 *Stellar Structure and Evolution* (New York: Gordon and Breach)

[Cra86]   Crane P *et al* 1986 *Astrophys. J.* **309** 822

[Cre79]   Crewther R *et al* 1979 *Phys. Lett.* B **88** 323

[Cre83]   Creutz M 1983 *Quarks, Gluons and Lattices* (Cambridge: Cambridge University Press)

[Cro86]   Cromar M W *et al* 1986 *Phys. Rev. Lett.* **56** 2561

[Cro93]   Cronin J W, Gibbs K G and Weekes T C 1993 *Ann. Rev. Nucl. Part. Sci.* **43** 883

[Cro96]   Crotts A P S 1996 *Preprint* astro-ph/9610067

[Cun91]   Cundy D C 1991 *Proc. NEUTRINO '90, Nucl. Phys. B (Proc. Suppl.)* **19** 227, eds J Panman and K Winter

[DaC92]   DaCosta G S 1992 The stellar population of galaxies, Brazil, 1991, eds B Barbuy and A Renzini (Dordrecht: Kluwer), *IAU Proc.* **149** 191

[DaC94]   DaCosta G S *et al* 1994 *Astrophys. J.* **424** L1

[Dan95]   Danevich F A *et al* 1995 *Phys. Lett.* B **344** 72

[Dar96]   Dar A and Shaviv G 1996 *Nucl. Phys. B (Proc. Suppl.)* **48** 335, *Proc. 4th Int. Workshop on Theoretical and Phenomenological Aspects of Underground Physics (TAUP'95), Toledo, Spain* eds A Morales, J Morales and J A Villar (Amsterdam: North-Holland)

[Dav64]   Davis R 1964 *Phys. Rev. Lett.* **12** 303

[Dav83]   Davis M and Peebles P J E 1983 *Astrophys. J.* **267** 465

[Dav84]   Davis R, Cleveland B T and Rowley J K 1984 *Proc. Conf. on Interactions Between Particle and Nuclear Physics (Steamboat Springs, 23–30 May 1984) AIP Conf. Proc.* **123** ed R E Mischke

[Dav89]   Davidson K, Kinman T D and Friedman S D 1989 *Astrophys. J.* **97** 1591

[Dav92a]   Davis M, Summers F J and Schlegel D 1992 *Nature* **359** 393

[Dav92b]   Davis R L *et al* 1992 *Phys. Rev. Lett.* **69** 1856

[Dav92c]   Davis P 1992 *Sky and Telescope* **1** 20

[Dav92d]   Davis P (ed) 1992 *The New Physics* (Cambridge: Cambridge University Press)

[Dav94a]   Davis R 1994 *Prog. Part. Nucl. Phys.* **32** 13

[Dav94b]   Davis R 1994 *Proc. 6th Workshop on Neutrino Telescopes, Venezia, Italy* ed M Baldo-Ceolin (Padova University)

[Dav96]   Davis R Jr 1996 *Nucl. Phys. B (Proc. Suppl.)* **48** 284,   *Proc. 4th Int. Workshop on Theoretical and Phenomenological Aspects of Underground Physics (TAUP'95), Toledo, Spain* eds A Morales, J Morales and J A Villar (Amsterdam: North-Holland)

[Daw95]   Dawson B R 1995 *Ann. NY Acad. Sci.* **759** 460, *Proc. 17th TEXAS Int. Symp. on Relativistic Astrophysics and Cosmology, Munich, Germany* eds N Böhringer, G E Morfill and J E Trümper

[Dea86]   Dearborn D S P, Schramm D N and Steigman G 1986 *Astrophys. J.* **302** 35

[deB94]   de Boer W 1994 *Prog. Nucl. Part. Phys.* **33** 201

[Dec92a]   Decamp D *et al* (ALEPH Collaboration) 1992 *Phys. Lett.* B **284** 151

[Dec92b]   Decamp D *et al* (ALEPH Collaboration) 1992 *Phys. Rep.* **216** 253

[Ded95]   Dedenko L G *et al* 1995 *JETP Lett.* **61** 241

[Dek90]   Dekel A, Bertschinger E and Faber S M 1990 *Astrophys. J.* **364** 349

[Dek92]   Dekel A 1993 *Proc. 4th Rencontres de Blois: Particle Astrophysics* eds G Fontaine and J Thanh Van (Paris: Editions Frontieres)

[Dek93]   Dekel A *et al* 1993 *Astrophys. J.* **412** 1

[Dek94]   Dekel A 1994 *Ann. Rev. Astron. Astrophys.* **32** 371

[deL86]   de Lapparant N, Geller M and Huchra J 1986 *Astrophys. J.* **302** L1

[Den90]   Denegri D, Sadoulet B and Spiro M 1990 *Rev. Mod. Phys* **62** 1

[Den97]   Denegri D 1997 *Proc. DARK '96, Proc. Int. Workshop on Dark Matter in Astro- and Particle Physics, Heidelberg 1996* eds H V Klapdor-Kleingrothaus and Y Ramachers (Singapore: World Scientific)

[deR96]   de Rujula A 1996 *Nucl. Phys. B (Proc. Suppl.)* **48** 514,   *Proc. 4th Int. Workshop on Theoretical and Phenomenological Aspects of Underground Physics (TAUP'95), Toledo, Spain* eds A Morales, J Morales and J A Villar (Amsterdam: North-Holland)

[Der83]   Derbin A V und Popeko L A 1983 *Sov. J. Nucl. Phys.* **38** 665

[Der93a]   Derbin A I *et al* 1993 *JETP Lett.* **57** 769

[Der93b]   Derrick M *et al* (ZEUS Collaboration) 1993 *Phys. Lett.* B **306** 173

[Der95]   Derrick M *et al* (ZEUS Collaboration) 1995 *Z. Phys.* C **65** 379

[Der95a]   Derrick M *et al* (ZEUS Collaboration) 1995 *Z. Phys.* C **65** 627

[Der96]   Derrick M *et al* (ZEUS Collaboration) 1996 *Preprint* DESY-96-076

[deS97]   de Santo A, Popov B *et al* (NOMAD Collaboration) 1997 *Proc. 32nd Rencontres des Moriond, Electroweak Interactions and Unified Theories*

[DeW67]   DeWitt B S 1967 *Phys. Rev.* **160** 1113

[DiC93]   DiCredico A *et al* (MACRO Collaboration) 1994 *Proc. 23rd Int. Conf. on Cosmic Rays, Calgary, Canada, 1993* eds R B Hicks, D A Leahy, D Venkatesan and R Edge (Singapore: World Scientific)

[Die93]   Diehl R *et al* 1993 *Astron. Astrophys. Suppl.* **97** 181

[Die95]   Diehl R 1995 *Ann. NY Acad. Sci.* **759** 384, *Proc. 17th TEXAS Int. Symp. on Relativistic Astrophysics and Cosmology, Munich, Germany* eds N Böhringer, G E Morfill and J E Trümper

[Dim86]   Dimopolous S, Starkman G and Lynn B 1986 *Phys. Lett.* B **168** 145

[Din81]   Dine M, Fischler W and Srednicki M 1981 *Phys. Lett.* B **104** 199

[Din85]   Dine M *et al* 1985 *Phys. Lett.* B **156** 55

[Din90]   Dine M 1990 *Ann. Rev. Nucl. Part. Sci.* **40** 145

[Din92] Dine M 1992 *Particles and Fields '91: Meeting of the Division of Particles and Fields of the APS, Vancouver, Canada* eds D Axen, D Bryman, M Comyn and R Edge (Singapore: World Scientific) 831

[Dir31] Dirac P A M 1931 *Proc. R. Soc. London* A **133** 60

[Dir37] Dirac P A M 1937 *Nature* **139** 323

[Doi85] Doi M, Kotani T and Takasugi E 1985 *Prog. Theor. Phys. Suppl.* **83** 1

[Dol81] Dolgov A D and Zeldovich Y B 1981 *Rev. Mod. Phys.* **53** 1

[Dol83] Dolgov A D 1983 *Proc. 1982 Nuffield Workshop 'The Very Early Universe'* eds G W Gibbons, S W Hawking and S T C Siklos (Cambridge: Cambridge University Press) 449

[Dol90] Doll P *et al* 1990 *KfK-Report* **4686** (Kernforschungszentrum Karlsruhe)

[Dol92] Dolgov A D 1992 *Phys. Rep.* **222** 309

[Dom87] Dombeck T *et al* 1987 *Phys. Lett.* B **194** 491

[Don92] Donoghue J F, Golowich F and Holstein B R 1992 *Dynamics of the Standard Model* (Cambridge: Cambridge University Press)

[Dra87] Dragon N, Ellwanger U and Schmidt M G 1987 *Prog. Nucl. Part. Phys.* **18** 1

[Dre83] Drell S D *et al* 1983 *Phys. Rev. Lett.* **50** 644

[Dre87] Dressler A *et al* 1987 *Astrophys. J.* **313** L37

[Dre91a] Dressler A and Faber S M 1991 *Astrophys. J.* **354** L45

[Dre91b] Dressler A 1991 *Nature* **350** 391

[Dre94] Dreiner H and Morawitz P 1994 *Nucl. Phys.* B **428** 31

[Dub91] Dubbers D 1991 *Prog. Part. Nucl. Sci.* **26** 173

[Dun92] Duncan D, Lambert D and Lemke D 1992 *Astrophys. J.* **401** 584

[Dus92] Duschl W J and Wagner S J (eds) 1992 *Physics of Active Galactic Nuclei* (Berlin: Springer)

[Eck96] Eckhart A and Genzel R 1996 *Nature* **383** 415

[Eck97] Eckhart A and Genzel R 1997 *Mon. Not. R. Astron. Soc.* **284** 576

[Efi88] Efimov N N *et al* 1988 *Catalogue of the Highest Energy Cosmic Rays, WDC-C2 for Cosmic Rays* **3** 1

[Efs82] Efstathiou G, Ellis R S and Carter D 1982 *Mont. Not. R. Astron. Soc.* **201** 975

[Efs83] Efstathiou G and Silk J 1983 *Fund. Cosmic Phys.* **9** 1

[Egg95] Eggert K 1995 *Nucl. Phys. B (Proc. Suppl.)* **38** 240, *Proc. 16th Int. Conf. on Neutrino Physics and Astrophysics, NEUTRINO '94* eds A Dar, G Eilam and M Gronau (Amsterdam: North-Holland)

[Ein17] Einstein A 1917 *Sitzungsberichte Preuß. Akad. Wiss.* **142**

[Eji92] Ejiri H *et al* 1992 *Nucl. Phys. B (Proc. Suppl.)* **28A** 219, *Proc. 4th Int. Workshop on Theoretical and Phenomenological Aspects of Underground Physics (TAUP'95), Toledo, Spain* eds A Morales, J Morales and J A Villar (Amsterdam: North-Holland)

[Ell87a] Elliot S R, Hahn A A and Moe M K 1987 *Phys. Rev. Lett.* **59** 1649, 2020

[Ell87b] Ellis J, Flores R A and Ritz S 1987 *Phys. Lett.* B **198** 393

[Ell84] Ellis J *et al* 1984 *Nucl. Phys.* B **238** 453

[Ell90] Ellis J *et al* 1990 *Phys. Lett.* B **245** 251

[Ell91a] Ellis J, Kelly S and Nanopoulos D V 1991 *Phys. Lett.* B **260** 131

[Ell91b] Ellis J 1991 *Int. School of Astroparticle Physics, Woodlands, Texas* eds D V Nanopoulos and R Edge (Singapore: World Scientific)

[Ell92]   Elliott S R *et al* 1992 *Phys. Rev.* C **46** 1535

[Ell93a]   Ellis J 1993 *Ann. NY Acad. Sci.* **688** 164

[Ell93b]   Ellis J and Flores R 1993 *Phys. Lett.* B **300** 175

[Ell93c]   Ellis J and Karliner M 1993 *Phys. Lett.* B **313** 131

[Ell93d]   Ellis R S 1993 *Sky surveys: Protostars to Protogalaxies* ed B T Stoifer, ASP Conf. Series **43** 165

[Err83]   Errede S M 1983 *Monopole '83* ed J L Stone (Nato ASI Series) (New York: Plenum) p 251

[Ewa92]   Ewan G T 1992 *Nucl. Instrum. Methods* A **314** 373

[Ewa95]   Ewan G T 1995 *Proc Weak and Electromagnetic Interactions in Nuclei (WEIN'95), Osaka* eds H Ejiri, T Kishimoto and T Sato (Singapore: World Scientific) p 647

[Fab76]   Faber S M and Jackson R E 1976 *Astrophys. J.* **204** 668

[Fab92]   Fabian A C and Barcons X 1992 *Ann. Rev. Astron. Astrophys.* **30** 429

[Fal94]   Falk T, Olive K and Srednicki M 1994 *Phys. Lett.* B **339** 248

[Fei88]   von Feilitzsch F 1988 in *NEUTRINOS* ed H V Klapdor (Berlin: Springer) p 1

[Fer34]   Fermi E 1934 *Z. Phys.* **88** 161

[Fer94]   Ferger P *et al* 1994 *Phys. Lett.* B **323** 95

[Fey65]   Feynman R P and Hibbs A H 1965 *Quantum Mechanics and Path Integrals* (New York: McGraw-Hill)

[Fic75]   Fichtel C E *et al* 1975 *Astrophys. J.* **198** 163

[Fic91]   Fich M and Tremaine S 1991 *Ann. Rev. Astron. Astrophys.* **29** 409

[Fic95]   Fichtel C E 1995 *Ann. NY Acad. Sci.* **759** 221

[Fid61]   Fidecaro M, Finocchiaro G and Giacomelli G 1961 *Nuovo Cimento* **22** 657

[Fil93]   Filippenko A V 1993 *Sky and Telescopy* **12** 30

[Fio95]   Fiorentini G, Kavanagh R W and Rolfs C 1995 *Z. Phys.* A **350** 289

[Fio96]   Fiorini E 1996 *Proc. NEUTRINO '96 (Helsinki, June 1996)*

[Fis81]   Fisher J R and Tully R B 1981 *Astrophys. J. Suppl.* **47** 119

[Fis92]   Fisher K B *et al* 1992 *Astrophys. J.* **402** 42

[Fis95]   Fisher K B *et al* 1995 *Astrophys. J. Suppl.* **100** 69

[Fis95a]   Fishman G J and Meegan C A 1995 *Ann. Rev. Astron. Astrophys.* **33** 415; Fishman G J 1995 *Ann. NY Acad. Sci.* **759** 232

[Fix96]   Fixsen D J *et al* 1996 *Preprint* astro-ph/9605054

[Fla75]   van Flandern T C 1975 *Mon. Not. R. Astron. Soc.* **170** 333

[Fla76]   van Flandern T C 1976 *Sci. Am.* **234** 44

[Fla96]   Flanz M, Paschos E A, Sarkar U and Weiss J 1996 *Phys. Lett.* B **389** 693

[Fon70]   Fonda L and Chirardi G C 1970 *Symmetry Principles in Quantum Physics* (New York: Dekker)

[Fon95]   Fonseca V *et al* (HEGRA Collaboration) 1995 *Proc. XXIV Int. Cosmic Rays Conf., Rome* 474

[For85]   Forman W, Jones C and Tucker W 1985 *Astrophys. J.* **293** 535

[For95]   Forty R 1995 *Proc. 27th Int. Conf. on High Energy Physics, Glasgow* eds P J Bussey and I G Knowles (Bristol: IOP Publishing) 171

[Fow60]   Fowler W A and Hoyle F 1960 *Ann. Phys.* **10** 280

[Fow72]   Fowler W A 1972 *Cosmology, Fusion and Other Matters* ed F Reines (Boulder, CO: Associated University Press) 67

[Fow75]   Fowler W A, Caughlan G R and Zimmerman B A 1975 *Ann. Rev. Astron. Astrophys.* **13** 69

[Fow78]   Fowler W A 1978 *Proc. Welch Foundation Conf. on Chemical Research* ed W D Milligan (Houston, TX: Houston University Press) 61

[Fra90]   Frampton P H 1990 *Phys. Rev. Lett.* **64** 619

[Fra97]   Frampton P H 1997 *Proc. Beyond the Standard Model V, Balestrand, Norway, May 1997, Preprint* hep-ph/9706220

[Fre74]   Freedman D Z 1974 *Phys. Rev.* D **9** 1389

[Fre77]   Freedman D Z, Schramm D N and Tubbs D L 1977 *Ann. Rev. Nucl. Part. Sci.* **27** 167

[Fre83]   Freese K, Turner M S and Schramm D N 1983 *Phys. Rev. Lett.* **51** 1625

[Fre86]   Freese K 1986 *Phys. Lett.* B **167** 295

[Fre93]   Freedman S J *et al* 1993 *Phys. Rev.* D **47** 811

[Fre96]   Freedman J W L 1996 *Preprint* astro-ph/9612024, *Proc. Critical Dialogues in Cosmology (Princeton, NJ, 1996)* ed N Turok (Singapore: World Scientific)

[Fri75]   Fritzsch H and Minkowski P 1975 *Ann. Phys.* **93** 193

[Fri86]   Fritschi M *et al* 1986 *Phys. Lett.* **173** 485

[Fri88]   Frieman J A, Haber H E and Freese K 1988 *Phys. Lett.* B **200** 115

[Fri91]   Fritschi M *et al* 1991 *Nucl. Phys.* B *(Proc. Suppl.)* **19** 205, *Proc. 14th Int. Conf. on Neutrino Physics and Astrophysics (NEUTRINO'90)* eds J Panman and K Winter (Amsterdam: North-Holland)

[Fuj80]   Fujikawa K and Shrock R E 1980 *Phys. Rev. Lett.* **45** 963

[Fuk87]   Fukugita M and Yanagida T 1987 *Phys. Rev. Lett.* **58** 1807

[Fuk90a]   Fukugita M and Yanagida T 1990 *Phys. Rev.* D **42** 1285

[Fuk90b]   Fukugita M, Futamase T and Kasai M 1990 *Mon. Not. R. Astron. Soc.* **246** 24

[Fuk94]   Fukuda Y *et al* (KAMIOKANDE Collaboration) 1994 *Phys. Lett.* B **335** 237

[Fuk96]   Fukuda Y *et al* (KAMIOKANDE Collaboration) 1996 *Phys. Rev. Lett.* **77** 1683

[Ful82a]   Fuller G M, Fowler W A and Newman M J 1982 *Astrophys. J.* **252** 715

[Ful82b]   Fuller G M, Fowler W A and Newman M J 1982 *Astrophys. J. Suppl.* **48** 289

[Ful82c]   Fuller G M 1982 *Astrophys. J.* **252** 741

[Ful93]   Fuller G M 1993 *Phys. Rep.* **227** 143 and Fuller G M 1992 *Astrophys. J.* **389** 517

[Ful95]   Fuller G M, Primack J R and Qian Y Z 1995 *Phys. Rev.* D **52** 656, 1288

[Gab95]   Gabrielse G *et al* 1995 *Phys. Rev. Lett.* **74** 3544

[Gai90]   Gaisser T K 1990 *Cosmic Rays and Particle Physics* (Cambridge: Cambridge University Press)

[Gai94]   Gaisser T K 1994 *Nucl. Phys.* B *(Proc. Suppl.)* **35** 209, *Proc. 3rd Int. Workshop on Theoretical and Phenomenological Aspects in Astroparticle and Underground Physics (TAUP'93), Assergi, Italy* eds C Arpesella, E Bellotti and A Bottino (Amsterdam: North-Holland)

[Gai94a]   Gaisser T K 1994 *Phil. Trans. Soc.* A **346** 75

[Gai95]   Gaisser T K, Halzen F and Stanev T 1995 *Phys. Rep.* **258** 173

[Gai96a]   Gaisser T K 1996 *Nucl. Phys. B (Proc. Suppl.)* **48** 405   *Proc. 4th Int. Workshop on Theoretical and Phenomenological Aspects of Underground Physics (TAUP'95), Toledo, Spain* eds A Morales, J Morales and J A Villar (Amsterdam: North-Holland)

[Gai96b]   Gaisser T K 1996 *Proc. NEUTRINO '96, Helsinki 1996*

[Gai97]   Gaitskell 1997 in *Proc. Int. Conf. on Particles Physics Beyond the Standard Model, 'BEYOND THE DESERT '97', Castle Ringberg, Germany* eds H V Klapdor-Kleingrothaus and H Päs (Bristol: IOP)

[Gaj92]   Gajewski W *et al* (IMB-Collaboration) 1992 *Nucl. Phys. B (Proc. Suppl.)* **28**, *Proc. 2nd Int. Workshop on Theoretical and Phenomenological Aspects of Underground Physics (TAUP'91), (Toledo, Spain)* (Amsterdam: North-Holland)

[Gal94]   Galeotti P *et al* (LVD Collaboration) 1994 *Nucl. Phys. B (Proc. Suppl.)* **35** 267, *Proc. 3rd Int. Workshop on Theoretical and Phenomenological Aspects in Astroparticle and Underground Physics (TAUP'93) Assergi, Italy* ed C Arpesella, E Belloti and A Bottino (Amsterdam: North Holland)

[Gal96]   Galeotti P *et al* (LVD Collaboration) 1996 *Proc. Int. Cosmic Rays Conf. ICRC '95, Rome, Italy, 1995*

[Gal97]   Galeazzi M *et al* 1997 *Phys. Lett.* B **398** 187

[Gam38]   Gamow G 1938 *Phys. Rev.* **53** 595

[Gam46]   Gamow G 1946 *Phys. Rev.* **70** 572

[Gan93]   Ganga K *et al* 1993 *Astrophys. J.* **410** L57

[Gan95]   Gandhi R *et al* 1995 *Preprint* hep-ph/9512364

[Gar91]   Gardner R D *et al* 1991 *Phys. Rev.* D **44** 622

[Gas97]   Gasperini M 1997 *Preprint* gr-ge/9707034

[Gat95]   Gates E I, Gynk G and Turner M S 1995 *Phys. Rev. Lett.* **74** 3724

[Gau86]   Gauthier A 1986 *Proc. XXIII Int. Conf. on High-Energy Physics (Berkeley)* (Singapore: World Scientific)

[Gav90]   Gavrin V N *et al* (SAGE Collaboration) 1990 *IAU Proc.* **121** *Inside the Sun, Versailles, France* eds G Berthomieu and M Cribier (Berlin: Kluwer) 201

[Geh90]   Gehrels N 1990 *Nucl. Instrum. Methods* A **292** 505

[Geh93]   Gehrels N and Chen W 1993 *Nature* **361** 706

[Gel64]   Gell-Mann M 1964 *Phys. Lett.* **8** 214

[Gel78]   Gell-Mann M, Ramond P and Slansky R 1978 *Supergravity* eds F van Nieuwenhuizen and D Freedman (Amsterdam: North Holland) 315

[Gel81]   Gelmini G B and Roncadelli M 1981 *Phys. Lett.* B **99** 411

[Gel88]   Gelmini G B 1988 in *Neutrinos* ed H V Klapdor (Berlin: Springer) p 309

[Gel89]   Geller M and Huchra J P 1989 *Science* **246** 897

[Gel91]   Gelmini G, Nussinov S and Peccei R D 1991 *Preprint* UCLA-91-TEP-15

[Gel92]   Gelmini G and Yanagida T 1992 *Phys. Lett.* B **294** 53

[Gel95]   Gelmini G and Roulet E 1995 *Rep. Prog. Phys.* **58** 1207

[Gen87]   Genzel R and Townes C H 1987 *Ann. Rev. Astron. Astrophys.* **25** 377

[Gen96]   Genzel R *et al* 1996 *Astrophys. J.* **472** 153

[Geo74]   Georgi H and Glashow S L 1974 *Phys. Rev. Lett.* **32** 438

[Geo75]   Georgi H 1975 *Particles and Fields* ed C E Carloso (New York: AIP)

[Geo97]	Georgadze A S, Klapdor-Kleingrothaus H V, Päs H and Zdesenko Yu G 1997 *Astroparticle Phys.* **7** 173
[Ger90]	Gerbier G *et al* 1990 *Phys. Rev.* D **42** 3211
[Ger94]	Gerbier G *et al* (BRPS Collaboration) 1994 *Nucl. Phys. B (Proc. Suppl.)* **35** 159, *Proc. 3rd Int. Workshop on Theoretical and Phenomenological Aspects in Astroparticle and Underground Physics (TAUP'93), Assergi, Italy* eds C Arpesella, E Bellotti and A Bottino (Amsterdam: North-Holland)
[Ger96]	Gervasio *et al* 1996 in *Double Beta Decay and Related Topics* eds H V Klapdor-Kleingrothaus and S Stoica (Singapore: World Scientific) 475
[Ger96a]	Gerdes D *et al* (CDF Collaboration) *Fermilab-conf. 96/342E*
[Gia71]	Giacconi R *et al* 1971 *Astrophys. J.* **230** 540
[Gia94]	Giammarchi M G *et al* 1994 *Nucl. Phys. B (Proc. Suppl.)* **35** 433, *Proc. 3rd Int. Workshop on Theoretical and Phenomenological Aspects in Astroparticle and Underground Physics (TAUP'93), Assergi, Italy* eds C Arpesella, E Bellotti and A Bottino (Amsterdam: North-Holland)
[Gib93]	Gibbons L K *et al* (E731 Collaboration) 1993 *Phys. Rev. Lett.* **70** 1203
[Gil91]	Gilmore G, Edvardson B and Nissen P 1991 *Astrophys. J.* **378** 17
[Gin64]	Ginzburg V I and Syrovatskii S I 1964 *The Origin of Cosmic Rays* (London: Pergamon)
[Gin80]	Ginzburg V L, Khazan Y M and Ptuskin V S 1980 *Astron. Space Sci.* **68** 295
[Gio91]	Giovanelli R and Haynes M 1991 *Ann. Rev. Astron. Astrophys.* **29** 499
[Gla61]	Glashow S L 1961 *Nucl. Phys.* **22** 579
[Goe94]	Goenner H 1994 *Einführung in die Kosmologie* (Berlin: Spektrum Akademischer)
[Gol62]	Goldstone J, Salam A and Weinberg S 1962 *Phys. Rev.* **127** 965
[Gol85]	Goldstein H 1985 *Klassische Mechanik* (Wiesbaden: Aula)
[Gol88]	Goldman I *et al* 1988 *Phys. Rev. Lett.* **60** 1789
[Gol92]	Goldwurm A *et al* 1992 *Astrophys. J.* **389** L79
[Gon96]	Gonin M *et al* (NA50 Collaboration) 1996 *Proc. Quark Matter '96 (Heidelberg, Germany)* eds P Braun-Munzinger, H J Specht, R Stock and H Stocker (Amsterdam: North-Holland), *Nucl. Phys.* A **610** 404
[Goo85]	Goodman M W and Witten E 1985 *Phys. Rev.* D **31** 3059
[Goo95a]	Goobar A and Perlmutter S 1995 *Astrophys. J.* **450** 14
[Goo95b]	Goodman M C (Soudan 2 Collaboration) 1995 *Nucl. Phys. B (Proc. Suppl.)* **38** 337, *Proc. 16th Int. Conf. on Neutrino Physics and Astrophysics, NEUTRINO '94* eds A Dar, G Eilam and M Gronau (Amsterdam: North-Holland)
[Gou92]	Gould A 1992 *Astrophys. J.* **388** 338
[Gou96]	Gough D O *et al* 1996 *Science* **272** 1281
[Gra96]	Gratta G *et al* (PALO VERDE Collaboration) 1996 *Proc. NEUTRINO '96 (Helsinki, June 1996)*
[Gre66]	Greisen K 1966 *Phys. Rev. Lett.* **16** 748
[Gre86]	Green M and Schwarz J 1986 *Phys. Lett.* B **151** 21
[Gre86a]	Greiner W and Müller B 1986 *Theoretische Physik Bd 8 Eichtheorie der Schwachen Wechselwirkung* (Frankfurt: Deutsch)

468   *References*

[Gre87]   Green M, Schwarz J and Witten E 1987 *Superstring Theory* vol I, II (Cambridge: Cambridge University Press)

[Gre89]   Greiner W and Schäfer A 1989 *Theoretische Physik Bd 10 Quantenchromodynamik* (Frankfurt: Deutsch)

[Gre91]   Gregory P C and Condon J J 1991 *Astrophys. J. Suppl.* **75** 1011

[Gre94]   Greife U *et al* 1994 *Nucl. Instrum. Methods* A **350** 326

[Gri75]   Grishchuk S P 1975 *Sov. Phys.–JETP* **40** 409

[Gri88]   Griest K 1988 *Phys. Rev.* D **38** 2357; 1988 *Phys. Rev.* D (Erratum) **39** 3802

[Gri90]   Griest K and Kamionkowski M 1990 *Phys. Rev. Lett.* **64** 615

[Gro73]   Gross D J and Wilczek F 1973 *Phys. Rev. Lett.* **30** 1343

[Gro76]   Groote H v, Hilf E R and Takahashi K 1976 *At. Data Nucl. Data Tables* **17** 418

[Gro84]   Grotz K, Klapdor H V and Metzinger J 1984 *Proc. Int. Symp. on Capture Gamma Ray Spectroscopy (September 1984, Knoxville, USA)* AIP Proc. No **125** 793

[Gro86a]   Groom D 1986 *Phys. Rep.* **140** 323

[Gro86b]   Grotz K and Klapdor H V 1986 *Nucl. Phys.* A **460** 395

[Gro86c]   Grotz K, Klapdor H V and Metzinger J 1986 *Phys. Rev.* C **33** 1263; 1986 *Astron. Astrophys.* **154** L1

[Gro89]   Grotz K and Klapdor H V 1989 *Die Schwache Wechselwirkung in Kern-, Teilchen- und Astrophysik* (Stuttgart: Teubner)

[Gro90]   Grotz K and Klapdor H V 1990 *The Weak Interaction in Nuclear, Particle and Astrophysics* (Bristol: Hilger)

[Gro93]   Gross D J 1993 *Ann NY Acad. Sci.* **688** 148

[Gru93]   Gruwe M *et al* (CHARM-II Collaboration) 1993 *Phys. Lett.* B **309** 463

[Gun78]   Gunn J E 1978 *Observational Cosmology, SAAS-FE lecture series, Geneva Observatory* vol 8 eds A Maeder, L Martinet and G Tammann

[Gun90]   Gunion G F, Haber H E, Kane G and Dawson S 1990 *The Higgs–Hunter Guide, Frontiers in Physics* vol 80 (London: Addison-Wesley)

[Gun94]   Gunn J E and Weinberg D H 1994 *Preprint* astro-ph/9412080 and in *Proc. Wide-field Spectroscopy in the Distant Universe, Cambridge, UK, 1994* (Singapore: World Scientific)

[Gün97]   Günter M *et al* (HEIDELBERG–MOSCOW Collaboration) 1997 *Phys. Rev.* D **55** 54

[Gup91]   Gupta S K 1991 *Astron. Astrophys.* **245** 141

[Gur76]   Gursey F, Ramond P and Sikivie P 1976 *Phys. Lett.* B **60** 177

[Gut81]   Guth A H 1981 *Phys. Rev.* D **23** 347

[Gut89]   Guth A H 1989 in *Bubbles, Voids and Bumps in Time: The New Cosmology* ed J Cornell (Cambridge: Cambridge University Press)

[Hab85]   Haber H E and Kane G L 1985 *Phys. Rep.* **117** 75

[Hab93]   Haber H E 1993 *Proc. 9th Int. Workshop on Recent Advances in the Superworld, Woodlands, TX* and *Preprint* hep-ph/9308209

[Hag90]   Hagmann C *et al* 1990 *Phys. Rev.* D **42** 1297

[Hag95]   Hagmann C *et al* 1995 *Preprint* astro-ph/9508013

[Hag96]   Hagmann C *et al* 1996 *Preprint* astro-ph/9607022

[Hal84]   Halzen F and Martin A D 1984 *Quarks and Leptons* (London: Wiley)

[Hal92]   Halpern J P and Holt S S 1992 *Nature* **357** 306

[Hal95]   Halzen F 1995 *Nucl. Phys. B (Proc. Suppl.)* **38** 472, *Proc. 16th Int. Conf. on Neutrino Physics and Astrophysics, NEUTRINO '94* eds A Dar, G Eilam and M Gronau (Amsterdam: North-Holland)

[Hal97]   Halzen F 1997 *Proc. Int. Workshop on Dark Matter in Astro- and Particle Physics, DARK '96 (16–20 September 1996, Heidelberg, Germany)* eds H V Klapdor-Kleingrothaus and Y Ramachers (Singapore: World Scientific)

[Hal97a]   Halzen F 1997 in *Proc. Int. Conf. on Particles Physics Beyond the Standard Model, 'BEYOND THE DESERT '97', Castle Ringberg, Germany* eds H V Klapdor-Kleingrothaus and H Päs (Bristol: IOP)

[Ham86]   Hampel W 1986 *Proc. Int. Symp. on Weak and Electromagnetic Interactions in Nuclei (WEIN '86) Heidelberg* ed H V Klapdor (Berlin: Springer) 718

[Ham93]   Hampel W 1993 *J. Phys. G: Nucl. Phys.* **19** 209

[Ham94]   Hampel W 1994 *Phil. Trans. R. Soc.* A **346** 3

[Ham95]   Hamuy M *et al* 1995 *Astron. J.* **109** 1

[Ham96]   Hamel L A *et al* 1996 *Preprint* hep-ex 9602004

[Ham96a]   Hampel W *et al* (GALLEX Collaboration) 1996 *Phys. Lett.* B **388** 384

[Har70]   Harrison E R 1970 *Phys. Rev.* D **1** 2726

[Har83]   Hartle J and Hawking S W 1983 *Phys. Rev.* D **28** 2960

[Har90]   Hara T *et al* 1990 *21st Int. Cosmic Ray Conf. (Adelaide, 1990)* ed R J Protheroe p 95

[Har91]   Harding A K 1991 *Phys. Rep.* **206** 327

[Har95]   Hartmann D H, Briggs M S and Pendleton G N 1995 *Ann. NY Acad. Sci.* **759** 434

[Har96]   Harvey J W *et al* 1996 *Science* **272** 1284

[Har96a]   Harris J W and Müller B 1996 *Preprint* hep-ph/9607022 and 1996 *Ann. Rev. Nucl. Part. Phys.* **46** 71

[Has73]   Hasert F *et al* 1973 *Phys. Lett.* B **46** 121

[Has91]   Hasinger G, Schmidt M and Trümper J 1991 *Astron. Astrophys.* **246** L2

[Has95]   Hasinger G 1995 *Ann. NY Acad. Sci.* **759** 200

[Has96]   Hasinger G, Aschenbach B and Trümper J 1996 *Preprint* astro-ph/9606149

[Has96a]   Hasinger G, Aschenbach B and Trümper J 1996 *Astron. Astrophys.* **312** 9

[Hat94a]   Hata N, Bludman S and Langacker P 1994 *Phys. Rev.* D **49** 3622

[Hat94b]   Hata N and Langacker P 1994 *Phys. Rev.* D **50** 632

[Hat95]   Hata N and Langacker P 1995 *Phys. Rev.* D **52** 420

[Hat97]   Hata N and Langacker P 1997 *Preprint* hep-ph/9705339

[Haw74]   Hawking S 1974 *Nature* **248** 30

[Hax87]   Haxton W C 1987 *Phys. Rev.* D **36** 2283

[Hax88]   Haxton W C 1988 *Phys. Rev. Lett.* **60** 768

[Hay94]   Hayashida N *et al* (AKENO Collaboration) 1994 *Phys. Rev. Lett.* **73** 3491

[Hay95]   Hayes J and Burrows A 1995 *Sky and Telescope* **8** 30

[He89]   He X G, McKellar B H J and Pakvasa S 1989 *Int. J. Mod. Phys.* A **4** 5011

[Hel96]   1996 *Proc. NEUTRINO '96 (Helsinki, June 1996)*

[Hel96a]   Hellemans A 1996 *Science* **272** 1264

[Hen91]   Henbest N 1991 *New Sci.* **132** 42

[Hen92]   Henbest N 1992 *New Sci.* **133** 25

[Her80]   Herrmann J 1980 *dtv-Atlas zur Astronomie* (München: Deutscher Taschenbuch)

470   *References*

[Her94]   1994 Proposal HERA-B, DESY/PRC 94-02
[Het87]   Hetherington D W *et al* 1987 *Phys. Rev.* C **36** 1504
[Heu76]   Heusch C A *et al* 1976 *Phys. Rev. Lett.* **37** 405, 409
[Hew68]   Hewish A *et al* 1968 *Nature* **217** 709
[Hig64]   Higgs P W 1964 *Phys. Lett.* **12** 252
[Hil72]   Hillas A M 1972 *Cosmic Rays* (Oxford: Pergamon)
[Hil76]   Hilf E R, Groote H v and Takahashi K 1976 *CERN-Report 76-13* 142
[Hil78]   Hillebrandt W 1978 *Space Sci. Rev.* **21** 639
[Hil79]   Hillebrandt W 1979 *Proc. 4th EPS General Conf.* 255
[Hil82]   Hillebrandt W 1982 *Phys. Bl.* **38** 189
[Hil84]   Hillas A M 1984 *Ann. Rev. Astron. Astrophys.* **22** 425
[Hil88]   Hillebrandt W 1988 in *NEUTRINOS* ed H V Klapdor (Berlin: Springer)
            285
[Hil90]   Hill F 1990 *IAU Proc.* **121,** *Inside the Sun, Versailles, France* eds G
            Berthomieu and M Cribier (Berlin: Kluwer) p 265
[Hil95]   Hill J E 1995 *Phys. Rev. Lett.* **75** 2654
[Him89]   Hime A and Simpson J J 1989 *Phys. Rev.* D **49** 1837
[Him91]   Hime A *et al* 1991 *Phys. Lett.* B **260** 381
[Him93]   Hime A 1993 *Phys. Lett.* B **299** 165
[Hin94]   Hindmarsh M B and Kibble T W B 1994 *Preprint* hep-ph/9411342
[Hin96]   Hinshaw G *et al* 1996 *Preprint* astro-ph/9601061
[Hin96a]   Hinshaw G *et al* 1996 *Astrophys. J* **464** L25
[Hip90]   Hipdon J C and Lingenfelter R E 1990 *Ann. Rev. Astron. Astrophys.* **28** 401
[Hir83]   Hirata K S *et al* (KAMIOKANDE Collaboration) 1983 *Phys. Rev. Lett.* **63**
            16
[Hir87]   Hirata K S *et al* (KAMIOKANDE Collaboration) 1987 *Phys. Rev. Lett.* **58**
            1490
[Hir88]   Hirata K S *et al* (KAMIOKANDE Collaboration) 1988 *Phys. Rev.* D **38**
            448
[Hir89]   Hirata K S *et al* (KAMIOKANDE Collaboration) 1989 *Phys. Lett.* B **220**
            308
[Hir89a]   Hirata K S *et al* (KAMIOKANDE Collaboration) 1989 *Phys. Rev. Lett.* **63**
            16
[Hir90]   Hirata K S *et al* (KAMIOKANDE Collaboration) 1990 *Phys. Rev. Lett.* **65**
            1297
[Hir91]   Hirata K S *et al* (KAMIOKANDE Collaboration) 1991 *Phys. Rev.* D **44**
            2241
[Hir92a]   Hirata K S *et al* (KAMIOKANDE Collaboration) 1992 *Phys. Lett.* B **280**
            146
[Hir92b]   Hirsch M, Staudt A and Klapdor-Kleingrothaus H V 1992 *At. Data Nucl.
            Data Tables* **51** 243
[Hir93]   Hirsch M, Staudt A, Muto K and Klapdor-Kleingrothaus H V 1993 *At.
            Data Nucl. Data Tables* **53** 165
[Hir95]   Hirsch M, Klapdor-Kleingrothaus H V and Kovalenko S 1995 *Phys. Bl.*
            51 418; 1995 *Phys. Rev. Lett.* **75** 17; 1995 *Phys. Lett.* B **352** 1
[Hir95a]   Hirsch M, Klapdor-Kleingrothaus H V and Kovalenko S G 1995 *Annual
            Report, Max Planck Institute für Physik, Heidelberg*

[Hir96a]   Hirsch M, Klapdor-Kleingrothaus H V, Kovalenko S G and Päs H 1996
            Phys. Lett. B **372** 8
[Hir96b]   Hirsch M, Klapdor-Kleingrothaus H V and Kovalenko S G 1996 *Phys. Rev.*
            D **53** 1329
[Hir96c]   Hirsch M, Klapdor-Kleingrothaus H V and Kovalenko S G 1996 *Phys. Lett.*
            B **372** 181
[Hir96d]   Hirsch M, Klapdor-Kleingrothaus H V and Panella O 1996 *Phys. Lett.* B
            374 7
[Hir96e]   Hirsch M, Klapdor-Kleingrothaus H V and Kovalenko S G 1996 *Phys. Lett.*
            B **378** 17
[Hir96f]   Hirsch M, Klapdor-Kleingrothaus H V and Kovalenko S G 1996 *Phys. Rev.*
            D **54** R4207
[Hir97a]   Hirsch M, Klapdor-Kleingrothaus H V and Kovalenko S G 1997 *Preprint*
            hep-ph/9707207, submitted to *Phys. Rev.* D
[Hir97b]   Hirsch M, Klapdor-Kleingrothaus H V and Kovalenko S G 1997 *Phys. Lett.*
            B **398** 311
[Hir97c]   Hirsch M, Klapdor-Kleingrothaus H V and Kovalenko S G 1997 *Phys. Lett.*
            B **403** 291
[Hoe91]   Hoell J and Priester W 1991 *Astron. Astrophys.* **251** L23
[Hol92a]   Holzschuh E 1992 *Rep. Prog. Phys.* **55** 851
[Hol92b]   Holzschuh E *et al* 1992 *Phys. Lett.* B **287** 381
[Hom92]   Homer G J *et al* 1992 *Z. Phys.* C **55** 549
[Hom96]   Homma H, Bender E, Hirsch M, Muto K, Klapdor-Kleingrothaus H V and
            Oda T 1996 *Phys. Rev.* C **54** 2972
[Hon94]   Hong J T *et al* (MACRO Collaboration) 1994 *Nucl. Phys. B (Proc. Suppl.)*
            35 261, *Proc. 3rd Int. Workshop on Theoretical and Phenomenological
            Aspects in Astroparticle and Underground Physics (TAUP'93), Assergi,
            Italy* eds C Arpesella, E Bellotti and A Bottino (Amsterdam: North-
            Holland)
[How80]   Howard W M and Möller P 1980 *At. Data Nucl. Data Tables* **25** 219
[Hu95]   Hu W, Sugiyama N and Silk J 1995 *Preprint* astro-ph/9504057
[Hua92]   Huang K 1992 *Quarks, Leptons and Gauge Fields* (Singapore: World
            Scientific)
[Hub29]   Hubble E 1929 *Proc. Nat. Acad.* **15** 168
[Hub90]   Huber M *et al* 1990 *Phys. Rev. Lett.* **64** 835
[Huc90]   Huchra J P *et al* 1990 *Astrophys. J. Suppl.* **72** 433
[Hwa90]   Hwa R C (ed) 1990 *The Quark-Gluon-Plasma, Advanced Series in High
            Energy Physics* vol. 6 (Singapore: World Scientific)
[Ibe75]   Iben I 1975 *Astrophys. J.* **196** 525, 549
[Ibe82]   Iben I and Renzini A 1982 *Astrophys. J. Lett.* **263** L188
[ICR95]   1995 *Proc. XXIV Int. Cosmic Rays Conf. (ICRC'95), Rome*
[Inc84]   Incandella J *et al* 1984 *Phys. Rev. Lett.* **53** 2067
[Ion94]   Ioannissyan A and Valle J W F 1994 *Phys. Lett.* B **332** 93
[Itz85]   Itzykson C and Zuber J B 1985 *Quantum Field Theory* (London: McGraw-
            Hill) International Edition
[Iud94]   Iudin A *et al* 1994 *Astron. Astrophys.* **284** L1
[Jac75]   Jackson J D 1975 *Classical Electrodynamics* (New York: Wiley)
[Jac82]   Jackson J D 1982 *Klassische Elektrodynamik* (Amsterdam: de Gruyter)

[Jam93]   James P A and Puxley P J 1993 *Nature* **363** 240
[Jan95]   Janka H T *et al* 1995 *Preprint* astro-ph/9507023 and *Phys. Rev. Lett.* **76** 2621
[Jan95a]   Janka H T and Müller E 1995 *Ann. NY Acad. Sci.* **759** 269
[Jan95b]   Jansen K E 1995 *Preprint* DESY 95-169 and *Preprint* hep-lat/9509018
[Jar89]   Jarlskog C (ed) 1989 *CP-Violation, Advanced Series in High Energy Physics* vol 3 (Singapore: World Scientific)
[Jef70]   Jeffrey P M and Anders E 1970 *Geochim. Cosmochim. Acta.* **34** 1175
[Jeo95]   Jeon H and Longo M J 1995 *Phys. Rev. Lett.* **75** 1443
[Jon85]   Jones B J T and Wyse R F 1985 *Astron. Astrophys.* **149** 144
[Jon89]   Jones L W *et al* 1989 *Z. Phys.* C **43** 349
[Jon93]   Jones M *et al* 1993 *Nature* **365** 320
[Jun95]   Jungmann G, Kamionkowski M and Griest K 1995 *Preprint* hep-ph/9506380 and 1996 *Phys. Rep.* **267** 195
[Kaf94]   Kafka T *et al* (Soudan 2 Collaboration) 1994 *Nucl. Phys. B (Proc. Suppl.)* **35** 427, *Proc. 3rd Int. Workshop on Theoretical and Phenomenological Aspects in Astroparticle and Underground Physics (TAUP'93), Assergi, Italy* eds C Arpesella, E Bellotti and A Bottino (Amsterdam: North-Holland)
[Kah86]   Kahana S 1986 *Proc. Int. Symp. on Weak and Electromagnetic Interaction in Nuclei (WEIN'86)* ed H V Klapdor (Heidelberg: Springer) 939
[Kak88]   Kaku M 1988 *Introduction to Superstrings* (Heidelberg: Springer)
[Kak93]   Kakita M *et al* (KAMIOKANDE Collaboration) 1993 *Proc. Int. Conf on High Energy Physics, (Dallas)* p 1187
[Kaj88]   Kajino T, Mathews G J and Fuller G M 1988 *Proc. Int. Symp. on Heavy Ion Physics and Nuclear Astrophysical Problems (21–23 July 1988, Tokyo)* (Singapore: World Scientific) p 51
[Kal21]   Kaluza T 1921 *Preuß. Acad. Wiss.* **K1** 966
[Kal97]   Kalinowski J, Rückl R, Spiesberger H and Zerwas P M 1997 *Preprint* hep-ph/9703288
[Kam97]   Kampert K H *et al* 1997 *Preprint* astro-ph/9703182
[Kan93]   Kane G L *et al* 1993 *Preprint* UM-TH-93-24
[Kan94]   Kane G L *et al* 1995 *Proc. Int. Conf. on Particles, Strings and Cosmology (PASCOS'94), Syracuse* ed K C Wali (River Edge, NJ) (Singapore: World Scientific) p 188
[Käp89]   Käppeler F, Beer H and Wisshak K 1989 *Rep. Prog. Phys.* **52** 945
[Käp91]   Käppeler F 1991 *Nuclei in the Cosmos* ed G Oberhummer (Heidelberg: Springer) 179
[Kaw87]   Kawakami H *et al* 1987 *Phys. Lett.* B **187** 198
[Kaw88a]   Kawakami H *et al* 1988 *J. Japan. Phys. Soc.* **57** 2873
[Kaw88b]   Kawano L, Schramm D N and Steigman G 1988 *Astrophys. J.* **327** 750
[Kaw91]   Kawakami H *et al* 1991 *Phys. Lett.* B **256** 105
[Kay89]   Kayser B, Gibrat-Debu F and Perrier F 1989 *Physics of Massive Neutrinos* (Singapore: World Scientific)
[Kei96]   Keil W *et al* 1996 *Preprint* astro-ph/9612222
[Kel93]   Kellerman K 1993 *Nature* **361** 134
[Ken83]   Kent S M and Sargent W L W 1983 *Astrophys. J.* **88** 692
[Key91]   You K *et al* 1991 *Phys. Lett.* B **265** 53

[Kib67]   Kibble T W B 1967 *Phys. Rev.* **155** 1554
[Kib76]   Kibble T W B 1976 *J. Phys. A: Math. Gen* **9** 1387
[Kif94]   Kifnne T *et al* (CANGAROO Collaboration) 1994 *Astrophys. J.* **438** L91
[Kim79]   Kim J 1979 *Phys. Rev. Lett.* **43** 103
[Kim87]   Kim J 1987 *Phys. Rep.* **150** 1
[Kim93]   Kim J C W and Pevzner A 1993 *Neutrinos in Physics and Astrophysics, Comtemporary Concepts in Physics vol 8* (New York: Harwood Academic)
[Kin89]   Kinoshita K *et al* 1989 *Phys. Lett.* B **228** 543
[Kin90]   Kinoshita T *et al* 1990 *Quantum Electrodynamics* ed T Kinoshita (Singapore: World Scientific)
[Kin92]   Kinoshita K *et al* 1992 *Phys. Rev.* D **46** R881
[Kin93]   Kinney A L 1993 *Ann. NY Acad. Sci.* **688** 195
[Kip90]   Kippenhahn R and Weigert A 1990 *Stellar Structure and Evolution* (Berlin: Springer)
[Kir68]   Kirsten T *et al* 1968 *Phys. Rev. Lett.* **20** 1300
[Kir78]   Kirsten T 1978 *The Origin of the Solar System* ed S F Dermott (London: Wiley)
[Kir81]   Kirshner R P *et al* 1981 *Astrophys. J.* **248** L57
[Kir83]   Kirshner R P *et al* 1983 *Astron. J.* **88** 1285
[Kir90]   Kirsten T *et al* (GALLEX Collaboration) 1990 *Inside the Sun, IAU Proc.* **121** eds G Berthomieu and M Cribier (Berlin: Kluwer) 187
[Kir93]   Kirsten T 1993 *Sterne und Weltraum* **1** 16
[Kir95]   Kirsten T 1995 *Ann. NY Acad. Sci.* **759** 21
[Kir96]   Kirk J 1996 *MPG Spiegel* **1** 16
[Kir96a]   Kirsten T 1996 *Proc. NEUTRINO '96 (Helsinki, June 1996)*
[Kir97]   Kirshner R 1997 *Sky and Telescope* **2** 35
[Kla79]   Klapdor H V and Wene C O 1979 *Astron. J.* **230** L113
[Kla81]   Klapdor H V *et al* 1981 *Z. Phys.* A **299** 213
[Kla82a]   Klapdor H V and Metzinger J 1982 *Phys. Lett.* B **112** 22
[Kla82b]   Klapdor H V 1982 *Phys. Bl.* **38** 182
[Kla83]   Klapdor H V 1983 *Prog. Nucl. Part. Phys.* **10** 131
[Kla84]   Klapdor H V, Metzinger J and Oda T 1984 *At. Data Nucl. Data Tables* **31** 81
[Kla85]   Klapdor H V 1985 *Fortschritte der Physik* **33** 1
[Kla86a]   Klapdor H V and Grotz K 1986 *Astrophys. J.* **304** L39
[Kla86b]   Klapdor H V 1986 *Prog. Part. Nucl. Phys.* **17** 419
[Kla87]   Klapdor H V 1987 *Proposal* MPI-Bericht MPI-H-1987-V17
[Kla88]   Klapdor H V and Metzinger J 1988 *Proc. Int. Conf. on Nuclear Data for Science and Technology* ed S Igarasi (Japan Atomic Energy Research Institute) 827
[Kla89]   Klapdor H V 1989 *Der Betazerfall der Atomkerne und das Alter des Universums, Rheinisch-Westf Akademie der Wissenschaften Vorträge N365* (Berlin: Westdeutscher) 73
[Kla91a]   Klapdor-Kleingrothaus H V 1991 *Proc. Conf. on Capture Gamma-ray Spectroscopy, (Pacific Grove, CA)* ed R W Hoff (AIP Conf. Proc. Series) **238** 870
[Kla91b]   Klapdor-Kleingrothaus H V 1991 *J. Phys. G: Nucl. Phys.* **17** 129, 537

[Kla91c]   Klapdor-Kleingrothaus H V 1991 *Proc. Conf. on Gamma-Ray Line Astrophysics, Paris-Saclay, 1990, AIP Conf. Proc.* **232** 464

[Kla91d]   Klapdor-Kleingrothaus H V 1991 *Nuclei in the Cosmos* ed G Oberhummer (Heidelberg: Springer) 199

[Kla92]   Klapdor-Kleingrothaus H V and Zuber K 1992 *Phys. Bl.* **48** 125

[Kla94]   Klapdor-Kleingrothaus H V 1994 *Prog Part. Nucl. Phys.* **32** 261

[Kla94a]   Klapdor-Kleingrothaus H V 1994 *MPG-Spiegel* **6** 6

[Kla95]   Klapdor-Kleingrothaus H V and Staudt A 1995 *Teilchenphysik ohne Beschleuniger* (Stuttgart: Teubner) and *Non-Accelerator Particle Physics* (Bristol: Institute of Physics)

[Kla95a]   Klapdor-Kleingrothaus H V 1995 *Proc. Weak and Electromagnetic Interactions in Nuclei (WEIN'95), (Osaka)* eds H Ejiri, T Kishimoto and T Sato (Singapore: World Scientific) 174

[Kla96]   Klapdor-Kleingrothaus H V 1996 *Proc. ECT Workshop on Double Beta Decay and Related Topics, Trento, Italy, 1995* eds H V Klapdor-Kleingrothaus and S Stoica (Singapore: World Scientific) 3

[Kla96a]   Klapdor-Kleingrothaus H V 1996 *Proc. NEUTRINO '96 (Helsinki, June 1996)*

[Kla97a]   Klages H O *et al* (KASKADE Collaboration) 1997 *Nucl. Phys. B (Proc. Suppl)* **52** 92

[Kla97b]   Klapdor-Kleingrothaus H V 1997 *Proc. Int. Conf. on Particle Physics Beyond the Standard Model, 'BEYOND THE DESERT '97'* eds H V Klapdor-Kleingrothaus and H Päs (Bristol: IOP)

[Kla97c]   Klapdor-Kleingrothaus H V and Ramachers Y 1997 *Proc. Int. Workshop on Dark Matter in Astro- and Particle Physics, DARK '96 (16–20 September 1996, Heidelberg, Germany)* eds H V Klapdor-Kleingrothaus and Y Ramachers (Singapore: World Scientific)

[Kla97d]   Klapdor-Kleingrothaus H V and Ramachers Y 1997 *Proc. Int. Conf. on Particle Physics Beyond the Standard Model, 'BEYOND THE DESERT '97', Castle Ringberg, Germany* eds H V Klapdor-Kleingrothaus and H Päs (Bristol: IOP)

[Kla97e]   Klapdor-Kleingrothaus H V and Päs H (eds) 1997 *Proc. Int. Conf. on Particle Physics Beyond the Standard Model, 'BEYOND THE DESERT '97'* (Bristol: IOP)

[Kla97f]   Klapdor-Kleingrothaus H V 1997 in *Neutrino Physics* ed K Winter 2nd edn (Cambridge: Cambridge University Press)

[Kla97g]   Klapdor-Kleingrothaus H V, Kudravtsev M I, Stolpovski V G, Svetilov S I, Melnikov V F and Krivosheina I 1997 *J. Moscow Phys. Soc.* **7** 41

[Kle26]   Klein K 1926 *Z. Phys.* **37** 895: 1926 *Nature* **118** 516

[Kle96]   Kleinfeller J 1996 *Proc. Neutrino '96 (Helsinki, June 1996)*

[Kli84]   Klinkhammer D and Manton N 1984 *Phys. Rev. D* **30** 2212

[Kli86]   Klimenko A A *et al* 1986 *Proc. Int. Symp. on Weak and Electromagnetic Interactions in Nuclei (WEIN'86), Heidelberg, Germany* ed H V Klapdor (Heidelberg: Springer) 701

[Kly92]   Klypin A and Melott A 1992 *Astrophys. J.* **399** 725

[Kne95]   Kneib J P *et al* 1995 *Preprint* astro-ph/950403; *Astron. Astrophys.* **303** 27

[Kob73]   Kobayashi M and Maskawa T 1973 *Prog. Theor. Phys.* **49** 652

[Kof93]   Kofman L A, Gnedin N Y and Bahcall N A 1993 *Astrophys. J.* **413** 1

[Kog93]	Kogut A *et al* 1993 *Astrophys. J.* **419** 1
[Kog96]	Kogut A *et al* 1996 *Astrophys. J.* **464** L5
[Köh95]	Köhler T 1995 *Proc. 27th Int. Conf on High Energy Physics, (Glasgow)* eds P J Busey and I G Knowles (Bristol: Institute of Physics) vol 2 p 801
[Kol84]	Kolb E W and Turner M S 1984 *Astrophys. J.* **286** 702
[Kol87]	Kolb E W, Stebbins A J and Turner M S 1987 *Phys. Rev.* D **35** 3598
[Kol89]	Kolb E W and Turner M S 1989 *Phys. Rev. Lett.* **62** 509
[Kol90]	Kolb E W and Turner M S 1990 *The Early Universe, Frontiers in Physics* vol 69 (Reading, MA: Addison Wesley)
[Kol91]	Kolb E W and Turner M S 1991 *Phys. Rev. Lett.* **67** 5
[Kol93]	Kolb E W and Turner M S 1993 *The Early Universe* (Reading, MA: Addison-Wesley)
[Kol95]	Kolatt T and Dekel A 1995 *Preprint* astro-ph/9512132
[Kol97]	Kolb S M, Hirsch M and Klapdor-Kleingrothaus H V 1997 *Phys. Rev.* D in press
[Koo87]	Koo D C and Kron R G 1987 *Observational Cosmology* eds A Hewitt, G Burbidge, L Z Fang and D Reidel (Dordrecht: Reidel) *IAU Symp. Proc.* **124** 383
[Kon96]	Konopelko A *et al* (HEGRA Collaboration) 1996 *Astropart. Phys.* **4** 199
[Kor89]	Kormendy J and Djorgovski S 1989 *Ann. Rev. Astron. Astrophys.* **27** 235
[Kos86]	Kosvintzev Y Y *et al* 1986 *JETP Lett.* **44** 571
[Kos89]	Kossakowski R *et al* 1989 *Nucl. Phys.* A **503** 473
[Kos92]	Koshiba M 1992 *Phys. Rep.* **220** 229
[Kra90a]	Krauss L M 1990 *Phys. Rev. Lett.* **64** 999
[Kra90b]	Krauss L M and Romanelli P 1990 *Astrophys. J.* **358** 47
[Kra91]	Krauss L M 1991 *Phys. Lett.* B **263** 441
[Kra95]	Krauss L M and Kernan P 1995 *Phys. Lett.* B **347** 347
[Kra96]	Kraan-Korteweg R C *et al* 1996 *Nature* **379** 519
[Kul92]	Kulessa A S and Lynden-Bell D 1992 *Mon. Not. R. Astron. Soc.* **255** 105
[Kun82]	Kunde V *et al* 1982 *Astrophys. J.* **263** 443
[Kün92]	Kündig W and Holzschuh E 1992 *Prog. Part. Nucl. Sci.* **32** 131
[Kur91]	Kurki-Suonio H *et al* 1991 *Astrophys. J.* **353** 406
[Kur92]	Kurfess J D *et al* 1992 *Astrophys. J.* **399** L137
[Kuz66]	Kuzmin V A 1966 *Sov. Phys. JETP* **22** 1050
[Kuz85]	Kuzmin V A, Rubakov V A and Shaposhnikov M E 1985 *Phys. Lett.* B **155** 36
[Kuz97]	Kuzmin V 1997 in *Proc. Int. Conf. on Particles Physics Beyond the Standard Model, 'BEYOND THE DESERT '97', Castle Ringberg, Germany* eds H V Klapdor-Kleingrothaus and H Päs (Bristol: IOP)
[Kwo81]	Kwon H *et al* 1981 *Phys. Rev.* D **24** 1097
[Lah88]	Lahav O, Rowan-Robinson M and Lynden-Bell D 1988 *Mon. Not. R. Astron. Soc.* **234** 677
[Lal94]	Lalanne D *et al* (NEMO Collaboration) 1994 *Nucl. Phys.* B *(Proc. Suppl.)* **35** 369 *Proc. 3rd Int. Workshop on Theoretical and Phenomenological Aspects in Astroparticle and Underground Physics (TAUP'93), Assergi, Italy* eds C Arpesella, E Bellotti and A Bottino (Amsterdam: North-Holland)
[Lam97]	Lamoreaux S K 1997 *Phys. Rev. Lett.* **78** 5

476   *References*

[Lan52]	Langer L M and Moffat R J D 1952 *Phys. Rev.* **88** 689
[Lan75]	Landau L D and Lifschitz E M 1975 *Lehrbuch der Theoretischen Physik Bd* **4a** 5 (Berlin: Academischer)
[Lan81]	Langacker P 1981 *Phys. Rep.* **72** 185
[Lan86]	Langacker P 1986 *Proc. Int. Symp. Weak and Electromagnetic Interactions in Nuclei, WEIN '86* ed H V Klapdor (Heidelberg: Springer) p 879
[Lan88]	Langacker P 1988 *Neutrinos* ed H V Klapdor (Heidelberg: Springer) p 71
[Lan91]	Langer N 1991 *Astron. Astrophys.* **243** 155
[Lan92]	Langacker P 1992 *Proc. Fourth Int. Workshop on Neutrino Telescopes (10– 13 March, 1992, Venezia, Italy)* ed M Baldo-Ceolin p 73
[Lan93a]	Langacker P 1993 *Ann. NY Acad. Sci.* **688** 34
[Lan93b]	Langacker P and Polonsky N 1993 *Phys. Rev.* D **47** 4028
[Lan94]	Langanke K 1994 *Proc. Solar Modeling Workshop, Seattle, WA*
[Lan95]	Langacker P 1995 *Preprint* hep-ph/9511207; 1995 *Proc. SUSY '95 (Palaiseau, France)* 199
[Las88]	Last J *et al* 1988 *Phys. Rev. Lett.* **60** 995
[Lat88]	Lattimer J and Cooperstein J 1988 *Phys. Rev. Lett.* **61** 24
[Lau94]	Lauer T R and Postman M 1994 *Astrophys. J.* **425** 418
[Laz92]	Lazarus D M *et al* 1992 *Phys. Rev. Lett.* **69** 2333
[Lee72]	Lee B W and Zinn-Justin J 1972 *Phys. Rev.* D **5** 3121
[Lee77a]	Lee B W and Weinberg S 1977 *Phys. Rev. Lett.* **39** 165
[Lee77b]	Lee B W and Shrock R E 1977 *Phys. Rev.* D **16** 1444
[Lee94]	Lee D G and Mohapatra R N 1994 *Phys. Lett.* B **329** 963
[Lee95]	Lee D G *et al* 1995 *Phys. Rev.* D **51** 229
[Lee95a]	Lee D G *et al* 1995 *Phys. Rev.* D **51** 1353
[Leu84]	Leung C N and Petcov S T 1984 *Phys. Lett.* B **145** 416
[Lev93]	Leventhal *et al* 1993 *Astrophys. J.* **405** 425
[Lid95]	Lidsey J E *et al* 1995 *Preprint* astro-ph/9508078
[Lid96]	Liddle A 1996 *Preprint* astro-ph/9612093, Lecture given at *Winter School 'From Quantum Fluctuations to Cosmological Structures (CASA '96), Casablanca, December 1996'*
[Lim88]	Lim C S and Marciano W J 1988 *Phys. Rev.* D **37** 1368
[Lin63]	Lindhard J *et al* 1963 *K. Dansk. Vidensk. Selsk. Mat. Fys. Medd.* **33** 10
[Lin76]	Linde A D 1976 *JETP Lett.* **23** 64
[Lin82]	Linde A D 1982 *Phys. Lett.* B **108** 389
[Lin84]	Linde A D 1984 *Rep. Prog. Phys.* **47** 925
[Lin87]	Lin J 1987 *Phys. Rev.* D **35** 3447
[Lin89]	Lingenfelter R E and Ramaty R 1989 *Astrophys. J.* **43** 686
[Lin90]	Lin A 1990 *Particle Physics and Inflationary Cosmology* (Boston, MA: Harvard University Press)
[Lin92]	Linsky J L *et al* 1992 *Astrophys. J.* **402** 694
[Lin93]	Lingenfelter R E, Chan K W and Ramaty R 1993 *Phys. Rep.* **227** 133
[Lin95]	Lin D N C, Jones B F and Klemola A R 1995 *Astrophys. J.* **439** 652
[Lin96]	Lin H *et al* 1996 *Preprint* astro-ph/9606055
[Lin96a]	Linde A 1996 *Preprint* astro-ph/9601004
[Lin96b]	Lineweaver C H 1996 *Preprint* astro-ph/9609034, *Proc. Morioud Conf. on Cosmic Microwave Background, 1996*
[Lob96]	Lobashev V M 1996 *Proc. NEUTRINO '96 (Helsinki, June 1996)*

[Loh81]	Lohrmann E 1981 *Hochenergiephysik* (Stuttgart: Teubner)
[Loh83]	Lohrmann E 1983 *Einführung in die Elementarteilchenphysik* (Stuttgart: Teubner)
[Loh86]	Loh E and Spillar E 1986 *Astrophys. J.* **307** L1
[Lon89]	Longair M S 1989 *Evolution of Galaxies—Astronomical Observation* eds I Appenzeller, H J Habing and P Lena *Lecture Notes in Physics* (Heidelberg: Springer) **333** 1
[Lon92, 94]	Longair M S 1992, 1994 *High Energy Astrophysics* (Cambridge: Cambridge University Press)
[Lop94]	Lopez J L, Nanopoulous D V and Zichichi A 1994 *Riv. Nuovo Cimento* **17** 2
[Lop94a]	Lopez J L *et al* 1994 *Phys. Rev.* D **50** 2164
[Lop96]	Lopez J L 1996 *Preprint* hep-ph/9601208
[Lop96a]	Lopez J L 1996 *Rep. Prog. Phys.* **59** 819
[Lor95]	Lorenz E 1995 *Ann. NY Acad. Sci.* **759** 472
[Lor96]	Lorenz E 1996 *Nucl. Phys. B (Proc. Suppl.)* **48** 391 *Proc. 4th Int. Workshop on Theoretical and Phenomenological Aspects of Underground Physics (TAUP'95), Toledo, Spain* eds A Morales, J Morales and J A Villar (Amsterdam: North-Holland)
[LoS87]	LoSecco J M *et al* (IMB Collaboration) 1987 *Phys. Lett.* B **188** 388
[Lou95]	Louis W 1995 *Nucl. Phys. B (Proc. Suppl.)* **38** 229, *Proc. 16th Int. Conf. on Neutrino Physics and Astrophysics, NEUTRINO '94* eds A Dar, G Eilam and M Gronau (Amsterdam: North-Holland)
[Lov96a]	Loveday J *et al* 1996 *Mon. Not. R. Astron. Soc.* **278** 1025
[Lov96b]	Loveday J 1996 *Preprint* astro-ph/9605028
[Low91]	Lowder D M *et al* 1991 *Nature* **353** 331
[LTD96]	*Proc. 6th Int. Workshop on Low Temperature Detectors (Beatenberg, Switzerland, 1995)* eds H R Ott and A Zehnder (Amsterdam: North-Holland), 1996 *Nucl. Instrum. Methods* A **370**(1)
[Lub80]	Lubimov V A *et al* 1980 *Phys. Lett.* B **94** 266
[Luc86]	Lucha W 1986 *Comment. Nucl. Part. Phys.* **16** 155
[Lüd54]	Lüders G 1954 *K. Dansk. Vidensk. Mat.-Fys. Medd.* **28** 5
[Lüd57]	Lüders G 1957 *Ann. NY Acad. Phys.* **2** 1
[Lyn90]	Lyne A G and Graham-Smith F 1990 *Pulsar Astronomy* (Cambridge: Cambridge University Press)
[Mac96]	McDonald A B 1996 *Nucl. Phys. B (Proc. Suppl.)* **48** 357, *Proc. 4th Int. Workshop on Theoretical and Phenomenological Aspects of Underground Physics (TAUP'95), Toledo, Spain* eds A Morales, J Morales and J A Villar (Amsterdam: North-Holland)
[Mag93]	Magneville C *et al* (EROS Collaboration) 1993 *Ann. NY Acad. Sci.* **688** 619
[Mag96]	Magueijo J *et al* 1996 *Preprint* astro-ph/9605047
[Mah84]	Mahoney W A *et al* 1984 *Astrophys. J.* **286** 578
[Mal93]	Malaney R A and Mathews G J 1993 *Phys. Rep.* **229** 145, 147
[Mam89]	Mampe W *et al* 1989 *Phys. Rev. Lett.* **63** 593
[Mam93]	Mampe W *et al* 1993 *JETP Lett.* **57** 82
[Man93]	Manchester R N 1993 *Ann. NY Acad. Sci.* **688** 331
[Mao92]	Mao S and Paczynski B 1992 *Astrophys. J.* **389** L13

[Mar92]   Marshak R E 1992 *Conceptional Foundations of Modern Particle Physics* (Cambridge: Cambridge University Press)

[Mar97]   Martel H and Weinberg S 1997 *Preprint* astro-ph/9701099

[Mat88]   Matz S M *et al* 1988 *Nature* **331** 416

[Mat90a]   Mather J C *et al* 1990 *Astrophys. J.* **354** L37

[Mat90b]   Mathews G J and Cowan J J 1990 *Nature* **345** 491

[Mat91]   Matsuki S and Yamamoto K 1991 *Phys. Lett.* B **263** 523

[Mat94]   Mather J C *et al* 1994 *Astrophys. J.* **420** 439

[Mat94a]   Matthews J M (ed) 1994 *High Energy Astrophysics: Models and Observations from MeV to EeV* (Singapore: World Scientific)

[May87]   Mayle R, Wilson J R and Schramm D N 1987 *Astrophys. J.* **318** 288

[McC92]   McCullough P R 1992 *Astrophys. J.* **390** 213

[McN87]   McNaught R M 1987 *IAU Circ. No* **4316**

[Mel90]   Melchiorri B and Melchiorri F 1990 *Proc. Int. School of Physics 'Enrico Fermi', Course CV* (Amsterdam: North Holland)

[Mes95]   Messous Y *et al* 1995 *Astropart. Phys* **3** 361

[Mey86]   Meyer H 1986 *Proc. Int. Symp. Weak and Electromagnetic Interactions in Nuclei, WEIN '86* ed H V Klapdor (Heidelberg: Springer) p 846

[Mey86a]   Meyer D M *et al* 1986 *Astrophys. J.* **308** L37

[Mey92]   Meyer B S *et al* 1992 *Astrophys. J.* **399** 656

[Mey94]   Meyer B S 1994 *Ann. Rev. Astron. Astrophys.* **32** 371

[Mic94]   Michael D G (MACRO Collaboration) 1994 *Nucl. Phys. B (Proc. Suppl.)* **35** 235 *Proc. 3rd Int. Workshop on Theoretical and Phenomenological Aspects in Astroparticle and Underground Physics (TAUP'93), Assergi, Italy* eds C Arpesella, E Bellotti and A Bottino (Amsterdam: North-Holland)

[Mig97]   Migliozzi P *et al* (CHORUS Collaboration) 1997 *Proc. Workshop 11th Les Rencontres de Physique de la Vallee D'aoste: Results and Perspectives in Particle Physics (La Thuile, Italy, 1997)*

[Mik86a]   Mikheyev S P and Smirnov A Y 1986 *Nuovo Cimento* C **9** 17

[Mik86b]   Mikheyev S P and Smirnov A Y 1986 *Proc. 12th Int. Conf. on Neutrino Physics and Astrophysics* eds T Kitugaki and H Yuta (Singapore: World Scientific)

[Mil83]   Milgram M 1983 *Astrophys. J.* **270** 365

[Mil96]   Milsztajn A 1996 *Proc. NEUTRINO '96 (Helsinki, June 1996)*

[Mis73]   Misner C, Thorne K and Wheeler J 1973 *Gravitation* (London: Freeman)

[Moe95]   Moe M K 1995 *Nucl. Phys. B (Proc. Suppl.)* **38** 36, *Proc. 16th Int. Conf. on Neutrino Physics and Astrophysics, NEUTRINO '94* eds A Dar, G Eilam and M Gronau (Amsterdam: North-Holland)

[Moh86]   Mohapatra R N 1986 *Phys. Rev.* D **34** 3457

[Moh86, 92] Mohapatra R N 1986 and 1992 *Unification and Supersymmetry* (Heidelberg: Springer)

[Moh88]   Mohapatra R N 1988 *Neutrinos* ed H V Klapdor (Heidelberg: Springer) p 117

[Moh89]   Mohapatra R N 1989 *Nucl. Instrum. Methods* A **284** 1

[Moh91]   Mohapatra R N and Pal P B 1991 *Massive Neutrinos in Physics and Astrophysics* (Singapore: World Scientific)

[Moh94]   Mohapatra R N 1994 *Prog. Part. Nucl. Phys.* **32** 187

[Moh95]   Mohapatra R N and Rasin A 1995 *Preprint* hep-ph/9511391; 1996 *Phys. Rev. Lett.* **76** 3490

[Moh96]   Mohapatra R N 1996 *Double Beta Decay and Related Topics* eds H V Klapdor-Kleingrothaus and S Stoica (Singapore: World Scientific) p 44

[Moh96a]   Mohapatra R N 1996 *Proc. Int. Workshop on Future Prospects of Baryon Instability Search in p-Decay and n–n̄ Oscillation Experiments (28–30 March, 1996, Oak Ridge)* eds S J Ball and Y A Kamyshkov (US Dept of Energy Publications) p 73

[Moh96b]   Mohapatra R N and Nussinov S 1997 *Preprint* UMD-PP-97-38

[Moh96c]   Mohapatra R N 1996 *Proc. NEUTRINO '96 (Helsinki, June 1996)*

[Mol95]   Molaro P, Primas F and Bonifacio P 1995 *Astron. Astrophys.* **295** L47

[Mon96]   Montaruli T 1996 *Nucl. Phys. B (Proc. Suppl.)* **48** 87, *Proc. 4th Int. Workshop on Theoretical and Phenomenological Aspects of Underground Physics (TAUP'95), Toledo, Spain* eds A Morales, J Morales and J A Villar (Amsterdam: North-Holland)

[Mon97]   Montaruli T 1997 *Proc. Int. Workshop on Dark Matter in Astro- and Particle Physics, DARK'96 (Heidelberg, 16–20 September, 1996)* eds H V Klapdor-Kleingrothaus and Y Ramachers (Singapore: World Scientific)

[Moo96]   Moorhead M E 1996 *Nucl. Phys. B (Proc. Suppl.)* **48** 378, *Proc. 4th Int. Workshop on Theoretical and Phenomenological Aspects of Underground Physics (TAUP'95), Toledo, Spain* eds A Morales, J Morales and J A Villar (Amsterdam: North-Holland)

[Mor91a]   Mori M *et al* (KAMIOKANDE Collaboration) 1991 *Phys. Lett.* B **270** 89

[Mor91b]   Morrison D R O 1992 *Proc. Int. Conf. on Joint Lepton Photon and Europhysics Conf. on High Energy Physics, Geneva, 1991* (Singapore: World Scientific) p 641

[Mor93]   Mori M *et al* 1993 *Phys. Rev.* D **48** 5505

[Mor93a]   Mori T 1993 *Proc. Int. Conf. XXVI on High Energy Physics, Dallas 1992* ed J R Sanford (Singapore: World Scientific) p 1321

[Mor95]   Morris D 1995 *Proc. 17th Texas Symp. on Rel. Astrophysics, München 1994* eds H Böhringer, G E Morfill and J E Trümper; *Ann. NY Acad. Sci.* **759** 397

[Möß93]   Mößbauer R 1993 *Nucl. Phys. B (Proc. Suppl.)* **31** 385

[MPG96]   1996 *MPG-Spiegel* **3** 8

[Mui65]   Muirhead H 1965 *The Physics of Elementary Particles* (Oxford: Oxford University Press)

[Mül85]   Müller B 1985 *The Physics of the Quark-Gluon-Plasma, Lecture Notes in Physics* **225** (Heidelberg: Springer)

[Mül91]   Müller D 1991 *Astrophys. J.* **374** 356

[Mul93]   Mulchaey J S *et al* 1993 *Astrophys. J.* **404** L9

[Mül95a]   Müller B 1995 *Rep. Prog. Phys.* **58** 611

[Mül95b]   Müller E 1995 *Sterne und Weltraum* **34** 350

[Mül95c]   Müller E and Janka H T 1995 *Ann. NY Acad. Sci.* **759** 368

[Mur91]   Muraki Y 1991 *Astrophys. J.* **373** 657

[Mur93]   Murthy R G and Wolfendale A W 1993 *Gamma-Ray Astronomy* (Cambridge: Cambridge University Press)

[Mur94]   Murayama H 1994 *Preprint* hep-ph/9410285; *Proc. 22nd Int. Symp. on Physics with High Energy Colliders (Tokyo)* eds S Yamada and T Ishi (River Edge, NJ: World Scientific) p 357

[Mus83]   Musset P, Price M and Lohrmann E 1983 *Phys. Lett.* B **128** 333

[Mut88]   Muto K and Klapdor H V 1988 *NEUTRINOS* ed H V Klapdor (Heidelberg: Springer) p 183

[Mut91]   Muto K, Bender E and Klapdor H V 1991 *Z. Phys.* A **39** 435

[Nac86]   Nachtmann O 1986 *Einführung in die Elementarteilchenphysik* (Berlin: Vieweg)

[Nak90]   Nakahata M 1990 *Tests of Fundamental Laws, 9th Moriond Workshop (1989)* eds O Fackler and J Tran Thanh Van (Gif sur Yvette: Edition Frontieres)

[Nar83]   Narlikar J V 1983 *Introduction to Cosmology* (Boston, MA: Jones and Bartlett)

[Nar91]   Narlikar J V and Padmanabhan T 1991 *Ann. Rev. Astron. Astrophys.* **29** 325

[Nar92]   Narayan R, Paczynski B and Piran T 1992 *Astrophys. J.* **395** L83

[Nar93]   Narlikar J V (ed) 1993 *Introduction to Cosmology* 2nd edn (Cambridge: Cambridge University Press)

[Ncs92]   Nesvizhevskii V V *et al* 1992 *Sov. Phys. JETP* **55** 84

[Ng92]   Ng Y J 1992 *Int. J. Mod. Phys.* D **1** 145

[Nil95]   Nilles H P 1995 *Preprint* hep-th/9511313; 1995 *Conf. on Gauge Theories, Applied Supersymmetry and Quantum Gravity (July 1995, Leuven, Belgium)* eds B de Wit, A Sevrin, M Tonin *et al* (Belgium: University Press)

[Nir92]   Nir Y and Quinn H 1992 *Ann. Rev. Nucl. Part. Sci.* **42** 211

[Noe18]   Noether E 1918 *Kgl. Ges. Wiss. Nachrichten. Math. Phys. Klasse* Göttingen p 235

[Nor87]   Norman E *et al* 1987 *Phys. Rev. Lett.* **58** 1403

[Nuc94]   Nucciotti A *et al* 1994 *Nucl. Phys. B (Proc. Suppl.)* B **35** 172, *Proc. 3rd Int. Workshop on Theoretical and Phenomenological Aspects in Astroparticle and Underground Physics (TAUP'93), Assergi, Italy* eds C Arpesella, E Bellotti and A Bottino (Amsterdam: North-Holland)

[Nug95]   Nugent P *et al* 1995 *Phys. Rev. Lett.* **75** 394

[Nus81]   Nussinov S, Glashow S L and Georgi H 1981 *Nucl. Phys.* B **193** 297

[Nut96]   McNutt J *et al* (NESTOR Collaboration) 1996 *Nucl. Phys. B (Proc. Suppl.)* **48** 469, *Proc. 4th Int. Workshop on Theoretical and Phenomenological Aspects of Underground Physics (TAUP'95), Toledo, Spain* eds A Morales, J Morales and J A Villar (Amsterdam: North-Holland)

[Obe92]   Oberauer L and von Feilitzsch F 1992 *Rep. Prog. Phys.* **55** 1093

[Oda94]   Oda T *et al* 1994 *At. Data Nucl. Data Tables* **56** 231

[Oga96]   Ogawa I, Matsuki S and Yamamoto K 1996 *Phys. Rev.* D **53** R1740

[Oku78]   Okun L B and Zeldovich Y B 1978 *Phys. Lett.* B **78** 597

[Oku82]   Okun L B 1982 *Leptons and Quarks* (Amsterdam: North-Holland)

[Oli87]   Olinto A V 1987 *Phys. Lett.* **192** 71

[Oli90a]   Olive K 1990 *Phys. Rep.* **190** 307

[Oli90b]   Olive K *et al* 1990 *Phys. Lett.* B **236** 454

[Oli95]   Olive K 1995 *Preprint* astro-ph/9512166

[Oli95a]   Olive K A and Steigman G 1995 *Appl. J. Suppl.* **97** 49

[Ono91] Ono Y and Suematsu H E 1991 *Phys. Lett.* B **271** 265

[Oor83] Oort H J 1983 *Ann. Rev. Astron. Astrophys.* **21** 373

[Ori85] Orito S and Yoshimura M 1985 *Phys. Rev. Lett.* **54** 2457

[Ori91] Orito S *et al* 1991 *Phys. Rev. Lett.* **66** 1951

[Ost73] Ostriker J P and Peebles P J E 1973 *Astrophys. J.* **186** 467

[Ost74] Ostriker J P, Peebles P J E and Yahil A 1974 *Astrophys. J.* **193** L1

[Oth96] Otha K and Takasugi E 1996 *Proc. ECT Workshop on Double Beta Decay and Related Topics, Trento, Italy, 1995* eds H V Klapdor-Kleingrothaus and S Stoica (Singapore: World Scientific) p 256

[Ott95] Otten E W 1995 Nucl. Phys. B (Proc. Suppl.) **38** 26, *Proc. 16th Int. Conf. on Neutrino Physics and Astrophysics, NEUTRINO '94* eds A Dar, G Eilam and M Gronau (Amsterdam: North-Holland) p 26

[Pac86] Paczynski B 1986 *Astrophys. J.* **304** 1

[Pac93] Paczynski B 1993 *Ann. NY Acad. Sci.* **688** 321

[Pac95] Paczynski B *et al* 1995 *Bull. Am. Astron. Soc.* **187** 1407

[Pac96] Paczynski B 1996 *Ann. Rev. Astron. Astrophys.* **34** 419

[Pad93] Padmanabhan T 1993 *Structure Formation in the Universe* (Cambridge: Cambridge University Press)

[Pag76] Page D N and Hawking S W 1976 *Astron. J.* **206** 1

[Pai96] Pain R *et al* 1996 *Preprint* astro-ph/9607034

[Pal93] Palamara O *et al* (MACRO Collaboration) 1993 *Proc. 23rd Int. Conf. on Cosmic Rays (Calgary, Canada)* eds R B Hicks, D Λ Leahy and D Venkatesan (Singapore: World Scientific)

[Pal97] Palmieri V 1997 *Proc. Int. Workshop on Dark Matter, DARK '96, Heidelberg* eds H V Klapdor-Kleingrothaus and Y Ramachers (Singapore: World Scientific)

[Pan91] Panagia N *et al* 1991 *Astrophys. J.* **380** L23

[Pan92] Panman J *et al* (CHORUS Collaboration) 1992 *Joint Int. Lepton Photon and Europhysics Conf. on High Energy Physics, Geneva 1991* eds K Potter and E Quercigh (Singapore: World Scientific) p 628

[Pan94] Panella O and Srivastava Y 1994 *Preprint* LPC-94-39

[Pan95] Panter M 1995 *XXIV Int. Conf. on Cosmic Rays, Rome* vol 1 958

[Pan96] Panella O 1996 *Proc. ECT Workshop on Double Beta Decay and Related Topics (Trento, Italy, 1995)* eds H V Klapdor-Kleingrothaus and S Stoica (Singapore: World Scientific) p 145

[Pan97] Panella O 1997 *Preprint*

[Par67] Partridge B and Peebles P J E 1967 *Astron. J.* **147** 868

[Par70] Parker E N 1970 *Astrophys. J.* **160** 383

[Par88] Partridge B 1988 *Rep. Prog. Phys.* **51** 674

[Par94] Parker P D 1994 *Proc. Solar Modeling Workshop (Seattle, WA)* eds A B Balantekin and J N Bahcall (Singapore: World Scientific)

[Par95] Partridge R B 1995 *3K: The Cosmic Microwave Background Radiation* (Cambridge: Cambridge University Press)

[Par97] van Paradijs J *et al* 1997 *Nature* **386** 686

[Pas94] Paschos E A and Zioutas K 1994 *Phys. Lett.* B **323** 367

[Pas96] Passalacqua L *et al* (ALEPH Collaboration) 1996 *Preprint* LNF-96/063

[Päs96]   Päs H *et al* (HEIDELBERG–MOSCOW Collaboration) 1996 *Proc. ECT Workshop on Double Beta Decay and Related Topics (Trento, Italy, 1995)* eds H V Klapdor-Kleingrothaus and S Stoica (Singapore: World Scientific) p 130

[Pat74]   Pati J C and Salam A 1974 *Phys. Rev.* D **10** 275

[Pau89]   Paul W *et al* 1989 *Z. Phys.* C **45** 25

[PDG94]   Review of Particle Properties 1994 *Phys. Rev.* D **50** 1413

[PDG96]   Review of Particle Properties 1996 *Phys. Rev.* D **54** 1

[Pec77]   Peccei R and Quinn H 1977 *Phys. Rev. Lett.* **38** 1440

[Pec89]   Peccei R 1989 *C P*-Violation *Advanced Series in High Energy Physics* vol 3 ed C Jarlskog (Singapore: World Scientific)

[Pec93]   Peccei R 1993 *Ann. NY Acad. Sci.* **688** 418

[Pec96]   Peccei R 1996 *Preprint* hep-ph/9606475

[Pee80]   Peebles P J E 1980 *The Large-Scale Structure of the Universe* (Princeton, NY: Princeton University Press)

[Pee84]   Peebles P J E 1984 *Astrophys. J.* **284** 439

[Pee88]   Peebles P J E and Ratra B 1988 *Astrophys. J.* **325** L17

[Pee93]   Peebles P J E 1993 *Principles of Physical Cosmology, Princeton Series in Physics* (Princeton, NJ: Princeton University Press)

[Pee95]   Peebles P J E 1995 *Principles of Physical Cosmology, Princeton Series in Physics* (Princeton, NJ: Princeton University Press)

[Pel92]   Pello R *et al* 1992 *Astron. Astrophys.* **266** 6

[Pen65]   Penzias A A and Wilson R W 1965 *Astrophys. J.* **142** 419

[Pen93]   Pendlebury J M 1993 *Ann. Rev. Nucl. Part. Phys.* **43** 687

[Per84]   Perkins D H 1984 *Ann. Rev. Nucl. Part. Phys.* **34** 1

[Per87]   Perkins D H 1987 *Introduction to High Energy Physics* (New York: Addison Wesley)

[Per88]   Perkins D H 1988 *Proc. IX Workshop on Grand Unification, (Aix-les-Bains)* (Singapore: World Scientific) p 170

[Per95]   Perlmutter S *et al* 1995 *Astrophys. J.* **440** L40

[Per95a]   Percival J W *et al* 1995 *Astrophys. J.* **446** 832

[Per96]   Perlmutter S *et al* 1996 *Preprint* astro-ph/9602122

[Per96a]   Perlmutter S *et al* 1996 *Preprint* astro-ph/9608192

[Pet94]   Petcov S T and Smirnov A Y 1994 *Phys. Lett.* B **322** 109

[Pet96]   Petcov S T 1996 *Proc. ECT Workshop on Double Beta Decay and Related Topics, Trento, Italy, 1995* eds H V Klapdor-Kleingrothaus and S Stoica (Singapore: World Scientific) p 195

[Pet96a]   Petry D *et al* (HEGRA Collaboration) 1996 *Preprint* astro-ph/9606159

[Pic95]   Pich A 1995 *Preprint* hep-ph/9505231; *Proc. European School of High Energy Physics (September 1994, Sorrento, Italy)* eds N Ellis and M B Gavela (CERN-95-04)

[Pie94]   Pierce M J *et al* 1994 *Nature* **371** 385

[Pil95]   Pildis R A, Bregman J N and Evrard A E 1995 *Preprint* astro-ph/9501004

[Pin88]   Pinto P A and Woosley S E 1988 *Nature* **333** 534

[Pin93]   Pinfold J L *et al* 1993 *Phys. Lett.* B **316** 407

[Pir94]   Piran T 1994 *Gamma-Ray Bursters AIP Conf. Proc.* **307** ed G Fishman (New York: AIP)

[Pir97]   Piro L *et al* 1997 *Preprint* astro-ph/9701080

[Pis94]   Piskilli P 1994 *Nucl. Phys. B (Proc. Suppl.)* **35** 191, *Proc. 3rd Int. Workshop on Theoretical and Phenomenological Aspects in Astroparticle and Underground Physics (TAUP'93), Assergi, Italy* eds C Arpesella, E Bellotti and A Bottino (Amsterdam: North-Holland)

[Plu86]   Plumien G, Müller B and Greiner W 1986 *Phys. Rep.* **134** 87

[Pod95]   Podsiadlowski P, Rees M and Ruderman M 1995 *Ann. NY Acad. Sci.* **759** 283

[Pol73]   Politzer H D 1973 *Phys. Rev. Lett.* **30** 1346

[Pol74]   Polyakov A 1974 *JETP Lett.* **20** 194

[Pon93]   Ponman T J and Bertram D 1993 *Nature* **363** 51

[Pra89]   Prantzos N *et al* 1989 *Supernovae, 10th Santa Cruz Summer Workshop on Astronomy and Astrophysics* ed S E Woosley (Heidelberg: Springer) p 630

[Pra96]   Prantzos N and Diehl R 1996 *Phys. Rep.* **267** 1

[Pre84]   Preskill J 1984 *Ann. Rev. Nucl. Part. Phys.* **34** 461

[Pre85]   Press W H and Spergel D N 1985 *Astrophys. J.* **296** 79

[Pre89]   Press W H and Spergel D N 1989 *Physics Today* **3** 29

[Pre92]   Press W H, Rybicki G B and Hewitt J N 1992 *Astrophys. J.* **385** 404, 416

[Pre93]   Pretzl K P 1993 *Europhys. News* **24** 167

[Pri71]   Price P B and Fleischer R L 1971 *Ann. Rev. Nucl. Sci.* **21** 295

[Pri83]   Price P B *et al* 1983 *Phys. Rev. Lett.* **52** 1265

[Pri86]   Price P B and Salomon M H 1986 *Phys. Rev. Lett.* **56** 1226

[Pri87]   Price P B, Guoxiao R and Kinoshita K 1987 *Phys. Rev. Lett.* **59** 2523

[Pri88]   Primack J, Seckel D and Sadoulet B 1988 *Ann. Rev. Nucl. Part. Phys.* **38** 751

[Pri97]   Primack J 1997 *Proc. Int. Workshop on Dark Matter in Astro- and Particle Physics, DARK '96, Heidelberg* eds H V Klapdor-Kleingrothaus and Y Ramachers (Singapore: World Scientific)

[Pun92]   Punch M *et al* 1992 *Nature* **358** 477

[Pur63]   Purcell E M *et al* 1963 *Phys. Rev.* **129** 2326

[Pur93]   Purcell W R *et al* 1993 *Astrophys. J.* **413** L85

[Qia93]   Qian Y Z, Fuller G-M, Mayle R W, Mathews G J, Wilson J R and Woosley S E 1993 *Phys. Rev. Lett.* **71** 1965

[Que95]   Quenby J J 1995 *Phys. Lett. B* **351** 70

[Qui83]   Quigg C 1983 *Gauge Theories of the Strong, Weak and Electromagnetic Interactions, Frontiers in Physics* vol 56 (New York: Addison-Wesley)

[Qui93]   Quirrenbach A 1993 *Sterne und Weltraum* **32** 98

[Raf88]   Raffelt G G and Seckel D 1988 *Phys. Rev. Lett.* **60** 1793

[Raf90]   Raffelt G G 1990 *Phys. Rep.* **198** 1

[Raf95]   Raffelt G G and Silk J 1995 *Preprint* hep-ph/9502306 and 1996 *Phys. Lett. B* **366** 429

[Raf95a]   Raffelt G G 1995 *Preprint* hep-ph/9502358 and *Proc. Moriond XXX Workshop on Dark Matter in Cosmology Clocks and Tests of Fundamental Laws* eds B Guiderdoni, G Greene, D Hinds and J Tran Thanh Van (Paris: Edition Frontieres) p 159

[Raf96]   Raffelt G 1996 *Stars as Laboratories for Fundamental Physics* (Chicago, IL: University of Chicago Press)

[Raf97]   Raffelt G 1997 Private communication

[Rag86]   Raghavan R S, Pakvasa S and Brown B A 1986 *Phys. Rev. Lett.* **57** 801

[Rag97]   Raghavan R S 1997 *Phys. Rev. Lett.* **78** 3618

[Ram77]   Ramaty R and Lingenfelder R E 1977 *Astrophys. J.* **213** L5

[Ram90]   Ramsey N F 1990 *Ann. Rev. Nucl. Part. Phys.* **40** 1

[Ram92]   Ramaty R 1992 *Workshop Compton Observ. Science, NASA CP-3137*

[Ram95]   Ramaty R and Lingenfelter R E 1995 *Preprint* astro-ph/9503045 and 1995 *The Analysis of Emission Lines* eds R E Williams and M Livio (Cambridge: Cambridge University Press)

[Rea92]   Readhead A C S and Lawrence C R 1992 *Ann. Rev. Astron. Astrophys.* **30** 653

[Ree94]   Reeves H 1994 *Rev. Mod. Phys.* **66** 193

[Ref64]   Refsdal S 1964 *Mon. Not. R. Astron. Soc.* **128** 307

[Ren91]   Renzini A 1991 *Observational Tests of Cosmological Inflation* eds T Shanks *et al* (Dordrecht: Kluwer) p 131

[Rep95]   Rephaeli Y 1995 *Ann. Rev. Astron. Astrophys.* **33** 541

[Res93]   Ressell M T *et al* 1993 *Phys. Rev.* D **48** 5519

[Res94]   Resvanis L K (NESTOR Collaboration) 1994 *Nucl. Phys. B (Proc. Suppl.)* **35** 294, *Proc. 3rd Int. Workshop on Theoretical and Phenomenological Aspects in Astroparticle and Underground Physics (TAUP'93), Assergi, Italy* eds C Arpesella, E Bellotti and A Bottino (Amsterdam: North-Holland)

[Reu91]   Reusser D *et al* 1991 *Phys. Lett.* B **255** 143

[Ric87]   Rich J, Lloyd Owen D and Spiro M 1987 *Phys. Rep.* **151** 239

[Rie95]   Riess A G, Press W H and Kirshner R P 1995 *Astrophys. J. Lett.* **438** 217

[Rob91]   Robertson R G H *et al* 1991 *Phys. Rev. Lett.* **67** 957

[Rob96]   Robson I 1996 *Workshop on Active Galactic Nuclei* (New York: Wiley)

[Röd72]   Röde B and Daniel H 1972 *Nuovo Cimento Lett.* **5** 139

[Rog73]   Rogerson J and York D 1973 *Astrophys. J.* **186** L95

[Rol88]   Rolfs C E and Rodney W S 1988 *Cauldrons in the Cosmos* (Chicago, IL: University of Chicago Press)

[Roo88]   Rood H J 1988 *Ann. Rev. Astron. Astrophys.* **26** 245

[Ros93]   Roszkowski L 1993 *Proc. XXIII Workshop: The Decay Properties of SUSY Particles (Erice, 1992)* eds L Cifarelli and V A Khoze (Singapore: World Scientific) (*Science and Culture Series: Physics* **6**) p 429

[Rot93]   Roth K C, Meyer D M and Hawkins I 1993 *Astrophys. J.* **413** L67

[Rou97]   Roulet E and Mollerach S 1997 *Phys. Rep.* **279** 67

[Row85a]   Rowley J K, Cleveland B T and Davis R 1985 *Proc. Solar Neutrinos and Neutrino Astronomy (Homestake, USA), AIP Conf. Proc.* **126** (New York: AIP)

[Row85b]   Rowan-Robinson M 1985 *The Cosmological Distance Ladder* (London: Freeman)

[Roy92]   Roy D P 1992 *Phys. Lett.* B **283** 270

[Rub82]   Rubakov V A 1982 *Nucl. Phys.* B **203** 311

[Rub96]   Rubbia C 1996 *Nucl. Phys. B (Proc. Suppl.)* **48** 172, *Proc. 4th Int. Workshop on Theoretical and Phenomenological Aspects of Underground Physics (TAUP'95), Toledo, Spain* eds A Morales, J Morales and J A Villar (Amsterdam: North-Holland)

[Rub96a]	Rubbia A 1996 *Proc. 7th Int. Workshop on Neutrino Telescopes (Venezia, Italy)* ed M Baldo-Ceolin (Padova: University of Padova Press)
[Rya90]	Ryan S *et al* 1990 *Astrophys. J.* **348** L57
[Sac67a]	Sachs R K and Wolfe A M 1967 *Astrophys. J.* **147** 73
[Sac67b]	Sacharov A D 1967 *JETP Lett* **6** 24
[Sac94]	Sackett P D *et al* 1994 *Nature* **370** 441
[Sad94]	Sadoulet B 1994 *Nucl. Phys. B (Proc. Suppl.)* **35** 117, *Proc. 3rd Int. Workshop on Theoretical and Phenomenological Aspects in Astroparticle and Underground Physics (TAUP'93), Assergi, Italy* eds C Arpesella, E Bellotti and A Bottino (Amsterdam: North-Holland)
[Sag93]	Saglia R P *et al* 1993 *Astrophys. J.* **403** 567
[Sah95]	Sahni V and Coles P 1995 *Phys. Rep.* **262** 1
[Sah97]	Sahu K C etal 1997 *Preprint* astro-ph/9705184, and 1997 *Nature* **387** 476
[Sal68]	Salam A 1968 *Proc. 8th Nobel Symposium, Hrsg. N. Svartholm* (Stockholm: Almquist and Wiskell)
[Sam83]	Samorski M and Stamm W 1983 *Astrophys. J. Lett.* **268** L17
[Sam94]	Samm D (DUMAND Collaboration) 1994 *Trends in Astroparticle-Physics* ed P C Bosetti (Stuttgart: Teubner) p 9
[San78]	Sandage A A 1978 *Astron. J.* **83** 904
[San88]	Santamaria R and Valle J W F 1988 *Phys. Rev. Lett.* **60** 397
[San90]	Sanders R H 1990 *Astron. Astrophys. Rev.* **2** 1
[San90a]	Sandage A and Cacciari C 1990 *Astrophys. J.* **350** 645
[San92]	Sandage A A *et al* 1992 *Astrophys. J. Lett.* **401** L7
[San94]	Sandage A R 1994 *The Carnegie Atlas of Galaxies* vols 1 and 2 (Washington, DC: Carnegie Institute of Washington)
[Sar80]	Sargent W L W *et al* 1980 *Astrophys. J. Suppl.* **42** 41
[Sar94]	Sarsa M L *et al* 1994 *Nucl. Phys. B (Proc. Suppl.)* **35** 154, *Proc. 3rd Int. Workshop on Theoretical and Phenomenological Aspects in Astroparticle and Underground Physics (TAUP'93), Assergi, Italy* eds C Arpesella, E Bellotti and A Bottino (Amsterdam: North-Holland)
[Sar96]	Sarkar S 1996 *Rep. Prog. Phys.* **59** 1493
[Sat85]	Satz H 1985 *Ann. Rev. Nucl. Part. Phys.* **35** 245
[Sat90]	Satz H 1990 *The Quark–Gluon-Plasma, Advanced Series in High Energy Physics* ed R C Hwa (Singapore: World Scientific) vol 6
[Sat91]	Sato N *et al* 1991 (KAMIOKANDE Collaboration) *Phys. Rev. D* **44** 2220
[Sau91]	Saunders W *et al* 1991 *Nature* **349** 32
[Sch83]	Schweizer F, Whitmore B C and Rubin V C 1983 *Astron. J.* **88** 909
[Sch85a]	Schreckenbach K *et al* 1985 *Phys. Lett. B* **160** 325
[Sch85b]	Schrempp B and Schrempp F 1985 *Phys. Bl.* **41** 335
[Sch87a]	Scherrer R, Applegate J and Hogan C 1987 *Phys. Rev. D* **35** 1151
[Sch87b]	Schwarzschild B 1987 *Physics Today* **40** 17
[Sch87c]	Schramm D N 1987 *Proc. XXII Recontre de Moriond: Starbursts and Galaxy Evolution* eds T Xuan Thuan, T Montmerle and J Tran Thanh Van (Gif sur Yvette: Edition Frontieres) p 125
[Sch89]	Schröder H 1989 *Proc. XXIV Int. Conf. on High Energy Physics, München 1988* eds R Kotthaus and J H Kühn (Heidelberg: Springer) p 370

[Sch90a]	Schramm D N 1990 *Supernovae, Jerusalem Winter School for Theoretical Physics, 1988* eds J C Wheeler, T Piran and S Weinberg (Singapore: World Scientific)
[Sch90b]	Schramm D N 1990 *Proc. La Thuile Workshop*
[Sch90c]	Schramm D N and Truran J W 1990 *Phys. Rep.* **189** 89
[Sch91a]	Schoenfelder V 1991 *Phys. Bl.* **47** 295
[Sch91b]	Schneider D P, Schmidt M and Gunn J E 1991 *Astron. J.* **102** 837
[Sch91c]	Schneps J 1991 *Neutrino Telescopes* ed M B Ceolin (Padova: Padova University Press) p 263
[Sch94]	Schoenfelder V 1994 *Sterne und Weltraum* **1** 28
[Sch95]	Schoenfelder V *et al* 1995 *Ann. NY Acad. Sci.* **759** 226; *Phys. in unserer Zeit* **6** 262
[Sch95a]	Schindler S 1995 *Preprint* astro-ph/9511086
[Sch97]	Schutz B 1997 *Proc. Int. Conf. on Particle Physics Beyond the Standard Model, 'BEYOND THE DESERT '97'* eds H V Klapdor-Kleingrothaus and H Päs (Bristol: IOP)
[Sch97a]	Schramm D N and Turner M S 1997 *Preprint* astro-ph/9706069
[Sco95]	Scott D, Silk J and White M 1995 *Science* **268** 829
[Sei90]	Seidel W *et al* 1990 *Phys. Lett.* B **236** 483
[Sei97]	Seidel W 1997 *Proc. Int. Workshop on Dark Matter in Astro- and Particle Physics, DARK '96, Heidelberg* eds H V Klapdor-Kleingrothaus and Y Ramachers (Singapore: World Scientific)
[Sex87]	Sexl R U and Urbantke H K 1987 *Gravitation und Kosmologie* (Mannheim: B I Wissenschaftsverlag) and 1995 (Heidelberg: Spectrum Academischer)
[Sha67]	Shane C D and Wirtanen C A 1967 *Publ. Lick Obs.* **22** 1
[Sha70]	Shapiro M M and Silberberg M 1970 *Ann. Rev. Nucl. Part. Sci.* **20** 323
[Sha83]	Shapiro S L and Teukolsky S A 1983 *Black Holes, White Dwarfs and Neutron Stars* (London: Wiley)
[Sha92]	Shaban N T and Stirling W J 1992 *Phys. Lett.* B **291** 281
[Sha95]	Shandarin S F *et al* 1995 *Phys. Rev. Lett.* **75** 7
[Sha95a]	Shaviv G 1995 *Nucl. Phys.* B *(Proc. Suppl.)* **38** 81, *Proc. 16th Int. Conf. on Neutrino Physics and Astrophysics, NEUTRINO '94* eds A Dar, G Eilam and M Gronau (Amsterdam: North-Holland)
[She85]	Shectman S 1985 *Astrophys. Suppl. J.* **57** 77
[She96]	Shectman S *et al* 1996 *Preprint* astro-ph/9604167 and 1996 *Astrophys. J.* **470** 172
[Shi80]	Shifman M A *et al* 1980 *Nucl. Phys.* B **166** 493
[Shu92]	Shutt T *et al* 1992 *Phys. Rev. Lett.* **69** 3425
[Sik83]	Sikivie P 1983 *Phys. Rev. Lett.* **51** 1415
[Sik95]	Sikivie P 1995 *XXX Renc. de Moriond: Dark Matter in Cosmology: Clocks and Tests of Fundamental Laws (Gif-sur-Yvette, Switzerland)* eds B Guiderdoni, G Greene, D Hinds and J Tran Thanh Van (Gif sur Yvette: Editions Frontières) p 149
[Sik97]	Sikivie P 1997 *Proc. Int. Workshop on Dark Matter in Astro- and Particle Physics, DARK '96, Heidelberg* eds H V Klapdor-Kleingrothaus and Y Ramachers (Singapore: World Scientific)
[Sil84]	Silk J and Srednicki M 1984 *Phys. Rev. Lett.* **53** 624

[Sil85]	Silk J, Olive K and Srednicki M 1985 *Phys. Rev. Lett.* **55** 257
[Sil93]	Silk J and Wyse R F G 1993 *Phys. Rep.* **231** 293
[Sim81]	Simpson J J 1981 *Phys. Rev.* D **24** 2971
[Sim83]	Simpson J A 1983 *Ann. Rev. Nucl. Part. Sci.* **33** 323
[Sim85]	Simpson J J 1985 *Phys. Rev. Lett.* **54** 1891
[Sim97]	Simkovic F *et al* 1997 *Phys. Lett.* B **393** 267
[Sin93]	Singh C P 1993 *Phys. Rep.* **236** 147
[Sky93]	1993 *Sky and Telescope* **2** 23
[Sky95]	1995 *Sky and Telescope* **8** 12
[Sky95a]	1995 *Sky and Telescope* **9** 10
[Sky95b]	1995 *Sky and Telescope* **9** 21
[Smi89]	Smith P F 1989 *Ann. Rev. Nucl. Part. Sci.* **39** 73
[Smi90]	Smith P F and Lewin J D 1990 *Phys. Rep.* **187** 203
[Smi93a]	Smith V P, Nissen P E and Lambert D L 1993 *Astrophys. J.* **408** 262
[Smi93b]	Smith M S, Kawano L H and Malaney R A 1993 *Astrophys. J. Suppl.* **85** 219
[Smi93c]	Smith D M *et al* 1993 *Astrophys. J.* **414** 165
[Smi96]	Smith P F *et al* 1996 *Phys. Lett.* B **379** 299
[Smi97]	Smirnov Yu 1997 *Proc. Int. Conf. on High Energy Physics, Warsaw 1996* and *Preprint* hep-ph/9611465
[Smo90]	Smoot G F *et al* 1990 *Astrophys. J.* **360** 685
[Smo92]	Smoot G F *et al* 1992 *Astrophys. J.* **396** L1
[Smo93]	Smoot G F 1993 *Proc. Int. Conf. on High Energy Physics, Dallas* ed J R Sanford (AIP Conf. Proc. Series) **272** p 1591
[Smo95]	Smoot G F 1995 *Preprint* astro-ph/9505139 and 1994 *DPF Summer Study on High Energy Physics: Particle and Nuclear Astrophysics and Cosmology in the Next Millenium, Snowmass 1994* p 547
[Smo97a]	Smoot G 1997 *Preprint* astro-ph/9705101
[Smo97b]	Smoot G 1997 *Preprint* astro-ph/9705135
[Sok89]	Sokolsky P 1989 *Introduction to Ultrahigh Energy Cosmic Ray Physics, Frontiers in Physics* vol 76 (London: Addison-Wesley)
[Sok92]	Sokolsky P, Sommers P and Dawson B R 1992 *Phys. Rep.* **217** 225
[Son94a]	Songaila A *et al* 1994 *Nature* **368** 599
[Son94b]	Songaila A *et al* 1994 *Nature* **371** 43
[Sou92]	D'Souza I A and Kalman C S 1992 *Preons, Models of Leptons, Quarks and Gauge Bosons as Composite Objects* (Singapore: World Scientific)
[Spe85]	Spergel D N and Press W H 1985 *Astrophys. J.* **294** 663
[Spi82]	Spite F and Spite M 1982 *Nature* **297** 483
[Spi88]	Spivak B V 1988 *JETP* **47** 267
[Spi93]	Spiering Ch 1993 *Phys. Bl.* **49** 871
[Spi96]	Spiering Ch 1996 *Nucl. Phys. B (Proc. Suppl.)* **48** 463, *Proc. 4th Int. Workshop on Theoretical and Phenomenological Aspects of Underground Physics (TAUP'95), Toledo, Spain* eds A Morales, J Morales and J A Villar (Amsterdam: North-Holland)
[Sre88]	Srednicki M, Watkins K and Olive K 1988 *Nucl. Phys.* B **310** 693
[Sta90a]	Staudt A, Muto K and Klapdor-Kleingrothaus H V 1990 *Europhys. Lett.* **13** 31

[Sta90b]	Staudt A, Bender E, Muto K and Klapdor-Kleingrothaus H V 1990 *At. Data Nucl. Data Tables* **44** 79
[Sta92a]	Staudt A and Klapdor-Kleingrothaus H V 1992 *Nucl. Phys.* A **549** 254
[Sta92b]	Starkman G D 1992 *Phys. Rev.* D **45** 476
[Sta96]	Stanev T 1996 *Nucl. Phys. B (Proc. Suppl.)* **48** 165 *Proc. 4th Int. Workshop on Theoretical and Phenomenological Aspects of Underground Physics (TAUP'95), Toledo, Spain* eds A Morales, J Morales and J A Villar (Amsterdam: North-Holland)
[Ste81]	Steigman G, Schramm D N and Gunn J E 1981 *Phys. Lett.* B **66** 454
[Ste84]	Stewart G C *et al* 1984 *Astrophys. J.* **278** 536
[Ste87]	Stephens S A and Golden R L 1987 *Space Sci. Rev.* **46** 31
[Ste88a]	Stebbins A 1988 *Astrophys. J.* **327** 584
[Ste88b]	Stefanov M A 1988 *JETP Lett.* **47** 1
[Ste91]	Steinberger J 1991 *Phys. Rep.* **203** 345
[Ste92]	Steigman G 1992 *Nucl. Phys. B (Proc. Suppl.)* **28A** 28, *Proc. TAUP'91, Toledo* eds A Morales, J Morales, J A Villar (Amsterdam: North-Holland)
[Ste96]	Stevenson E, Goldman T and McKellar B J H 1996 *Preprint* hcp-ph/9603392
[Ste96a]	Steinhardt P J 1996 *Proc. SUSY'95* (Gif sur Yvette: Editions Frontieres) p 552
[Sti89]	Stix M 1989 *The Sun* (Heidelberg: Springer)
[Sto84]	Stone J L (ed) 1984 *MONOPOLE '83* NATO-ASI Series vol 111 (New York: Plenum)
[Sto96]	Stone J L 1996 *Nucl. Phys. B (Proc. Suppl.)* **48** 453, *Proc. 4th Int. Workshop on Theoretical and Phenomenological Aspects of Underground Physics (TAUP'95), Toledo, Spain* eds A Morales, J Morales and J A Villar (Amsterdam: North-Holland)
[Str64]	Streater R F and Wightman A S 1964 *PCT, Spins and Statistics, and all that* (New York: Benjamin)
[Str88]	Strauss M A and Huchra J P 1988 *Astron. J.* **95** 1602
[Str89]	Streitmatter R E *et al* 1989 *Adv. Space Sci.* **9** 65
[Str93]	Strauss M A *et al* 1993 *Astrophys. J.* **397** 95
[Sug95]	Sugiyama N 1995 *Astrophys. J. Suppl.* **100** 281
[Sun80]	Sunyaev R A and Zeldovich Y B 1980 *Ann. Rev. Astron. Astrophys.* **18** 537
[Sun91]	Sunyaev R A *et al* 1991 *Astrophys. J.* **383** L49
[Sun92]	Suntzeff N B *et al* 1992 *Astrophys. J. Lett.* **384** 33
[Sut92]	Sutton C 1992 *Spaceship Neutrino* (Cambridge: Cambridge University Press)
[Sut96]	Sutherland W *et al* (MACHO Collaboration) 1996 *Preprint* astro-ph/9611059 and 1997 *Int. Workshop on the Identification of Dark Matter (IDM'96) Sheffield, UK* (Singapore: World Scientific)
[Suz94]	Suzuki Y 1994 *Nucl. Phys. B (Proc. Suppl.)* **35** 273, *Proc. 3rd Int. Workshop on Theoretical and Phenomenological Aspects in Astroparticle and Underground Physics (TAUP'93), Assergi, Italy* eds C Arpesella, E Bellotti and A Bottino (Amsterdam: North-Holland)

[Suz95] Suzuki Y 1995 *Nucl. Phys. B (Proc. Suppl.)* **38** 54, *Proc. 16th Int. Conf. on Neutrino Physics and Astrophysics, NEUTRINO '94* eds A Dar, G Eilam and M Gronau (Amsterdam: North-Holland)

[Suz96] Suzuki Y 1996 *Proc. NEUTRINO '96, Helsinki*

[Suz97] Suzuki Y 1997 *Proc. 4th Int. Conf. on Solar Neutrinos, Heidelberg, April 1997*

[Swo82] Swordy S *et al* 1982 *Nucl. Instrum. Methods* **193** 591

[Sym81] Symbalisty E M D and Schramm D N 1981 *Rep. Prog. Phys.* **44** 293

[Tak96] Takasugi E 1996 *Proc ECT Workshop on Double Beta Decay and Related Topics, Trento, Italy, 1995* eds H V Klapdor-Kleingrothaus and S Stoica (Singapore: World Scientific) p 165

[Tak97] Takita M 1997 in *Proc. Int. Conf. on Particle Physics Beyond the Standard Model, 'BEYOND THE DESERT '97', Castle Ringberg, Germany* eds H V Klapdor-Kleingrothaus and H Päs (Bristol: IOP)

[Tam86] Tammann G A *et al* 1986 *Proc. WEIN '86, Int. Symp. on Weak and Electromagnetic Interactions in Nuclei* ed H V Klapdor (Berlin: Springer) 1016

[Tam96] Tammann G A *et al* 1996 *Preprint* astro-ph/9603076

[Tat95] Tata X 1995 *Preprint* hep-ph/9510287

[Tat97] Tata X 1997 *Preprint* hep-ph/9706307

[Tau93] Taubes G 1993 *Science* **259** 177

[Tau94] Taubes G 1994 *Science* **263** 1682

[Tau95] TAUP'95 1995 *Proc. 4th Int. Workshop on Theoretical and Phenomenological Aspects of Underground Physics, Toledo, Spain* eds A Morales, J Morales and J A Villar (Amsterdam: North-Holland), *Nucl. Phys. B (Proc. Suppl.)* **48**

[Tau96] Taubes G 1996 *Science* **272** 1431

[Tau96a] Taubes G 1996 *Science* **273** 1492

[Tay83] Taylor G N *et al* 1983 *Phys. Rev. D* **28** 2705

[Tay86] Taylor J H and Stinebring D R 1986 *Ann. Rev. Astron. Astrophys.* **24** 285

[Tay92] Taylor A N and Rowan-Robinson M 1992 *Nature* **359** 396

[Tay94] Taylor J H 1994 *Rev. Mod. Phys.* **66** 711

[Teg95] Tegmark M 1995 *Preprint* astro-ph/9511148

[Teg96] Tegmark M 1996 *Astrophys. J.* **464** L35

[Teg96a] Tegmark M *et al* 1996 *Astrophys. J.* **474** L77

[Thi83] Thielemann F K, Metzinger J and Klapdor H V 1983 *Z. Phys. A* **309** 301; 1983 *Astron. Astrophys.* **123** 162

[Thi90] Thielemann F K *et al* 1990 in *Supernovae* ed S E Woosley (Berlin: Springer) p 609

[t'Ho71] 't Hooft G 1971 *Nucl. Phys. B* **33** 173

[t'Ho72] 't Hooft G and Veltman M 1972 *Nucl. Phys. B* **50** 318

[t'Ho74] 't Hooft G 1974 *Nucl. Phys. B* **79** 276

[t'Ho76] 't Hooft G 1976 *Phys. Rev. D* **14** 3432

[Tho89] Thomas P 1989 *Mon. Not. R. Astron. Soc.* **238** 1319

[Tho95] Thorne K 1995 *Ann. NY Acad. Sci.* **759** 127

[Thr92] Thron J L *et al* (Soudan2 Collaboration) 1992 *Phys. Rev. D* **46** 4846

[Tim95] Timmes F X, Woosley S E and Weaver T A 1995 *Astrophys. J. Suppl. Ser.* **98** 617

490   *References*

[Ton93]   Tonry J L 1993 *Ann. NY Acad. Sci.* **688** 113
[Tot92]   Totsuka Y *et al* (KAMIOKANDE Collaboration) 1992 *Nucl. Phys. B (Proc. Suppl.)* **28A** 67, *Proc 2nd Int. Workshop on Theoretical and Phenomenological Aspects of Underground Physics, TAUP'91, Toledo, Spain* eds A Morales, J Morales and J A Villar (Amsterdam: North-Holland)
[Tot96]   Totsuka Y *et al* 1996 *Nucl. Phys. B (Proc. Suppl.)* **48** 547, *Proc. 4th Int. Workshop on Theoretical and Phenomenological Aspects of Underground Physics (TAUP'95), Toledo, Spain* eds A Morales, J Morales and J A Villar (Amsterdam: North-Holland)
[Tou92]   Toutain T and Frölich C 1992 *Astron. Astrophys.* **257** 287
[Tri82]   Trimble V 1982 *Rev. Mod. Phys.* **54** 1183
[Tri83]   Trimble V 1983 *Rev. Mod. Phys.* **55** 511
[Tri97]   Tripathy D N and Mishra S 1997 *Preprint* gr-ge/9705024v2
[Trü90]   Trümper J 1990 *Phys. Bl.* **46** 137
[Trü93]   Trümper J 1993 *Ann. NY Acad. Sci.* **688** 260 and *Science* **260** 1769
[Trü97]   Trümper J 1997 Private communication
[Tuc85]   Tucker W and Giacconi R 1985 *The X-ray Universe* (Harvard, MA: Harvard University Press)
[Tuc96]   Tucker D L *et al* 1996 *Mon. Not. R. Astron. Soc.* (*Preprint* astro-ph/9611206)
[Tue90]   Tueller J *et al* 1990 *Astrophys. J.* **351** L41
[Tul77]   Tully R B and Fisher J R 1977 *Astron. Astrophys.* **54** 661
[Tur82]   Turner M S, Parker E N and Bogdan T J 1982 *Phys. Rev.* D **26** 1296
[Tur86]   Turok N and Brandenberger R 1986 *Phys. Rev.* D **33** 2175
[Tur87]   Turner M S 1987 *Phys. Rev. Lett.* **59** 2489
[Tur88]   Turck-Chieze S *et al* 1988 *Astrophys. J.* **335** 415
[Tur90a]   Turner E L 1990 *Astrophys. J.* **365** L43
[Tur90b]   Turner M S 1990 *Phys. Rep.* **197** 1
[Tur92]   Turner E L and Ikeuchi S 1992 *Astrophys. J.* **389** 478
[Tur92a]   Turner M S 1992 *Trends in Astroparticle Physics* eds D Cline and R Peccei (Singapore: World Scientific) p 3
[Tur93a]   Turck-Chieze S *et al* 1993 *Phys. Rep.* **230** 57
[Tur93b]   Turck-Chieze S and Lopes I 1993 *Astrophys. J.* **408** 347
[Tur93c]   Turner M S 1993 *New Scientist* **137** 31
[Tur96]   Turck-Chieze S 1996 *Nucl. Phys. B (Proc. Suppl.)* **48** 350, *Proc. 4th Int. Workshop on Theoretical and Phenomenological Aspects of Underground Physics (TAUP'95), Toledo, Spain* eds A Morales, J Morales and J A Villar (Amsterdam: North-Holland)
[Tur96a]   Turner M S 1996 *Preprint* astro-ph/9602050
[Tur97]   Turner M S 1997 *Preprint* astro-ph/9704062
[Tur97a]   Turner M S 1997 *Phys. Rev.* D **55** 435
[Tyt94]   Tytler D and Fan X 1994 *Bull. Am. Astr. Soc.* **26** 1424
[Tyt96]   Tytler D *et al* 1996 *Nature* **381** 207; *Preprints* astro-ph/9603070, astro-ph/9612121
[Uda92]   Udalski A *et al* (OGLE Collaboration) 1992 *Act. Astron.* **42** 253
[Uda94]   Udalski A *et al* (OGLE Collaboration) 1994 *Act. Astron.* **44** 165
[Uns92]   Unsöld A and Baschek B 1992 *Der Neue Kosmos* (Heidelberg: Springer)

[Ush81]	Ushida N *et al* 1981 *Phys. Rev. Lett.* **47** 1694
[Ush86]	Ushida N *et al* 1986 *Phys. Rev. Lett.* **57** 2897
[Uso91]	Uson J M, Bagri D S and Cornwell T J 1991 *Phys. Rev. Lett.* **67** 3328
[Vac85]	Vachaspati T and Vilenkin A 1985 *Phys. Rev.* D **31** 3052
[Vac86]	Vachaspati T 1986 *Phys. Rev. Lett.* **57** 1655
[Vac91]	Vacanti G *et al* 1991 *Astrophys. J.* **377** 467
[Val90]	Valentijn E A 1990 *Nature* **346** 153
[Vau48]	de Vaucouleur G 1948 *Ann. Astrophys.* **11** 247
[Vau78]	de Vaucouleur G and Pence W D 1978 *Astron. J.* **83** 1163
[Ver87]	Vergados J D 1987 *Phys. Lett.* B **184** 55
[Vid94]	Vidyakin G S *et al* 1994 *JETP Lett.* **59** 364
[Vil85]	Vilenkin A 1985 *Phys. Rep.* **121** 263
[Vil87]	Vilenkin A 1987 *Sci. Am.* **12** 52
[Vil88]	Vilenkin A 1988 *Phys. Rev.* D **37** 888
[Vol86]	Voloshin M, Vysotskii M and Okun L B 1986 *Sov. Phys. JETP* **64** 446 and *Sov. J. Nucl. Phys.* **44** 440
[Vui93]	Vuilleumier J C *et al* 1993 *Phys. Rev.* D **48** 1009
[Wag67]	Wagoner R, Fowler W A and Hoyle F 1967 *Astrophys. J.* **148** 3
[Wal79]	Walsh D, Carswell R F and Weymann R J 1979 *Nature* **270** 381
[Wal91]	Walker T P *et al* 1991 *Astrophys. J.* **376** 51
[Wat92]	Watson R A *et al* 1992 *Nature* **357** 660
[Wax95]	Waxman E 1995 *Phys. Rev. Lett.* **75** 386
[Wea80]	Weaver T A and Woosley S E 1980 *Ann. NY Acad. Sci.* **336** 335
[Web74]	Webber W R and Lezniak J A 1974 *Astron. Space Sci.* **30** 361
[Wee86]	Weedman D W 1986 *Quasar Astronomy* (Cambridge: Cambridge University)
[Wee88]	Weekes T C 1988 *Phys. Rep.* **160** 1
[Wei37]	von Weizsäcker C F 1937 *Z. Phys.* **38** 176
[Wei67]	Weinberg S 1967 *Phys. Rev. Lett.* **19** 1264
[Wei72]	Weinberg S 1972 *Gravitation and Cosmology* (New York: Wiley)
[Wei74]	Weinberg S 1974 *Phys. Rev.* D **9** 3357
[Wei78]	Weinberg S 1978 *Phys. Rev. Lett.* **40** 223
[Wei79]	Weinberg S 1979 *Phys. Rev. Lett.* **42** 850
[Wei89]	Weinberg S 1989 *Rev. Mod. Phys.* **61** 1
[Wei93]	Weinheimer C *et al* 1993 *Phys. Lett.* B **300** 210
[Wei96]	Weinberg S 1996 *Preprint* astro-ph/9610044
[Wen76]	Wene C O and Johansson S A E 1976 *Proc. 3rd Int. Conf. on Nuclei far from Stability, Cargese* CERN-Report 76-13 584
[Wes74]	Wess J and Zumino B 1974 *Nucl. Phys.* B **70** 39
[Wes86, 90]	West P 1986, 1990 *Introduction to Supersymmetry* 2nd edition (Singapore: World Scientific)
[Wes87]	Wess J 1987 *Phys. Bl.* **1** 2
[Wet94]	Wetterich C 1994 *Preprint* hep-th/9408025
[Whe68]	Wheeler J A 1968 *Batelle Rencontres* eds C M DeWitt and J A Wheeler (New York: Benjamin) 242
[Whe73]	Whelan J and Iben I 1973 *Astrophys. J.* **186** 1007

[Whe90]   Wheeler J C 1990 *Supernovae, Jerusalem Winter School for Theoretical Physics, 1989* vol 6 eds J C Wheeler, T Piran and S Weinberg (Singapore: World Scientific) p 1

[Whi94]   White M, Scott D and Silk J 1994 *Ann. Rev. Astron. Astrophys.* **32** 319

[Wil78]   Wilczek F 1978 *Phys. Rev. Lett.* **40** 279

[Wil86]   Wilson J R *et al* 1986 *Ann. NY Acad. Sci.* **470** 267

[Wil87]   Wilkerson J F *et al* 1987 *Phys. Rev. Lett.* **58** 2023

[Wil90]   Willick J A 1990 *Astrophys. J.* **351** L5

[Wil93]   Wilson T 1993 *Sterne und Weltraum* **3** 164

[Wil94]   Wilson J R and Rood H J 1994 *Ann. Rev. Astron. Astrophys.* **32** 191

[Wil94a]   Willmer C N A *et al* 1994 *Astrophys. J.* **437** 560

[Win87]   Winget D E *et al* 1987 *Astrophys. J.* **315** L77

[Win91]   Winter K (ed) 1991 *Neutrino Physics* (Cambridge: Cambridge University Press)

[Win92]   Winstein B 1992 *Proc. Joint Int. Symp. on Lepton–Photon and Europhys. Conf. on High Energy Physics, Geneva, 1991* eds S Hegarty, K Potter and E Quercigh (Singapore: World Scientific) p 186

[Win95]   Winter K 1995 *Nucl. Phys. B (Proc. Suppl.)* **38** 211, *Proc. 16th Int. Conf. on Neutrino Physics and Astrophysics, NEUTRINO '94* eds A Dar, G Eilam and M Gronau (Amsterdam: North-Holland)

[Wit84]   Witten E 1984 *Phys. Rev. D* **30** 272

[Wit85]   Witten E 1985 *Phys. Lett. B* **155** 151

[Wit96]   Witten E 1996 *Physics Today* **49** 24

[Wol78]   Wolfenstein L 1978 *Phys. Rev. D* **17** 2369

[Wol81]   Wolfenstein L 1981 *Phys. Lett. B* **107** 77

[Wol85]   Wolfsberg K 1985 *Solar Neutrinos and Neutrino Astronomy, Homestake USA, AIP Conf. Proc.* **126** eds M L Cherry, W A Fowler and K Lande (New York: AIP) p 196

[Wol86]   Wolfenstein L 1986 *Ann. Rev. Nucl. Part. Sci.* **36** 137

[Wol93]   Wolfe A M 1993 *Ann. NY Acad. Sci.* **688** 281

[Won94]   Wong C Y 1994 *Introduction to High Energy Heavy-Ion Collisions* (Singapore: World Scientific)

[Woo82]   Woosley S E and Weaver T A 1982 *Supernovae: A Survey of Current Research, Proc. NATO Advanced Study Institute, Cambridge* eds M J Rees and R J Sonteham (Dordrecht: Reidel) p 79

[Woo86a]   Woosley S E and Weaver T A 1986 *Ann. Rev. Astron. Astrophys.* **24** 205

[Woo86b]   Woosley S E 1986 *Nucleosynthesis and Chemical Evolution (Santa Fe, Lecture Notes)* eds B Hauch, A Maeder and G Meynet p 1; Geneva Observatory 1986; *Ann. Rev. Astron. Astrophys.* **24** 205

[Woo88]   Woods M *et al* (E731 Collaboration) 1988 *Phys. Rev. Lett.* **60** 1695

[Woo88a]   Woosley S E and Phillips M M 1988 *Science* **240** 750

[Woo90a]   Woosley S E (ed) 1990 *Supernovae* (Heidelberg: Springer)

[Woo90b]   Woosley S E *et al* 1990 *Astrophys. J.* **356** 272

[Woo92]   Woosley S E and Hoffman R D 1992 *Astrophys. J.* **395** 202

[Woo93]   Woosley S E 1993 *Astrophys. J.* **405** 273

[Woo94]   Woosley S E *et al* 1994 *Astrophys. J.* **433** 229

[Woo95]   Woosley S E 1995 *Ann. NY Acad. Sci.* **759** 446

[Woo95a]   Woosley S E *et al* 1995 *Ann. NY Acad. Sci.* **759** 352

[Woo95b]	Woosley S E *et al* 1995 *Ann. NY Acad. Sci.* **759** 388
[Woo97]	Woosley S E *et al* 1997 *Preprint* astro-ph/9705146
[Wri92]	Wright E L *et al* 1992 *Astrophys. J.* **396** L13
[Wri94a]	Wright E L *et al* 1994 *Astrophys. J.* **420** 450
[Wri94b]	Wright E L *et al* 1994 *Astrophys. J.* **436** 443
[Wu57]	Wu S *et al* 1957 *Phys. Rev.* **105** 1413
[Wu95]	Wu X P 1995 *Preprint* astro-ph/9512110
[Wue89]	Wuensch W *et al* 1989 *Phys. Rev.* D **40** 3153
[Yah77]	Yahil A, Tammann A and Sandage A 1977 *Astrophys. J.* **217** 903
[Yam83]	Yamada S 1983 *Proc. Int. Conf. on Lepton and Photon Interactions at High Energies (Ithaca, NY, 1983)* eds D G Cassel and D L Kreinick (Ithaca, NY: Cornell University Press) p 525
[Yan54]	Yang C N and Mills R 1954 *Phys. Rev.* **96** 191
[Yan78]	Yanagida T 1978 *Prog. Theor. Phys.* B **135** 66
[Yan79]	Yang J *et al* 1979 *Astrophys. J.* **227** 697
[Yan84]	Yang J *et al* 1984 *Astrophys. J.* **281** 493
[Yok83]	Yokoi K, Takahashi K and Arnould M 1983 *Astron. Astrophys.* **117** 65
[You91]	You K *et al* 1991 *Phys. Lett.* B **265** 53
[Yps94]	Ypsilantis T 1994 *Proc. Int. School on Cosmological Dark Matter (Valencia, Spain, 1993)* eds J W F Valle and A Perez (Singapore: World Scientific)
[Yps96]	Ypsilantis T 1996 *Europhys. News* **27** 97
[Zac86]	Zacek G *et al* 1986 *Phys. Rev.* D **34** 2621
[Zac94]	Zacek V 1994 *Nuovo Cimento* A **104** 291
[Zat66]	Zatsepin G T and Kuzmin V A 1966 *JETP Lett.* **4** 53
[Zat95]	Zatsepin V I 1995 *J. Phys. G: Nucl. Part. Phys.* **21** L31
[Zei91]	Zeilik M 1991 *Astronomy* 6th edn (New York: Wiley)
[Zel67]	Zeldovich Y 1967 *JETP Lett.* **6** 316
[Zel70]	Zeldovich Y 1970 *Astron. Astrophys.* **5** 84
[Zel72]	Zeldovich Y 1972 *Mon. Not. R. Astron. Soc.* **160** 1
[Zhi80]	Zhitnitsky A R 1980 *Sov. J. Nucl. Phys.* **31** 260
[Zub93]	Zuber K and Klapdor-Kleingrothaus H V 1993 *Phys. Bl.* **49** 125
[Zub96]	Zuber K 1996 *Preprint* hep-ph/9605405 and 1997 *Phys. Rev.* D **56** 1816
[Zub97]	Zuber K 1997 *Proc. 4th Int. Solar Neutrino Conf. (Heidelberg, April 1997)*
[Zwe64]	Zweig G 1964 *CERN-Berichte Th-401 and Th-412*
[Zwi68]	Zwicky F *et al* 1968 *Catalog of Galaxies and Clusters of Galaxies* vols 1–6 (Pasadena)
[Zwi83]	Zwirner F 1983 *Phys. Lett.* B **132** 103

Index